Lecture Notes in Computer Sc

Jesús Labarta Kazuki Joe
Toshinori Sato (Eds.)

High-Performance Computing

6th International Symposium, ISHPC 2005
Nara, Japan, September 7-9, 2005
and First International Workshop
on Advanced Low Power Systems, ALPS 2006
Revised Selected Papers

Volume Editors

Jesús Labarta
Universidad Politecnica de Catalunya
Spain
E-mail: jesus@ac.upc.edu

Kazuki Joe
Nara Women's University
Japan
E-mail: joe@ics.nara-wu.ac.jp

Toshinori Sato
Kyushu University
Japan
E-mail: toshinori.sato@computer.org

Library of Congress Control Number: 2007943157

CR Subject Classification (1998): D.1, D.2, F.2, E.4, G.1-4, J.1-2, J.6, I.6

LNCS Sublibrary: SL 1 – Theoretical Computer Science and General Issues

ISSN 0302-9743
ISBN-10 3-540-77703-2 Springer Berlin Heidelberg New York
ISBN-13 978-3-540-77703-8 Springer Berlin Heidelberg New York

Springer is a part of Springer Science+Business Media

springer.com

Printed in Germany

Typesetting: Camera-ready by author, data conversion by Scientific Publishing Services, Chennai, India
Printed on acid-free paper SPIN: 12216628 06/3180 5 4 3 2 1 0

Preface

This is the joint post-proceedings of the 6th International Symposium on High Performance Computing (ISHPC-VI) and the First International Workshop on Advanced Low Power Systems 2006 (ALPS2006). The post-proceedings also contain the papers presented at the Second HPF International Workshop: Experiences and Progress (HiWEP2005) and the Workshop on Applications for PetaFLOPS Computing (APC2005), which are workshops of ISHPC-VI.

ISHPC-VI, HiWEP2005 and APC2005 were held in Nara, Japan during September 7–9, 2005. Fifty-eight papers from 11 countries were submitted to ISHPC-VI. After the reviews of the submitted papers, the ISHPC-VI Program Committee selected 15 regular (12-page) papers for oral presentation. In addition, several other papers with favorable reviews were recommended for poster presentation, and 14 short (8-page) papers were also selected. Twenty-eight papers out of 29 ISHPC-VI papers are contained in the post-proceedings. HiWEP2005 and APC2005 received eight and ten submissions, with six and eight papers being accepted for oral presentation after reviews, respectively. All the HiWEP2005 and APC2005 papers are included in the post-proceedings.

ALPS2006 was held in Cairns, Australia on July 1, 2006 in conjunction with the ACM 20th International Conference on Supercomputing. The number of submitted papers was 15, and eight papers were accepted for oral presentation. The post-proceedings contain six of the eight papers.

Initially, the ISHPC-VI Program Committee tried to publish on-site proceedings, but could not make it because of the very tight schedule. After ISHPC-VI, the ISHPC-VI Organizing Chair discussed with the General Chair of ALPS2006 (Hironori Nakajo) the possibility of a joint post-proceedings. After ALPS2006, we made a proposal of having joint post-proceedings to Springer, and it was accepted as an LNCS volume. Then we organized the post-proceedings committee, which is based on the PCs of ISHPC-VI, HiWEP2005, APC2005 and ALPS2006, to ask all authors to submit their papers to the post-proceedings because it took a long time since their paper presentations. All submitted papers were re-reviewed and most papers were accepted for the post-proceedings.

The post-proceedings consist of six sections. The first section is of ISHPC-VI award papers. The ISHPC-VI Program Committee selected a distinguished paper award and a best student paper award. The distinguished paper award was given for "Multiple Stream Prediction" by Oliverio J. Santana. The best student paper award was given for "Enhanced Loop Coalescing: A Compiler Technique for Transforming Non-Uniform Iteration Spaces" by Arun Kejariwal. The second and the third sections are of ISHPC-VI regular and short papers, respectively. The second section on ISHPC-VI regular papers contains many architecture-related papers, while the third section on ISHPC-VI short papers has more variety such as HPC applications or visualization. The fourth section of HiWEP2005 shows the recent trend in HPF research. The fifth section on APC2005 includes HPC applications which require Peta scale supercomputers. The last section on ALPS2006 includes low-power processor-related papers, which is a key issue for the future innovative HPC processors and systems.

We hope that the post-proceedings are of significant interest to many readers. Last but not least, we thank the members of the Program Committee and the referees for all the hard work that made the post-proceedings possible.

October 2007

Jesús Labarta
Kazuki Joe
Toshinori Sato

Organization

ISHPC–VI Executive Committee

Conference Chair	Hideo Aiso (Tokyo University of Technology)
Program Chair	Jesús Labarta (Technical University of Catalonia)
Program Co-chairs	Hironori Nakajo (Tokyo University of Agriculture and Technology)
	Utpal Banerjee (Intel)
	Theodore S. Papatheodorou (University of Patras)
Organizing Chair	Kazuki Joe (Nara Women's University)
Local Arrangements Chair	Noriyuki Fujimoto (Osaka University)
Workshop Chairs	Yasuo Okabe (Kyoto University)
	Masahiro Fukuda (Japan Aerospace Exploration Agency)

ISHPC–VI Program Committee Members

Hamid R. Arabnia (University of Georgia)
Nicholas Carter (University of Illinois at Urbana-Champaign)
Claudia Dinapoli (National Research Council)
Rudolf Eigenmann (Purdue University)
Ophir Frieder (Illinois Institute of Technology)
Mario Furnari (National Research Council)
Dennis Gannon (Indiana University)
Kyle Gallivan (Florida State University)
Steve Lumetta (University of Illinois at Urbana-Champaign)
Allen Malony (University of Oregon)
Trevor Mudge (University of Michigan)
Alex Nicolau (University of California Irvine)
Constantine Polychronopoulos (University of Illinois at Urbana-Champaign)
Eleftherios Polychronopoulos(University of Patras)
Mateo Valero (Technical University of Catalonia)
Alex Veidenbaum (University of California Irvine)
Harry Wijshoff (Leiden University)

HiWEP2005 Program Committee

Program Chair	Hitoshi Sakagami (National Institute of Fusion Science)
PC members	PC members: Yoshiki Seo (NEC)
	Hidetoshi Iwashita (Fujitsu)

Kunihiko Watanabe (Earth Simulator Center)
Masahiro Fukuda (Japan Aerospace Exploration Agency)
Mitsuhisa Sato (University of Tsukuba)

APC2005 Program Committee

Program Chair — Masahiro Fukuda (Japan Aerospace Exploration Agency)

PC members — Hidehiko Hasegawa (University of Tsukuba)
Masanori Kanazawa (Kyoto University)
Ryo Nagai (Nagoya University)
Naoki Hirose (Japan Aerospace Exploration Agency/Asian Technology Information Program)
Shinji Hioki (Tezukayama University)

ALPS2006 Program Committee

Program Chair — Toshinori Sato (Kyushu University)

PC Members — David Albonesi (Cornell University)
Pradip Bose (IBM T.J. Watson Research Center)
David Brooks (Harvard University)
Naehyuck Chang (Seoul National University)
Pai Chou (University of California Irvine)
Nikil Dutt (University of California Irvine)
Farzan Fallah (Fujitsu Labs of America)
Pierfrancesco Foglia (University of Pisa)
Masahiro Goshima (University of Tokyo)
José González (Intel Barcelona Research Center)
Kenji Kise (Tokyo Institute of Technology)
Tadahiro Kuroda (Keio University)
José F. Martínez (Cornell University)
Vasily Moshnyaga (Fukuoka University)
Hiroshi Nakashima (Toyohashi University of Technology)
Vijaykrishnan Narayanan (Pennsylvania State University)
Sri Parameswaran (University of New South Wales)
Mitsuhisa Sato (University of Tsukuba)
Youngsoo Shin (Korea Advanced Institute of Science and Technology)
Hiroyuki Tomiyama (Nagoya University)

Sponsoring Institutions[1]

Information Processing Society Japan Kansai Branch
Intel (r) Corporation
Fujitsu Limited
HPF Promoting Consortium
Nara Convention Breau
Advance Soft Corporation

[1] These institutions have supported ISHPC-VI.

Table of Contents

High Performance Computing

I ISHPC-VI Award Papers

II ISHPC-VI Regular Papers

III ISHPC-VI Short Papers

V APC2005

VI ALPS2006

Multiple Stream Prediction

Oliverio J. Santana[1], Alex Ramirez[1,2], and Mateo Valero[1,2]

[1] Departament d'Arquitectura de Computadors
Universitat Politècnica de Catalunya
Barcelona, Spain
[2] Barcelona Supercomputing Center
Barcelona, Spain
{osantana,aramirez,mateo}@ac.upc.edu

Abstract. The next stream predictor is an accurate branch predictor that provides stream level sequencing. Every stream prediction contains a full stream of instructions, that is, a sequence of instructions from the target of a taken branch to the next taken branch, potentially containing multiple basic blocks. The long size of instruction streams makes it possible for the stream predictor to provide high fetch bandwidth and to tolerate the prediction table access latency. Therefore, an excellent way for improving the behavior of the next stream predictor is to enlarge instruction streams.

In this paper, we provide a comprehensive analysis of dynamic instruction streams, showing that there are several kinds of streams according to the terminating branch type. Consequently, focusing on particular kinds of stream is not a good strategy due to Amdahl's law. We propose the multiple stream predictor, a novel mechanism that deals with all kinds of streams by combining single streams into long virtual streams. We show that our multiple stream predictor is able to tolerate the prediction table access latency without requiring the complexity caused by additional hardware mechanisms like prediction overriding, also reducing the overall branch predictor energy consumption.

Keywords: microarchitecture, branch prediction, access latency, instruction stream.

1 Introduction

High performance superscalar processors require high fetch bandwidth to exploit all the available instruction-level parallelism. The development of accurate branch prediction mechanisms has provided important improvements in the fetch engine performance. However, it has also increased the fetch architecture complexity. Our approach to achieve high fetch bandwidth, while maintaining the complexity under control, is the stream fetch engine [11,16].

This fetch engine design is based on the next stream predictor, an accurate branch prediction mechanism that uses instruction streams as the basic prediction unit. We call stream to a sequence of instructions from the target of a taken

J. Labarta, K. Joe, and T. Sato (Eds.): ISHPC 2005 and ALPS 2006, LNCS 4759, pp. 1–16, 2008.

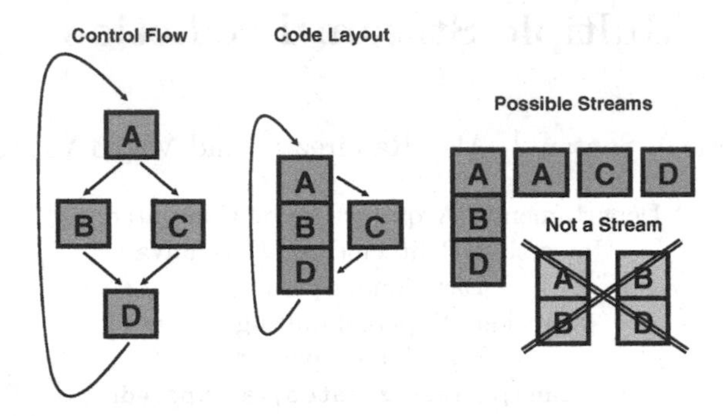

Fig. 1. Example of instruction streams

branch to the next taken branch, potentially containing multiple basic blocks. Figure 1 shows an example control flow graph from which we will find the possible streams. The figure shows a loop containing an if-then-else structure. Let us suppose that our profile data shows that $A \rightarrow B \rightarrow D$ is the most frequently followed path through the loop. Using this information, we lay out the code so that the path $A \rightarrow B$ goes through a not-taken branch, and falls-through from $B \rightarrow D$. Basic block C is mapped somewhere else, and can only be reached through a taken branch at the end of basic block A.

From the resulting code layout we may encounter four possible streams composed by basic blocks ABD, A, C, and D. The first stream corresponds to the sequential path starting at basic block A and going through the frequent path found by our profile. Basic block A is the target of a taken branch, and the next taken branch is found at the end of basic block D. Neither the sequence AB, nor the sequence BD can be considered streams because the first one does not end in a taken branch, and the second one does not start in the target of a taken branch. The infrequent case follows the taken branch at the end of A, goes through C, and jumps back into basic block D.

Although a fetch engine based on streams is not able to fetch instructions beyond a taken branch in a single cycle, streams are long enough to provide high fetch bandwidth. In addition, since streams are sequentially stored in the instruction cache, the stream fetch engine does not need a special-purpose storage, nor a complex dynamic building engine. However, taking into account current technology trends, accurate branch prediction and high fetch bandwidth is not enough. The continuous increase in processor clock frequency, as well as the larger wire delays caused by modern technologies, prevent branch prediction tables from being accessed in a single cycle [1,7]. This fact limits fetch engine performance because each branch prediction depends on the previous one, that is, the target address of a branch prediction is the starting address of the following one.

A common solution for this problem is prediction overriding [7,19]. A small and fast predictor is used to obtain a first prediction in a single cycle. A slower but more accurate predictor provides a new prediction some cycles later, overriding the first prediction if they differ. This mechanism partially hides the branch predictor access latency. However, it also causes an increase in the fetch architecture complexity, since prediction overriding requires a complex recovery mechanism to discard the wrong speculative work based on overridden predictions.

An alternative to the overriding mechanism is using long basic prediction units. A stream prediction contains enough instructions to feed the execution engine during multiple cycles [16]. Therefore, the longer a stream is, the more cycles the execution engine will be busy without requiring a new prediction. If streams are long enough, the execution engine of the processor can be kept busy during multiple cycles while a new prediction is being generated. Overlapping the execution of a prediction with the generation of the following prediction allows to partially hide the access delay of this second prediction, removing the need for an overriding mechanism, and thus reducing the fetch engine complexity.

Since instruction streams are limited by taken branches, a good way to obtain longer streams is removing taken branches through code optimizations. Code layout optimizations have a beneficial effect on the length of instruction streams [16]. These optimizations try to map together those basic blocks that are frequently executed as a sequence. Therefore, most conditional branches in optimized code are not taken, enlarging instruction streams. However, code layout optimizations are not enough for the stream fetch engine to completely overcome the need for an overriding mechanism [17].

Looking for novel ways of enlarging streams, we present a detailed analysis of dynamic instruction streams. Our results show that most of them finalize in conditional branches, function calls, and return instructions. As a consequence, it would seem that these types of branches are the best candidates to apply techniques for enlarging instruction streams. However, according to Amdahl's law, focusing on particular branch types is not a good approach to enlarge instruction streams. If we focus on a particular type of stream, the remainder streams, which do not benefit from the stream enlargement, will limit the achievable performance improvement. This leads to a clear conclusion: the correct approach is not focusing on particular branch types, but trying to enlarge all dynamic streams. In order to achieve this, we present the multiple stream predictor, a novel predictor that concatenates those streams that are frequently executed as a sequence. This predictor does not depend on the type of the branch terminating the stream, making it possible to generate very long virtual streams.

The remainder of this paper is organized as follows. Section 2 describes previous related work. Section 3 presents our experimental methodology. Section 4 provides an analysis of dynamic instruction streams. Section 5 describes the multiple stream predictor. Section 6 evaluates the proposed predictor. Finally, Section 7 presents our concluding remarks.

2 Related Work

The prediction table access latency is an important limiting factor for current fetch architectures. The processor front-end must generate the fetch address in a single cycle because this address is needed for fetching instructions in the next cycle. However, the increase in processor clock frequency, as well as the slower wires in modern technologies, cause branch prediction tables to require multi-cycle accesses [1,7].

The trace predictor [5] is a latency tolerant mechanism, since each trace prediction is potentially a multiple branch prediction. The processor front-end can use a single trace prediction to feed the processor back-end with instructions during multiple cycles, while the trace predictor is being accessed again to obtain a new prediction. Overlapping the prediction table access with the fetch of instructions from a previous prediction allows to hide the branch predictor access delay. Our next stream predictor has the same ability [17], since a stream prediction is also a multiple branch prediction able to provide enough instructions to hide the prediction table access latency.

Using a fetch target queue (FTQ) [12] is also helpful for taking advantage of this fact. The FTQ decouples the branch prediction mechanism and the instruction cache access. Each cycle, the branch predictor generates the fetch address for the next cycle, and a fetch request that is stored in the FTQ. Since the instruction cache is driven by the requests stored in the FTQ, the fetch engine is less likely to stay idle while the predictor is being accessed again.

Another promising idea to tolerate the prediction table access latency is pipelining the branch predictor [6,20]. Using a pipelined predictor, a new prediction can be started each cycle. Nevertheless, this is not trivial, since the outcome of a branch prediction is needed to start the next prediction. Therefore, each branch prediction can only use the information available in the cycle it starts, which has a negative impact on prediction accuracy. In-flight information could be taken into account when a prediction is generated, as described in [20], but this also involves an increase in the fetch engine complexity. It is possible to reduce this complexity in the fetch engine of a simultaneous multithreaded processor by pipelining the branch predictor and interleaving prediction requests from different threads each cycle [2]. Nevertheless, analyzing the accuracy and performance of pipelined branch predictors is out of the scope of this work.

A different approach is the overriding mechanism described by Jimenez et al. [7]. This mechanism provides two predictions, a first prediction coming from a fast branch predictor, and a second prediction coming from a slower, but more accurate predictor. When a branch instruction is predicted, the first prediction is used while the second one is still being calculated. Once the second prediction is obtained, it overrides the first one if they differ, since the second predictor is considered to be the most accurate. A similar mechanism is used in the Alpha EV6 [3] and EV8 [19] processors.

The problem of prediction overriding is that it requires a significant increase in the fetch engine complexity. An overriding mechanism requires a fast branch predictor to obtain a prediction each cycle. This prediction should be stored for being compared with the main prediction. Some cycles later, when the main prediction is generated, the fetch engine should determine whether the first prediction is correct or not. If the first prediction is wrong, all the speculative work done based on it should be discarded. Therefore, the processor should track which instructions depend on each prediction done in order to allow the recovery process. This is the main source of complexity of the overriding technique.

Moreover, a wrong first prediction does not involve that all the instructions fetched based on it are wrong. Since both the first and the main predictions start in the same fetch address, they will partially coincide. Thus, the correct instructions based on the first prediction should not be squashed. This selective squash will increase the complexity of the recovery mechanism. To avoid this complexity, a full squash could be done when the first and the main predictions differ, that is, all instructions depending on the first prediction are squashed, even if they should be executed again according to the main prediction. However, a full squash will degrade the processor performance and does not remove all the complexity of the overriding mechanism. Therefore, the challenge is to develop a technique able to achieve the same performance than an overriding mechanism, but avoiding its additional complexity, which is the objective of this work.

3 Experimental Methodology

The results in this paper have been obtained using trace driven simulation of a superscalar processor. Our simulator uses a static basic block dictionary to allow simulating the effect of wrong path execution. This model includes the simulation of wrong speculative predictor history updates, as well as the possible interference and prefetching effects on the instruction cache. We feed our simulator with traces of 300 million instructions collected from the SPEC 2000 integer benchmarks[1] using the *reference* input set. To find the most representative execution segment we have analyzed the distribution of basic blocks as described in [21].

Since previous work [16] has shown that code layout optimizations are able to enlarge instruction streams, we present data for both a baseline and an optimized code layout. The baseline code layout was generated using the Compaq C V5.8-015 compiler on Compaq UNIX V4.0. The optimized code layout was generated with the Spike tool shipped with Compaq Tru64 Unix 5.1. Optimized code generation is based on profile information collected by the Pixie V5.2 tool using the *train* input set.

3.1 Simulator Setup

Our simulation setup corresponds to an aggressive 8-wide superscalar processor. The main values of this setup are shown in Table 1. We compare our stream fetch architecture with three other state-of-the-art fetch architectures: a fetch architecture using an interleaved BTB and a 2bcgskew predictor [19], the fetch target buffer (FTB) architecture [12] using a perceptron predictor [8], and the trace cache fetch architecture using a trace predictor [5]. All these architectures use an 8-entry fetch target queue (FTQ) [12] to decouple branch prediction from

[1] We excluded *181.mcf* because its performance is very limited by data cache misses, being insensitive to changes in the fetch architecture.

Table 1. Configuration of the simulated processor

fetch, rename, and commit width	8 instructions
integer and floating point issue width	8 instructions
load/store issue width	4 instructions
fetch target queue	8 entries
instruction fetch queue	32 entries
integer, floating point, and load/store issue queues	64 entries
reorder buffer	256 entries
integer and floating point registers	160

L1 instruction cache	64/32 KB, 2-way, 128 byte block, 3 cycle latency
L1 data cache	64 KB, 2-way, 64 byte block, 3 cycle latency
L2 unified cache	1 MB, 4-way, 128 byte block, 16 cycle latency
main memory latency	350 cycles
maximum trace size	32 instructions (10 branches)
filter and main trace caches	128 traces, 4-way associative

the instruction cache access. We have found that larger FTQs do not provide additional performance improvements.

Our instruction cache setup uses wide cache lines, that is, 4 times the processor fetch width [11], and 64KB total hardware budget. The trace fetch architecture is actually evaluated using a 32KB instruction cache, while the remainder 32KB are devoted to the trace cache. This hardware budget is equally divided into a filter trace cache [13] and a main trace cache. In addition, we use selective trace storage [10] to avoid trace redundancy between the trace cache and the instruction cache.

3.2 Fetch Models

The stream fetch engine [11,16] model is shown in Figure 2.a. The stream predictor access is decoupled from the instruction cache access using an FTQ. The stream predictor generates requests, composed by a full stream of instructions, which are stored in the FTQ. These requests are used to drive the instruction cache, obtain a line from it, and select which instructions from the line should be executed. In the same way, the remainder three fetch models use an FTQ to decouple the branch prediction stage from the fetch stage.

Our interleaved BTB fetch model (iBTB) is inspired by the EV8 fetch engine design described in [19]. This iBTB model decouples the branch prediction mechanism from the instruction cache with an FTQ. An interleaved BTB is used to allow the prediction of multiple branches until a taken branch is predicted, or until an aligned 8-instruction block is completed. The branch prediction history is updated using a single bit for prediction block, which combines the outcome of the last branch in the block with path information [19]. Our FTB model is similar to the one described in [12] but using a perceptron branch predictor [8] to predict the direction of conditional branches. Figure 2.b shows a diagram representing these two fetch architectures.

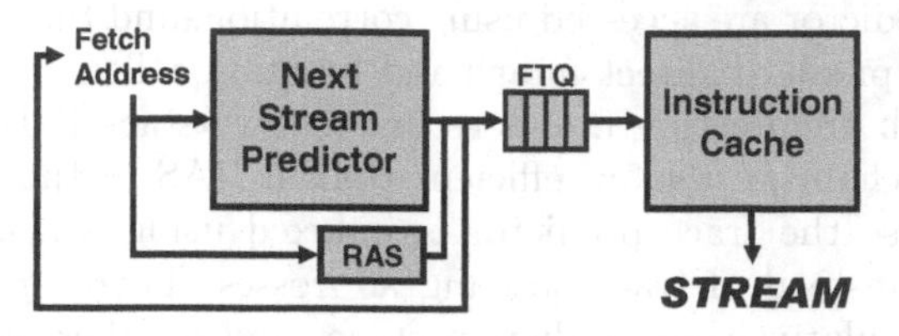

(a) The stream fetch engine

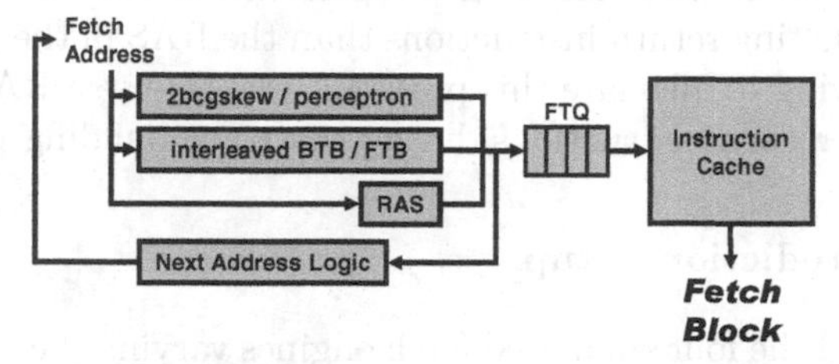

(b) Fetch engine using an iBTB/FTB and a decoupled conditional branch predictor

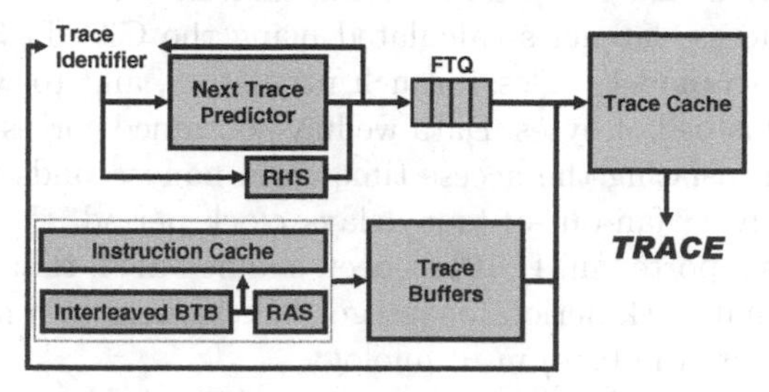

(c) Trace cache fetch architecture using a next trace predictor

Fig. 2. Fetch models evaluated

Our trace cache fetch model is similar to the one described in [14] but enhanced using an FTQ [12] to decouple the trace predictor from the trace cache, as shown in Figure 2.c. Trace predictions are stored in the FTQ, which feeds the trace cache with trace identifiers. An interleaved BTB is used to build traces in the case of a trace cache miss. This BTB uses 2-bit saturating counters to predict the direction of conditional branches when a trace prediction is not available. In addition, an aggressive 2-way interleaved instruction cache is used to allow traces to be built as fast as possible. This mechanism is able to obtain up to a full cache line in a cycle, independent of PC alignment.

The four fetch architectures evaluated in this paper use specialized structures to predict return instructions. The iBTB, the FTB, and the stream fetch architecture use a return address stack (RAS) [9] to predict the target address of return instructions. There are actually two RAS, one updated speculatively in prediction stage, and another one updated non-speculatively in commit stage, which is used to restore the correct state in case of a branch misprediction. The iBTB and FTB fetch architectures also use a cascaded structure [15] to improve the prediction accuracy of the rest of indirect branches. Both the stream predictor and the trace predictor are accessed using correlation, and thus they are already able to correctly predict indirect jumps and function calls.

The trace fetch architecture uses a return history stack (RHS) [5] instead of a RAS. This mechanism is more efficient than a RAS in the context of trace prediction because the trace predictor is indexed using a history of previous trace identifiers instead of trace starting addresses. There are also two RHS, one updated speculatively in prediction stage, and another one updated non-speculatively in commit stage. However, the RHS in the trace fetch architecture is less accurate predicting return instructions than the RAS in the rest of evaluated architectures. Trying to alleviate this problem, we also use a RAS to predict the target address of return instructions during the trace building process.

3.3 Branch Prediction Setup

We have evaluated the four simulated fetch engines varying the size of the branch predictor from small and fast tables to big and slow tables. We use realistic prediction table access latencies calculated using the CACTI 3.0 tool [22]. We modified CACTI to model tagless branch predictors, and to work with setups expressed in bits instead of bytes. Data we have obtained corresponds to 0.10μm technology. For translating the access time from nanoseconds to cycles, we assumed an aggressive 8 fan-out-of-four delays clock period, that is, a 3.47 GHz clock frequency as reported in [1]. It has been claimed in [4] that 8 fan-out-of-four delays is the optimal clock period for integer benchmarks in a high performance processor implemented in 0.10μm technology.

We have found that the best performance is achieved using three-cycle latency tables [17]. Although bigger predictors are slightly more accurate, their increased access delay harms processor performance. On the other hand, predictors with a lower latency are too small and achieve poor performance. Therefore, we have chosen to simulate all branch predictors using the bigger tables that can be accessed in three cycles. Table 2 shows the configuration of the simulated predictors. We have explored a wide range of history lengths, as well as DOLC index [5] configurations, and selected the best one found for each setup. Table 2 also shows the approximate hardware budget for each predictor. Since we simulate the larger three cycle latency tables[2], the total hardware budget devoted to each predictor is different. The stream fetch engine requires less hardware resources because it uses a single prediction mechanism, while the other evaluated fetch architectures use some separate structures.

Our fetch models also use an overriding mechanism [7,19] to complete a branch prediction each cycle. A small branch predictor, supposed to be implemented using very fast hardware, generates the next fetch address in a single cycle. Although being fast, this predictor has low accuracy, so the main predictor is used to provide an accurate back-up prediction. This prediction is obtained three

[2] The first level of the trace and stream predictors, as well as the first level of the cascaded iBTB and FTB, is actually smaller than the second one because larger first level tables do not provide a significant improvement in prediction accuracy.

Table 2. Configuration of the simulated branch predictors

iBTB fetch architecture (approx. 95KB)		
2bcgskew predictor	interleaved BTB	1-cycle predictor
four 64K entry tables 16 bit history (bimodal 0 bits)	1024 entry, 4-way, first level 4096 entry, 4-way, second level DOLC 14-2-4-10	64 entry gshare 6-bit history 32 entry, 1-way, BTB
FTB fetch architecture (approx. 50KB)		
perceptron predictor	FTB	1-cycle predictor
256 perceptrons 4096x14 bit local and 40 bit global history	1024 entry, 4-way, first level 4096 entry, 4-way, second level DOLC 14-2-4-10	64 entry gshare 6-bit history 32 entry, 1-way, BTB
Stream fetch architecture (approx. 32KB)		
next stream predictor		1-cycle predictor
1024 entry, 4-way, first level 4096 entry, 4-way, second level DOLC 16-2-4-10		32 entry, 1-way, spred DOLC 0-0-0-5
Trace fetch architecture (approx. 80KB)		
next trace predictor	interleaved BTB	1-cycle predictor
2048 entry, 4-way, first level 4096 entry, 4-way, second level DOLC 10-4-7-9	1024 entry, 4-way, first level 4096 entry, 4-way, second level DOLC 14-2-4-10	32 entry, 1-way, tpred DOLC 0-0-0-5 perfect BTB override

cycles later and compared with the prediction provided by the single-cycle predictor. If both predictions differ, the new prediction overrides the previous one, discarding the speculative work done based on it. The configuration of the single-cycle predictors used is shown in Table 2.

4 Analysis of Dynamic Instruction Streams

Fetching a single basic block per cycle is not enough to keep busy the execution engine of wide-issue superscalar processors during multiple cycles. In this context, the main advantage of instruction streams is their long size. A stream can contain multiple basic blocks, whenever only the last one ends in a taken branch. This makes it possible for the stream fetch engine to provide high fetch bandwidth while requiring low implementation cost [11,16].

However, having high average length does not involve that most streams are long. Some streams could be long, providing high fetch bandwidth, while other streams could be short, limiting the potential performance. Therefore, in the search for new ways of improving the stream fetch engine performance, the distribution of dynamic stream lengths should be analyzed.

Figure 3 shows an histogram of dynamic streams classified according to their length. It shows the percentage of dynamic streams that have a length ranging from 1 to 30 instructions. The last bar shows the percentage of streams that are longer than 30 instructions. Data is shown for both the baseline and the optimized

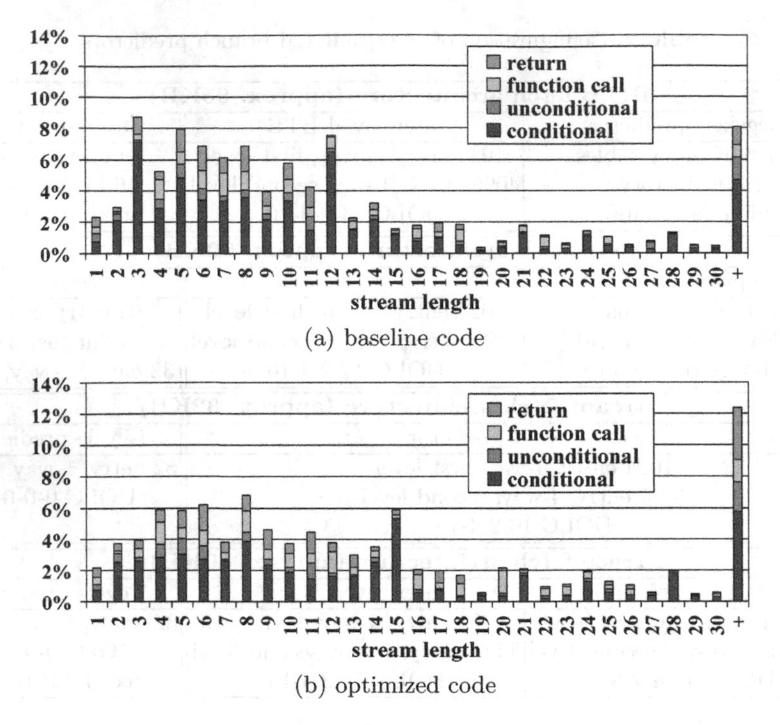

Fig. 3. Histograms of dynamic streams classified according to their length and the terminating branch type. The results presented in these histograms are the average of the eleven benchmarks used.

code layout. In addition, streams are divided according to the terminating branch type: conditional branches, unconditional branches, function calls, and returns.

Using the baseline code layout, most streams are shorter than the average length: 70% of the dynamic streams have 12 or less instructions. Using the optimized code layout, the average length is higher. However, most streams are still shorter than the average length: 70% of the dynamic streams have 15 or less instructions. Therefore, in order to increase the average stream length, research should be focused in those streams that are shorter than the average length. For example, if we consider an 8-wide execution core, research effort should be devoted to enlarge streams shorter than 8 instructions. Using optimized codes, the percentage of those streams is reduced from 40% to 30%. Nevertheless, there is still room for improvement.

Most dynamic streams finish in taken conditional branches. They are 60% when using the baseline code and 52% when using the optimized code. The percentage is lower in the optimized codes due to the higher number of not taken conditional branches, which never finish instruction streams. There also is a big percentage of streams terminating in function calls and returns. They are 30% of all dynamic streams in the baseline code. The percentage is larger in the optimized code: 36%. This happens because code layout optimizations are mainly focused on conditional branches. Since the number of taken conditional branches is lower, there is a higher percentage of streams terminating in other types of branches, although the total number is similar.

5 Multiple Stream Prediction

According to the analysis presented in the previous section, one could think that, in order to enlarge instruction streams, the most promising field for research are conditional branches, function calls, and return instructions. However, we have found that techniques for enlarging the streams finalizing in particular branch types achieve poor results [18]. This is due to Amdahl's law: although these techniques enlarge a set of instructions streams, there are other streams that are not enlarged, limiting the achievable benefit. Therefore, we must try to enlarge not particular stream types, but all instruction streams. Our approach to achieve this is the multiple stream predictor.

5.1 The Multiple Stream Predictor

The next stream predictor [11,16], which is shown in Figure 4.a, is a specialized branch predictor that provides stream level sequencing. Given a fetch address, i.e., the current stream starting address, the stream predictor provides the current stream length, which indicates where is the taken branch that finalizes the stream. The predictor also provides the next stream starting address, which is used as the fetch address for the next cycle. The current stream starting address and the current stream length form a fetch request that is stored in the FTQ. The fetch requests stored in the FTQ are then used to drive the instruction cache.

Actually, the stream predictor is composed by two cascaded tables: a first level table indexed only by the fetch address, and a second level table indexed using path correlation. A stream is only introduced in the second level if it is not accurately predicted by the first level. Therefore, those streams that do not need correlation are kept in the first level, avoiding unnecessary aliasing. In order to generate a prediction, both levels are looked up in parallel. If there is a second level table hit, its prediction is used. Otherwise, the prediction of the first level table is used. The second level prediction is prioritized because it is supposed to be more accurate than the first level due to the use of path correlation.

The objective of our multiple stream predictor is predicting together those streams that are frequently executed as a sequence. Unlike the trace cache, the instructions corresponding to a sequence of streams are not stored together in a special purpose buffer. The instruction streams belonging to a predicted sequence are still separate streams stored in the instruction cache. Therefore, the multiple stream predictor does not enable the ability of fetching instructions beyond a taken branch in a single cycle. The benefit of our technique comes from grouping predictions, allowing to tolerate the prediction table access latency.

Figure 4.b shows the fields required by a 2-stream multiple predictor. Like the original single stream predictor, a 2-stream predictor requires a single tag field,

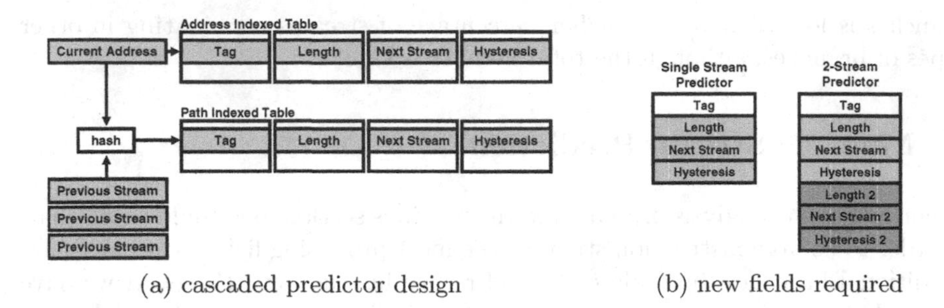

(a) cascaded predictor design (b) new fields required

Fig. 4. The next stream predictor

which corresponds to the starting address of the stream sequence. However, the rest of the fields should be duplicated. The tag and length fields determine the first stream that should be executed. The target of this stream, determined by the next stream field, is the starting address of the second stream, whose length is given by the second length field. The second next stream field is the target of the second stream, and thus the next fetch address.

In this way, a single prediction table lookup provides two separate stream predictions, which are supposed to be executed sequentially. After a multiple stream prediction, every stream belonging to a predicted sequence is stored separately in the FTQ, which involves that using the multiple-stream predictor does not require additional changes in the processor front-end. Extending this mechanism for predicting three or more streams per sequence would be straightforward, but we have found that sequences having more than two streams do not provide additional benefit.

5.2 Multiple Stream Predictor Design

Providing two streams per prediction needs duplicating the prediction table size. In order to avoid a negative impact on the prediction table access latency and energy consumption, we only store multiple streams in the first-level table of the cascaded stream predictor, which is smaller than the second-level table. Since the streams belonging to a sequence are supposed to be frequently executed together, it is likely that, given a fetch address, the executed sequence is always the same. Consequently, stream sequences do not need correlation to be correctly predicted, and thus keeping them in the first level table does not limit the achievable benefit.

In order to take maximum advantage of the available space in the first level table, we use hysteresis counters to detect frequently executed stream sequences. Every stream in a sequence has a hysteresis counter associated to it. All hysteresis counters behave like the counter used by the original stream predictor to decide whether a stream should be replaced from the prediction table [16]. When the predictor is updated with a new stream, the corresponding counter is increased if the new stream matches with the stream already stored in the selected entry. Otherwise, the counter is decreased and, if it reaches zero, the whole predictor entry is replaced with the new data, setting the counter to one. If the decreased counter does not reach zero, the new data is discarded. We have found that 3-bit hysteresis counters, increased by one and decreased by two, provide the best results for the multiple stream predictor.

When the prediction table is looked up, the first stream is always provided. However, the second stream is only predicted if the corresponding hysteresis counter is saturated, that is, if the counter has reached its maximum value. Therefore, if the second hysteresis counter is not saturated, the multiple stream predictor provides a single stream prediction as it would be done by the original stream predictor. On the contrary, if the two hysteresis counters are saturated, then a frequently executed sequence has been detected, and the two streams belonging to this sequence are introduced in the FTQ.

6 Evaluation of the Multiple Stream Predictor

Our multiple stream predictor is able to provide a high amount of instructions per prediction. Figure 5 shows an histogram of instructions provided per prediction. It shows the percentage of predictions that provide an amount of instructions ranging from 1 to 30 instructions. The last bar shows the percentage of predictions that provide more than 30 instructions. Data are shown for both the baseline and the optimized code layout. In addition, data are shown for the original single-stream predictor, described in [11,16], and a 2-stream multiple predictor.

The main difference between both code layouts is that, as can be expected, there is a lower percentage of short streams in the optimized code. Besides, it is clear that our multiple stream predictor efficiently deals with the most harmful problem, that is, the shorter streams. Using our multiple stream predictor, there is an important reduction in the percentage of predictions that provide a small number of instructions. Furthermore, there is an increase in the percentage of predictions that provide more than 30 instructions, especially when using optimized codes. The lower number of short streams points out that the multiple stream predictor is an effective technique for hiding the prediction table access latency by overlapping table accesses with the execution of useful instructions.

Figure 6 shows the average processor performance achieved by the four evaluated fetch architectures, for both the baseline and the optimized code layout. We have evaluated a wide range of predictor setups and selected the best one found for each evaluated predictor. Besides the performance of the four fetch engines using overriding, the performance achieved by the trace cache fetch architecture and the stream fetch engine not using overriding is also shown. In the latter case, the stream fetch engine uses a 2-stream multiple predictor instead of the original single-stream predictor.

The main observation from Figure 6 is that the multiple stream predictor without overriding provides a performance very close to the original single-stream predictor using overriding. The performance achieved by the multiple stream predictor without overriding is enough to outperform both the iBTB and the

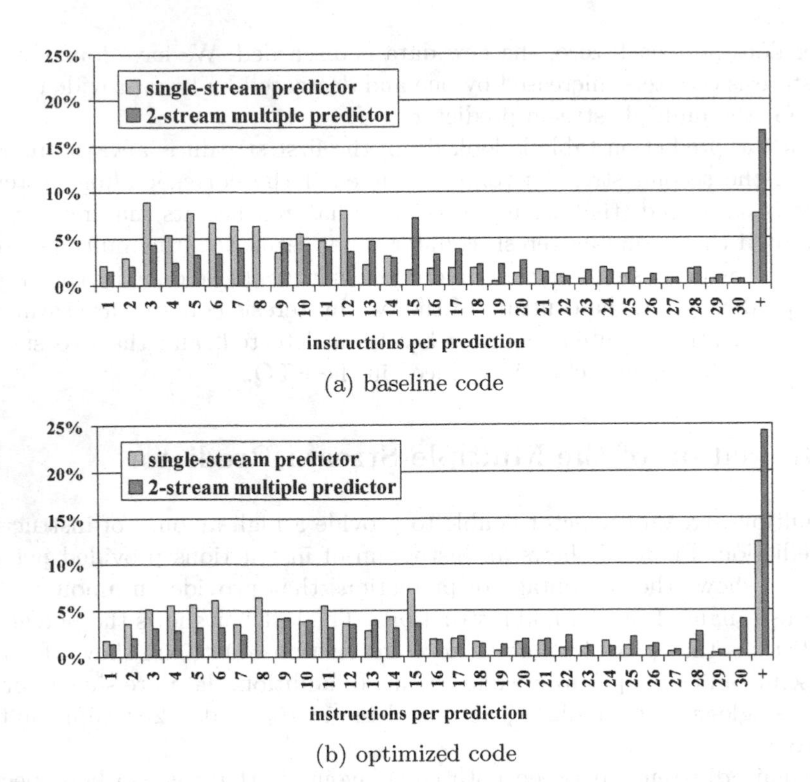

(a) baseline code

(b) optimized code

Fig. 5. Histograms of dynamic predictions, classified according to the amount of instructions provided, when using a single-stream predictor and a 2-stream multiple predictor. The results presented in these histograms are the average of the eleven benchmarks used.

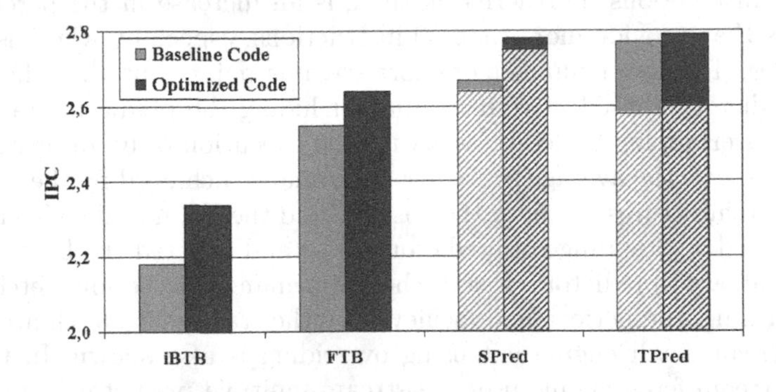

Fig. 6. Processor performance when using (full bar) and not using (shadowed bar) overriding

FTB fetch architectures, even when they do use overriding. The performance of the multiple stream predictor without overriding is also close to a trace cache using overriding, while requiring lower complexity.

It should be taken into account that this improvement is achieved by increasing the size of the first level table. Fortunately, the tag array is unmodified and no additional access port is required. We have checked using CACTI [22] that the increase in the predictor area is less than 12%, as well as that the prediction table access latency is not increased. Moreover, our proposal not only avoids the need for a complex overriding mechanism, but also reduces the predictor energy consumption. Although the bigger first level table consumes more energy per access, it is compensated with the reduction in the number of prediction table accesses. The ability of providing two streams per prediction causes 35% reduction in the total number of prediction table lookups an updates, which leads to 12% reduction in the overall stream predictor energy consumption.

7 Conclusions

Current technology trends create new challenges for the fetch architecture design. Higher clock frequencies and larger wire delays cause branch prediction tables to require multiple cycles to be accessed [1,7], limiting the fetch engine performance. This fact has led to the development of complex hardware mechanisms, like prediction overriding [7,19], to hide the prediction table access delay.

To avoid this increase in the fetch engine complexity, we propose to use long instruction streams [11,16] as basic prediction unit, which makes it possible to hide the prediction table access delay. If instruction streams are long enough, the execution engine can be kept busy executing instructions from a stream during multiple cycles, while a new stream prediction is being generated. Therefore, the prediction table access delay can be hidden without requiring any additional hardware mechanism.

In order to take maximum advantage of this fact, it is important to have streams as long as possible. We achieve this using the multiple stream predictor, a novel predictor design that combines frequently executed instruction streams into long virtual streams. Our predictor provides instruction streams long enough for allowing a processor not using overriding to achieve a performance close to a processor using prediction overriding, that is, we achieve a very similar performance at a much lower complexity, also requiring less energy consumption.

Acknowledgements

This work has been supported by the Ministry of Education of Spain under contract TIN–2004–07739–C02–01, the HiPEAC European Network of Excellence, the Barcelona Supercomputing Center, and an Intel fellowship.

References

1. Agarwal, V., Hrishikesh, M.S., Keckler, S.W., Burger, D.: Clock rate versus IPC: The end of the road for conventional microarchitectures. In: 27th Intl. Symp. on Computer Architecture (2000)

2. Falcón, A., Santana, O.J., Ramirez, A., Valero, M.: Tolerating branch predictor latency on SMT. In: 5th Intl. Symp. on High Performance Computing (2003)
3. Gwennap, L.: Digital 21264 sets new standard. Microprocessor Report 10(14) (1996)
4. Hrishikesh, M.S., Jouppi, N.P., Farkas, K.I., Burger, D., Keckler, S.W., Shivakumar, P.: The optimal useful logic depth per pipeline stage is 6-8 fo4. In: 29th Intl. Symp. on Computer Architecture (2002)
5. Jacobson, Q., Rotenberg, E., Smith, J.E.: Path-based next trace prediction. In: 30th Intl. Symp. on Microarchitecture (1997)
6. Jimenez, D.A.: Reconsidering complex branch predictors. In: 9th Intl. Conf. on High Performance Computer Architecture (2003)
7. Jimenez, D.A., Keckler, S.W., Lin, C.: The impact of delay on the design of branch predictors. In: 33rd Intl. Symp. on Microarchitecture (2000)
8. Jimenez, D.A., Lin, C.: Dynamic branch prediction with perceptrons. In: 7th Intl. Conf. on High Performance Computer Architecture (2001)
9. Kaeli, D., Emma, P.: Branch history table prediction of moving target branches due to subroutine returns. In: 18th Intl. Symp. on Computer Architecture (1991)
10. Ramirez, A., Larriba-Pey, J.L., Valero, M.: Trace cache redundancy: red & blue traces. In: 6th Intl. Conf. on High Performance Computer Architecture (2000)
11. Ramirez, A., Santana, O.J., Larriba-Pey, J.L., Valero, M.: Fetching instruction streams. In: 35th Intl. Symp. on Microarchitecture (2002)
12. Reinman, G., Austin, T., Calder, B.: A scalable front-end architecture for fast instruction delivery. In: 26th Intl. Symp. on Computer Architecture (1999)
13. Rosner, R., Mendelson, A., Ronen, R.: Filtering techniques to improve trace cache efficiency. In: 10th Intl. Conf. on Parallel Architectures and Compilation Techniques (2001)
14. Rotenberg, E., Bennett, S., Smith, J.E.: A trace cache microarchitecture and evaluation. IEEE Transactions on Computers 48(2) (1999)
15. Santana, O.J., Falcón, A., Fernández, E., Medina, P., Ramirez, A., Valero, M.: A comprehensive analysis of indirect branch prediction. In: 4th Intl. Symp. on High Performance Computing (2002)
16. Santana, O.J., Ramirez, A., Larriba-Pey, J.L., Valero, M.: A low-complexity fetch architecture for high-performance superscalar processors. ACM Transactions on Architecture and Code Optimization 1(2) (2004)
17. Santana, O.J., Ramirez, A., Valero, M.: Latency tolerant branch predictors. In: Intl. Workshop on Innovative Architecture for Future Generation High-Performance Processors and Systems (2003)
18. Santana, O.J., Ramirez, A., Valero, M.: Techniques for enlarging instruction streams. Technical Report UPC-DAC-RR-2005-11, Departament d'Arquitectura de Computadors, Universitat Politècnica de Catalunya (2005)
19. Seznec, A., Felix, S., Krishnan, V., Sazeides, Y.: Design tradeoffs for the Alpha EV8 conditional branch predictor. In: 29th Intl. Symp. on Computer Architecture (2002)
20. Seznec, A., Fraboulet, A.: Effective ahead pipelining of instruction block address generation. In: 30th Intl. Symp. on Computer Architecture (2003)
21. Sherwood, T., Perelman, E., Calder, B.: Basic block distribution analysis to find periodic behavior and simulation points in applications. In: 10th Intl. Conf. on Parallel Architectures and Compilation Techniques (2001)
22. Shivakumar, P., Jouppi, N.P.: CACTI 3.0: an integrated cache timing, power and area model. Technical Report 2001/2, Western Research Laboratory (2001)

Enhanced Loop Coalescing: A Compiler Technique for Transforming Non-uniform Iteration Spaces

Arun Kejariwal[1], Alexandru Nicolau[1], and Constantine D. Polychronopoulos[2]

[1] Center for Embedded Computer Systems
University of California at Irvine
Irvine, CA 92697, USA
arun_kejariwal@computer.org, nicolau@cecs.uci.edu
http://www.cecs.uci.edu/

[2] Center for Supercomputing Research and Development
University of Illinois at Urbana-Champaign
Urbana, IL 61801, USA
cdp@csrd.uiuc.edu
http://www.csrd.uiuc.edu/

Abstract. Parallel nested loops are the largest potential source of parallelism in numerical and scientific applications. Therefore, executing parallel loops with low run-time overhead is very important for achieving high performance on parallel computers. Guided self-scheduling (GSS) has long been used for dynamic scheduling of parallel loops on shared memory parallel machines and for efficient utilization of dynamically allocated processors. In order to minimize the synchronization (or scheduling) overhead in GSS, loop coalescing has been proposed as a restructuring technique to transform nested loops into a single loop. In other words, coalescing "flattens" the iteration space in lexicographic order of the indices of the original loop. Although coalescing helps reduce the run-time scheduling overhead, it does not necessarily minimize the makespan, i.e., the maximum finishing time, especially in situations where the execution time (workload) of iterations is not uniform as is often the case in practice, e.g., in control intensive applications. This can be attributed to the fact that the makespan is directly dependent on the workload distribution across the flattened iteration space. The latter in itself depends on the order of coalescing of the loop indices. We show that coalescing (as proposed) can potentially result in large makespans. In this paper, we present a loop permutation-based approach to loop coalescing, referred to as enhanced loop coalescing, to achieve near-optimal schedules. Several examples are presented and the general technique is discussed in detail.

1 Introduction

Advances in silicon technology has enabled the development of petascale computing systems [1,2,3,4,5,6,7]. Applications that require petaflop speed include (but are not limited to) simulation of physical and artificial phenomenon [8],

J. Labarta, K. Joe, and T. Sato (Eds.): ISHPC 2005 and ALPS 2006, LNCS 4759, pp. 17–32, 2008.

medicine (reconstruction of 3-d images from 2-d images), business (modeling of transportation and economic systems), computational biology applications such as protein folding, modeling of integrated Earth systems, remote sensing data assimilation, simulation of aerodynamics, hydrodynamics [9], transient dynamics [10], molecular dynamics [11,12] gas turbine engines et cetera. In order to meet the high performance requirements of such applications and efficiently exploit the parallel processing power of such petaflop architectures [13,14,15], advanced algorithms and compilers need to be developed for program parallelization and optimization and scheduling.

It has been shown that parallel nested loops account for the greatest percentage of parallelism in high performance computing applications [16]. Thus, it is very important to optimize the execution of such loops on parallel processor systems [17,18]. Guided self-scheduling has long been used for dynamic scheduling (as in `DOALL` Cray scheduling) of nested loops on shared memory parallel systems [19]. In this context GSS is widely used as it allows for efficient utilization of dynamically allocated processors[1] and minimizes run-time scheduling overhead. In order to make nested loops amenable for GSS, *loop coalescing* has been proposed as a compiler restructuring technique for shared memory supercomputers [20]. Loop coalescing transforms multiply nested `DOALL` loops into singly nested loops. The transformation maps a multidimensional iteration space into a single dimensional iteration space through a set of mappings that express the indices or "co-ordinates" of the multidimensional iteration space as a function of a single index (the transformation is further discussed in Section 3). The "flattening" of the iteration space simplifies the loop structure, thereby facilitating efficient dynamic scheduling by reducing the associated scheduling overhead. However, loop coalescing does not account for the workload distribution across the iteration space – iterations typically have variable workloads (execution times) due to both algorithmic and systemic variations (operating system, interference with other programs).

In this paper we present a novel technique for transforming, specifically coalescing (or flattening), iteration spaces with non-uniform workloads to minimize scheduling overhead and the makespan. For the same, we compute the workload gradient (to characterize the workload distribution) across the iteration space. The gradient can either be computed at the loop boundaries (in a nested loop) or at the iteration-level. Subsequently, the iteration space is coalesced along an axis corresponding to the maximum gradient. Finally, the transformed loop is dynamically scheduled on a parallel multicomputer using *guided self-scheduling.*

The rest of the paper is organized as follows: Next, we present the terminology used in the rest of the paper. In Section 3 we present a brief overview of *loop coalescing* and *guided self-scheduling* (GSS). A formal description of the problem we address in this paper is presented in Section 4. Section 5 presents a motivating example. Next, in Section 6 we discuss our approach for transforming iteration spaces with non-uniform workload distribution. A case study is presented in

[1] High performance systems are often time shared across jobs (of a single user) or users owing to their high cost. This calls for dynamic allocation of processors.

Section 7. In Section 8 we discuss previous work. Finally, we conclude with directions for future research in Section 9.

2 Terminology

Our loop model consists of a non-perfectly nested `DOALL` loop with fixed loop bounds, as shown in Figure 1. Further, our model also supports nested conditionals at each level of the nested loop. The **index variables** of the individual loops are $i_1, i_2, \ldots, i_n$ and they constitute an **index vector** $\mathbf{i} = \langle i_1, i_2, \ldots, i_n \rangle$. An **iteration** is an instance of the index vector $\mathbf{i}$. The set of iterations of a loop nest $\mathbf{L}$ is an **iteration space** $\Gamma = \{\mathbf{i}\}$. Let $\mathbf{N}$ denote the total number of iterations in Γ. Assuming normalized indices, $\mathbf{N}$ is given by:

$$\mathbf{N} = \prod_{k=1}^{n} N_k$$

where, N_k is the upper bound of index variable i_k. An iteration space is said to have *uniform* workload distribution if all the iterations in Γ have equal execution times (or workloads). However, in the presence of conditionals the iterations tend to have different workloads. Such an iteration space is said to have *non-uniform* workload distribution. Let $\mathcal{W}(\mathbf{i})$ denote the average workload (obtained by profiling the application with multiple training sets) of an iteration $\mathbf{i}$ and $\mathbf{W}$ denote the total workload.

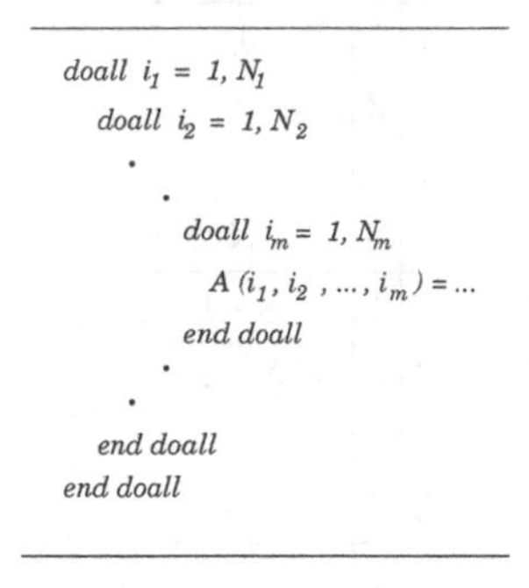

Fig. 1. Our loop model

3 Background

In this section we present a brief overview of *loop coalescing* and *guided self-scheduling.*

3.1 Loop Coalescing

In [20], Polychronopoulos proposed *loop coalescing* as a restructuring technique for transforming nested parallel loops. Loop coalescing manipulates the loop subscripts such that there exists a one-to-one mapping between the array subscripts of the original and the transformed loop. In other words, the transformation determines a mapping between the subscripts of the original loop and the single subscript of the restructured loop. An example of loop coalescing is illustrated in Figure 2.

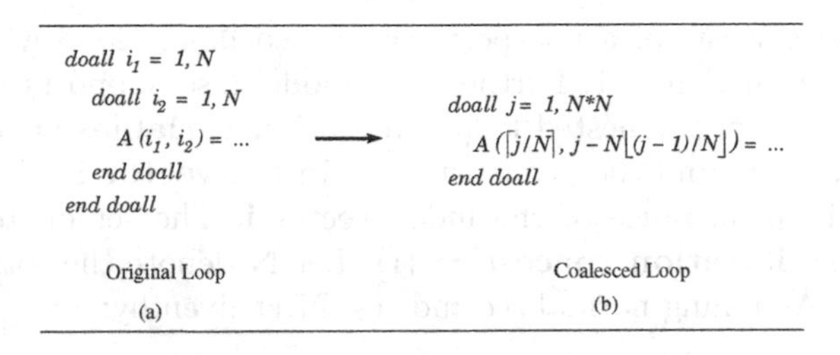
doall i1 = 1, N
doall i2 = 1, N
A (i1, i2) = ...
end doall
end doall
doall j= 1, N*N
A (⌈j/N⌉, j – N⌊(j – 1)/N⌋) = ...
end doall
Original Loop
(a)
Coalesced Loop
(b)

Fig. 2. Example of Loop Coalescing

Table 1 shows the index values for the original and the coalesced loop. Note that in case of non-unit strides, the lower and upper loop bounds are normalized w.r.t. the stride before coalescing.

Table 1. Index Mapping for the example shown in Figure 2

i_1	i_2	j
1	1	1
1	2	2
.	.	.
1	N	N
2	1	N+1
2	2	N+2
.	.	.
2	N	2N
.	.	.
.	.	.
N	1	(N-1) N+1
N	2	(N-1) N+2
.	.	.
N	N	N*N

Observe that the first subscript i_1 of the original loop is transformed into an expression involving j, i.e.,

$$i_1 \rightarrow f(j)$$

where f is an integer-valued function such that the value of $f(j)$ is incremented by 1 each time j assumes a value of the form $wN+1$, for $w \in \mathbb{Z}^+$. Similarly, i_2 is transformed such that

$$i_2 \rightarrow g(j)$$

$g(j)$ assumes the successive values $1, 2, \ldots, N$ but its values wrap around each time $f(j)$ becomes $wN+1$. The mapping functions for the example shown in Figure 2(b) is given by:

$$f(j) = \left\lceil \frac{j}{N} \right\rceil \quad \text{and} \quad g(j) = j - N \left\lfloor \frac{j-1}{N} \right\rfloor$$

It is easy to verify that the mappings satisfy the properties mentioned above. Observe that the mapping functions f and g follow a regular pattern. In [20], Polychronopoulos showed that loop coalescing can also applied to nested loops with unequal loop bounds. The general array subscript transformation for a loop of the form shown in Figure 1 is given by the following theorem.[2]

Theorem 1. *[20] Let* $\mathbf{L}$, $\mathbf{L}'$ *represent the original and the coalesced loop respectively. An array reference of the form* $A(i_1, i_2, \ldots, i_m)$ *in* $\mathbf{L}$ *can be uniquely expressed by an equivalent array reference* $A(f_1(j), f_2(j), \ldots, f_m(j))$ *in* $\mathbf{L}'$ *where*

$$f_k(j) = \left\lceil \frac{j}{\Pi_{p=k+1}^{m} N_p} \right\rceil - N_k \left\lfloor \frac{j-1}{\Pi_{p=k}^{m} N_p} \right\rfloor \quad \textit{for} \quad k = 1, 2, \ldots, m \qquad (1)$$

From Equation 1, we observe that the transformation for the outermost loop i_1 is $\left\lceil \frac{j}{\Pi_{p=2}^{m} N_p} \right\rceil$. In case of equal loop bounds, the mapping can be simplified as follows:

$$f_k(j) = \left\lceil \frac{j}{N^{m-k}} \right\rceil - N_k \left\lfloor \frac{j-1}{N^{m-k+1}} \right\rfloor \quad \text{for} \quad k = 1, 2, \ldots, m$$

Loop coalescing reduces the amount of synchronization (in other words, run-time overhead) associated with the execution of parallel loops. The transformation can be viewed as flattening of the iteration space along an axis corresponding to the outermost loop.

3.2 Guided Self-Scheduling

Polychronopoulos and Kuck proposed the *guided self-scheduling* (GSS) technique for dynamic scheduling of parallel loops [21]. Given a nested loop, GSS coalesces the loop to obtain a single index $j = 1, \ldots, \mathbf{N}$. At each scheduling step p, GSS allocates Λ number of iterations (also referred to as a *chunk*) to an idle processor, where

$$\Lambda_p = \left\lceil \frac{R_i}{\mathbf{P}} \right\rceil; \quad R_{p+1} \leftarrow R_p - \Lambda_p$$

[2] The transformation is also applicable for non-perfectly nested parallel loops.

Table 2. Example of Guided Self-Scheduling

Chunk Sizes	# of Sync. Points
10 5 3 1 1	5

$R_1 = \mathbf{N}$ and $\mathbf{P}$ is the number of processors. For example, given $\mathbf{N} = 20$ and $\mathbf{P} = 2$, the chunk sizes and the number of synchronization points corresponding to the schedule obtained by GSS is given by:

From Table 2 we observe that towards the end of a GSS(1) schedule, chunks are comprised of single iterations. Scheduling single iterations can potentially incur large scheduling overhead. Therefore, the chunk size is restricted to $\Lambda_{\min}$ (and is pre-specified). The modified expression for Λ is given by

$$\Lambda_p = \max\left(\Lambda_{\min}, \left\lceil \frac{R_i}{\mathbf{P}} \right\rceil\right) \tag{2}$$

The range of iterations assigned to the p-th processor is given by $[\mathbf{N} - R_p + 1, \ldots, \mathbf{N} - R_p + \Lambda_p]$. GSS models loops with conditionals as pseudo-tree. The execution time of the loop body is assumed to be equal to the execution time of the shortest path in the tree. GSS minimizes the scheduling overhead as processors perform scheduling by themselves at run-time, in contrast to (traditional) dynamic scheduling wherein processors are scheduled by the operating system or by a global control unit.

4 Problem Statement

In this section we present a formal description of scheduling iteration spaces (of parallel nested loops) with non-uniform workloads. We are given an iteration space $\Gamma = \{\mathbf{i}\}$ (comprised of $\mathbf{N}$ independent iterations), each iteration having $\mathcal{W}(\mathbf{i})$ workload and a set of $\mathbf{P}$ identical processors. A schedule in this case can be thought as a partition $\mathcal{S} = \langle S_1, S_2, \ldots, S_{\mathbf{P}} \rangle$ of Γ into $\mathbf{P}$ disjoint sets. The q-th processor, for $1 \leq q \leq \mathbf{P}$, executes the iterations in S_q. The makespan (or maximum finishing time) of the schedule is given by

$$f(\mathcal{S}) = \max_{1 \leq q \leq \mathbf{P}} \ell(S_q)$$

where for any $X \subseteq \Gamma$, $\ell(X)$ is defined to $\sum_{\mathbf{i} \in X} \mathcal{W}(\mathbf{i})$.[3]

The above formalization is analogous to the classical bin-packing problem or the job-shop scheduling problem. Note that unlike the job-scheduling problem where the jobs can be arbitrarily reordered, in our case reordering of iterations of a parallel (nested) loop is restricted to loop boundaries (in other words, iterations are reordered at loop-level (via loop interchange). The restriction is introduced so as to minimize the run-time scheduling overhead and the increase in code

[3] The notation used in this section is the one used by Coffman et al. [22].

size. It is well known that these problems are NP-complete [23]. Therefore, we seek efficient algorithms to achieve "near-optimal" schedules.

In the next section we present a motivating example to present an intuitive idea of approach. Subsequently, we discuss our approach in detail.

5 A Motivating Example

Consider the iteration space shown in Figure 3(a). The iteration space shows non-uniform workload distribution of a doubly-nested DOALL loop with $N_1 = N_2 = 5$. Assuming 2 processors and $\Lambda_{\min} = 3$, the chunk sizes corresponding to a GSS schedule is given by:

Since iterations of a DOALL loop are independent (by definition), the iteration space can be flattened along any axis of the iteration space. Let us flatten the iteration space shown in Figure 3 along the i_1 and i_2 axis, shown in Figures 3(b) and 3(c) respectively. Observe that the flattened iteration spaces have *different* workload profiles. Assuming processor 1 is available at $t = 0$ and processor 2 is available at $t = 5$, the GSS schedules corresponding to the iteration spaces of Figures 3(b) and 3(c) is shown in Figures 4(a) and 4(b) respectively. The numbers

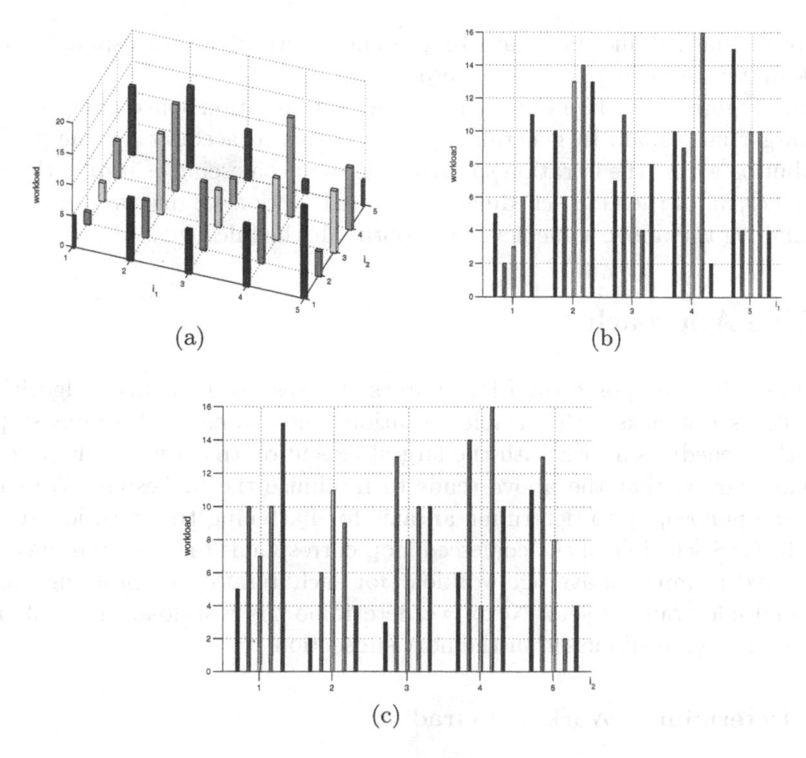

Fig. 3. a) Iteration space with non-uniform workload distribution; b) Flattened iteration space along i_1; and c) Flattened iteration space along i_2

Table 3. Chunk sizes corresponding to the self-schedule of the iteration space shown in Figure 3

Chunk Sizes	# of Sync. Points
13 6 3 3	4

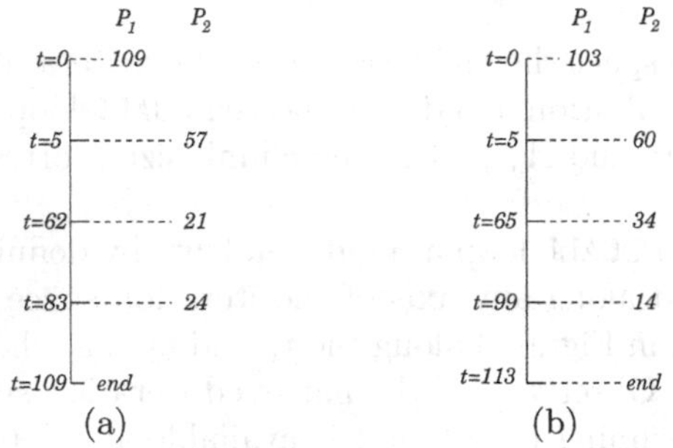

Fig. 4. a) GSS schedule for the iteration space of Figure 3(b); b) GSS schedule for the iteration space of Figure 3(c)

next to the dashed lines in Figure 4 represent the workload corresponding to the chunk of iterations mapped onto a processor.

From Figure 4 we observe that flattening the iteration space along i_2 results in a larger makespan. In general, it is imperative to account for the workload distribution across the iteration space during loop coalescing as it directly affects the makespan of the GSS schedule. In the next section we discuss our approach for handling the above to achieve near-optimal schedules.

6 The Approach

In [24], Graham proposed the LPT (Largest Processing Time first) algorithm to schedule a set of tasks with variable execution times. At each scheduling step, the algorithm schedules a task with the largest execution time on an idle processor. Graham showed that the above tends to minimize the makespan. We employ the same philosophy to determine an axis for flattening the iteration space so that the GSS schedule of the coalesced loop corresponds to minimum makespan. First, we determine an average workload for each iteration by profiling the loop with multiple training sets. Next, we determine the workload gradient of the iteration space, as discussed in the next subsection.

6.1 Determining Workload Gradient

We define a workload gradient, denoted by $\nabla(\Gamma)$, to characterize the variation in workload across the iteration space. To determine the workload gradient along an axis corresponding to a given loop index i_k, first the iteration space is flattened (symbolically) in lexicographical order of the loop indices excluding i_k. Subsequently, the gradient is computed either at a coarse-grain or at a fine-grain granularity as discussed below:

- *Coarse-grain:* The coarse-grain workload is computed as follows: Compute the average workload of the set of iterations corresponding to $i_k = 1, 2, \ldots, N_k$. Next, the workload gradient along i_k, denoted by $\nabla(\Gamma, i_k)$, is computed by summing the difference in successive average values. For example, the coarse-grain workload gradient of the iteration space shown in Figure 3(b) is $\nabla(\Gamma, i_1) = -3.2$ and for the iteration space shown in Figure 3(c) is $\nabla(\Gamma, i_2) = 1.8$.
- *Fine-grain:* In this case, $\nabla(\Gamma)$ is computed as a cumulative sum of the difference in workloads of successive iterations of the coalesced (along i_k) loop. For example, the fine-grain workload gradient of the iteration spaces shown in Figures 3(b) and 3(c) is $\nabla(\Gamma, i_1) = -1.0$ and $\nabla(\Gamma, i_2) = -11.0$ respectively.

In context of GSS, fine-grain workload gradient is not a good measure to determine the variation in workload across the iteration space; in contrast, the coarse-grain approach better captures the workload gradient as GSS allocates a chunk of iterations (to an idle processor) at each scheduling step. However, the chunks in GSS do not necessarily correspond to $i_k = 1, 2, \ldots, N_k$. Therefore, we propose to compute the workload gradient between the chunks of a GSS schedule. Recall that both the chunk sizes and the iterations constituting the chunks can be determined at compile-time. For example, given $\Lambda_{\min} = 3$ and $\mathbf{P} = 2$, the chunk sizes for the iteration space shown in Figure 3(a) are given in Table 3. The workload of the chunks are 109, 57, 21, 24 respectively for coalescing done along i_1 and 103, 60, 34, 14 respectively for coalescing done along i_2. The workload gradient for the two cases is $\nabla(\Gamma, i_1) = 85$ and $\nabla(\Gamma, i_2) = 83$. Since, $\nabla(\Gamma, i_1) > \nabla(\Gamma, i_2)$ we coalesce the loop along i_1. In the next subsection, we formally outline the algorithm for *enhanced loop coalescing*.

6.2 The Transformation

A formal description of our approach is presented in Algorithm 1. As discussed earlier, the prime objective of the algorithm is to reorder the iterations in non-increasing order of their workloads, i.e., $\mathcal{W}(\mathbf{i}) \geq \mathcal{W}(\mathbf{j})$ for $\mathbf{i} < \mathbf{j}$ (based on the LPT philosophy). In this context, reordering corresponds to flattening of the iteration space and is restricted along an axis corresponding to maximum workload gradient.

First, the workload distribution across the iteration space is determined. For the same, the loop is profiled with multiple training sets. The workload of each iteration is an average of the workloads corresponding to the different training sets. Then, given a fixed number of processors $\mathbf{P}$, Algorithm 1 determines the chunk sizes for GSS. Next, the algorithm determine an axis $i*$ corresponding to

Algorithm 1. Enhanced Loop Coalescing

Input : An N-dimensional non-uniform iteration space, Γ and **P** processors.
Output : A flattened iteration space.

Determine the workload distribution across Γ via profiling

Compute the chunk sizes using the GSS algorithm (refer to Section 3.2)

/* Determine the axis for flattening the iteration space */
for each iterator i_k, where $1 \leq k \leq n$ **do**
  Determine $\nabla(\Gamma, i_k)$
end for
$i* \leftarrow \max_{1 \leq k \leq n} \nabla(\Gamma, i_k)$

Interchange the loop corresponding to the index $i*$ with the outermost loop

Coalesce the loop along the (new) outermost loop

maximum workload gradient (refer to Section 6.1). Subsequently, the outermost loop and the loop corresponding to index $i*$ are interchanged. Note that for the loop model shown in Figure 1 loop interchange is always valid as there do not exist any loop carried dependences [25]. Finally, the transformed loop is coalesced along the (new) outermost loop. Note that Algorithm 1 does not guarantee chunks (of a GSS schedule corresponding to the coalesced loop) with monotonically decreasing workloads.

The maximum workload gradient obtained in Algorithm 1 may not correspond to a global maximum. In order to achieve the same, we propose an iterative approach wherein Algorithm 1 is applied for each permutation of the loop indices. The permutation corresponding to the highest workload gradient is chosen for coalescing the loop. Arguably, the exhaustive approach is expensive w.r.t. compilation time. However, it improves program performance.

6.3 Analysis

The problem of obtaining a minimum makespan GSS schedule of a parallel nested loop is identical to that scheduling a set of tasks $\mathbf{T} = \{T_1, T_2, \ldots, T_r\}$ with an empty partial order [26] and a function $\mu : T \rightarrow (0, \infty)$.[4] Once a processor begins to execute a task T_j, it works without interruption until the completion of that task, requiring $\mu(T_j)$ units of time. In [24] Graham proposed the LPT algorithm wherein an idle processor always executes the longest remaining unexecuted task. In our case, a task (obtained from Algorithm 1) corresponds to a chunk of iterations; the number of iterations per task is governed by Equation 2. Note that the workload gradient between the tasks need not be monotonically decreasing. In addition, the tasks, though independent, cannot be reordered. However, in the

[4] The notation in this subsection is the same as in [24].

best case, i.e., when the workload gradient between the tasks is monotonically decreasing, the bound on the performance of GSS is given by[5]

$$1 \leq \frac{t(GSS)}{t(OPT)} \leq \frac{(4\mathbf{P}-1)}{3\mathbf{P}} \quad \text{for } \mathbf{P} \geq 1$$

6.4 Hybrid Loops

Enhanced loop coalescing can also applied to hybrid loops. A loop is *hybrid* if it contains combination of `DOALL`s, `DOACROSS`s and serial loops. In such cases enhanced loop coalescing can be applied to transform only the `DOALL`s of the hybrid loop. Only the subscripts of array references that correspond to the `DOALL`s are transformed in this case. The indices (subscripts) of any serial or `DOACROSS` loop are left unchanged.

In a similar fashion, enhanced loop coalescing can be applied to non-perfectly nested loops. The subscript transformations remain the same, but care must be taken to assure correct execution of loop bodies at different nesting levels. In such cases, loop bodies are executed conditionally. Likewise, enhanced loop coalescing can be applied to multiway loops subject to direction vector of the flow dependence between the loops at the same level.

7 Case Study

In this section, we illustrate the behavior of our approach with the help of an example. Consider the iteration space shown in Figure 5(a). Note that the workload is non-uniformly distributed across the iteration space. The workload of each iteration is obtained by profiling the doubly-nested `DOALL` loop of image reconstruction (another execution profile of the same loop is available in [28]). Assuming $\mathbf{P} = 2$, the chunk sizes corresponding to a GSS schedule are given in Table 3. Next, we determine the workload for each chunk and workload gradient (summarized in Table 4) assuming coalescing along each (i_1 and i_2) axis (the flattened iteration spaces are shown in Figures 5(b) and 5(c) respectively). Note that the successive chunks of the loop coalesced along i_1 do not have monotonically decreasing workloads.

Table 4. Workload of each chunk and the workload gradient corresponding to the iteration spaces shown in Figures 5(b) and 5(c)

Axis	Workload of each chunk	$\nabla(\Gamma)$
i_1	97 64 10 45	52
i_2	88 78 33 17	71

[5] Bounds on LPT schedules for uniform processors are further discussed in [27].

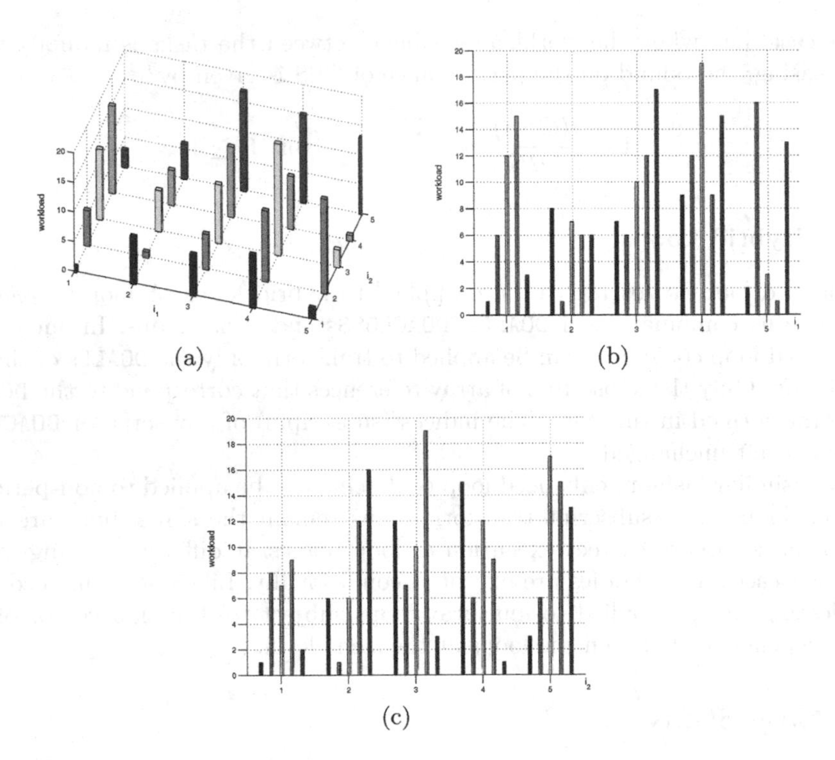

Fig. 5. a) An example non-uniform iteration space b) Flattened iteration space along i_1; and c) Flattened iteration space along i_2

Assuming that the processors P_1, P_2 are available at $t = 0$ and $t = 5$ respectively, the GSS schedules corresponding to the loops coalesced along i_1 and i_2 are shown in Figures 6(a) and 6(b) respectively. From Figure 6(a) we observe that processor P_1 is assigned a large (97 units) amount of work at $t = 0$. As a consequence, the remaining chunks are mapped onto processor P_2. The last chunk (with 45 units of work) is scheduled on P_2 at $t = 79$, thus resulting in a makespan of 124. Observe that P_1 remains idle for the period $97 \leq t \leq 124$.[6] The large idle time of P_1 can be attributed to large workload of the last chunk. Recall that chunks with small workloads towards the end of the schedule help balance load amongst different processors. On the other hand, from Figure 6(b) we note that P_1 is allocated 88 units of work at time $t = 0$, thus avoiding allocation of large chunks of work at the beginning of the schedule. It facilitates scheduling of the last chunk (with 17 units of work) on P_1, thereby balancing the workload between the two processors. The latter is evident from the fact that in this case the makespan is 116. Furthermore, the idle time of P_1 is limited to $105 \leq t \leq 116$.

Observe that unlike the example shown in Section 5 where the loop was coalesced along i_1, in this case the loop is coalesced along i_2.

[6] Recall that GSS generates a non-preemptive schedule.

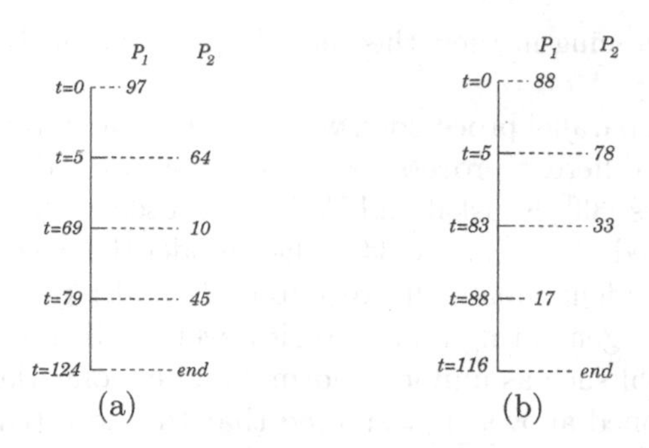

Fig. 6. a) GSS schedule for the iteration space of Figure 5(b); b) GSS schedule for the iteration space of Figure 5(c)

8 Related Work

Static scheduling of parallel nested loops can be solved optimally in polynomial time. In [29], Polychronopoulos et al. proposed an algorithm, called *OPTAL*, for the same. In case of singly nested loops with uniform iteration spaces, the obvious one-step processor assignment is also the optimal one, the optimal distribution is clearly the one that assigns $\lceil \mathbf{N}/\mathbf{P} \rceil$ iterations to each processor. It would be therefore desirable to have, if possible, if possible, parallel programs with singly nested loops. Thus, for those loops that can be restructured the optimal processor assignment problem becomes simple. Moreover, the processor assignment of the transformed (coalesced) loops is generally better than the optimal assignments of the original loops.

Loop coalescing resembles *loop collapsing* [30]. However, the latter is different from the former in both its purpose and mechanism. Loop collapsing is a memory related transformation that collapses doubly nested loops only, to single loops by transforming two dimensional arrays into vectors. An example of loop collapsing is illustrated in Figure 7.

```
doall i1 = 1, N
  doall i2 = 1, M                             doall i1 = 1, N*M
    A (i1, i2) = B (i1, i2) + 2    ---->        A (i1) = B (i1) + 2
  end doall                                   end doall
end doall
```

Fig. 7. Example of Loop Collapsing

Loop collapsing is used to create long vectors for efficient execution on memory-to-memory SIMD systems. No subscript manipulation is attempted in loop collapsing, which by the applicable only to double perfectly nested `DOALL`s. The use of loop coalescing for MIMD architectures is discussed in [31]. Tabirca et al. employ loop coalescing in their theoretical approach for dynamic scheduling of loops [32].

Self-scheduling of parallel processors, where successive iterations are allocated and executed on to different processors one by one, was first used on Denelcor HEP multiprocessors [33]. Kruskal and Weiss proposed *static chunking* of iterations during the scheduling process [34]. They model the execution times of the iterations as independent identically distributed (i.i.d.) random variables and possessing a moment generating function. However, their model is restrictive to IFR distributions [35] such as uniform, normal and exponential. For the class of distributions mentioned above, they showed that the expected completion time is given by $\mathbf{N}\mu/\mathbf{P} + \sigma\sqrt{2\mathbf{N}\ln\mathbf{P}/\mathbf{P}}$, for $\mathbf{N} \gg \mathbf{P}\log\mathbf{P}$. Although static chunking reduces the synchronization overhead, it has a greater potential for load imbalance than self-scheduling as processors finish within K iterations of each other in the worst case, where K is the chunk size. Arguably, one can randomly assign chunks of iterations to the processors; however, Lucco showed that the random assignment is more efficient than dynamic methods only when $\sigma \ll \mu$, i.e., for a uniform workload distribution or if the scheduling overhead is much greater than μ. In [36], Tang and Yew proposed a scheme for self-scheduling of multiple nested parallel loops. Fang et al. proposed an approach for self-scheduling general parallel nested loops in [37].

9 Conclusions

Since parallel loops account for the greatest percentage of parallelism in numerical programs, the efficient scheduling of such loops is vital to program and system performance. Minimizing scheduling overhead in conjunction with maximal exploitation of the available parallelism is critical for achieving high speedup on supercomputers. In this paper, we presented a loop restructuring technique, called *enhanced loop coalescing*, to transform nested loops with non-uniform iterations spaces into a single loop. The iteration space is flattened along an axis corresponding maximum workload gradient. The transformation achieves dual benefits:

- ✦ Reduction of synchronization operations during parallel execution
- ✦ Minimization of the makespan of a GSS schedule

The "complicated" index expressions introduced by coalescing do not pose a performance bottleneck as they are evaluated only once per processor. If the iteration blocks assigned to each processor are large enough (which is indeed the case in context of GSS), the time spent for subscript calculation should be negligible.

References

1. Petaflops Computing. `http://www.aeiveos.com/~bradbury/petaflops`
2. IBM Blue Gene. `http://www.research.ibm.com/bluegene/`

3. Brockman, J., Kogge, P., Thoziyoor, S., Kang, E.: PIM Lite: On the road towards relentless multithreading in massively parallel systems. Technical Report 03-01, Department of Computer Science, University of Notre Dame (2003)
4. Dongarra, J.J., Walker, D.W.: The quest for petascale computing. Computing in Science and Engineering 3(3), 32–39 (2001)
5. Bailey, D.H.: Onward to petaflops computing. Communications of the ACM 40(6), 90–92 (1997)
6. Kogge, P.M., Bass, S.C., Brockman, J.B., Chen, D.Z., Sha, E.: Pursuing a Petaflop: Point designs for 100 TF computers using PIM technologies. In: Proceedings of the 6th Symposium on the Frontiers of Massively Parallel Computation, pp. 88–97 (October 1996)
7. Sterling, T., Messina, P., Smith, P.H.: Enabling Technologies for Petaflops Computing. MIT Press, Cambridge (1995)
8. Lou, J., Farrara, J.: Performance analysis and optimization on the UCLA parallel atmospheric general circulation model code. In: Proceedings of the 1996 ACM/IEEE conference on Supercomputing, Pittsburgh, PA, p. 14 (1996)
9. Plimpton, S., Hendrickson, B., Attaway, S., Swegle, J., Vaughan, C., Gardner, D.: Transient dynamics simulations: parallel algorithms for contact detection and smoothed particle hydrodynamics. In: Proceedings of the 1996 ACM/IEEE conference on Supercomputing, Pittsburgh, PA (1996)
10. Plimpton, S., Attaway, S., Hendrickson, B., Swegle, J., Vanghan, C.: Parallel transient dynamics simulations. J. Parallel Distrib. Comput. 50(1-2), 104–122 (1998)
11. Taiji, M., Narumi, T., Ohno, Y., Futatsugi, N., Suenaga, A., Takada, N., Konagaya, A.: Protein Explorer: A petaflops special-purpose computer system for molecular dynamics simulations. In: Proceedings of the 2003 ACM/IEEE conference on Supercomputing (2003)
12. Almasi, G.S., Caşcaval, C., Casta nos, J.G., Denneau, M., Donath, W., Eleftheriou, M., Giampapa, M., Ho, H., Lieber, D., Moreira, J.E., Newns, D., Snir, M., Warren Jr., H.S.: Demonstrating the scalability of a molecular dynamics application on a petaflop computer. In: Proceedings of the 15th International conference on Supercomputing, Sorrento, Italy, pp. 393–406 (2001)
13. Wallach, S.: Petaflop architectures. In: Proceedings of the Second Conference on Enabling Technologies for Petaflops Computing (February 1999)
14. Gao, G.R., Likharev, K.K., Messina, P.C., Sterling, T.L.: Hybrid technology multithreaded architecture. In: Proceedings of the 6th Symposium on the Frontiers of Massively Parallel Computation, Annapolis, MD (1996)
15. Sterling, T.L., Zima, H.P.: Gilgamesh: A multithreaded processor-in-memory architecture for petaflops computing. In: Proceedings of the 2002 ACM/IEEE conference on Supercomputing, Baltimore, MD, pp. 1–23 (2002)
16. Kuck, D., Sameh, A.H., Cytron, R., Veidenbaum, A., Polychronopoulos, C.D., Lee, G., McDaniel, T., Leasure, B.R., Beckman, C., Davies, J.R.B, Kruskal, C.P.: The effects of program restructuring, algorithm change and architecture choice on program performance. In: Proceedings of the 1984 International Conference on Parallel Processing, pp. 129–138 (August 1984)
17. Polychronopoulos, C.D., Kuck, D.J., Padua, D.A.: Utilizing multidimensional loop parallelism on large scale parallel processor systems. IEEE Transactions on Computers 38(9), 1285–1296 (1989)
18. Petersen, P., Padua, D.: Machine-independent evaluation of parallelizing compilers. Technical Report 1173, Center for Supercomputing Research and Development, University of Illinois at Urbana-Champaign (1992)

19. Rudolph, D.C., Polychronopoulos, C.D.: An efficient message-passing scheduler based on guided self scheduling. In: Proceedings of the 3rd international conference on Supercomputing, Crete, Greece, pp. 50–61 (1989)
20. Polychronopoulos, C.: Loop coalescing: A compiler transformation for parallel machines. In: Proceedings of the 1987 International Conference on Parallel Processing, pp. 235–242 (August 1987)
21. Polychronopoulos, C.D., Kuck, D.J.: Guided self-scheduling: A practical scheduling scheme for parallel supercomputers. IEEE Transactions on Computers 36(12), 1425–1439 (1987)
22. Coffman Jr., E.G., Garey, M.R., Johnson, D.S.: An application of bin-packing to multiprocessor scheduling. SIAM Journal of Computing 7(1), 1–17 (1978)
23. Garey, M., Johnson, D.: Computers and Intractability, A Guide to the Theory of NP-Completeness. W. H. Freeman and Co., New York (1979)
24. Graham, R.L.: Bounds on multiprocessing timing anomalies. SIAM Journal of Applied Mathematics 17(2), 416–428 (1969)
25. Banerjee, U.: A theory of loop permutations. In: Gelernter, D., Nicolau, A., Padua, D. (eds.) Languages and Compilers for Parallel Computing, MIT Press, Cambridge (1990)
26. Kelley, J.L.: General Topology. D. van Nostrand Company Inc., Princeton (1955)
27. Gonzalez, T.F., Ibarra, O.H., Sahni, S.: Bounds for LPT schedules on uniform processors. SIAM Journal of Computing 6(1), 155–166 (1977)
28. Lucco, S.: A dynamic scheduling method for irregular parallel programs. In: Proceedings of the SIGPLAN 1992 Conference on Programming Language Design and Implementation, San Francisco, CA, pp. 200–211 (1992)
29. Polychronopoulos, C., Kuck, D.J., Padua, D.A.: Execution of parallel loops on parallel processor systems. In: Proceedings of the 1986 International Conference on Parallel Processing, pp. 519–527 (August 1986)
30. Padua, D.A., Wolfe, M.J.: Advanced compiler optimizations for supercomputers. Communications of the ACM 29(12), 1184–1201 (1986)
31. O'Keefe, M.T., Dietz, H.G.: Loop coalescing and scheduling for barrier mimd architectures. IEEE Transactions on Parallel and Distributed Systems 4(9), 1060–1064 (1993)
32. Tabirca, T., Freeman, L., Tabirca, S., Yang, L.T.: Feedback guided dynamic loop scheduling; a theoretical approach. In: International Conference on Parallel Processing Workshops, Valencia, Spain, pp. 115–121 (2001)
33. Lusk, E.L., Overbeek, R.A.: Implementation of monitors with macros: A programming aid for the HEP and other parallel processors. TR ANL-83-97, Argonne National Laboratory (December 1983)
34. Kruskal, C.P., Weiss, A.: Allocating independent subtasks on parallel processors. IEEE Transactions on Software Engineering 11(10), 1001–1016 (1985)
35. Barlow, R.E., Proschan, F.: Statistical Theory of Reliability and Life Testing. Holt Rinehart & Winston Inc. (1975)
36. Tang, P., Yew, P.C.: Processor self-scheduling for multiple nested parallel loops. In: Proceedings of the 1986 International Conference on Parallel Processing, pp. 528–535 (August 1986)
37. Fang, Z., Tang, P., Yew, P.-C., Zhu, C.-Q.: Dynamic processor self-scheduling for general parallel nested loops. IEEE Transactions on Computers 39(7), 919–929 (1990)

Folding Active List for High Performance and Low Power

Yuichiro Imaizumi and Toshinori Sato*

Kyushu Institute of Technology
680-4, Kawazu, Iizuka, 820-8502 Japan
toshinori.sato@computer.org
http://www.slrc.kyushu-u.ac.jp/~tsato

Abstract. Out-of-order processors schedule instructions dynamically in order to exploit instruction level parallelism. It is necessary to increase instruction window size for improving instruction scheduling capability. In addition, current trend of exploiting thread-level parallelism requires further large instruction window. However, it is difficult to increase the size, because the instruction window is one of the dominant deciding processor cycle time and power consumption. This paper proposes a large instruction window, focusing on power-aware active list with large capacity. Restricting allocation and commitment policies, we achieve both high performance and low power. Simulation results show that our proposed active list significantly boosts processor performance with slight degradation from the traditional unrealistic active list.

Keywords: Out-of-order processors, instruction window, instruction-level parallelism, thread-level parallelism, active list.

1 Introduction

High performance superscalar processors require large instruction window, especially to tolerate long latency memory operations. In addition, recent emergence of simultaneous multithreading[8,11] increases requirements of large instruction window, because an increasing number of instructions from multiple threads share the instruction window. Hence, large instruction window is often necessary to exploit thread level parallelism as well as to exploit instruction level parallelism, while small hardware structures are required to achieve high clock frequency and low power. To achieve the goal, researchers have proposed novel techniques for performance enhancement[1,2,5,10,16] and for power reduction[1,9,13]. Particularly important structures are a recovery mechanism for deep speculation, a large instruction issue queue, a large register file, and a large active list. This paper focuses on the last one: the active list.

Large monolithic active lists will diminish clock frequency and consume much power. To improve processor performance with maintaining its power consumption, we propose to construct a large active list with a number of small active

* Currently with Kyushu University, Japan.

J. Labarta, K. Joe, and T. Sato (Eds.): ISHPC 2005 and ALPS 2006, LNCS 4759, pp. 33–42, 2008.

lists. We divide a large active list into small pieces of active lists and fold them. By keeping only active portions in high speed mode, power consumption is kept equivalent with that of the conventional small active list.

The rest of this paper is structured as follows: Section 2 summarizes the background of this study. Section 3 proposes schemes to realize a large instruction window. Section 4 introduces our evaluation environment. Section 5 presents simulation results. Section 6 summarizes related works. And last, Section 7 concludes this paper.

2 Terminology

Several definitions are given here to simplify future references in this paper, using MIPS R10000's instruction window shown in Fig. 1[17]. Superscalar processors fetch multiple instructions per cycle. Following instruction fetch, the instructions are decoded and dispatched into the instruction window. We use the term dispatch to indicate the process of placing the instructions into the instruction window. In order to eliminate anti- and output-dependences, out-of-order processors perform register renaming. There are two common ways to implement the register renaming. One is using a separated renaming registers which are usually constructed by a reorder buffer. The other combines the renaming registers with architected registers in a single register file. We focus on the latter case, based on R10000[17]. Hence, the instruction window consists of instruction issue queue, a buffer maintaining program order such as the reorder buffer, and register files. Following R10000, we use an active list for maintaining program order.

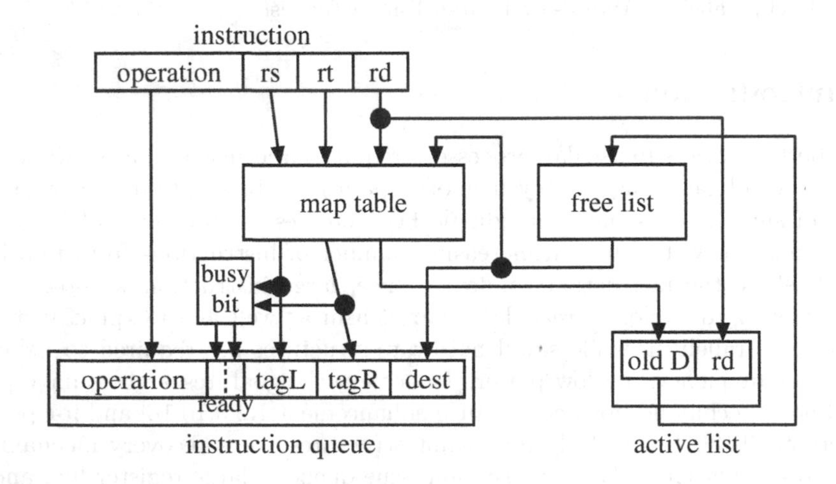

Fig. 1. Instruction window

The instructions remain in the instruction issue queue until their operands have been ready. Once their dependences have been resolved, instruction issue logic schedules the instructions and then issues them into functional units. The instruction issue queue entries containing the issued instructions are deallocated so that new instructions may be dispatched. We use the term issue to move the instructions from the instruction issue queue to the functional units, where they are executed. After completion of execution, the instructions still wait in the instruction window until their preceding instructions have been retired from the instruction window. When the instructions reach the head of the instruction window, they are retired from it. The instructions may be completed out-of-order but are retired in-order.

The register mapping hardware mainly consists of three structures – a map table, the active list, and a free list. By means of the map table, each logical register is mapped into a physical register. The destination register is mapped to a free physical register which is supplied by the free list, while operand registers are translated into the last mapping assigned to them. The old destination register is kept in the active list. When an instruction is retired, the old destination register which is allocated by the previous instruction with the same logical destination register is freed and is placed in the free list. The translated operand registers are held in the instruction queue as tags which are used for Tomasulo's algorithm. Busy bit table contains a bit indicating whether each physical register contains a valid value. It is used for initializing ready bits in the instruction issue queue for ready operands.

3 Large Instruction Window

While there are a lot of proposals for realizing large instruction windows, we propose an alternative technique. In order to realize large instruction window, the followings are required:

- Deep speculation
- Large instruction issue queue
- Large register file
- Large active list

For each requirement, we propose to utilize the following techniques respectively.

- Selective checkpointing[2,5]
- Waiting instruction buffer[10]
- Speculative register release
- Folded active list

While the following sections explain them shortly, this paper focuses on the folded active list.

3.1 Selective Checkpointing

We can achieve deep speculation by providing a lot of checkpoints. However, every checkpoint requires a plenty of hardware storage. Hence, in order to reduce the number of checkpoints, we utilize to selectively make checkpoints for predicted branches[2]. Only for branches predicted with low confidence[7], we make checkpoints. This enables deep speculation with relatively small number of checkpoints.

3.2 Waiting Instruction Buffer

When long latency operations, such as memory operations at cache misses, stall in the pipeline, processor performance is seriously degraded since even ready instructions can not be dispatched into the instruction issue queue. In order for the long latency operations to release the instruction issue queue, we utilize the waiting instruction buffer[10]. Such instructions move to the waiting instruction buffer from the instruction issue queue, and then succeeding ready instructions can be dispatched into the instruction issue queue. This effectively increases instruction issue queue capacity. A drawback in the waiting instruction buffer is that it requires a large active list. To satisfy the requirement, we will propose a large active list in the following section.

3.3 Speculative Register Release

As explained in the previous section, out-of-order processors utilize register renaming to remove write-after-read and write-after-write hazards. Register renaming requires a large number of physical registers. A physical register is allocated when every instruction is decoded and renamed. Even when the instruction is committed, its associated physical register is not released. It is only released when another instruction who has the same logical register for its destination is committed. This is required for mispredicted branch recovery, while this increases the life time of the register, resulting in the increase in physical register requirement. In order to reduce the number of physical registers, we propose to speculatively release renamed registers with the help of the degree-of-use prediction[4]. The degree-of-use predictor tells us how many consumers will appear for each producer. When the number of consumers matches the predicted value, we speculatively release the producer's renamed register. When mispredicted, processor rolls back to the nearest checkpoint.

3.4 Folded Active List

Large monolithic active lists will diminish clock frequency and consume much power. In order to realize a large active list with keeping its speed and power, we propose to construct it with a number of small active lists. We divide a large active list into small pieces of active lists and fold them. The tail of each small active list is logically connected with the head of the next active list.

Figure 2 shows this folded active list. We call the small active list *sublist*. While this structure resembles banking, its operations are different. It also looks like the distributed reorder buffer[9,13], however, the folded active list is logically a single active list. The key characteristic of the folded active list is that only one sublist is active each for allocation and commitment respectively. That is

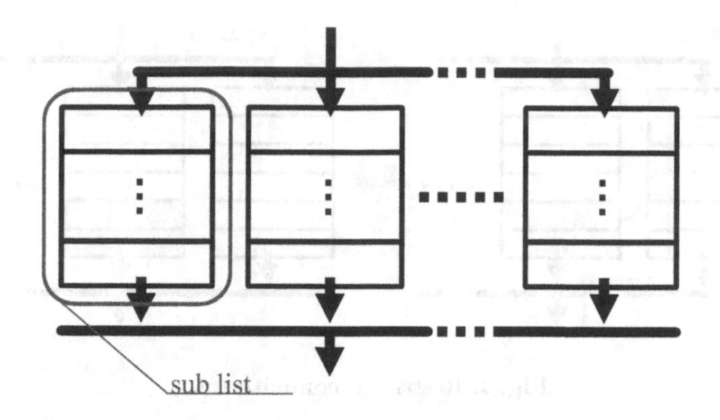

Fig. 2. Folded active list

at most two sublists are active at the time. Please note that there are not any read operations of operand values nor write operations of execution results in the case of the active lists. This is the different characteristic from the reorder buffers. Figure 3 shows how the next new entry is allocated. If the tail of the current sublist is allocated, the current sublist becomes inactive for write. The next sublist becomes active and its head is allocated for the coming instruction. Figure 4 shows an example of the commit operation. In this example, we assume the processor has 4-instruction-wide commit width. Even so, in the folded active list, only two instructions in the current sublist can be committed at the time, and the remaining two instructions are committed at the succeeding cycle after the next sublist is active.

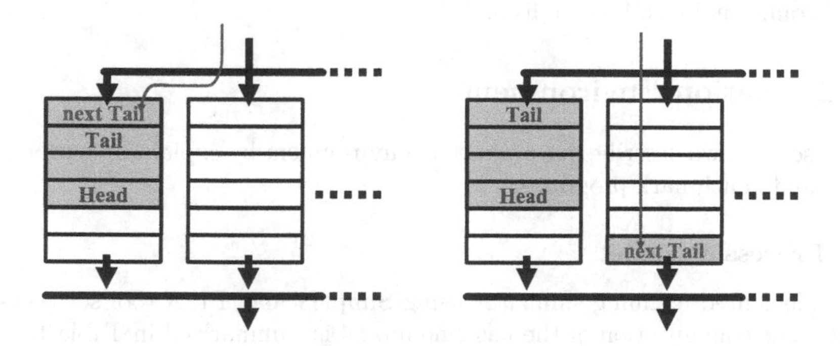

Fig. 3. Restricted allocation

The characteristic that at most two sublists are active at the time keeps high clock frequency and low power. As shown in Fig. 5, it is possible to reduce power supply voltage to inactive sublists. It is also possible to raise threshold voltage to inactive sublist by modulating body bias. This does not diminish processor performance due to the restriction in allocation and commitment. Thus, while

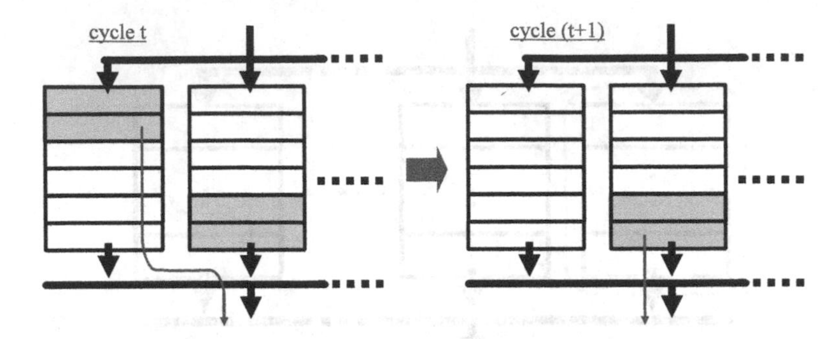

Fig. 4. Restricted commitment

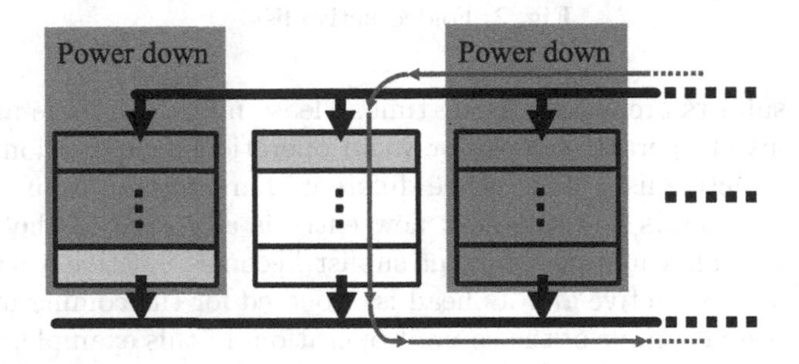

Fig. 5. Selective activation

the folded active list has a large capacity, its power consumption is comparable to the traditional small active lists.

4 Evaluation Environment

In this section, we describe our evaluation environment by explaining a processor model and benchmark programs.

4.1 Processor Model

We implemented a timing simulator using SimpleScalar/PISA tool set (version 3.0)[3]. The configuration of the baseline model is summarized in Table 1.

4.2 Benchmark Programs

The SPEC2000 benchmark suite is used in this study. Table 2 lists the benchmarks and the input sets. We focus on floating-point applications, because we are interested in long latency tolerance. We use the object files distributed at the SimpleScalar LLC web page. For each program, 1 billion instructions are skipped before actual simulation begins. Each program is executed to completion or for 1 billion instructions. We count only committed instructions.

Table 1. Processor configuration

OoO Core	32-entry instruction issue queue, 32-entry active list, 32-entry load/store queue, 4-wide decode, issue, and retirement
Branch Predictor	16K-entry gshare, 2K 4-way BTB, 64-entry RS, 8-cycle branch misprediction penalty
FUs	4 iALUs, 2 iMUL/iDIVs, 4 fALUs, 2 fMUL/fDIVs
Memory System	32K 4-way L1 I/D caches, 32-byte line size, 3-cycle hit, unified 256K 4-way L2 cache, 32-byte line size, 15-cycle hit, 1,000-cycle memory access, 2 cache ports

Table 2. Benchmark programs

Program	Input set
171.swim	swim.in
177.mesa	mesa.in mesa.ppm
179.art	c756hel.in a10.img
183.equake	inp.in
188.ammp	ammp.in
301.aspi	–

5 Results

This section presents simulation results. We focus on processor performance. Power efficiency will be improved as explained in Section 3.4. Detailed evaluation on power is remained for the future study.

First, we evaluate how processor performance is improved when we can increase the capacity of the active list with the folded active list. Figure 6 shows the percent increase in processor performance when a 32-entry active list is replaced by a 2K-entry folded active list. The instruction issue queue still has 32 entries. Each sublist is a 32-entry active list, and hence the folded active list evaluated here has 64 sublists. We use instructions per cycle (IPC) as a metric for this evaluation. We can see significant performance improvement, except for `188.ammp`. This is because `188.ammp` does not have the potential of performance improvement. Even when we use the ideal monolithic 2K-entry active list, we do not find any improvement. Even when we remove `179.art` as an exceptional result (over 200% improvement!), an average improvement of 24% is achieved. Since the folded active list maintains clock frequency, this IPC improvement turns out to be net performance improvement.

Next, we evaluate how the folding affects processor performance. Figure 7 shows the percent performance loss when we compare the 64x32-entry folded active list with the monolithic but unrealistic 2K-entry active list. We find little performance loss of between 1.3% and 7.5%, except for `188.ammp`, where we find the completely same results for both the folded and monolithic active lists. This

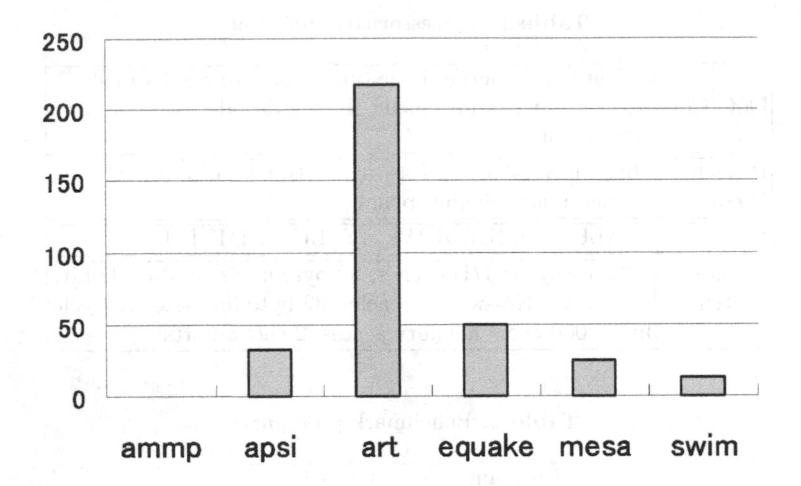

Fig. 6. %Performance improvement over 32-entry monolithic active list

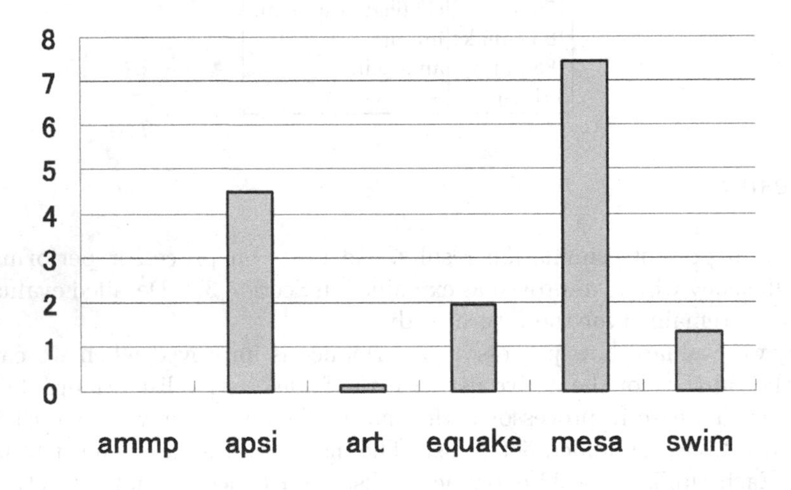

Fig. 7. %Performance loss under 2K-entry monolithic active list

performance loss is not serious, because the monolithic active list diminishes clock frequency and thus net performance for both models will be equivalent. In addition, the monolithic active list consumes much higher power than the folded active list, which exploits selective activation.

6 Related Work

Checkpointing for large instruction window is proposed by Cristal et al.[5] and Akkary et al.[2]. This enables to realize a large virtual reorder buffer using a small one and to reduce instruction issue queue entry. Checkpoints are created at long latency loads or at low-confidence branches. We follow their studies and determine to take checkpoints at branches because it is shown that branches are a good place for taking checkpoints[5]. Cherry proposed by Martinez et al.[12] also uses checkpointing. However, its purpose is not increasing instruction window but early register release.

Early register release was originally proposed by Moudgill et al.[15]. This uses counters to hold the number of reads. It unfortunately does not support precise exceptions, and this is very severe for modern speculative processors. Monreal et al.[14] and Ergin et al.[6] propose schemes to implement precise exception for early register release. Monreal et al.[14] extends the active list. Ergin et al.[6] uses the Checkpointed Register File bit cell, which is a special register file bit cell and essentially builds a shadow register file.

In order to effectively increase instruction issue queue capacity, a slow lane instruction queue[5], the waiting instruction buffer[10], and a slice processing unit[16] are proposed. The aim is shared by all schemes and is to tolerate long latency memory operations with keeping the instruction issue queue small. Every blocking instruction is moved from the instruction issue queue to a secondary buffer, releasing its instruction issue queue entry for other short latency operations. A good survey on schemes to implement a physically large instruction issue queue is found in [1].

Kucuk et al.[9] and Monferrer et al.[13] propose the distributed reorder buffer for power and temperature reduction. In contrast, the folded active list is proposed to improve processor performance under the constraint of keeping its power consumption.

7 Concluding Remarks

Modern superscalar processors rely on dynamic instruction scheduling for aggregating high performance. Instruction window size is an important factor for exploiting instruction level parallelism. Especially to tolerate long latency operations, large instruction windows are required. In this paper, we proposed a large instruction window, which consists of the selective checkpointing scheme, the waiting instruction buffer, the speculative register release scheme, and the folded active list. The large and fast active list is enabled by folding a conventional large active list, which will diminish clock frequency and consume much power. Based on detailed simulations, we found that the folded active list boosts processor performance by up to more than 200% with the expectation of speed and power, which are equivalent with those of the conventional small active list.

Acknowledgments

This work was partially supported by PRESTO program from Japan Science and Technology Agency, and by a Grants-in-Aid for Scientific Research #16300019 from Japan Society for the Promotion of Science.

References

1. Abella, J., Canal, R., Gonzalez, A.: Power- and Complexity-Aware Issue Queue Designs. IEEE Micro 23(5) (September 2003)
2. Akkary, H., Rajwar, R., Srinivasan, S.T.: Checkpoint Processing and Recovery: Towards Scalable Large Instruction Window Processors. In: 36th International Symposium on Microarchitecture (December 2003)
3. Burger, D., Austin, T.M.: The SimpleScalar Tool Set, Version 2.0. ACM SIGARCH Computer Architecture News 25(3) (1997)
4. Butts, J.A., Sohi, G.: Characterizing and Predicting Value Degree of Use. In: 35th International Symposium on Microarchitecture (November 2002)
5. Cristal, A., Ortega, D., Llosa, J., Valero, M.: Kilo-instruction Processors. In: 5th International Symposium on High Performance Computing (October 2003)
6. Ergin, O., Balkan, D., Ponomarev, D., Ghose, K.: Increasing Processor Performance Through Early Register Release. In: 22nd International Conference on Computer Design (October 2004)
7. Jacobsen, E., Rotenberg, E., Smith, J.E.: Assigning Confidence to Conditional Branch Predictions. In: 29th International Symposium on Microarchitecture (December 1996)
8. Kalla, R., Sinharoy, B., Tendler, J.: Simultaneous Multi-threading Implementation in POWER5 – IBM's Next Generation POWER Microprocessor. Hot Chips 15 (August 2003)
9. Kucuk, G., Ergin, O., Ponomarev, D., Ghose, K.: Distributed Reorder Buffer Schemes for Low Power. In: 21st International Conference on Computer Design (October 2003)
10. Lebeck, A.R., Li, T., Rotenberg, E., Koppanalil, J., Patwardhan, J.: Large, Fast Instruction Window for Tolerating Cache Misses. In: 29th International Symposium on Computer Architecture (May 2002)
11. Marr, D.T., Binns, F., Hill, D.L., Hinton, G., Koufaty, D.A., Miller, J.A., Upton, M.: Hyper-Threading Technology Architecture and Microarchitecture. Intel Technology Journal 6(1) (February 2002)
12. Martinez, J.F., Renau, J., Huang, M., Prvulovic, M., Torrellas, J.: Cherry: Checkpointed Early Resource Recycling in Out-of-order Microprocessors. In: 35th International Symposium on Microarchitecture (November 2002)
13. Monferrer, P.C., Magklis, G., Gonzalez, J., Gonzalez, A.: Distributing the Frontend for Temperature Reduction. In: 11th International Symposium on High-Performance Computer Architecture (February 2005)
14. Monreal, T., Vinals, V., Gonzalez, A., Valero, M.: Hardware Schemes for Early Register Release. In: 31st International Conference on Parallel Processing (August 2002)
15. Moudgill, M., Pingali, K., Vassiliadis, S.: Register Renaming and Dynamic Speculation: an Alternative Approach. In: 26th International Symposium on Microarchitecture (December 1993)
16. Srinivasan, S.T., Rajwar, R., Akkary, H., Gandhi, A., Upton, M.: Continual Flow Pipelines. In: 11th International Conference on Architectural Support for Programming Languages and Operating Systems (October 2004)
17. Yeager, K.C.: The MIPS R10000 Superscalar Microprocessor. IEEE Micro 6(2) (April 1996)

Reducing Misspeculation Penalty in Trace-Level Speculative Multithreaded Architectures

Carlos Molina[1], Jordi Tubella[2], and Antonio González[2,3]

[1] Dept. Eng. Informàtica i Matemàtiques, Universitat Rovira i Virgili, Tarragona - Spain
[2] Dept. d'Arquitectura de Computadors, Universitat Politècnica de Catalunya, Barcelona - Spain
[3] Intel Barcelona Research Center, Intel Labs-UPC, Barcelona - Spain
carlos.molina@urv.net,antonio@ac.upc.edu,jordit@ac.upc.edu

Abstract. Trace-Level Speculative Multithreaded Processors exploit trace-level speculation by means of two threads working cooperatively. One thread, called the speculative thread, executes instructions ahead of the other by speculating on the result of several traces. The other thread executes speculated traces and verifies the speculation made by the first thread. Speculated traces are validated by verifying their live-output values. Every time a trace misspeculation is detected, a thread synchronization is fired. This recovery action involves flushing the pipeline and reverting to a safe point in a program, which results in some performance penalties. This paper proposes a new thread synchronization scheme based on the observation that a significant number of instructions whose control and data are independent of the mispredicted instruction. This scheme significantly increases the performance potential of the architecture at less cost. Our experimental results show that the mechanism cuts the number of executed instructions by 8% and achieves on average speed-up of almost 9% for a collection of SPEC2000 benchmarks.

1 Introduction

Data dependences are one of the most important hurdles that limit the performance of current microprocessors. Two techniques have so far been proposed to avoid the serialization caused by data dependences: data value speculation [12] and data value reuse [23]. Both techniques exploit the high percentage of repetition in the computations of conventional programs. Speculation predicts a given value as a function of past history. Value reuse is possible when a given computation has already been made exactly. Both techniques can be considered at two levels: the instruction level and the trace level. The difference is the unit of speculation or reuse: an instruction or a dynamic sequence of instructions.

Reusing instructions at trace level means that the execution of a large number of instructions can be skipped in a row. More importantly, as these instructions do not need to be fetched, they do not consume fetch bandwidth. Unfortunately, trace reuse introduces a live-input test that it is not easy to handle. Especially

J. Labarta, K. Joe, and T. Sato (Eds.): ISHPC 2005 and ALPS 2006, LNCS 4759, pp. 43–55, 2008.

complex is the validation of memory values. Speculation may overcome this limitation but it introduces a new problem: penalties due to a misspeculation. Trace-level speculation avoids the execution of a dynamic sequence of instructions by predicting the set of live-output values based, for instance, on recent history. There are two important issues with regard to trace-level speculation. The first of these involves the microarchitecture support for trace speculation and how the microarchitecture manages trace speculation. The second involves trace selection and data value speculation techniques.

Recently, several thread-level speculation techniques [3], [7], [16], [20] have been explored to exploit parallelism in general-purpose programs. We lay on the same trend and focus on Trace-Level Speculative Multithreaded Architecture (TSMA) [14], which is tolerant to misspeculations in the sense that it does not introduce significant trace misprediction penalties and does not impose any constraint on the approach to building or predicting traces. This paper extends the previous TSMA microarchitecture with a novel verification engine that significantly improves performance. This new engine reduces the number of thread synchronizations and the penalties due to misspeculations. The main idea is that it does not throw away execution results of instructions that are independent of the mispredicted speculation, which reduces the number of instructions fetched and executed again.

The rest of this paper is organized as follows. Section 2 describes in detail the microarchitecture assumed to exploit trace-level speculation. Section 3 analyses the percentage of useful computation that is lost in recovery actions. Section 4 presents the novel verification engine. Section 5 analyses the performance of the processor with the proposed engine. Section 6 reviews related work and Section 7 summarizes our main conclusions and outlines future work.

2 Trace-Level Speculative Multithreaded Architecture (TSMA)

2.1 Trace-Level Speculation with Live-Output Test

Trace-level speculation is a dynamic technique (although the compiler may help) that requires a live-input or live-output test. The microarchitecture presented in this section focuses on the following approach: trace-level speculation with live-output test. This approach was introduced by Rotenberg et al [16], [17], [18] as the underlying concept behind Slipstream Processors. This approach is supported by means of a couple of threads (a speculative thread and a non-speculative one) working cooperatively to execute a sequential code.

Let us consider the program of Figure 1.a that is composed by three pieces of sequential code or traces. Figure 1.b shows the execution of the program from the point of view of code and Figure 1.c shows the execution of the program from the point of view of time. The speculative thread executes instructions and speculates on the result of whole traces. The non-speculative thread verifies instructions that are executed by the speculative thread and executes speculated traces. Each thread maintains its own state but only the state of the non-speculative thread is guaranteed to be correct. Communication between threads is done by means of a buffer that contains the executed instructions by the speculative thread. Once the non-speculative thread executes the speculated trace, instruction validation begins. This is done by verifying that source operands match the non-speculative state and updating the state with the new result. If validation does not succeed, recovery actions are required.

Note that speculated traces are validated by verifying their live-output values. Live- output values are those that are produced and not overwritten within the trace. The advantage with this approach is that only live-output values that are used are verified. Moreover, verification is fast because instructions consumed from the buffer have their operands ready (see trace3 execution and validation in Figure 1.c). Finally, speed-up is obtained when both threads execute instructions at the same time and validation does not produce a misspeculation, which implies to set some recovery actions (see trace2 and trace3 execution in Figure 1.c).

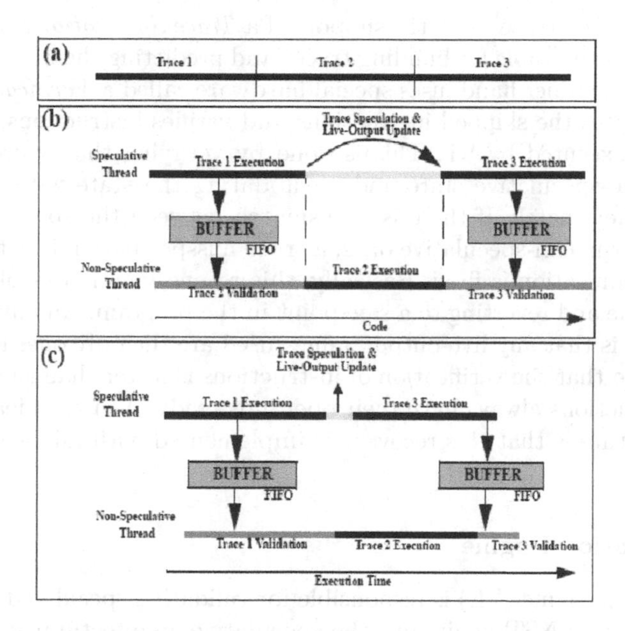

Fig. 1. Trace-lebel speculation with live-output test (a)program (b)point of view of code (c)point of view of time

2.2 Microarchitecture

A TSMA processor can simultaneously execute a couple of threads (a speculative one and a non-speculative one) that cooperate to execute a sequential code. The speculative thread is in charge of trace speculation. The non-speculative thread is in charge of validating the speculation. This validation is performed in two stages: (1) executing the speculated trace and (2) validating instructions executed by the speculative thread. Speculated traces are validated by verifying their live-output values. Live-output values are those that are produced and not overwritten within the trace. In the rest of the paper we will use the terms ST and NST to refer to the speculative thread and the non- speculative thread, respectively. Note that ST runs ahead of NST.

Both threads maintain their own architectural state by means of their associated architectural register file and a memory hierarchy with some special features. NST provides the correct and non-speculative architectural state, while ST works on a speculative architectural state. Note that each thread maintains its own state but that only the state of NST is guaranteed to be correct.

Additional hardware is required for each thread. ST stores its committed instructions to a special FIFO queue called *Look-Ahead Buffer*. NST executes the skipped instructions and verifies instructions in the look-ahead buffer executed by ST. Note that verifying instructions is faster than executing them because instructions always have their operands ready. In this way, NST catches ST up quickly.

ST speculates on traces with the support of a *Trace Speculation Engine* (TSE). This engine is responsible for building traces and predicting their live-output values. NST, on the other hand, uses special hardware called a *Verification Engine.* The NSTexecutes the skipped instructions and verifies instructions in the look-ahead buffer executed by ST. This is done by verifying that source operands match the non-speculative state and by updating the state with the new result in case they match. If there is a mismatch between the speculative source operands and the non-speculative ones, a trace misspeculation is detected and a thread synchronization is fired. Basically, this recovery action involves flushing the ST pipeline and reverting to a safe point in the program. An advantage with this approach is that any live-output values used are the only ones that are verified. Note also that the verification of instructions is faster than their execution because instructions always have their operands ready. A critical feature of this microarchitecture is that this recovery is implemented with minor performance penalties.

2.3 Verification Engine

The verification engine (VE) is responsible for validating speculated instructions and, together with NST, maintains the speculative architectural state. Instructions to be validated are stored in the look-ahead buffer. Verification involves testing source values of the instruction with the non-speculative architectural state. If they match, the destination value of the instruction can be updated in the non-speculative architectural state (register file or memory).

Memory operations require special considerations. First, the effective address is verified and, after this validation, store instructions update memory with the destination value. On the other hand, loads check whether the value of the destination register matches the non-speculative memory state. If it does, the destination value is committed to the register file.

This engine is independent of both threads but works cooperatively with NST to maintain the correct architectural state.

3 Thread Synchronization Analysis

Traces are identified by an initial and a final point in the dynamic instruction stream. They can be built according to different heuristics: basic blocks, loop bodies, etc [8], [9], [11]. Live-output values of a trace can be predicted in several ways, including with conventional value predictors such as last value, stride, context-based and hybrid schemes [13], [22].

Unfortunately, speculation accuracy decreases when the traces are large because they have a huge number of live-output values that have to be predicted. As outlined above, if source values of the instructions in the look ahead buffer do not match the non- speculative architectural state, a thread synchronization is required in the original TSMA architecture. This involves emptying all the ST structures and reverting to a safe point in the program. Note that, a misspeculation in one instruction causes younger instructions to be discarded from the look ahead buffer, though some may be correctly executed. Consider, for instance, a speculative trace in which just a single live-output value of the whole set is incorrectly predicted. Only the instructions dependent on the mispredicted one will be incorrectly executed by ST. In this section we analyse the number of correctly executed instructions that are squashed when a thread synchronization is fired. See Section 5.1 for details of the experimental framework.

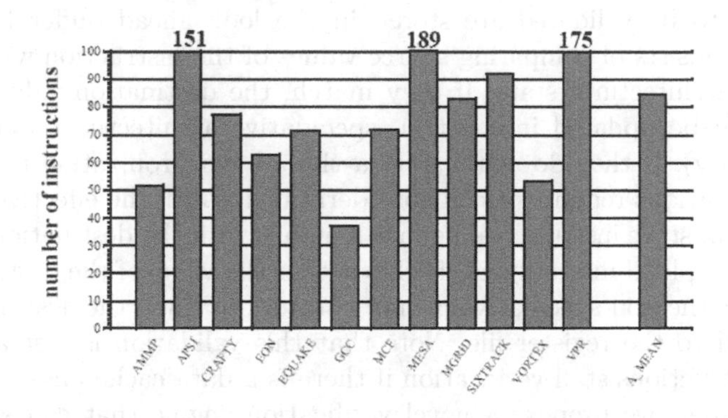

Fig. 2. Number of squashed instructions in each thread synchronization

Figure 2 shows the number of instructions that are squashed from the look ahead buffer every time a thread synchronization was fired. Note that the number of discarded instructions is significant for all benchmarks. On average, up to 80 instructions were squashed from the look ahead buffer in each thread synchronization irrespective of weather they were correctly or incorrectly executed.

Figure 3 shows on average the percentage of squashed instructions from the look ahead buffer that were correctly executed by ST was over 20Combined with the previous results, this means that on average 16 instructions that were correctly executed were discarded every time a thread synchronization was performed. This led us to reconsider thread synchronizations in order to try to avoid this waste of activity and reduce the number of fetched and executed instructions.

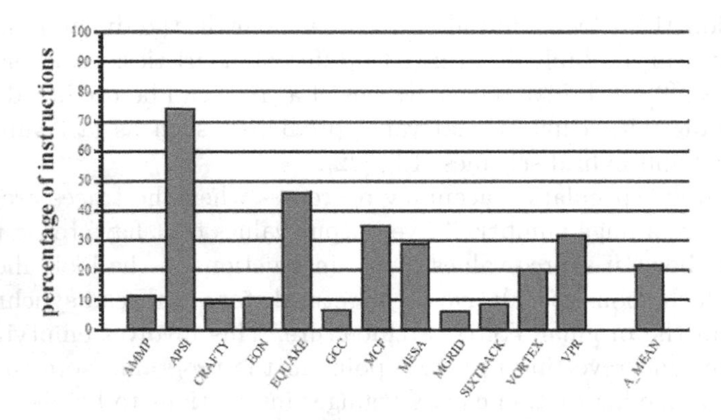

Fig. 3. Percentage of the squashed instructions that were correctly executed

4 Novel Verification Engine

The conventional verification engine is in charge of validating speculated instructions. Together with NST, it maintains the speculative architectural state. Instructions to be validated are stored in the look ahead buffer by ST. The verification consists of comparing source values of the instruction with the non-speculative architectural state. If they match, the destination value of the instruction can be updated in the non- speculative architectural state (register file or memory). If they do not match, a thread synchronization is performed. Memory operations require special considerations. First, the effective address is verified. Then, store instructions update memory with the destination value. On the other hand, load instructions check whether the value of the destination register matches the non-speculative memory state. If it does, the destination value is committed to the register file. Note that this validation is fast and simple. Memory instructions stall verification if there is a data cache miss.

In this paper, we propose a novel verification engine that can significantly improve the performance potential of the architecture. The underlying concept is based on the idea that a misspeculation in one instruction does not necessarily causes valid work from sequential younger computations to be aborted. Thread synchronization can therefore be delayed or even avoided. Below we describe how this new verification engine behaves depending on the type of instruction that is validated:

Branch instructions: These operations do not have an explicit destination value. Implicitly, they modify the program counter according to the branch direction that is taken or not taken. The idea is to validate the branch target instead of the source values. So, if source values are incorrectly predicted but the direction of the branch is correct, a thread synchronization is not fired.

Load instructions: First, the effective address is verified. If validation fails, the correct effective address is computed. Therefore, load instructions do not check whether the value of the destination register matches the non-speculative memory state. Simply, the destination value obtained from memory is committed to the register file. Note that an additional functional unit is required in order to compute the effective address.

Store instructions: As with load instructions, the effective address is first verified. If validation fails, store instructions update memory with the destination value obtained from the non-speculative architectural state, instead of the value obtained from the instruction. Note that only one functional unit is required to compute the effective address.

Arithmetic instructions: As with the conventional engine, the verification of arithmetic operations involves comparing the source operands of the instruction with the non-speculative architectural state. If they match, the destination value of the instruction can be committed to the register file. If they do not, the verification engine re-executes the instruction with values from the non-speculative state. In this case, verification is stalled and instructions after the re-executed one cannot be validated until the next cycle. Moreover, to maintain a high validation rate, this re- execution is only considered for single-cycle latency instructions. An additional functional unit is required in order to re-execute the instruction.

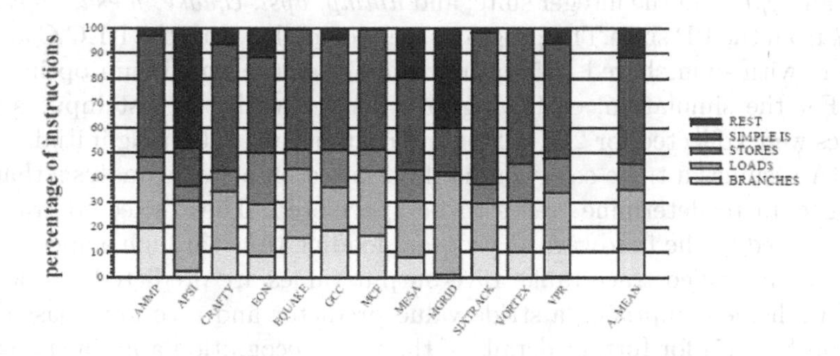

Fig. 4. Type of incorrect speculated instructions

Note that only branch instructions with a wrong target and non-single-latency instructions with wrong source operands fire a synchronization.

Figure 4 shows the breakdown of instructions in the look-ahead buffer that fail validation for the original validation engine. From bottom to top the categories are: branch instructions, load instructions, store instructions, instructions with single-cycle execution latency and, finally, the rest of the instructions. Note that branches, memory operations and instructions with a single-cycle latency account for 90% of the total incorrectly executed instructions. This means that there is a huge potential benefit for the new verification engine. Also, simulation results show that on average just 1% of the instructions inserted in the look-ahead buffer are incorrectly predicted. This suggests that the new verification engine may not need to re-execute many instructions, so the validation rate will not be greatly affected. Therefore, for the novel TSMA we assume that the number of functional units is the same as for the conventional one. Finally, we assume that the maximum number of instructions validated per cycle is the same and that no more than one instruction is re-executed per cycle.

5 Performance Evaluation

In this section we describe the experimental framework assumed in this paper, analyse the performance of the novel engine and compare it with the conventional one.

5.1 Experimental Framework

The TSMA simulator is built on top of the Simplescalar Alpha toolkit [4]. Table 1 shows the parameters of the baseline conventional microarchitecture. The TSMA assumes the same resources as the baseline configuration except for the issue queue, reorder buffer and logical register mapping table, which are replicated for each thread. It also has some new structures, which are shown in Table 2. The following Spec2000 benchmarks were randomly chosen: *crafty, eon, gcc, mcf, vortex,* and *vpr* from the integer suite; and *ammp, apsi, equake, mesa, mgrid,* and *sixtrack* from the FP suite.The programs were compiled with the DEC C and F77 compilers with -non_shared -O5 optimization flags (i.e. maximum optimization level). For the simulation, each program was run with the test input set and statistics were collected for 250 million instructions after skipping initializations.

TSMA assumes a trace selection method based on a static analysis that uses profiling data to determine traces to be speculated. These selected traces are communicated to the hardware at program loading time through a special hardware structure called trace table. Live-output values are predicted by means of a hybrid scheme comprising a stride value predictor and a context-based value predictor. See [15] for further details of the trace recognition approach based on profile guided heuristics.

5.2 Analysis of Results

The main objective of this section is to show that the number of thread synchronizations is lower when the new verification engine is used.

Figure 5 plots the percentage of thread synchronizations against the number of trace speculations.

Figure 6 shows the speed-up of TSMA over the baseline architecture. The first bar in each figure represents TSMA with the conventional engine and the second bar represents TSMA with the new verification engine. Our results show that

Table 1. Parameters of the baseline microarchitecture

Instruction fetch	4 instructions per cycle.
Branch predictor	2048-entry bimodal predictor
Instruction issue/ commit	Out-of-order issue, 4 instructions committed per cycle, 64-entry reorder buffer, loads execute only after all the preceding store addresses are known, store-load forwarding
Arch. Registers	32 integer and 32 FP
Functional units	4 integer ALUs, 4 load/store units, 4 FP adders, 2 integer mult/div, 2 FP mult/div
FU latency/repeat rate	int ALU 1/1, load/store 1/1, int mult 3/1, int div 20/19, FP adder 2/1, FP mult 4/1, FP div 12/12
Instruction cache	16 KB, direct-mapped, 32-byte block, 6-cycle miss latency
Data cache	16 KB, 2-way set-associative, 32-byte block, 6-cycle miss latency
Second Level Cache	Shared instruction & data cache, 256 KB, 4-way set-associative, 32-byte block, 100-cycle miss latency

Table 2. Parameters of TSMA additional structures

Speculative data cache	1 KB, direct-mapped, 8-byte block
Verification engine	Up to 8 instructions verified per cycle. Memory instructions stall verification for L1 misses. Only one instruction may be re-executed per cycle.
Trace speculation engine	128 history table, 4-way set-associative,.
Look ahead buffer	512 entries

the average speed-up for TSMA was 27 with the conventional verification engine. As expected, these speed-ups were significant for all the benchmarks despite a thread synchronization rate close to 30%.

On the other hand, the number of thread synchronizations was about 10% lower (from 30% to 20%) with the new verification engine than with the conventional scheme. Note that this engine did not always fires a thread synchronization to handle a miss trace speculation. It also provided a higher speed up (close to 38%), which implies that the average performance improvement was 9%. Note that the performance of most benchmarks improved significantly. Only benchmarks such as ammp, apsi or mgrid, whose misspeculation with the traditional verification engine was negligible, hardly improved since thread synchronizations were already low with the original verification engine.

These results demonstrates the tolerance to misspeculations of the proposed microarchitecture and encourage further work to develop more aggressive trace prediction mechanisms. Note that the novel verification engine opens up a new area of investigation i.e. aggressive trace predictor mechanisms that do not need to accurately predict all live output values.

Figure 7 shows the reduction in executed instructions with the new verification engine. On average, this reduction is almost 8%. Note that this also reduces memory pressure since these instructions do not need to be fetched all together.

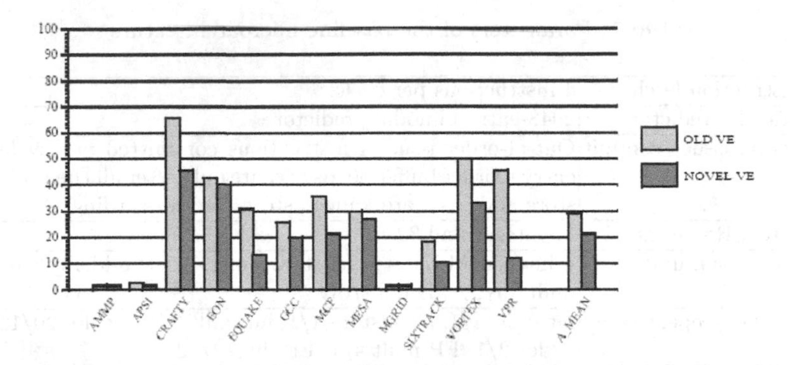

Fig. 5. Percentage of thread synchronization

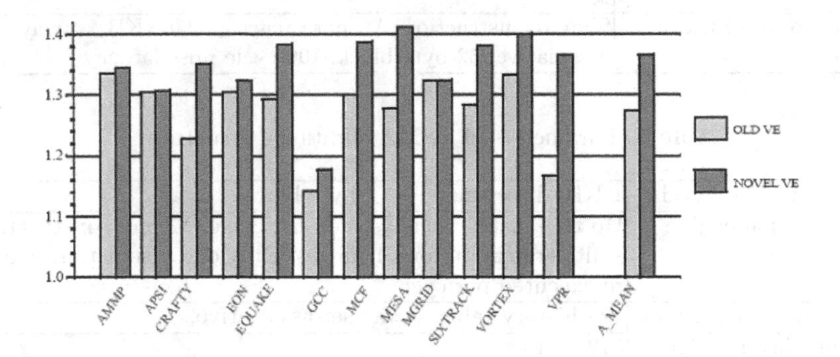

Fig. 6. Speed-up

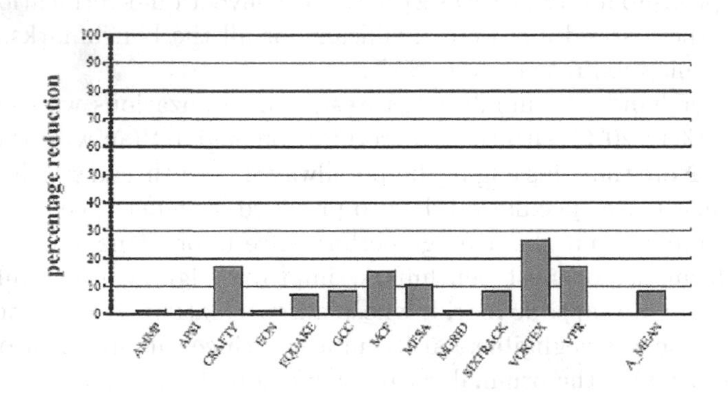

Fig. 7. Reduction in executed instructions

Again, benchmarks whose percentage of synchronization was negligible experienced a very small reduction in executed instructions. For the other benchmarks, on the other hand, the number of executed instructions and the number of thread synchronizations decreased, which led to significant speed-ups.

6 Related Work

Several techniques for reducing recovery penalties caused by speculative execution have been proposed. Instruction reissue, or selective squashing, was first proposed in . The idea is to retain instructions dependent on a predicted instruction in the issue queue until the prediction is validated. If the prediction is wrong, all dependent instructions are issued again. This technique trades the cost of squashing and re-fetching instructions for the cost of keeping instructions longer in the issue queue. A similar scheme that focused on load instructions was presented in [10]. The performance of the instruction reissue was investigated thoroughly in [25]. A practical and simple implementation of instruction reissue based on a slight modification of the register update unit was proposed in [21].

Squash reuse has also been proposed as a way to reduce branch miss-speculation penalty. This concept was first introduced in [23]. These authors proposed a table-based technique for avoiding the execution of an instruction that has previously been executed with the same inputs. As well as squash reuse, they also cover general reuse. A different implementation based on a centralized window environment was proposed in [6]. These authors also introduced the idea of dynamic control independence and showed how it can be detected and used in an out-of-order superscalar processor to reduce the branch misprediction penalty. Finally, register integration [19] has also been proposed as a simple and efficient implementation of squash reuse. This mechanism allows speculative results to remain in the physical register file after the producer instruction is squashed. They can later be reused through a modified renaming scheme.

The concept of dynamic verification was introduced in [17]. The proposed AR_SMT processor employs a time redundant technique that allows some transient errors to be tolerated. Slipstream processors [16] dynamically avoid the execution of a program's non essential computations. These authors suggested creating a shorter version of the original program by removing ineffectual computation. Using dynamic verification to reduce the burden of verification in complex microprocessor designs is covered in [5].

Several thread-level speculation techniques have been examined to exploit parallelism in general-purpose programs [1], [2], [13], [24]. Other recent studies have also focused on speculative threads. The pre-execution of critical instructions by means of speculative threads is proposed in several studies [3], [7], [20]. Critical instructions, such as mispredicted branches or loads that miss in cache are used to construct traces called slices that contain the subset of the program that relates to that instruction.

7 Conclusions and Future Work

In this paper, we have proposed a novel hardware technique to enhance the *Trace Level Speculative Multithreaded Architecture (TSMA)*. This hardware improvement focuses on the verification engine of the TSMA. The idea is to avoid the re-execution of instructions even when source values are incorrectly predicted.

Instead of firing a thread synchronization that wastes useful computations, the correct value is re-computed and used to update the architectural state. The new engine reduces the number of thread synchronizations and the penalty due to misspeculations. This avoids discarding instructions that are independent of a mispredicted one, thus reducing the number of fetched and executed instructions and cutting energy consumption and contention for execution resources.

Simulation results of the novel verification engine of the *Trace-Level Speculative Multithreaded Architecture* show that it can significantly improve performance without increasing complexity. These results encourage further work to develop more aggressive speculation schemes based on the idea that not all live-output values need to be highly predictable. This is also motivated by the relatively low penalty of misspeculations achieved by the *Trace-Level Speculative Multithreaded Architecture.*

Finally, TSMA processor can simultaneously execute a couple of threads that cooperate to execute a sequential code. To ensure the correctness of the architectural state, ST may only speculate a new trace when the look-ahead buffer is empty. This means that TSMA has only a single unverified trace speculation at any given time. Future work includes to modify the architecture in order to allow multiple unverified traces while maintaining the relatively low penalty of misspeculations. Future areas for investigation also include generalising the architecture to multiple threads in order to perform sub-trace speculation during the validation of a trace that has been speculated.

Acknowledgments

This work has been partially supported by the Ministry of Education and Science under grants TIN2004-07739-C02-01 and TIN2004-03072, the CICYT project TIC2001-0995-C02-01, Feder funds, and Intel Corporation. The research described in this paper has been developed using the resources of the European Center for Parallelism of Barcelona and the resources of the Robotics and Vision Group of Tarragona.

References

1. Ahuja, P.S., Skadron, K., Martonosi, M., Clark, D.W.: Multipath Execution: Opportunities and Limits. In: Proceedings of the International Symposium on Supercomputing (1998)
2. Akkary, H., Driscoll, M.: A Dynamic Multithreaded Processor. In: Proceedings of the 31st Annual International Symposium on Microarchitecture (1998)
3. Balasubramonian, R., Dwarkadas, S., Albonesi, D.: Dynamically Allocating Processor Resources between Nearby and Distant ILP. In: Proceedings of the 28th International Symposium on Computer Architecture (2001)
4. Burger, D., Austin, T.M., Bennet, S.: Evaluating Future Microprocessors: The SimpleScalar Tool Set. Technical Report CS-TR-96-1308. Univ. of Wisconsin (1996)
5. Chaterjee, S., Weaver, C., Austin, T.: Efficient Checker Processor Design. In: Proceedings of the 33rd Annual International Symposium on Microarchitecture (2000)

6. Chou, Y., Fung, J., Shen, J.: Reducing Branch Missprediction Penalties Via Dynamic Control Independence Detection. In: Proceedings of International Conference on Supercomputing (1999)
7. Collins, J., Wang, H., Tullsen, D., Hughes, C., Lee, Y., Lavery, D., Shen, J.: Speculative Precomputation: Long-range Prefetching of Delinquent Loads. In: Proceedings of the 28th International Symposium on Computer Architecture (2001)
8. Connors, D.A., Hwu, W.W.: Compiler-Directed Dynamic Computation Reuse: Rationale and Initial Results. In: Proceedings of the 32nd Annual International Symposium on Microarchitecture (1999)
9. Gonzalez, A., Tubella, J., Molina, C.: Trace Level Reuse. In: Proceedings of the International Conference on Parallel Processing (1999)
10. Gonzalez, J., Gonzalez, A.: Speculative Execution via Address Prediction and Data Prefetching. In: Proceedings of the 11th International Conf. on Supercomputing (1997)
11. Huang, J., Lilja, D.: Exploiting Basic Block Value Locality with Block Reuse. In: Proceedings of the 5th International Symposium on High-Performance Computer Architecture (1999)
12. Lipasti, M.H.: Value Locality and Speculative Execution, Ph.D. Dissertation, department of Electrical and Computer Engineering, Carnegie Mellon Univ. (April 1997)
13. Marcuello, P., Tubella, J., Gonzalez, A.: Value Prediction for Speculative Multithreaded Architectures. In: Proceedings of the 32th Annual International Symposium on Microarchitecture (1999)
14. Molina, C., Tubella, J., Gonzalez, A.: Trace-Level Speculative Multithreaded Architecture. In: Procs of the International Conference on Computer Design (2002)
15. Molina, C., Tubella, J., Gonzalez, A.: Compiler Analysis to Support Trace-Level Speculative Multithreaded Architectures. In: Proceedings of the 9th Annual Workshop on Interaction between Compilers and Computer Architectures (2005)
16. Purser, Z., Sundaramoorthy, K., Rotenberg, E.: A Study of Slipstream Processors. In: Proceedings of the 33rd International Symposium on Microarchitecture (2000)
17. Rotenberg, E.: AR-SMT: A Microarchitectural Approach to Fault Tolerance in Microprocessors. In: Procs of the 29th Fault-Tolerant Computing Symposium (1999)
18. Rotenberg, E.: Exploiting Large Ineffectual Instruction Sequences. Technical Report, North Carolina State University (November 1999)
19. Roth, A., Sohi, G.S.: Register Integration: A Simple and Efficient Implementation of Squash Reuse. In: Proceedings of the 33rd International Symposium on Microarchitecture (2000)
20. Roth, A., Sohi, G.: Speculative Data-Driven Multithreading. In: Proceedings of the 7th International Symposium on High-Performance Computer Architecture (2001)
21. Sato, T., Arita, I.: Comprehensive Evaluation of an Instruction Reissue Mechanism. In: Proceedings of the 5th International Symposium on Parallel Architectures, Algorithms and Networks (2000)
22. Sazeides, Y., Smith, J.E.: The Predictability of Data Values. In: Proceedings of the 30th International Symposium on Microarchitecture (1997)
23. Sodani, A., Sohi, G.S.: Dynamic Instruction Reuse. In: Proceedings of the 24th International Symposium on Computer Architecture (1997)
24. Tullsen, D.M., Eggers, S.J., Levy, H.M.: Simultaneous Multithreading: Maximizing on- chip Parallelism. In: Proceedings of the 22th Annual International Symposium on Computer Architecture (1995)
25. Tyson, G.S., Austin, T.M.: Improving the Accuracy and Performance of Memory Communication Through Renaming. In: Proceedings of the 30th Annual Symposium on Microarchitecture (1997)

Exploiting Execution Locality with a Decoupled Kilo-Instruction Processor

Miquel Pericàs[1,2], Adrian Cristal[2], Ruben González[1], Daniel A. Jiménez[3], and Mateo Valero[1,2]

[1] Computer Architecture Department, Technical University of Catalonia (UPC)
Jordi Girona, 1-3, Mòdul D6 Campus Nord, 08034 Barcelona (SPAIN)
{mpericas,adrian,gonzalez,mateo}@ac.upc.edu
[2] Barcelona Supercomputing Center (BSC)
Jordi Girona, 29, Edifici Nexus-II Campus Nord, 08034 Barcelona (SPAIN)
[3] Department of Computer Science, The University of Texas at San Antonio (UTSA)
Science Building, One UTSA Circle, San Antonio, TX 78249-1644 (USA)
djimenez@acm.org

Abstract. Overcoming increasing memory latency is one of the main problems that microprocessor designers have faced over the years. The two basic techniques introduced to mitigate latencies are *caches* and *out-of-order* execution. However, neither of these solutions is adequate for hiding off-chip memory accesses in the order of 200 cycles or more. Theoretically, increasing the size of the instruction window would allow much longer latencies to be hidden. But scaling the structures to support thousands of in-flight instructions would be prohibitively expensive.

However, the distribution of instruction issue times under the presence of L2 cache misses is highly correlated. This paper describes this phenomenon of *Execution Locality* and shows how it can be exploited with an inexpensive microarchitecture consisting of two linked cores. This *Decoupled Kilo-Instruction Processor* (D-KIP) is very effective in recovering lost potential performance. Extensive simulations show that speed-ups of up to 379% are possible for numerical benchmarks thanks to the exploitation of impressive degrees of Memory-Level Parallelism (MLP) and the execution of independent instructions in the shadow of L2 misses.

1 Introduction

The memory wall problem [1] is one of the main causes for the low instructions-per-cycle (IPC) rates that current architectures are able to achieve. In general, to overcome memory latencies, two very successful techniques were introduced in high performance microarchitectures: Memory Caches and Out-of-Order Processing.

Caches [2,3] exploit the locality in data access patterns exhibited by programs, giving the cost advantage of a large and cheap main memory with the low latency of a small and expensive memory. This technique has proven very successful and is used in all current microprocessors.

J. Labarta, K. Joe, and T. Sato (Eds.): ISHPC 2005 and ALPS 2006, LNCS 4759, pp. 56–67, 2008.

Out-of-Order execution allows instructions in the instruction window to execute in dataflow order. The required hardware is expensive and must be designed to be small enough so as to not impair the cycle time. Out-of-order execution allows the processor to continue executing while an earlier instruction is blocked. However, the technique is limited by the size of the instruction queues and, for the case of memory access latencies, this technique is not sufficient.

Recently, much research has proposed replacing the most critical microarchitectural structures with distributed structures that can scale much farther. Kilo-Instruction processors are such an approach. They provide an effective solution for overcoming the memory wall problem. A Kilo-Instruction processor does not need to stall under the presence of a cache miss. This has two benefits:

- Distant Load Misses can be pre-executed. This results in higher exploitation of MLP.
- Independent instructions can be executed in the shadow of a L2 Cache miss. This allows the architecture to take profit of distant parallelism.

However, in its current incarnation, the Kilo-Instruction Processor still has shortcomings in terms of its complexity. For example, the management of registers still lacks a low-complexity solution. It would be desirable to obtain simpler schemes that do not involve large amounts of complexity.

This paper proposes a new complexity-effective implementation of the KILO based on the novel concept of *Execution Locality.* Execution locality is a concept derived from data access locality and memory hierarchies. The idea is that instructions can be grouped at runtime into clusters depending on their `decode→issue` distance. Instructions that issue soon are said to have high execution locality. This allows building a kilo-instruction processor using a processor hierarchy in which different processors handle instructions belonging to different locality groups. The architecture presented in this paper proposes using two superscalar processors linked by instruction/register buffers (see Figure 1). The first processor (*Cache Processor*, CP) is small and fast, and it executes instructions that have high execution locality. The second processor (*Memory Processor*, MP) can be simple and wide, and it executes the remaining low-locality instructions. The proposal reuses some concepts recently introduced to scale the processor structures but reorganizes the processor architecture in a novel fashion resulting in a decoupled processor that combines scalability, sustained complexity, and high performance.

The main contribution of this paper is twofold:

1. The introduction of the concept of *Execution Locality* and its evaluation.
2. The proposal of a *Decoupled Kilo-Instruction Processor*, a low-complexity architecture designed to exploit Execution Locality.

This paper is organized as follows. In Sect. 2 a motivation study is presented that will provide background to the possible improvements in execution. Section 3 will provide a discussion of execution locality along with an initial evaluation of execution behavior. Using these concepts it will be possible to begin a description

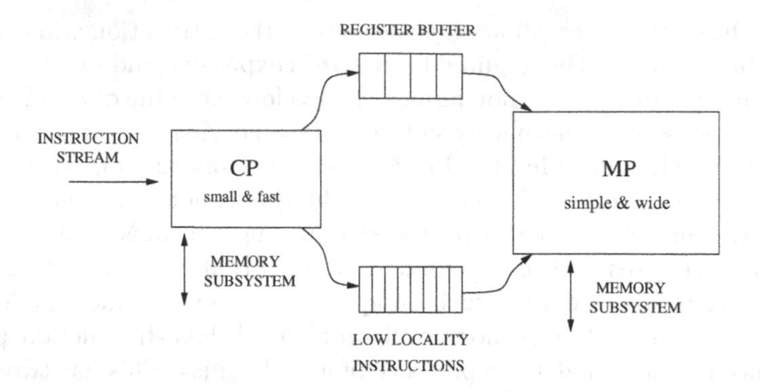

Fig. 1. 2-Level Decoupled Processor

of the decoupled Kilo-Instruction Processor in Sect. 4. The paper continues with a description of the simulation framework and the evaluation results in Sect. 5. Sections 6 and 7 complete the proposal with a summary of related works and the conclusions, respectively.

2 Motivation

To evaluate the impact of the memory wall, a workload consisting of all benchmarks of the SPECFP2000 benchmark suite is run on a simulator that models a typical 4-way out-of-order processor. The processor configurations are such that they are constrained only by the size of the ROB. As branch predictor the perceptron predictor [4] is used. Several processor configurations with ROB sizes from 32 to 4096 entries are evaluated. For each configuration, six different memory subsystems are tested. Three of these subsystems contain ideal caches. The IPC results are shown in Fig. 2. The sizes of the caches are 32KB for the first-level cache and 512KB for the second-level cache. Associativity is 4 in both cases and the access latencies are 1 and 10 cycles, respectively.

The figure shows how for numerical codes the possible gains of using a large-window processor are very large. Instructions are almost never discarded due to misspeculation and many independent instructions can be executed in the shadow of a cache miss, as predicted by [5]. This allows a large instruction window to much better hide the memory latencies. It can be seen that, at 4096 in-flight instructions, the processor configuration is almost insensitive to the memory latency. Only for the large 1000-cycle latency can a substantial IPC drop be observed. The reason is that the ROB is too small to hide latencies derived from two-level load chains, which is not the case for the 400-cycle configuration, where the cumulative latency of 800 cycles can still be hidden. The idea here is that performance close to perfect memory is feasible if an instruction window that supports thousands of in-flight instructions can be built.

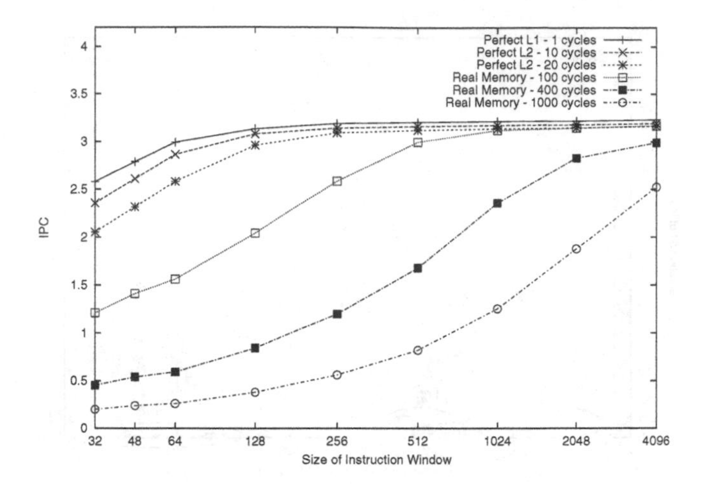

Fig. 2. Impact of ROB size and Memory Subsystem on SPECFP

3 Execution Locality

In a modern microprocessor, execution is data-driven. Once an instruction has all its operands computed, this instruction will execute within a small amount of cycles. Therefore, it should be expected that a large correlation exists between the number of cycles an instruction waits until *issue* and the presence of long-latency events in its dependency tree.

Figure 3 depicts the distribution of cycles that instructions remain in the instruction queues in a 4-way processor with 400 cycles of main memory access latency. These values have been obtained for all committed instructions in SPECFP. The distribution shows that most instructions (about 70%) do not depend on outstanding L2 cache misses while the remaining do depend directly on one or multiple L2 cache misses. A large part of these (11%) depend on a single cache miss and remain around 400 cycles in the queues. A smaller quantity (4%) depend on two caches misses. The remaining 15% of instructions belong either to issue clusters that depend on more than two L2 cache misses or are irregular instructions whose issue times do not belong clearly to one cluster or another.

This figure illustrates the major concept that this paper introduces: the *Execution Locality*. Similarly to data accesses, it is possible to classify all instructions using locality concepts. The phenomenon of execution locality classifies instructions based on their issue latency. Those instructions that will issue soon have a high degree of execution locality, while those that depend on long-latency memory accesses are classified as having low execution locality.

As will now be seen, execution locality can be used to reorganize the internal microarchitecture in a very efficient way such that a different core processes instructions belonging to different execution locality groups.

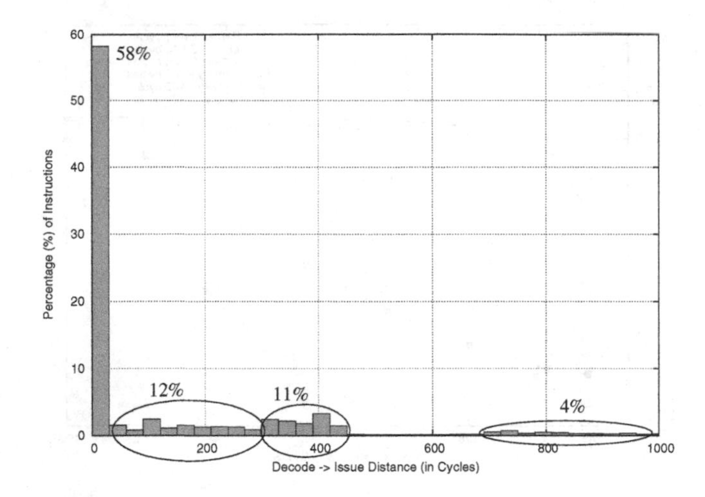

Fig. 3. Average distance between decode and issue for SPECFP

4 A Decoupled Kilo-Instruction Processor

The concepts of *execution locality* and the analysis presented in the previous section enable the efficient implementation of a *Kilo-Instruction Processor* using a heterogeneous decoupled dual core approach, a simplified version of which has been already shown in Fig. 1.

Some techniques used in the Kilo-Instruction processor are used in the proposal. They will be briefly summarized in Section 6. The next section will focus on the new concepts introduced by the Decoupled Kilo-Instruction Processor (D-KIP).

4.1 Heterogeneous Dual Core for Kilo-Instruction Processing

This section describes the basic microarchitecture of a D-KIP. The D-KIP is the first proposal designed from the ground up to provide an efficient solution that exploits *execution locality.* Here we will be giving a conceptual description of the D-KIP. A more detailed description is outside the scope of this paper.

Using the concepts of decoupling and execution locality this paper proposes a new and interesting approach to kilo-instruction processing: a two-level processor hierarchy. This proposal is, to the best of our knowledge, the first heterogeneous dual-core architecture aimed at single-thread performance, and the first to exploit execution locality concepts.

The D-KIP consists of two simple superscalar cores linked by a long-latency instruction buffer (LLIB). The LLIB is a FIFO structure that temporarily holds instructions that have low execution locality (ie, depend on a miss) while they are waiting for the L2 cache miss to complete. In combination both processors act as a Kilo-Instruction Processor. The first core will be referred to as the *cache processor* (CP) and the second core is the *memory processor* (MP). The reason to call the first processor *cache processor* is that its structures are sized to deal only with latencies resulting from cache hits, no further. A more detailed microarchitecture is shown in Fig. 4. There are actually many possible implementations. We will present only the most straight-forward implementation here.

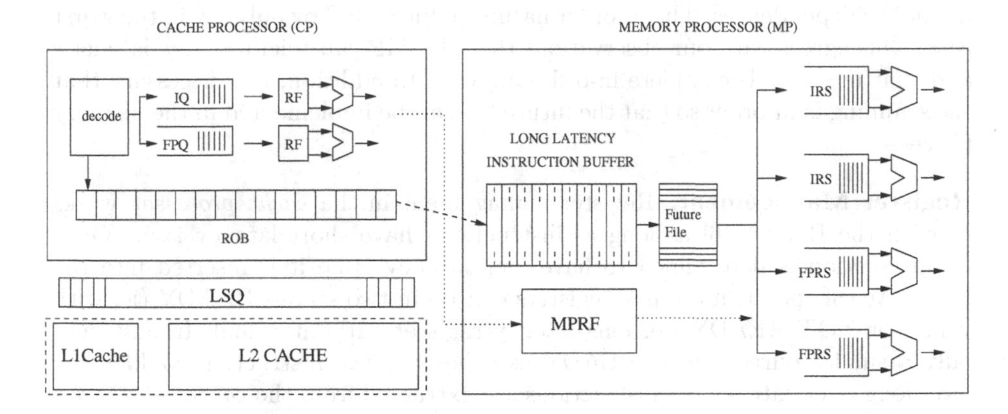

Fig. 4. Microarchitecture of a decoupled Kilo-Instruction Processor

The CP uses a conventional microarchitecture with a merged register file and separate instruction queues, similar to the MIPS R10000 [6]. This first-level processor is intended to process operations with short issue latency. All long issue-latency instructions are to be executed in the memory processor (MP). Therefore, the CP can be designed with a perfect L2 cache in mind. This means that all structures in the processor can be quite small. The resulting processor is simpler than current generation processors and can be very fast. The CP also contains a ROB structure to perform simple rollback of mis–predicted branches.

If an instruction depends on a L2 cache miss then it is classified as long-latency and sent to the memory processor.

Instructions that have executed in the CP are removed from the ROB without later consideration after they reach the head of the ROB. This means that all the information related to this instruction is discarded. This is not incorrect because once an instruction leaves the CP ROB it is associated with a checkpoint. In case recovery has to start, the architecture will rollback to this previous checkpoint and re-execute the instructions. As a consequence, there is no necessity for a ROB structure in the MP, which instead relies on checkpoints [7].

The memory processor (MP) requires a somewhat more innovative design, but the microarchitecture is basically a reservation station based back–end with the addition of some special structures to implement a Future File and the long-latency instruction buffer (LLIB).

Instructions at the head of the CP ROB are inserted into the LLIB when they have long issue-latency. The LLIB is a buffer that stores sequences of dependent instructions in program order. The load that starts a sequence is itself not stored in the LLIB, but in the Load/Store processor. Once a missing load returns to the MP, if this load is the oldest off-chip access in-flight, then the following instructions until the next unfinished load in the LLIB are awakened and sent to the reservation stations to be executed. To simplify the LLIB and the register management, the LLIB structure is strictly in-order. It is thus implemented as a buffer that stores the instruction opcode and the necessary data to compute and honor the dependences. The in-order nature reduces IPC by only 5% in the worst case. This agrees with our observations that the MP can tolerate large latencies and that most loads complete in-order anyway. In addition, it is necessary that the scanning is in-order so that the future file can be implemented in the memory processor.

Register Management. Register management in the *cache processor* works as with the R10000 [6] as long as instructions have short latency issue. Once an instruction is determined to have long-latency issue it is inserted into the LLIB. At this point, its source registers can be in two states: READY (ie, with value) or NOT READY (ie, long-latency register without value). In any case, only a single source can be READY as otherwise the instruction could never have long issue-latency. As instructions are extracted from the *cache processor*'s ROB, READY registers are retrieved and inserted into a dedicated register file, the MPRF. It is important to understand that no register sharing exists within the MPRF. Thus, two instructions in the LLIB that source the same READY register will see different entries once they are in the LLIB. This dramatically simplifies the register management, because all registers in the MPRF now have a *single consumer* and can be freed right after they are read.

NOT READY registers are inserted together with the instructions into the LLIB using only the logical register descriptor. Their purpose is to maintain the dependencies between the instructions in the LLIB.

5 Evaluation

5.1 Simulation Infrastructure

The decoupled kilo-instruction processor is evaluated using an execution-driven simulator consisting of the simplescalar-3.0 [8] front–end (executing Alpha ISA) and a new back-end that has been rewritten from scratch to simulate kilo-instruction processors. The focus of this work is on numerical and memory intensive applications. The workload is composed of all 14 benchmarks of the SPECFP2000 benchmark suite. The benchmarks have been compiled with *cc* version DEC C V5.9-008 on Digital UNIX V4.0 and using the `-O2` optimization level. Of each benchmark we simulate 200 million of representative instructions chosen using SimPoint [9].

Several architectures are simulated:

1. **BASE-256:** This baseline is a speculative out-of-order superscalar processor. It features structures (ROB, IQ, RF) that are somewhat larger than todays most aggressive implementations. The parameters are summarized in Table 1. The sizes are so that it is only constrained by the size of the ROB.

2. **BASE-92:** A smaller and cheaper baseline is also used. It is also limited only by the ROB which in this case has only 92 entries. This configuration has the same parameters as the CP in the decoupled kilo-instruction model. The remaining parameters are also shown in Table 1
3. **KILO:** A first-generation Kilo-Instruction Processor as described in [7] is simulated. The pseudo-ROB of this processor has 92 entries and the Slow Lane Instruction Queue has 1024 entries. These parameters make it more similar to the D-KIP model.
4. **D-KIP:** An implementation of the Decoupled Kilo-Instruction Processor is included. The parameters of the simulated D-KIP are shown in Table 2.

Table 1. Parameters of the two baseline configurations: BASE-256 and BASE-92

Fetch/Issue/Commit Width	4 instructions/cycle
Branch Predictor	Perceptron
I-L1 size	32 KB, 1 cycle latency
D-L1 size	32 KB, 4-way, 2 rd/wr ports, 1 cycle latency
D-L2 size	512 KB, 4-way, 2 rd/wr ports, 10 cycle latency
Memory Width / Latency	32 bytes / 400 cycles
Reorder Buffer Size	256 (BASE-256) / 92 (BASE-92)
[Integer/FP] Physical Registers	288 (BASE-256)/ 124 (BASE-92)
Load/Store Queue	256 / 92 entries
[Integer/FP] Queue Size	256 entries (BASE-256) / 92 entries (BASE-92)
Integer & FP Functional Units	4 Adders / 1 Multiplier

The *Cache Processor* of the D-KIP configuration has a 92-entry ROB. This size is a result of the design guidelines presented in section 4. A 92-entry ROB is enough to recover most of the lost performance due to memory when using a configuration with a perfect L2 cache. This value is extrapolated from the IPC analysis shown in Fig. 2.

The amount of registers in the MPRF is as large as the LLIB. Having 1024 registers may seem excessive, but due to the regular nature of insertion/extraction in the LLIB it can be implemented as a banked memory structure where each bank has a single read/write port. The cost of this is very small.

On the other hand, the size of the instruction queues may seem a bit large. These sizes have been chosen so that the ROB size limits the number of in-flight instructions. The real utilization of these instruction queues is likely to be quite small, almost never exceeding 32-40 instructions.

5.2 Instruction Level Parallelism

Using the same set of simulation and configuration parameters that has just been presented, the IPC achieved by all configurations has been measured for all benchmarks of SPECFP2000. The result is shown in Fig. 5. The last column represents the average IPC for the whole benchmark set.

Table 2. Parameters of the decoupled KILO processor

Common parameters	
ICache (L1) size	32 KB, 1 cycle latency
DCache (L1) size	32 KB, 4-way, 2 rd/wr ports, 1 cycle latency
Unified Cache (L2) size	512 KB, 4-way, 2 rd/wr ports, 10 cycle latency
Memory Width / Latency	32 bytes / 400 cycles
Load/Store Queue	unlimited
First Level Processor	
Fetch/Issue/Commit Width	4 instructions/cycle
Branch Predictor	Perceptron
Ports to the Register File	8 Read & 4 Write
Reorder Buffer Size	92
Integer/FP Physical Registers	128 / 128
Integer/FP Queue	92 entries / 92 entries
Integer & FP Functional Units	4 Adders / 1 Multiplier
Second Level Processor	
LLIB Size	1024 entries
LLIB extraction rate	4 instructions/cycle
Number of Checkpoints	16 stack entries
Registers in L2RF	1024 entries
Integer & FP Functional Units	4 Adders / 1 Multiplier

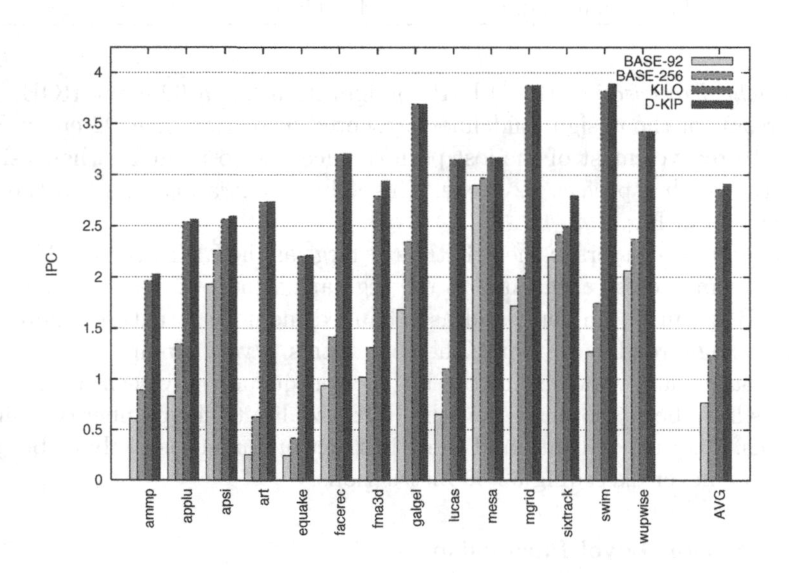

Fig. 5. IPC values for SPECFP2000

The speed-ups are very impressive for this set of numerical benchmarks. As a result of the accurate branch prediction in numerical codes most of the instructions in the large instruction window are useful for the execution and are not discarded. This results in an average speed-up of 236% compared to BASE-256 and an impressive 379% when compared to BASE-92. The memory-bound benchmarks are the ones that obtain the largest speed-ups. This should be no surprise as memory-bound applications are those that would spend a higher number of cycles stalled due to unavailability of ROB slots. The D-KIP with its non-blocking ROB mechanism provides tremendous speed-ups for these applications.

The decoupled model achieves a small speed-up of 1.02 versus the traditional KILO model [7]. This is thanks to the separate issue queues and additional functional units. However, the goal of decoupling is not to beat the KILO model, but to perform comparable with the benefit of sustainable design complexity.

Overall the FP performance is similar to that achieved by BASE-92 with an ideal L2 cache with a latency between 10 and 20 cycles (see Fig. 2).

6 Related Work

Processor behavior in the event of L2 Cache misses has been studied in detail by Karkhanis *et al.*. [5]. They showed that many independent instructions can be fetched and executed in the shadow of a cache miss. This observation has fueled the development of microarchitectures able to overcome the memory-wall problem [1].

Many suggestions have been proposed for overcoming the ROB size and management problem. Cristal *et al.* propose virtualizing the ROB by using a small sequential ROB combined with multicheckpointing [10,7,11]. Akkary *et al.* have also introduced a checkpointing approach [12] which consists in taking checkpoints on low-confidence branches.

The instruction queues have also received much attention recently. The *Waiting Instruction Buffer* (WIB) [13] is a structure that holds all the instructions dependent on a cache miss until its data returns from memory. The *Slow Lane Instruction Queue* (SLIQ) [7] is similar in concept to the WIB but is designed as an integral component of an overall kilo-instruction microarchitecture. In addition, it contains a pseudo-ROB structure to detect which instructions are long-latency and which are not. Recently, Akkary *et al.* have proposed the *Continual Flow Pipelines* (CFP) [14] architecture in which they propose en efficient implementation of a two-level instruction queue.

Several proposals try to improve the efficiency of physical register allocation. *Virtual Registers* [15] is a technique to delay the allocation of physical register until the issue stage. On the other hand, *Early Release* [16] tries to release registers earlier. An aggressive technique consists in combining both approaches. This technique is known as *Ephemeral Registers* [17].

Finally, several techniques have been proposed to attack the scalability problem of the load/store queues. There are several recent proposals for dealing with the scalability problems of store queues, including two-level queue proposals [12,18] or partitioned queues [19].

The main difference between these proposals and the technique presented so far is that the D-KIP proposes a complete solution from the point of view of *Execution Locality.*

Decoupling is an important technique that allows simplification of the design of loosely coupled microarchitectures while increasing its performance. Smith first proposed decoupling the memory system from the processor in the Decoupled Access-Execute computer architecture [20].

7 Conclusions

This paper has presented a new concept in program execution termed *execution locality.* This concept describes the execution behavior of instructions inside the instruction window. It has been shown how, based on the instruction issue times, instructions can be grouped into clusters. Based on this observation a new implementation of kilo-instruction processors called the *Decoupled Kilo-Instruction Processor* has been proposed. This processor consists of three subprocessors linked by queues: the *Cache Processor*, the *Memory Processor* and a *Load/Store processor.* The decoupling offers a complexity-effective approach to the design of kilo-instruction processors.

Extensive simulations conducted using the SPECFP2000 benchmarks have shown that this decoupled architecture fully exploits the possibilities of kilo-instruction processing. It even wins 2% of IPC when executing numerical codes. Overall, this represents a speedup of 379% compared to a single-windowed processor with the same structures as the *Cache Processor* when running numerical codes. But the important observation here is that this speed-up is obtained with only a small increase in the complexity of the design.

Acknowledgments

This work has been supported by the Ministry of Science and Technology of Spain under contract TIN–2004–07739–C02–01 and the HiPEAC European Network of Excellence under contract IST-004408. Daniel A. Jiménez is supported by NSF Grant CCF–0545898.

References

1. Wulf, W.A., McKee, S.A.: Hitting the memory wall: Implications of the obvious. Computer Architecture News (1995)
2. Wilkes, M.V.: Slave memories and dynamic storage allocation. IEEE Transactions on Electronic Computers, 270–271 (1965)
3. Smith, A.J.: Cache memories. ACM Computing Surveys 14(3), 473–530 (1982)

4. Jimenez, D.A., Lin, C.: Dynamic branch prediction with perceptrons. In: Proc. of the 7th Intl. Symp. on High Performance Computer Architecture, pp. 197–206 (2001)
5. Karkhanis, T., Smith, J.E.: A day in the life of a data cache miss. In: Proc. of the Workshop on Memory Performance Issues (2002)
6. Yeager, K.C.: The MIPS R10000 superscalar microprocessor. IEEE Micro 16, 28–41 (1996)
7. Cristal, A., Ortega, D., Llosa, J., Valero, M.: Out-of-order commit processors. In: Proc. of the 10th Intl. Symp. on High-Performance Computer Architecture (2004)
8. Austin, T., Larson, E., Ernst, D.: Simplescalar: an infrastructure for computer system modeling. IEEE Computer (2002)
9. Perelman, E., Hamerly, G., Biesbrouck, M.V., Sherwood, T., Calder, B.: Using SimPoint for accurate and efficient simulation. In: Proc. of the Intl. Conf. on Measurement and Modeling of Computer Systems (2003)
10. Cristal, A., Valero, M., Gonzalez, A., LLosa, J.: Large virtual ROBs by processor checkpointing. Technical report (2002), Technical Report number UPC-DAC-2002-39 (2002)
11. Cristal, A., Santana, O.J., Martinez, J.F., Valero, M.: Toward kilo-instruction processors. ACM Transactions on Architecture and Code Optimization (TACO), 389–417 (2004)
12. Akkary, H., Rajwar, R., Srinivasan, S.T.: Checkpoint processing and recovery: Towards scalable large instruction window processors (2003)
13. Lebeck, A.R., Koppanalil, J., Li, T., Patwardhan, J., Rotenberg, E.: A large, fast instruction window for tolerating cache misses. In: Proc. of the 29th Intl. Symp. on Computer Architecture (2002)
14. Srinivasan, S.T., Rajwar, R., Akkary, H., Gandhi, A., Upton, M.: Continual flow pipelines. In: Proc. of the 11th Intl. Conf. on Architectural Support for Programming Languages and Operating Systems (2004)
15. Gonzalez, A., Valero, M., Gonzalez, J., Monreal, T.: Virtual registers. In: Proc. of the 4th Intl. Conf. on High-Performance Computing (1997)
16. Moudgill, M., Pingali, K., Vassiliadis, S.: Register renaming and dynamic speculation: an alternative approach. In: Proc. of the 26th. Intl. Symp. on Microarchitecture, pp. 202–213 (1993)
17. Cristal, A., Martinez, J., LLosa, J., Valero, M.: Ephemeral registers with multicheckpointing. Technical report(2003), Technical Report number UPC-DAC-2003-51, Departament d'Arquitectura de Computadors, Universitat Politecnica de Catalunya (2003)
18. Park, I., Ooi, C.L., Vijaykumar, T.N.: Reducing design complexity of the load/store queue. In: Proc. of the 36th Intl. Symp. on Microarchitecture (2003)
19. Sethumadhavan, S., Desikan, R., Burger, D., Moore, C.R., Keckler, S.W.: Scalable hardware memory disambiguation for high ILP processors. In: Proc. of the 36th Intl. Symp. on Microarchitecture (2003)
20. Smith, J.E.: Decoupled access/execute computer architectures. In: Proc. of the 9th annual Intl. Symp. on Computer Architecture (1982)

Decoupled State-Execute Architecture

Miquel Pericàs[1,2], Adrián Cristal[2], Ruben González[1], Alex Veidenbaum[3], and Mateo Valero[1,2]

[1] Computer Architecture Department, Technical University of Catalonia (UPC)
Jordi Girona, 1-3, Mòdul D6 Campus Nord, 08034 Barcelona (SPAIN)
{mpericas,adrian,gonzalez,mateo}@ac.upc.edu
[2] Computer Sciences, Barcelona Supercomputing Center (BSC)
Jordi Girona, 29, Edifici Nexus-II Compus Nord, 08034 Barcelona (SPAIN)
[3] Department of Computer Science, University of California (UCI)
3019 Donald Bren Hall, Irvine, CA 92697-3435 (USA)
alexv@ics.uci.edu

Abstract. The majority of register file designs follow one of two well–known approaches. Many modern high-performance processors (POWER4 [1], Pentium4 [2]) use a *merged* register file that holds both architectural and rename registers. Other processors use a *Future File* (eg, Opteron [3]) with rename registers kept separately in reservation stations. Both approaches have issues that may limit their application in future microprocessors. The merged register file scales poorly in terms of power-performance while the Future File has to pay a large penalty due on branch mis–prediction recovery. In addition, the Future File requires the use of the less scalable mechanism of reservation stations.

This paper proposes to combine the best aspects of the traditional Future File architecture with those of the merged physical register file. The key point is that the new architecture separates the processor state, in particular the registers, and the execution units in the pipeline back–end. Therefore it is called *Decoupled State-Execute Architecture*. The resulting register file can be accessed in the pipeline front–end and has several desirable properties that allow efficient application of several optimizations, most notably the register file banking and a novel writeback filtering mechanism. As a result, only a 1.0% IPC degradation was observed with aggressive banking and the energy consumption was lowered by the new *writeback filtering* technique. Together, the two optimizations remove approximately 80% of the energy consumed in register file data array.

1 Introduction

Memory structures in microprocessors are one of the main sources of energy consumption [4]. As a consequence, great care has to be taken when designing structures that require large amounts of memory. One of such structures is a large register file of modern, out-of-order processors which use register renaming [5].

There are several approaches to intermediate value storage in registers in such dynamically scheduled architectures. One alternative is to implement a so-called

J. Labarta, K. Joe, and T. Sato (Eds.): ISHPC 2005 and ALPS 2006, LNCS 4759, pp. 68–78, 2008.

merged register file, an approach first followed by the ES/9000 [6] and further developed by the R10000 [7]. The merged register file has to supply values for both the computation and the mis–prediction recovery. It is typically accessed after an instruction is scheduled to execute, even if source operand values were available much earlier. As a result, this file needs to be both large and heavily multiported, increasing its energy consumption.

The alternative is to use a Future File. In this approach the *future file*, of size equal to the logical register file, is kept in the pipeline front–end while the rename registers correspond to storage in reservation stations. The future file contains the most recent values assigned to logical registers. The use of future registers is thus quite energy efficient. However, in the case of a branch mis–prediction, the architectural state must be recovered using the architectural register file at commit. With today's large memory latencies this approach can suffer a large IPC loss.

The Future File approach can be improved by providing direct access to registers required for recovery. The architecture proposed here uses a single register file containing all physical registers but located in the front end. Mis–prediction recovery can thus be done using a rename map stack, which check–points the rename map on each instruction that may require recovery.

The new register file is called the *Front-end Physical Register File* or FPRF. As source operand registers of an instruction are renamed, it can be determined if a source register has a computed value. The front–end physical register file is only read in this case, significantly reducing its access frequency. The remaining source operand values come directly from executing instructions via reservation stations which are also required in this architecture.

With the register file in the front end, the new architecture is called *Decoupled State-Execute Architecture* or DSE. Due to lower access frequency to the front–end register file it can be large but very more energy efficient. And because mis–prediction recovery is now fast, the DSE has better power-performance characteristics than the traditional approaches.

The new register file organization is more amenable to two important optimizations. Register file *banking* can be easily implemented, both due to the fact that registers are accessed in the front end and to the reduced access frequency. Also, an optimization to filter *unnecessary* writebacks into the register file can be performed efficiently in the DSE.

2 Related Work

The body of related work on register file design optimization is large.

Many papers have proposed to reorganize the register file architecture to reduce the number of ports and thus the energy [8, 9, 10]. Other techniques have been used to re–organize the register file. For instance, it is possible to *distribute* the register file based on the significance of register values [11]. Multilevel register files have also been proposed to reduce latency and save energy [12, 13, 14]. Clustered register files [15, 16, 17] have been used for the same reasons.

The Future File was proposed by Pleszkun and Smith in their 1985 work on precise exceptions [18]. The original proposal only provided operands to instructions via a logical register file in the front-end, hence the name *Future File*. More recent proposals for Future File design are capable of reading operands from both the Future File and an additional *architectural Register File*, which stores committed values [19]. This is specially useful after an exception/misprediction, when a precise instruction state needs to be recovered, as it avoids having to reconstruct the register state from the ROB.

3 The Decoupled State-Execute Architecture

This section describes the DSE in more detail. The DSE pipeline attempts to provide an instruction with source operands as early as possible. Similar to the Future File, it provides available source register values in the pipeline front–end. However, in the DSE approach the registers are accessed after being remapped to a large physical register space. This has two implications:

1. Access to computed values in the front–end needs to be delayed until the rename stage has completed
2. The number of registers in the front–end can be much larger than in the Future File

Figure 1 shows the DSE microarchitecture.

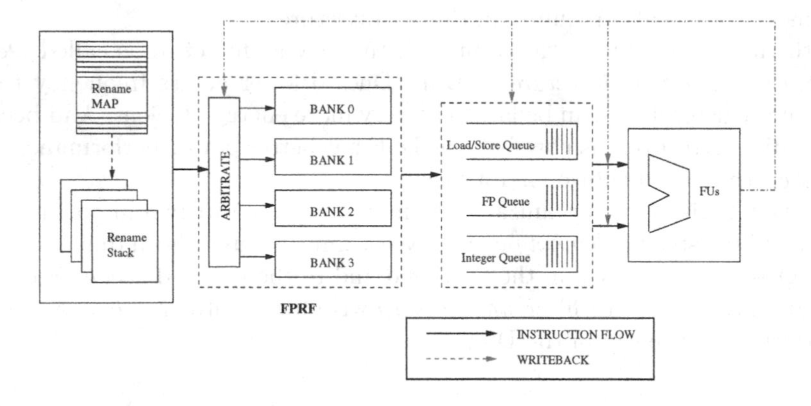

Fig. 1. The Decoupled State-Execute Architecture

Figure 2 shows the DSE pipeline. The total pipeline length of the DSE Microarchitecture is one stage longer than it would have been without the Front–End Physical Register File. The FPRF access in the front–end requires two stages: arbitrate and operand read. The former stage is necessary to implement register file banking.

The source register designators are checked in the arbitration stage for bank access conflicts in the same cycle. When a conflict is detected, all stages before the FPRF stage stall.

The DSE architecture maintains a bit for each logical register to mark if it contains a computed value. A FPRF register read operation is initiated if a desired source operand value is computed (available). The arbitration stage (ARB) logic checks if an N-instruction FPRF access has conflicts. The front-end stalls all stages prior to ARB in case of conflicts. Arbitration priority is given to older instructions to make sure that the front-end does not dispatch instructions out-of-order.

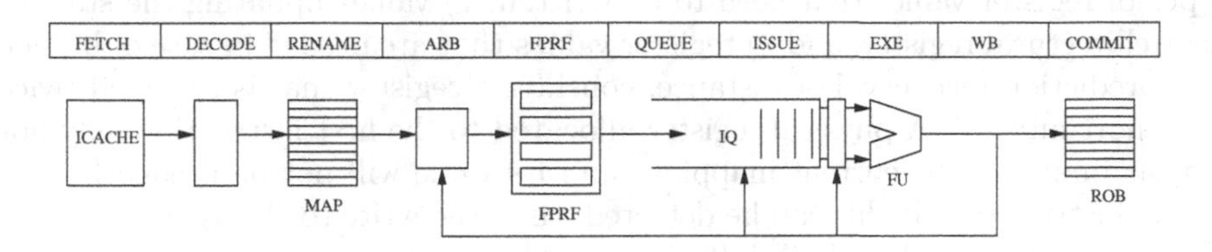

Fig. 2. The Pipeline of the DSE Architecture

Next, an instruction and the available operand values it read from the FPRF are inserted in an appropriate reservation station. A single, centralized reservation station can be used for all instructions, which can be implemented as an instruction queue in which the available source operand values are stored in the queue's payload RAM. This is the *Queue* stage.

As mentioned above, The DSE architecture has a lower access rate to the FPRF compared to a standard architecture accessing a back-end physical register file since only some source operands are available at this point in instruction execution. The number of integer operands obtained from the FPRF was observed to be near 40% of all required integer operands, while for floating point operands this number decreases to around 20% (for the SPEC2000 benchmarks and the architecture described in Sec. 4). This has the potential to reduce the number of banks as well as ports per bank in the FPRF. Note however, that there need to be at least two read ports per bank so that an instruction can obtain both operands from the same bank without stalling.

In the event of a branch mis–prediction the DSE architecture behaves exactly like the MIPS R10000. The processor aborts all instructions along the mis–predicted path, restores register mapping from the branch stack and starts fetching instructions from the correct path.

3.1 Read Sharing

Accesses to the same logical register often appear several times in a short instruction sequence. For instance, code that manipulates objects on the stack normally sources the stack pointer register many times. Such register accesses cannot be distributed among different banks. When they appear there is a high probability that a register access will result in a bank conflict.

Such conflicts in the FPRF access can be reduced by using a technique known as read sharing [13]. Read sharing allows multiple reads of a same register to share a single local port among the concurrently issued instructions. The impact of Read sharing will be evaluated for the FPRF architecture.

3.2 Writeback Filtering

Some of the values written to physical registers may never be used in the future. Such values do not actually need to be written to the register file. There are two types of register values that need to be written: 1) values updating the state of an architectural register, and 2) register values that are needed in case of branch mis–prediction recovery. For instance, consider a register that is renamed twice in a short interval. A physical register allocated to the first instruction may not appear in any of the current mappings and its value will not be needed by any future instructions. If this can be detected, then the write to this register can be eliminated. This is *Writeback Filtering.*

To implement Writeback Filtering the processor needs to check mapped registers in all rename checkpoints plus the current mapping and decide if a register write–back is necessary. Checkpoints need to be taken at all instructions that may cause a replay. There are many such instructions but the vast majority are conditional branches and load operations. Registers that are not referenced anywhere are candidates to be filtered out during writeback. Writeback filtering can be integrated in the renaming logic to detect short-lived registers. A short-lived register is a register that is not referenced by any checkpoint in the rename stack (including the current rename map). This can be detected by computing an OR of all the rename maps.

In practice, the number of loads that need to be replayed (due to ordering violations) is very small while the total number of loads is very large. Checkpointing all loads is thus expensive and very inefficient. The DSE architecture reduces the number of checkpoints by associating loads to older checkpoints. In case of a replay the restart is from an older point in execution but this is very infrequent and has almost no impact on performance.

These older checkpoints will usually correspond to branches, although it is possible that a load is the oldest operation in the processor and a previous branch in the instruction slice have committed already. To avoid check-pointing any loads, the architecture keeps an additional checkpoint associated to the last branch that has committed. This increases the size of the rename stack by one entry instead of a large increase that storing all mappings due to loads would have lead to.

4 Experimental Setup

The FPRF architecture was evaluated using a modified execution driven simulator based on SimpleScalar [20]. The simulator executes binaries compiled for the

Alpha ISA. The entire SPEC2000 suite was compiled using the Compaq/Digital *cc* compiler with the "-O2" optimization level. 100 million committed instructions are selected using SimPoint [21] and simulated.

Results for the DSE architecture are presented and compared to the baseline out-of-order microarchitecture as well as to the banked, multi–ported register file described in [22]. The common parameters of all architectures are shown in Table 1.

Table 1. Common architecture parameters for all configurations

Parameter	Value
Fetch/Issue/Commit Width	4 instructions/cycle
Branch Predictor	Combined bimodal + 2-level
I-L1 size	32 KB, 4-way, 1 cycle latency
D-L1 size	32 KB, 4-way, 2 rd/wr ports, 2 cycle latency
D-L2 size	256 KB, 4-way, 2 rd/wr ports, 11 cycle latency
Memory Bus Width	32 bytes
Ports to the Register File	8 Read & 4 Write
Reorder Buffer Size	128
Memory latency	100
Integer Physical Registers	160
FP Physical Registers	160
Load/Store Queue	128 entries
Integer Queue	32 entries
FP Queue	32 entries
Integer Functional Units	4 (latency 1)
FP Functional Units	4 (latency 2)

To evaluate the DSE architecture the 5 configurations described below were studied (they are summarized in Table 2).

1. *NON-BANKED* baseline configuration: a processor with a fully-ported (8Rd/4Wr), centralized physical register file without banking. It has a four-stage front-end and a five-stage back-end pipeline with operand access after ISSUE.
2. *NON-BANKED-WBF*: *NON-BANKED* configuration with writeback filtering.
3. *NON-BANKED-LONG*: *NON-BANKED* configuration but with an additional stage in the front-end. This model has the same branch mis–prediction penalty as the DSE. Introducing this model allows us to evaluate how much IPC our proposal loses just due to banking stalls in the front-end.
4. *DSE-BANKED*: the DSE architecture with the FPRF which has 8 banks and uses read sharing. Each bank has 2 read and 2 write ports. The banked register file is similar to the one in Tseng et al. [22].
5. *DSE-BANKED-WBF*: the DSE architecture with the *DSE-BANKED* configuration and writeback filtering.

Table 2. Configuration summary

Configuration	#Banks	Read Ports per Bank	Write Ports per Bank	Pipeline Length	Writeback Filtering
NON-BANKED	1	Unlimited	Unlimited	9	NO
NON-BANKED-WBF	1	Unlimited	Unlimited	9	YES
NON-BANKED-LONG	1	Unlimited	Unlimited	10	NO
DSE-BANKED	8	2	2	10	NO
DSE-BANKED-WBF	8	2	2	10	YES

In addition to these five configurations the model described in [22] was also implemented. Our implementation performed around 2% better than the original one as reported in that paper.

5 Performance Evaluation

This section reports on two major metrics of interest: IPC and Energy consumption. *Writeback Filtering* will be further analyzed in Sec. 5.3.

5.1 IPC

The amount of additional instruction level parallelism delivered by the DSE architecture is the focus of this section. For simplicity, only SPECINT and SPECFP average IPC for all configurations is shown in Fig. 3.

The figure shows that the slowdown due to banking or the use of the FPRF is very small. Only a 1.12% average slowdown is observed for SPECINT. For SPECFP the losses are even lower, less than 0.85%. The reason for small slowdowns is that the penalty due to the addition of two new stages in the front-end

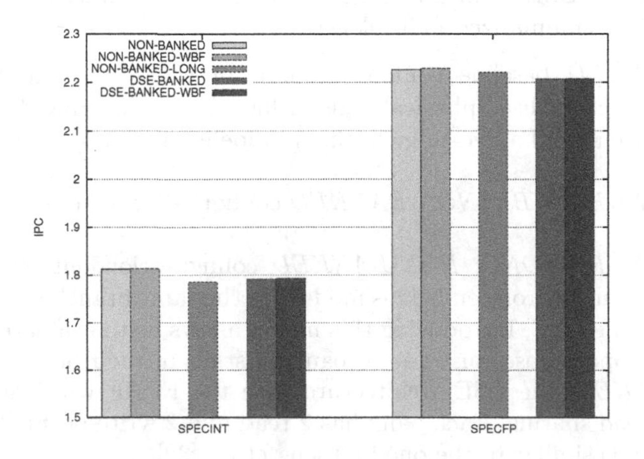

Fig. 3. Average IPC

of the architecture is largely hidden by the latency tolerance of the out-of-order execution back-end. The addition of an arbitration stage in the back end has a higher impact on IPC. Our implementation of a "standard" banked register file (per [22]) resulted in an average IPC loss of over 2%.

The *writeback filtering* technique has little impact on performance. The difference is less than 0.1%. Writeback filtering prevents writes in the writeback stage, which only reduces traffic on result buses and write ports. These are not a bottleneck in the simulated architectures.

Finally, the figure shows that the banked architectures are very close to the non-banked architectures with the same pipeline length. This implies that, even with 2 read and 2 write ports per bank, the number of conflicts is very small. The number of conflicts in FPRF access has been measured. For the reads to the FPRF in SPECINT, 0.18% of all accesses to the FPRF resulted in a conflict. This is approximately 1 conflict for every 600 accesses. For SPECFP this number is a bit larger, 0.5%. The reason it is higher for SPECFP is that most FP instructions read two sources while many integer operations read only a single source.

5.2 Energy Consumption

One of the main benefits of banking is that it reduces the energy consumption in the register file. This section evaluates the energy requirements of data arrays in the FPRF. The energy consumption in the register file is modeled per Rixner et al [23]. The model shows that an access in the FPRF with 160 registers, 8Rd/4Wr ports, and no banking consumes 4.32 times more energy compared to an access in the banked FPRF (8banks, 2Rd/2Wr ports per bank).

The total FPRF access energy is then computed using the total number of FPRF accesses obtained in simulations. The results are shown in Fig. 4 averaged over all SPEC2000 benchmarks. The impact of *banking* and *writeback filtering* on the FPRF energy consumption is clearly visible: the two techniques combined reduce the energy by 81.7%. Banking alone reduces the energy by 76%. The writeback filtering alone reduces it by about 20%. The latter has a smaller impact than banking because writeback filtering can only remove energy due to writes while banking reduces the energy on both reads and writes.

In this discussion on Energy Consumption we have not analyzed components other than the FPRF. But the FPRF is not the only source of registers in this architecture. The simulated DSE architecture features a centralized reservation station with 32 entries. This structure requires four write ports (driven by CAMs) and a series of read ports used to issue the instructions. Energy-wise it can be expensive. However, the instruction queue can also be implemented as a set of distributed reservation stations, a scheme in which each reservation station is attached to a single functional unit. This scheme adds a little complexity and requires somewhat more busses to drive operands around, but its use of very small structures (less number of entries, less read/write ports) allows it to be very small and have little energy consumption. For example, if four reservation

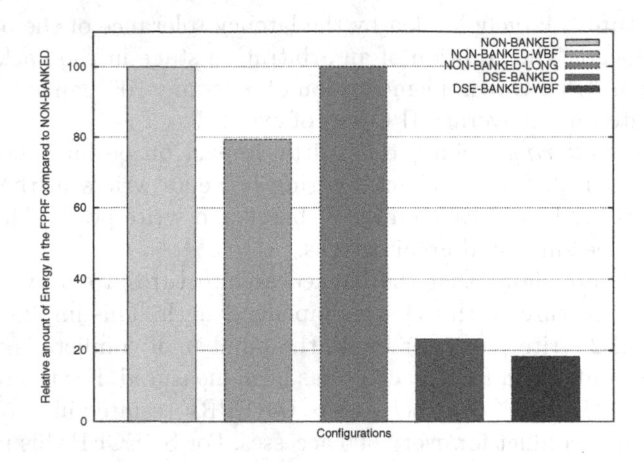

Fig. 4. Relative Energy Consumption

stations are being used, giving each one 8 entries and 2 write ports / 1 read port is still enough to support highly parallel execution.

5.3 Writeback Filtering

Fig. 5 shows percentage of writebacks that were actually filtered in the banked FPRF averaged over all SPECINT and all SPECFP benchmarks. For SPECINT writeback filtering reduces the number of writebacks by approximately 29%. In SPECFP benchmarks, the number of integer writes is reduced by 26% and 48% for floating–point writes. The results show that more FP registers are not part of the state when they are written back and can be filtered out.

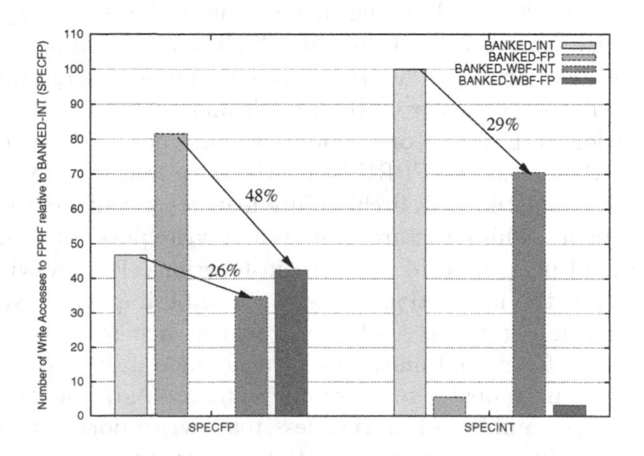

Fig. 5. The impact of Writeback Filtering in the FPRF

6 Conclusions

This paper proposes the Decoupled State-Execute Architecture (DSE) which combines the best features of the Future File and the centralized physical register file. The DSE places the physical register file in the processor front–end. This architecture decouples the processor state in registers from the execution back-end. It has been shown that this architecture allows significant power-performance improvement: a very small IPC loss and a large energy reduction in register file access are achieved. The main reasons are the move of register access arbitration to the front–end and a large reduction in register file access frequency (only values available early are read).

Two optimizations are applied to the DSE architecture: register file banking with read sharing and writeback filtering. Banking results in a minimal IPC loss, considerably less than in a previous proposal where the physical register file is in the back-end. This is due to a) lower access frequence and fewer conflicts and b) reduced impact of arbitration when performed in the front end.

Acknowledgments

This work has been supported by the Ministry of Science and Technology of Spain under contract TIN–2004–07739–C02–01 and the HiPEAC European Network of Excellence under contract IST-004408. This work was supported in part by the National Science Foundation under grant CNS–0220069.

References

1. Tendler, J., Dodson, S., Fields, S., Le, B.S.H.: Power4 system microarchitecture. IBM Journal of Research and Development 46(1) (2002)
2. Hinton, G., Sager, D., Upton, M., Boggs, D., Carmean, D., Kyker, A., Roussel, P.: The microarchitecture of the Pentium 4 processor. Intel Technology Journal (2001)
3. Keltcher, C., McGrath, K., Ahmed, A., Conway, P.: The AMD Opteron processor for multiprocessor servers. IEEE Micro 23, 66–76 (2003)
4. Gowan, M.K., Biro, L.L., Jackson, D.B.: Power considerations in the design lf the Alpha 21264. In: Proc. of the 35th Design Automation Conference (1998)
5. Tomasulo, R.M.: An efficient algorithm for exploiting multiple arithmetic units. IBM Journal of Research and Development, 25–33 (January 1967)
6. Liptay, J.: Design of the IBM Enterprise System/9000 high-end processor. IBM Journal of Research and Development 36(4) (July 1992)
7. Yeager, K.C.: The MIPS R10000 superscalar microprocessor. IEEE Micro 16, 28–41 (1996)
8. Zyuban, V., Kogge, P.: The energy complexity of register files. In: Intl. Symp. on Low Energy Electronics and Design, pp. 305–310 (1998)
9. Park, I., Powell, M.D., Vijaykumar, T.: Reducing register ports for higher speed and lower energy. In: Proc. of the 35th Annual Intl. Symposium on Microarchitecture (December 2002)

10. Kim, N.S., Mudge, T.: Reducing register ports using delayed write-back queues and operand pre-fetch. In: Proc. of the 17th ACM Intl. Conf. on Supercomputing (June 2003)
11. Gonzalez, R., Cristal, A., Ortega, D., Veidenbaum, A., Valero, M.: A content aware integer register file organisation. In: Proc. of the 31th Intl. Symp. on Computer Architecture (2004)
12. Cruz, J., Gonzez, A., Valero, M., Topham, N.: Multiple-banked register file architecture. In: Proc. of the 27th Intl. Symp. on Computer Architecture, pp. 316–325 (2000)
13. Balasubramonian, R., Dwarkas, S., Albonesi, D.: Reducing the complexity of the register file in dynamic superscalar processors. In: Proc of the 34th Intl. Symp. on Microarchitecture (2001)
14. Zalamea, J., Llosa, J., Ayguad, E., Valero, M.: Two-level hierarchical register file organization for VLIW processors. In: Proc of the 33th Intl. Symp. on Microarchitecture (MICRO-33), pp. 137–146 (2000)
15. Palacharla, S., Jouppi, N., Smith, J.: Complexity-effective superscalar processors. In: Proc. of the 24th Intl. Symp. on Computer Architecture (1997)
16. Kessler, R.: The Alpha 21264 microprocessor. IEEE MICRO 19 (March 1999)
17. Seznec, A., Toullec, E., Rochecouste, O.: Register write specialization register read specialization: a path to complexity-effective wide-issue superscalar processors. In: Proc. of the 35th Intl. Symp. on Microarchitecture, pp. 383–394 (2002)
18. Smith, J.E., Pleszkun, A.R.: Implementation of precise interrupts in pipelined processors. In: Proc. of the 12th Intl. Symp. on Computer Architecture, pp. 34–44 (1985)
19. Johnson, M.: Superscalar Microprocessor Design. Prentice-Hall, Englewood Cliffs (1990)
20. Austin, T., Larson, E., Ernst, D.: Simplescalar: an infrastructure for computer system modeling. IEEE Computer (2002)
21. Perelman, E., Hamerly, G., Biesbrouck, M.V., Sherwood, T., Calder, B.: Using SimPoint for accurate and efficient simulation. In: Proc. of the Intl. Conf. on Measurement and Modeling of Computer Systems (2003)
22. Tseng, J., Asanovic, K.: Banked multiported register files for high-frequency superscalar microprocessors. In: Proc. of the 30th Annual Intl. Symp. on Computer Architecture (2003)
23. Rixner, S., Dally, W.J., Khailany, B., Mattson, P.R., Kapasi, U.J., Owens, J.D.: Register organization for media processing. In: Proc. of the 6th Intl. Symp. on High Performance Computer Architecture, pp. 375–386 (2000)

A Scalable Methodology for Computing Fault-Free Paths in InfiniBand Torus Networks*

J.M. Montañana, J. Flich, A. Robles, and J. Duato

Dept. of Computer Engineering (DISCA,UPV)
Camino de Vera, 14, 46021–Valencia, Spain
jmontana@gap.upv.es

Abstract. Currently, clusters of PCs are considered as a cost-effective alternative to large parallel computers. In these systems the interconnection network plays a key role. As the number of elements increases in these systems, the probability of faults increases dramatically. Moreover, in some cases, it is critical to keep the system running even in the presence of faults. Therefore, an effective fault-tolerant strategy is needed.

InfiniBand (IBA) is a new standard interconnect suitable for clusters. Unfortunately, most of the fault-tolerant routing strategies proposed for massively parallel computers cannot be applied to IBA because routing and virtual channel transitions are deterministic, which prevent packets from avoiding the faults. A possible approach to provide fault-tolerance in IBA consists of using several disjoint paths between every source-destination pair of nodes and selecting the appropriate path at the source host. However, to this end, a routing algorithm able to provide enough disjoint paths, while still guaranteeing deadlock-freedom, is required. In this paper we address this issue, proposing a scalable fault-tolerant methodology for IBA Torus networks. Results show that the proposed methodology scales and supports up to $(2n-1)$-faults for n-dimensional tori when using 2 VLs (virtual lanes) and 4 SLs (service levels) regardless of the network size. Additionally the methodology is able to support up to 3 faults for 2D tori with 2 VLs and only 3 SLs.

1 Introduction

Over the recent years there is a trend in using clusters of PCs for building large systems. Some examples are cluster-based Internet portal servers like AOL, Google, Amazon or Yahoo. Also, clusters of PCs are currently being considered as a cost-effective alternative for small and large-scale parallel computing. Each time, more cluster-based systems are included in the top500 list of supercomputers [2]. As an example, the Abe system [1] with 2,400 quad-core Intel Xeon 64 2.3 GHz processors (9,600 cores) is in the eighth position.

In these systems, the interconnection network plays a key role in the performance achieved. In fact, clusters are being built by using high-end interconnection networks

* This work has been jointly supported by the Spanish MEC and European Commission FEDER funds under grants "Consolider Ingenio-2010 CSD2006-00046" and "TIN2006-15516-C04-0X"; and by JCC de Castilla-La Mancha under grant PBC-05-007-2.

J. Labarta, K. Joe, and T. Sato (Eds.): ISHPC 2005 and ALPS 2006, LNCS 4759, pp. 79–92, 2008.

like Quadrics [10], InfiniBand [6], and Myrinet [4]. Among them, InfiniBand (IBA) is a standard interconnect technology for interconnecting processor nodes and I/O nodes, thus building a system area network (SAN). The InfiniBand Architecture (IBA) is designed around a switch-based interconnect technology with high-speed serial point-to-point links connecting multiple independent and clustered hosts and I/O devices. Therefore, this interconnect technology is suitable to build large clusters. As an example, the Abe system uses InfiniBand.

Often, clusters are arranged on regular network topologies when the performance is the primary concern. Low dimensional tori (2D and 3D) are one of the most widely used topologies in commercial parallel computers. Furthermore, recent proposals, such as Alpha 21364 [9] and BlueGene/L [3], use 2D and 3D tori, respectively.

In many cluster-based systems is critical to keep the system running even in the presence of faults. These systems use a very large number of components (processors, switches, and links). Each individual component can fail, and thus, the probability of failure of the entire system increases. Although switches and links are robust, they are working close to their technological limits, and therefore they are prone to faults. Increasing clock frequency leads to a higher power dissipation, and a higher heating could lead to premature faults. So, fault-tolerant mechanisms in cluster-based systems are becoming a key issue.

Most of the fault-tolerant routing strategies proposed in the literature for massively parallel computers are not suitable for clusters (see chapter 6 of [5] for a description of some of the most interesting approaches). This is because they often require certain hardware support that is not provided by current commercial interconnect technologies [4,6]. Other strategies rely on the use of adaptive routing. However, they cannot be applied to IBA because routing is deterministic, which prevents packets from circumventing the faulty components found along their paths. Also, some of these routing strategies need to perform dynamic virtual channel transitions when the packet is blocked due to a fault. However, virtual channels in IBA cannot be selected at routing time.

In IBA, routing and virtual channel (in IBA they are referred to as Virtual Lanes, VLs) selection is performed based on the destination local ID (DLID) and the service level (SL) fields of the packet header. These two fields are computed at the source node and do not change along the path. As a consequence, a possible way to provide fault-tolerance in IBA would be to have several alternative paths between every source-destination pair, selecting one of them at the source host. In particular, to tolerate n faulty components it will be necessary to provide $n + 1$ disjoint paths. From a practical point of view, an appropriate methodology should be able to support a reasonable number of failures according to the network size. Obviously, the switch with the lowest degree will bound the number of possible disjoint paths in the last analysis, which will limit, in turn, the maximum number of faults that can be tolerated at the same time. For instance, in a 2D Torus network only four disjoint paths can be computed for every source-destination pair.

IBA provides a mechanism supported by hardware, referred to as Automatic Path Migration (APM)[6], which may be used for selecting among the available disjoint paths. According to this mechanism, at connection setup time, the source node is given two sets of path information for each destination, one for the primary path and one for the alternate path. APM provides a fast mechanism for migration from the primary to the alternate path when a faulty component is detected. Once path migration is accomplished, the alternate path is converted into the new primary path. Therefore, the subnet manager could reload the alternate path variables with new ones and re-enable the APM mechanism.

2 Motivation

In [8] we presented a methodology referred to as TFTR (Transition-based Fault Tolerant Routing) to compute disjoint paths for an InfiniBand network. In particular, TFTR computes four disjoint paths for every source-destination pair in a 2D Torus network. To get the maximum flexibility in computing disjoint paths, the methodology relies on virtual channel transitions for some paths. To do so, the methodology computes an underlying up*/down* tree [11] and then, the paths that use illegal up*/down* transitions are enforced to switch to a new (increasing) virtual channel.

The methodology requires the use of only two virtual channels to obtain four disjoint paths for every pair of nodes in a 2D Torus. Unfortunately, the methodology requires an unbounded number of SLs. This is because, according to IBA specs [6], the SL assigned to a packet can not be changed by the switches, and this may cause mapping conflicts (see Section 3 for details). Mapping conflicts are solved by using a different path (if any) or using a new SL (not causing a mapping conflict). Taking into account that in InfiniBand the maximum number of SLs is 16 and they are used for other purposes (mainly QoS), the methodology does not scale. For instance, the methodology requires using 2 VLs and 7 SLs for a 10×10 Torus network. Therefore, only two traffic classes could be used at most in conjunction with this fault-tolerant routing methodology, which would significantly limit the QoS capabilities of IBA technology. It has to be noted that, as network size increases, the length of paths also increases. Therefore more mapping conflicts arise, leading to an increase in the number of SLs required. This is a serious problem for IBA and requires an effective solution.

When computing disjoint paths for fault-tolerance issues in a 2D Torus network with InfiniBand, it would be desirable to obtain four disjoint paths for every pair of nodes in such a way that the number of resources used (VLs and SLs) is the same and as low as possible independently of the network size.

In this paper we take on such a challenge. To this end, we need to apply a methodology to compute disjoint paths completely different from that applied in [8], introducing novel concepts, models and computation strategies. In particular, we propose a new methodology referred as SPFTR (Scalable Pattern-based Fault Tolerant Routing). SPFTR will compute four disjoint paths for a 2D Torus network by requiring only the use of 2 VLs and 4 SLs regardless of the network size. Thus, it will guarantee the existence of up to four QoS channels and, at the same time, will tolerate up to 3 faults. Moreover, the methodology will generate a knowledge database of route patterns, which will allow the computation of the disjoint routing paths in a time-efficient manner for any network size. In particular, the computation cost of the methodology will be $O(N^2)$, where N is the number of nodes in the system. Also, the methodology will be easily extended to higher dimensional Torus networks, without increasing the number of resources required.

Notice that the methodology does not depend on the hardware used for detecting failures. Also, it does not depend on the way failures are notified.

The rest of the paper is organized as follows. In Section 3, mapping conflicts in InfiniBand will be explained. In Section 4, SPFTR will be presented for a 6×6 Torus network. In Section 5, the methodology will be extended to larger 2D and higher dimensional Torus networks. In Section 6, the methodology will be evaluated in terms of fault-tolerance, performance and cost. Finally, in Section 7, some conclusions will be drawn.

3 Mapping Conflicts in InfiniBand

In InfiniBand, every switch has a forwarding table where the output port for each destination is provided. The destination ID is located at the packet header and does not change along the path. Therefore, IBA routing is a kind of source routing with the routing info distributed.

Up to 15 data virtual lanes (VLs) can be used in IBA. Virtual lane selection is based on the use of service levels (SLs). By means of SLtoVL mapping tables located on every switch, SLs are used to select the proper VL at each switch. This table returns, for a given input port and a given SL, the VL to be used at the corresponding output port. For this, the SL is placed at the packet header and it cannot be changed by the switches. Therefore, we should also assign the proper SL to be used for a given path.

However, the fact of fixing a path with a unique SL can lead to a mapping conflict. It occurs when two packets labelled with the same SL enter a switch through the same input port, and they need to be routed to the same output port but through different VLs. The problem is that the SLtoVL mapping table does not use the input VL in order to determine the output VL. Figure 1.(a) shows an example. At switch R a mapping conflict arises as it is not possible to distinguish both paths because they are labelled with the same SL. It has to be noted that this problem arises only when there are paths with different VLs. For example, path A uses VL1 and path B uses VL0 until switch Q.

A mapping conflict can be solved only by using different service levels (SLs) for each path causing the mapping conflict. However, this often leads to an excessive number of SLs. Another solution is to use an alternative path that does not cause a mapping conflict. However, obtaining such alternative path strongly depends on the flexibility provided by the applied routing algorithm, on the available network resources (VLs), and the strategy applied to obtain SLtoVL tables.

4 Description of SPFTR

In this section we will describe SPFTR. For the sake of simplicity and ease of understanding we will apply the methodology to a 6×6 Torus network. In the next section we will extend the methodology to larger networks and n-dimensional Torus networks. The aim of this methodology is to compute the forwarding and SLtoVL tables for IBA Torus networks in such a way that they provide the maximum number of disjoint routing paths ($2n$) between every pair of nodes in a n-dimensional Torus in order to tolerate up to $2n - 1$ failures. The methodology must deal at the same time with different issues:

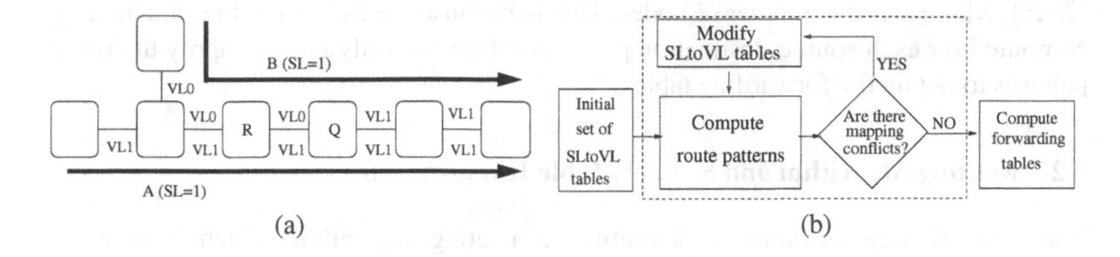

Fig. 1. (a) Mapping conflict example, (b) and steps followed by SPFTR

- It must ensure that any combination of the computed disjoint paths ($2n$ for every source-destination pair) does not lead to deadlock. Deadlock-freedom is ensured by the routing algorithm used.
- It must use only 4 SLs and 2 VLs, regardless of network size. This is achieved by computing in an appropriate manner the SLtoVL tables.
- It must be scalable. For this end, network regions whose switches require the same content for their SLtoVL tables will be defined.
- It must be simple enough in order to be applied in a time-efficient manner to larger networks. For this aim, we will obtain a set of route patterns to be used as templates to define the set of final disjoint paths.

All these issues have direct dependencies between them, and this fact makes obtaining the methodology quite challenging. The choice of the routing algorithm will influence the route patterns that can be used. At the same time, the route patterns used will influence the definition of the SLtoVL tables. In the other way, the definition of SLtoVL tables will influence the route patterns that can be applied.

4.1 Steps Followed by SPFTR

Figure 1.(b) shows the three steps followed by the methodology. Firstly, SLtoVL tables are initialized according to some initial rules: two virtual lanes will be used and transitions among virtual lanes will be done in increasing order (first VL0, then VL1) in order to avoid deadlocks.

In the second step, we look for a set of route patterns that accomplishes the restrictions imposed by the SLtoVL tables. This step will try to get paths as short as possible. The methodology performs as many iterations as required in order to get the final set of paths without introducing mapping conflicts. For this, once route patterns are computed, mapping conflicts are searched. If there is at least a mapping conflict, then the set of SLtoVL tables are changed accordingly (and manually), and a new iteration is performed.

Obviously, the computational cost of this procedure is not bounded and could become very high. However, this methodology takes advantage of the fact that once all the routes are successfully computed, it generates a unique definition of SLtoVL tables and route patterns that can be used in any 2D Torus regardless of its size. Applying the methodology to larger networks (once it was successfully applied to a small 2D Torus) will exhibit a low computational cost. In particular, the computational cost will be

$O(n^2)$, where n is the number of nodes. This is because the SLtoVL tables are already computed for each source-destination pair. Therefore, we only have to apply the route patterns to obtain the forwarding tables.

4.2 Routing Algorithm and SLtoVL Table Initialization

The methodology assumes a deterministic routing algorithm which guarantees deadlock-freedom by building an acyclic channel dependency graph (CDG). This is achieved by enforcing some routing restrictions at the switch level, which prevents packets from traversing some consecutive links (forbidden transitions).

To do this, an underlying deadlock-free routing algorithm will be used. This routing algorithm will indicate where the forbidden transitions are placed. In order to increase the routing flexibility when looking for disjoint paths, the applied routing scheme will traverse some forbidden transitions. The routing scheme will carry out a virtual lane transition each time a forbidden transition is traversed. As commented, the proposed methodology will use only 2 VLs. This means that every routing path can traverse at most one forbidden transition (i.e. only one virtual lane transition can be carried out by a packet at most). In addition, virtual lanes must be used in an ordered way (for example, first VL0, then VL1), in order to avoid cycles.

Thus, the key factor on defining our routing scheme is the selection of the appropriate underlying routing algorithm. This algorithm should be selected in such a way that it guarantees obtaining four disjoint paths for every pair of nodes and also it allows the scalability of the paths (the route patterns are valid regardless of the network size).

For this, we have started from up*/down* routing algorithm [11]. Up*/down* is based on an assignment of directions ("up" and "down") to the links. In our case the labeling of links will be slightly different from that performed by the original up*/down* routing. This assignment of directions is performed by building a certain spanning tree from the network graph. To do this, the node in the corner is selected as the root. Unlike original up*/down* routing, which builds a spanning tree from the complete network graph, the proposed strategy proceeds to remove all the wraparound links of the Torus before building the BFS spanning tree. Once the BFS spanning tree is built, the wraparound links are added, and then we impose a particular set of routing restrictions in order to break cyclic channel dependencies in the CDG. In particular, we have identified all the possible cycles that can be formed and proceeded to remove them by imposing the corresponding routing restrictions. The distribution of the routing restrictions can be seen in Figure 2.a. Deadlock-freedom is guaranteed by verifying that the resulting CDG is acyclic.

The main advantage of this algorithm is that, in most places, the orientation of the routing restrictions is the same, as can be seen in Figure 2.(a), which definitely will contribute to ease the computation process of SLtoVL tables carried out by the methodology [1].

Notice that this algorithm has not been designed with the aim of providing minimal paths between every pair of nodes nor minimizing the number of routing restrictions.

[1] We tried other routing algorithms, such as e-cube and up*/down*. However they were not able to provide a scalable methodology using the minimum number of SLs.

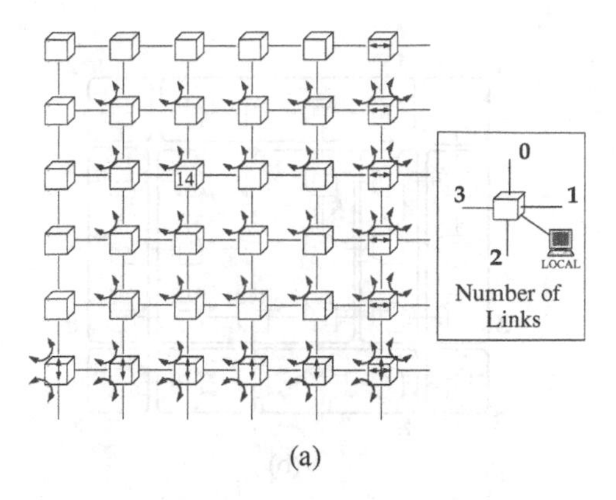

In Port	Out Port	SL	VLOut
0	1	X	N/D
0	2	X	N/D
0	3	X	**1**
1	X	X	N/D
2	X	X	N/D
3	0	X	**1**
3	1	X	N/D
3	2	X	N/D
local	X	X	N/D
X	local	X	N/D

(b)

Fig. 2. Underlying routing algorithm applied: (a) Imposed routing restrictions, (b) SLtoVL table initialization for switch 14. N/D means Not Defined

Instead, it is only used as an underlying routing to guarantee deadlock-freedom of the final routing scheme and provide the symmetry required to simplify the methodology. Notice that this scheme will be able to provide minimal paths by traversing some forbidden transitions and carrying out the corresponding virtual channel transition.

As commented above, the first step of the methodology is to initialize the SLtoVL tables. In particular, SLtoVL tables must initially contain those entries corresponding to link transitions (In Port - Out Port) that require performing in turn a virtual lane transition (i.e., transitions forbidden by the underlying routing algorithm). For instance, Figure 2.(b) shows the initial SLtoVL table corresponding to a switch in the centre of the network. The numbering of the links[2] is shown in Figure 2.(a). Specifically, the entries corresponding to link transitions 0 - 3 and 3 - 0 will be forced to use VL1 (shown in bold face in Figure 2.(b)) because both link transitions are forbidden, as can be seen in Figure 2.(a).

However, note that setting the SLtoVL table in such a way does not guarantee on its own deadlock-freedom. In order to enforce deadlock-freedom packets using these entries should enter the switch exclusively through VL0. This fact must be taken into account when computing the routing paths. The rest of table entries are not enforced in this step by the routing algorithm. The methodology will update those entries accordingly in the next step.

4.3 Network Regions and Route Patterns

The second step of the methodology aims at computing a set of route patterns. Route patterns constitute a kind of templates that can be used to obtain all the disjoint routing paths between every pair of nodes. A route pattern is defined by a sequence of movements (e.g., "go to next switch on the left" or "go to the left until the column of destination") together with a SL. This sequence of movements must be compatible with the routing scheme applied in order to guarantee deadlock-freedom. Moreover, the SL must be selected in such a way that mapping conflicts are avoided.

[2] This numbering will be used throughout the rest of the paper.

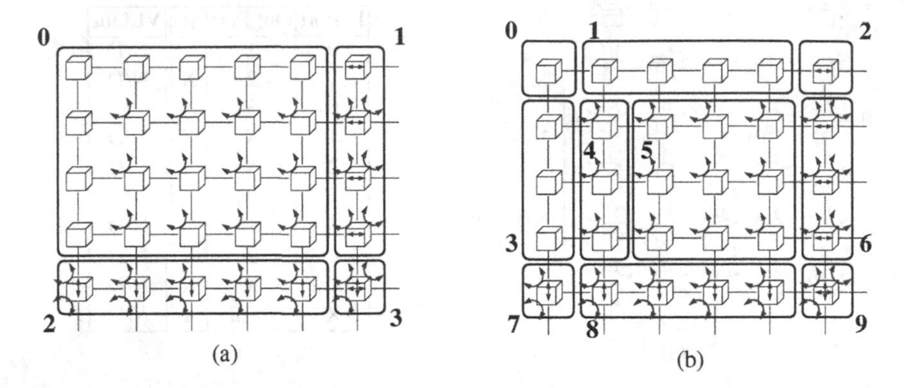

Fig. 3. (a) SPFTR regions (b) A-SPFTR regions in a 2D Torus Networks

Additionally, the methodology defines network regions. A network region will be formed by neighbour switches with the same SLtoVL table definition. Network regions are defined taking into account two conditions. The first one is that all the switches in a region must have the same routing restrictions (this is obvious as they must have the same SLtoVL table). And, second, in order to help the methodology to scale, the same distribution of regions should be kept regardless of network size. Figure 3.(a) shows the final regions defined for the 6×6 Torus network. As can be observed, 4 regions have been defined so that fulfil the conditions referred above. Each of them has a different SLtoVL table for every switch within the region.

The main advantage of using regions is that it minimizes the number of required route patterns. Route patterns depend on the regions where source and destination are located (the same or different) and their relative positions (i.e., they may be located either in the same row/column or in different rows and/or columns). To generate the route patterns, we try to obtain four disjoint paths (as short as possible) between every pair of nodes. To this end, it is necessary to establish the path followed by the pattern (sequence of movements) and the SL to be used. The former is accomplished by following the restrictions imposed by the applied routing algorithm (required VL transitions were already introduced in the initial SLtoVL tables), thus guaranteeing deadlock-freedom. The latter is carried out according to the current SLtoVL table entries and the bounded number of SLs used (4 SLs), checking that mapping conflicts are not introduced.

If it is not possible to obtain a set of valid route patterns, SLtoVL tables must be slightly updated (manually) and a new try to obtain route patterns will be carried out. Thus, an iterative process is performed with the two steps until a valid set of route patterns is found.

Each route pattern must be assigned a SL, checking the corresponding entries in the SLtoVL tables to verify that mapping conflicts are not introduced.

Table 1 shows the final SLtoVL table for every region. These tables have been computed taking into account the route patterns applied for every source-destination pair. The contents of the final SLtoVL tables are the same for all the switches belonging to the same region. According to the defined regions and the possible relative positions of every pair of nodes, 153 route patterns have been obtained.

Table 1. SLtoVL tables for every region in 2D Torus network (values imposed by routing restrictions in each region are written in bold face, X means any value)

		Region 0	Region 1	Region 2	Region 3
In	Out	SL0-1-2-3	SL0-1-2-3	SL0-1-2-3	SL0-1-2-3
0	1	0100	**1111**	0100	**1111**
0	2	0100	0100	**1111**	**1111**
0	3	**1111**	**1111**	**1111**	**1111**
1	0	1101	**1111**	1101	**1111**
1	2	0101	0101	1101	0101
1	3	1101	**1111**	1101	**1111**
loc.	X	0000	0000	0000	0000
X	loc.	1111	1111	1111	1111

		Region 0	Region 1	Region 2	Region 3
In	Out	SL0-1-2-3	SL0-1-2-3	SL0-1-2-3	SL0-1-2-3
2	0	1101	1101	**1111**	**1111**
2	1	0101	1101	0101	0101
2	3	1101	1101	**1111**	**1111**
3	0	**1111**	**1111**	**1111**	**1111**
3	1	0100	**1111**	0100	**1111**
3	2	0100	0100	**1111**	**1111**
loc.	loc.	0000	0000	0000	0000

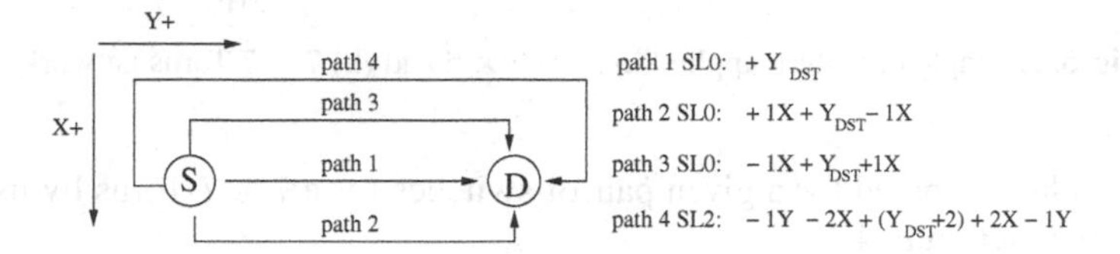

Fig. 4. Example of route pattern

To illustrate the process of computing the routing paths from the route patterns, without lose of generality, let us consider a pair of switches located in the same row in region 0 (the relative position of the switches and the regions will determine the pattern to use). Then, the patterns defined for computing the paths can be shown in Figure 4. The computed routing paths in a 6×6 Torus network can be shown in Figure 5.(a).

As an additional result from the methodology, we have successfully obtained an alternative set of route patterns and SLtoVL table definition that only requires 2 VLs and 3 SLs. This has been achieved by sacrificing the scalability in terms of higher dimensional Torus network, thus being only valid for 2D tori. However, scalability in terms of number of nodes is guaranteed (see next section). The resulting algorithm will be referred to as A-SPFTR (Asymmetric scalable patter-based fault tolerant routing). The regions defined are shown in Figure 3.(b).

5 Extending the Methodology

In this Section we will show how the methodology can be applied to other networks. In a first effort, the methodology will be extended to larger 2D Torus networks, thus scaling the methodology. In a second effort, we will extend it to 3D Torus networks.

A larger 2D Torus network can be viewed as a 6×6 Torus network with additional rows and/or columns of switches. These new rows or columns can be placed in the middle of the Torus network (taking as a reference Figure 3.(a)). Therefore, the new components will belong to regions 0 and 2 in the case of a column or to regions 0 and 1 in the case of a row. By doing this, the distribution of regions will be the same. Therefore, SLtoVL tables of new switches will be already defined. Also, the route patterns to use for every new switch to all the destinations have been already computed as each switch can use the route patterns of one of its neighbours. As an example, Figure 5.(b)

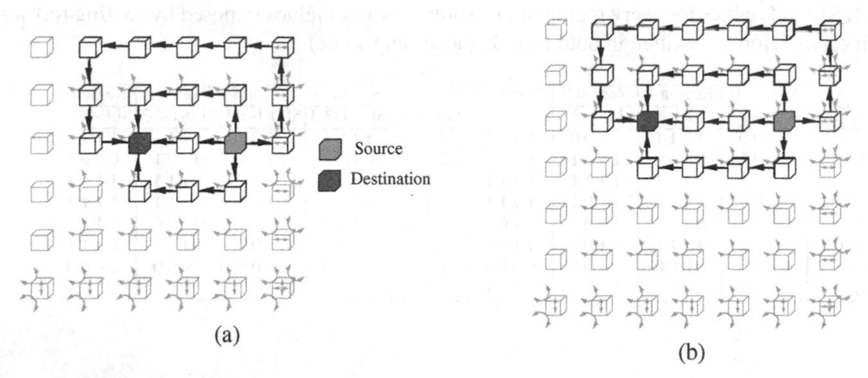

Fig. 5. Example of pattern applied in a (a) 6×6 and (b) 7×7 Torus network

shows the paths computed for a given pair of switches for a 7×7 Torus by using the patterns shown in Figure 4.

Notice also that A-SPFTR (2 VLs and 3 SLs) also scales. In this case, new rows could be added along regions 3, 4, 5, and 6, and new columns along regions 1, 5, and 8, maintaining the same region structure shown in Figure 3.(b).

In order to extend the methodology to n-dimensional Torus, we will use the same routing algorithm described in Section 4.2. It will define the routing restrictions (Figure 6.(a) shows an example for a $3 \times 3 \times 3$ Torus network). In particular, for every plane in the 3D Torus we can find an orientation at which all restrictions are allocated in the same positions as in the 2D Torus shown in Figure 2.(a).

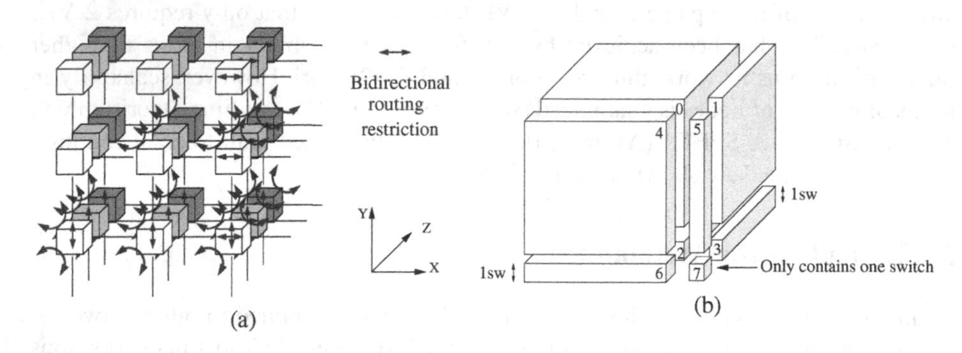

Fig. 6. (a) Frontal view of routing restrictions in the plain X-Y of a 3D Torus. (b) SPFTR Regions in a 3D Torus

Then we can define the SLtoVL table (defining the output VL for each pair of links), considering that each pair of links can be allocated in a plane and using the SLtoVL tables for the corresponding 2D plane, as shown in Figure 6.(a). By doing this in all 2D planes, all the entries for the SLtoVL tables (for the 3D case) will be filled.

Notice that every pair of links which are in the same direction will be included in two planes. Therefore, the SLtoVL values for them must be the same in both planes. That is, SLtoVL tables must be compatible plane by plane. The final regions in a 3D Torus can be shown in Figure 6.(b).

In the case of a 3D Torus the number of route patterns for each plane will be the same, then we will have 3 times the quantity of patterns defined for a plane (153 patterns). Additionally it is needed to add the route patterns for the pair of nodes not included in the same plane. Finally, the total amount of route patterns is 726.

6 Evaluation

In this section, we will evaluate SPFTR and A-SPFTR. Notice that both models are scalable in size but only SPFTR is scalable in dimensions too. For comparison purposes we will also evaluate the up*/down* routing scheme. First, we will present the evaluation model, describing all the simulation parameters and the Torus networks we have used. Finally, we will present the evaluation results.

6.1 Evaluation Model

We will evaluate the routing algorithms in 2D and 3D Torus networks. We have analyzed 2D tori with different sizes from 16 switches (4x4) up to 400 switches (20x20). Also 3D tori will be analyzed, from 64 switches (4x4x4) up to 512 switches (8x8x8).

In the analysis, we will only consider faults of links connecting switches. Note that a switch failure can be viewed as if all its links had failed. Moreover, a failure in a link connecting a host to a switch does not change the topology. The number of disjoint paths depends on the minimum degree of any switch in the network. Therefore, at maximum, there will be 4 disjoint paths in a 2D Torus (6 in a 3D Torus). Hence, at most, 3 faults can be tolerated (5 in a 3D Torus). The methodology will also be evaluated in terms of network performance. In particular, network performance degradation due to faults will be analyzed when applying the proposed methodology. The network performance is evaluated using the simulation tool and simulation parameters presented in [7]. The simulator models an IBA network, following the IBA specifications [6].

6.2 Evaluation Results

Fault Tolerance. We define a *singular case* as the fault combination that can not be tolerated by the routing algorithm, while still maintaining the network connected. In other words, those fault combinations for which the routing algorithm is not able to obtain a valid path for a particular source-destination pair.

We will evaluate the fault tolerance degree analyzing which combinations of faults are supported and which are not (singular case), for different amounts of link failures.

Figure 7 shows the percentages of singular cases when using up*/down*, SPFTR, (results for A-SPFTR are similar to SPFTR) for different number of faults. As can be observed, SPFTR (same happens for A-SPFTR) do not present any singular case up to any combination of $(2n - 1)$ faults in n-dimensional tori, as it is able to provide $2n$

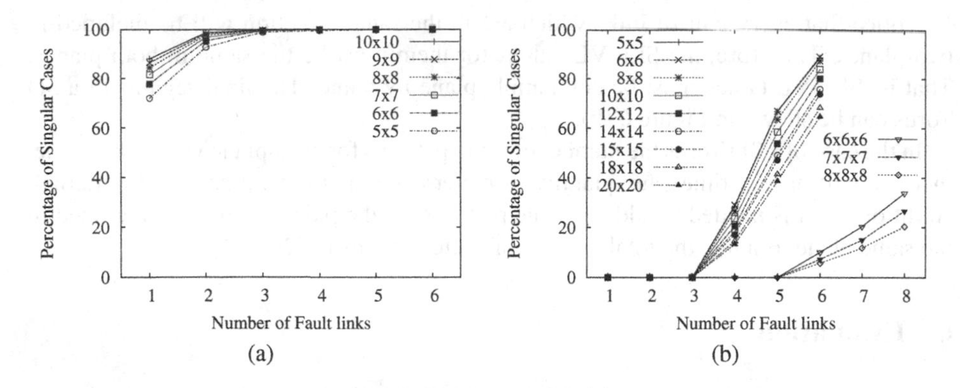

Fig. 7. Singular cases for (a) Up*/Down* and (b) SPFTR in 2D and in 3D (Maximum of 100.000 fault combination evaluated at each point when number of fault combinations is higher than 100.000)

disjoint paths. Therefore, the methodology is $(2n-1)$-fault tolerant with 2 VLs and 4 SLs for n-dimensional tori (SPFTR) or with 2 VLs and 3 SLs for 2D tori (A-SPFTR). On the other hand, we can see that up*/down* does not tolerate even a single failure.

From four faults and beyond we can see that none of the routing methods is able to tolerate all the failure combinations. Indeed, for 4 faults in the network, 30% of failure combinations are not tolerated in the worst case (5x5 Torus network). However, this is a reasonable fault-tolerance degree taking into account the analyzed network sizes, and that the mean time between failures is much greater than the mean time to repair.

However, notice that the up*/down* routing algorithm has a pretty worse behaviour. Practically, all fault combinations, even with one fault, are not tolerated by this algorithm. Also, notice that when a singular case arises the only solution will be to launch a network reconfiguration process to compute new routing tables.

Length of Paths. Figures 8.(b) and 8.(c) shows the average length of the shortest paths for the SPFTR, A-SPFTR (for 2D tori) and up*/down* routing, also showing the average topological distance between switches. When considering only the shortest path for every pair of switches (the common case in the absence of failures), SPFTR and A-SPFTR achieve, on average, shorter paths than up*/down*. Additionally, as network sizes increases, the difference between the average path length and the average topological distance slightly increases. This is due to the fact that only one forbidden transition is allowed (2 VLs are used) and some route patterns would need to take longer paths to reach destination. Also, in Figures 8.(b) and 8.(c) we can see for 2D and 3D tori, respectively, the average lengths of the shortest paths and the average length when the complete set of alternative paths is considered.

Performance degradation with faults. Figure 8.(a) shows the performance degradation suffered by the network in the presence of faults when the SPFTR algorithm is used (similar results have been obtained for A-SPFTR). In the presence of faults every source node will use the shortest disjoint path that does not traverse any faulty link.

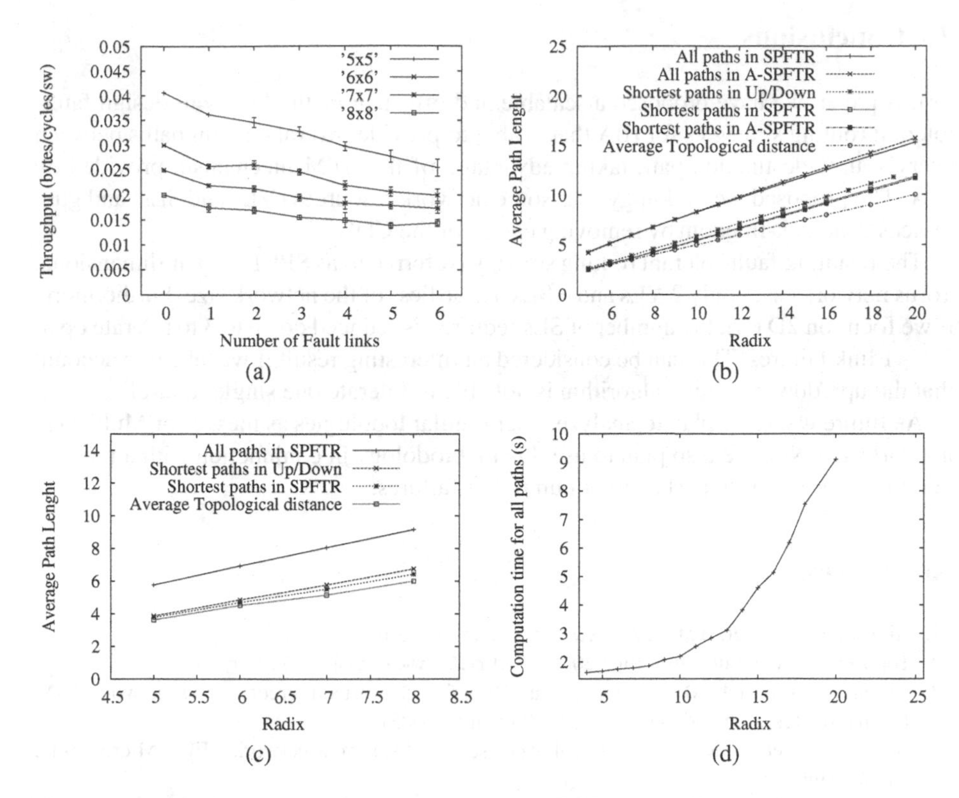

Fig. 8. (a) Degradation of performance with faults for SPFTR and A-SPFTR. (b) Average length of paths in 2D Torus. (c) Average length of paths in 3D Torus. (d) Computation Cost in 2D Torus. Radix stands for the number of switches per dimension.

For a particular number of faults, different random combinations of faults are injected. For every combination of faults the network is simulated, obtaining its throughput and displaying the average results. Error bars are shown for every number of faults.

As can be observed, throughput decreases as the number of faults increases. However, it can be noticed that performance degradation is relatively lower in larger networks. For instance, with 6 faulty links, throughput decreases in SPFTR up to 28 % for the 4×4 Torus and a 23 % for the 8×8 Torus. This is because the same number of faulty links affects a lower percentage of paths in larger networks. For up*/down* no results can be plotted as practically it does not tolerate any failure combination.

Computation Time. Finally, Figure 8.(d) shows the computation time of SPFTR (for A-SPFTR same results are obtained) for different 2D tori with different sizes. As can be observed, the computation time grows quadratically with the number of switches in the network. Notice that only the computation of forwarding tables is required. This is because SLtoVL tables and route patterns are already computed (tables for 2D Torus in Table 1). As an example, all routing information for a 20x20 Torus network was computed in less than 10 seconds using an Intel Xeon 3.06 GHz.

7 Conclusions

In this paper, we have proposed a scalable and effective methodology to design fault-tolerant routing strategies for IBA that is able to provide several disjoint paths between every source-destination pair, taking advantage of the APM mechanism provided by IBA. The proposed methodology uses some network resources (VLs and SLs) and guarantees deadlock-freedom by removing cycles in the CDG.

The resulting fault-tolerant routing strategy (referred to as SPFTR) on n-dimensional Torus networks uses only 2 VLs and 4 SLs, regardless of the network size. Furthermore, if we focus on 2D tori, the number of SLs required is reduced down to 3 to tolerate up to $2n - 1$ link failures. This can be considered an interesting result if we take into account that the up*/down* routing algorithm is not able to tolerate one single failure.

As future work, we plan to analyze other regular topologies as meshes or Multistage networks (MINs). We also plan to use this methodology in conjunction with a reconfiguration process to support a larger number of failures.

References

1. Abe supercomputer, `http://www.ncsa.uiuc.edu/`
2. Top500 supercomputer list (June 2007), `http://www.top500.org`
3. Adiga, N., Blumrich, M., Chen, D., et al.: Blue Gene/L torus interconnection network. IBM Journal of Research and Development 49 (March 2005)
4. Boden, N.J., et al.: Myrinet: A Gigabit-per-second local area network. IEEE Micro 15(1), 29–36 (1995)
5. Duato, J., Yalamanchili, S., Ni, L.: Interconnection Networks. An Engineering Approach. Morgan Kaufmann, San Francisco (2003)
6. Architecture. Specification Release 1.0, InfiniBand Trade Association^{TM} (October 2004)
7. Lysne, O., et al.: Simple Deadlock-Free Dynamic Network Reconfiguration. In: 11th Int. Conference on High Performance Computing (HiPC), December 19-22, 2004, Bangalore, India (2004)
8. Montañana, J.M., Flich, J., Robles, A., Duato, J.: A Transition-Based Fault-Tolerant Routing Methodology for InfiniBand networks. In: IPDPS 2004. Proc. of the 2004 Int. Parallel and Distributed Processing Symp, IEEE Computer Society Press, Los Alamitos (2004)
9. Mukherjee, S., Bannon, P., Lang, S., Spink, D.W.A.: The Alpha 21364 network architecture. IEEE MICRO (January-February 2002)
10. Petrini, F., et al.: The quadrics network (qsnet): High-performance clustering technology. In: HotI 2001. Proceedings of the 9th IEEE Hot Interconnects, Palo Alto, California (August 2001) (original version), IEEE Micro January-February 2002 (extended version)
11. Schroeder, M.D., et al.: Autonet: A high-speed, self-configuring local area network using point-to-point links. Journal on Selected Areas in Comm. 9(8) (October 1991)

Using a Way Cache to Improve Performance of Set-Associative Caches*

Dan Nicolaescu, Alexander Veidenbaum, and Alexandru Nicolau

Department of Computer Science
University of California Irvine
{dann,alexv,nicolau}@ics.uci.edu

Abstract. Modern high–performance out–of–order processors use L1 caches with increasing degree of associativity to improve performance. Higher associativity is not always feasible for two reasons: it increases cache hit latency and energy consumption. One of the main reasons for the increased latency is a multiplexor delay to select one of the lines in a set. The multiplexor is controlled by a *hit* signal, which means that tag comparison needs to be completed before the multiplexor can be enabled. This paper proposes a new mechanism called *Way Cache* for setting the multiplexor ahead of time in order to reduce the hit latency. The same mechanism allows access to only one of the tag stores and only one corresponding data store per cache access, which reduces the energy consumption. Unlike way prediction, the Way Cache always contains correct way information - but has misses. The performance of Way Cache is evaluated and compared with Way Prediction for data and instruction caches. The performance of the Way Cache is also evaluated in the presence of a Cached Load/Store Queue, an integrated L0 cache-Load/Store Queue which significantly reduces the number of accesses to the L1 cache.

1 Introduction

The increasing gap between processor and memory speeds requires further advances in cache design to minimize processor stalls due to slower memory. Increasing cache associativity is one way to improve the average performance of the memory system and thus the overall processor performance. Both high-performance and embedded microprocessors have been steadily increasing the associativity of their first-level data and instruction caches.

Examples of high-associtivty L1 caches include the Transmeta Crusoe [1] and Motorola's MPC7450 [2] in the embedded domain and high-performance Pentium–M[3] processors, which all use an 8–way set–associative L1 data cache. The Digital StrongArm and Intel XScale embedded processors have even implemented 32–way set associative caches.

* This work was supported in part by the National Science Foundation under grant NSF CCR-0311738.

J. Labarta, K. Joe, and T. Sato (Eds.): ISHPC 2005 and ALPS 2006, LNCS 4759, pp. 93–104, 2008.

The increased associativity poses several challenges for processor designers: an increased access time as well as increased energy consumption. The reason for both of these is the timing of a cache access.

An access in a set–associative cache is initiated before the cache "way" containing the requested address is known. All cache ways in a set are therefore accessed in parallel to reduce the access latency, a critical design parameter. The timing of the access is as follows: first, a simultaneous access to all N independent tag arrays of an N-way set–associative cache is performed, followed by N tag comparisons to generate the hit/miss signal. Access to N data arrays[1] is initiated at the same time as the tag access to minimize latency. An N-input multiplexor is used next to select the desired data array. The multiplexor is controlled by the tag comparison results and its presence thus delays the data availability. Accessing N tag and N data arrays even though tag/data may only reside in at most one array is also the reason for energy inefficiency.

Several mechanisms (discussed in more detail in Section 2) have been proposed as a way to reduce latency and/or save energy. One such mechanism, way prediction, has been used in high-performance processors [4,5]. The predictor records the most recently used way for each set and predicts that this way is going to be accessed again next. Only one way for both tags and data needs to be accessed. Furthermore, the output multiplexor can be set up in parallel with tag/data access. A successful tag comparison confirms that the predicted way contains the requested data, otherwise a standard, parallel N-way lookup is performed next. Misprediction thus leads to additional delay and energy consumption. In addition, predictor access delay needs to be hidden.

This paper presents a different mechanism that allows a cache way containing the desired data to be known before intiating the cache access. This is accomplished via a *Way Cache* (*WC*), a very small cache of recently observed address-way mappings. A hit in the WC guarantees that the data is in the cache way the WC supplied. A miss in the WC can occur for two reasons: a) the data is not in the cache and the access is going to be a miss, or b) the data is in the cache but the WC does not have a mapping. In either case a standard, N-way cache access is initiated. The main advantage of the new approach is that it requires a single access to the cache and cannot have mis-predictions. In addition, it may be possible to integrate the WC into the load/store queue (LSQ).

The proposed approach should work well because of locality. That is, a cache line being accessed is very likely to have been accessed in the recent past. This paper shows that the Way Cache can be implemented effectively for both data and instruction Level–1 (L1) caches and that, in addition to reducing access latency it significantly reduces the L1 cache energy consumption. Its effectivness increases with cache associtivity thus making it very suitable for next-generation caches. It is also shown to have higher performance than way prediction for high-associativity caches because the *Way Cache* does not have mis–predictions and the associated additional cache access delays.

[1] CAM-based implementations, such as StrongArm and Xscale, access only the desired data after the CAM tag comparison is completed.

The rest of this paper is organized as follows. Section 2 discusses related work. Section 3 describes the Way Cache and how it operates. The design of the Way Cache is described in Section 4. The simulation setup and benchmarks used are described in Section 5. The Way Cache is evaluated in Section 6 and compared with the performance of a way predictor in Section 6.1. Section 6.2 examines the performance of Way Cache in the presence of Cached Load/Store Queue (CLSQ). CLSQ allows both load and store data to be kept in the LSQ after corresponding instructions retire. It is thus a type of L0 cache integrated in the LSQ with little additional hardware cost. CLSQ significantly reduces the number of L1 cache accesses and alters their reference patterns. Thus it is interesting to examine its effect on the Way Cache.

2 Related Work

Several approaches, both hardware and software, have been proposed to reduce the delay and/or energy consumption of set–associative caches.

A phased cache [6] avoids the associative lookup to the data store by first accessing all N ways of the tag store and then only accessing the desired data way after the tag comparison completes and indicates the correct way for the data. This technique has the undesirable consequence of increasing the cache access latency and has a significant impact on performance. It is not suitable for high-performance L1 caches, but has been successfully applied to L2 caches [7] without significant performance impact.

A way–prediction scheme (mentioned above) was first implemented in the R10000 processor [8] (also described in [6]). It uses a predictor with an entry for each set in the cache. Each predictor entry stores the most recently used way for the cache set. On an access only the way returned by the predictor is accessed. In case of an incorrect prediction the access is restarted, accessing all the cache ways in parallel. Also, given the required predictor size for large modern L1 caches, it may increase the cache latency even for correct predictions.

Way prediction in L1 I-cache has been implemented by adding a pointer to the next line/way to each I-cache line [5] or in the branch target buffer [9], but these mechanisms do not work well in D-caches.

A mixed hardware–software approach was presented in [10]. Tag checks are avoided by having a compiler output special load/store instructions that use tags from a previous load. This approach requires changes to the compiler and the ISA and adds hardware complexity.

3 Way Caching

The goal of this work is to determine before or at the start of a cache access exactly which way needs to be accessed. This will allow for an energy reduction as well as reducing the access latency. The latter is possible because the signal that enables the data output multiplexor (MUX) will be generated very early in the access and the multiplexor delay reduced. (Note that this may or may not have an impact on the cycle time.)

This "way determination" is accomplished by a *Way Cache* (*WC*). The *Way Cache* is accessed prior to an L1 cache access using a CPU address. It supplies the way number where the desired data resides if the *WC* tag matches the address. *WC* access can be overlapped with the LSQ lookup on the data side or the BTB lookup on the instruction side. A block diagram of the Way Cache integrated in the data cache system is shown in Figure 1.

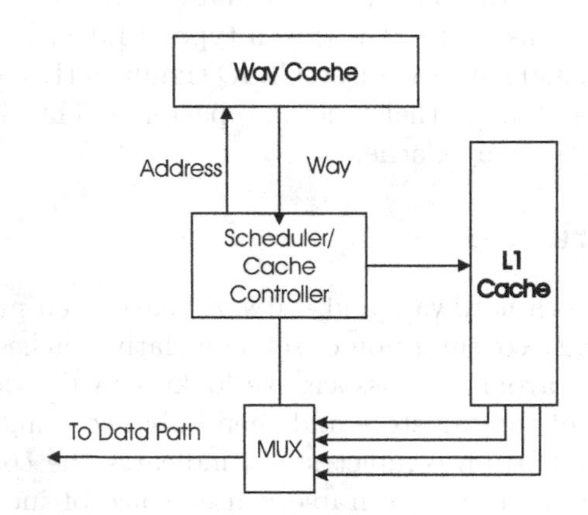

Fig. 1. The Way Cache in the cache hierarchy

The L1 cache address is sent to the Way Cache at the same time it is sent to the load/store queue for lookup. On a *Way Cache* hit the *WC* returns the corresponding way number. This way number is used to access only the designated way in both the tag and data stores.

On a Way Cache miss a standard N–way associative cache lookup is performed with all N ways accessed in parallel. At the same time a new entry is created in the WC containing the line address and empty way field. The L1 cache controller updates the entry's way field when it becomes known and marks the entry valid. All valid addresses in the Way Cache have been previously seen by the L1 cache and the way information stored in the WC is always correct. This is one of the main differences between the Way Cache and previously proposed systems using prediction.

Cache lines are accessed repeatedly due to both spatial and temporal locality and therefore the Way Cache can have a small number entries and still cover a significant number of accesses. Being small is important for the Way Cache because it allows for a fast access and small energy overhead. Since the Way Cache lookup is done in parallel with the LSQ access the Way Cache and the LSQ should be of similar size and latency to avoid adding extra latency to the cache access.

4 The Way Cache Design

The proposed Way Cache design is shown in Figure 2. The Way Cache is a small cache in which an entry contains a line address, way number, and a valid bit.

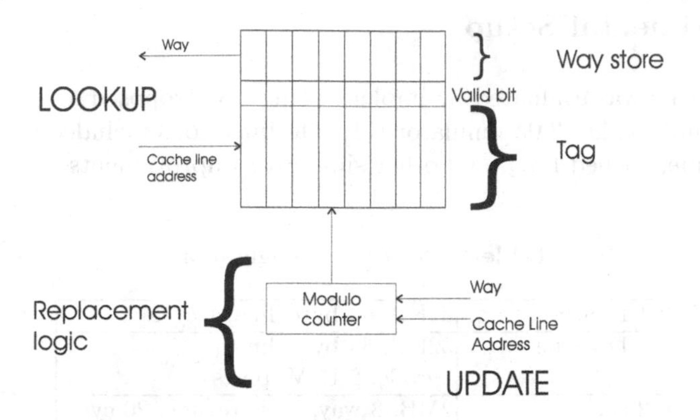

Fig. 2. The Way Cache structure

The address serves as the tag and the WC is assumed to be fully associative (due to its small size),

The Way Cache is accessed in two cases: to lookup a way for an address, and to add (write) a new address–way pair (an operation similar to an "update" in a predictor).

The WC lookup is like any cache read: given a cache line address the *WC* returns a way number for a matching tag (provided the entry is valid). The Way Cache is written to in two cases: on an L1 cache miss or *WC* miss and L1 cache hit. On a *WC* miss a new *WC* entry is immediately allocated, the line address is written to the tag part and the corresponding way number is recorded when the L1 cache controller determines it. The valid bit is set to true as soon as the way number is available. If the *WC* is full, the newly allocated entry replaces the oldest entry in the *WC*. The implementation used in this paper determines the oldest entry by using a modulo counter.

The *WC* design must address the issue of its coherence with the L1 cache on L1 line replacement or invalidation. The corresponding *WC* entry (if it exists) needs to be marked invalid in such cases. Another approach is to allow the L1 access to proceed using the *WC*–supplied way and to have a cache miss occur when the L1 cache tag comparison is performed. Since the way that the *WC* points to was the only place the line could have been found at, the L1 tag miss–match indicates a miss and does not require the search of all other cache ways. The *WC* will be updated when the line is allocated again. This is a more energy–efficient approach than the invalidation and it is used in the design presented here.

One last comment on the WC design. On the data side the WC performs a tag compare of a CPU line address which is only a few bits shorter than the LSQ address tag compare. Since both are performed at the same time after the address becomes available, the CAM structure can possibly be largely shared. (43-logN-3) bits of tag can be "shared" assuming a 43b physical address, N-way associativity, and a 64Byte line. Way number may then be recorded in the LSQ to be used when the entry is issued to the cache.

5 Experimental Setup

The Way Cache was evaluated by implementing the proposed architecture in a modified SimpleScalar–3.0c simulator [11]. Modifications included support for a longer pipeline, cached LSQ, and other simulator improvements.

Table 1. Processor configuration

L1 I–cache	32KB, 64 byte/line, 1 cycle
L1 D–cache	32KB, 64 byte/line, 3 cycle, 2 R/W ports
L2 cache	2MB, 8 way, 64 byte/line, 20 cycle
Issue	4 way out-of-order
Branch predictor	64K entry g–share, 4K-entry BTB
Reorder buffer	256 entry
Load/Store Queue	64 entry
Arithmetic Units	4 integer, 4 floating point units
Complex Units	2 INT, 2 FP multiply/divide units
Pipeline	15 stages

The architecture modeled was an aggressive 64–bit high–performance architecture. The SPECINT2000 benchmark suite was used in performance evaluation (see Table 2). The benchmarks were compiled with the -O4 flag using the Compaq compiler targeted for the Alpha 21264 processor. The benchmarks were fast–forwarded for 500 million instruction, then fully simulated for 5 billion instructions.

The memory hierarchy latencies are 3/20/200 cycles for the L1/L2/memory accesses, respectively. The details of the processor and system model are given in Table 1.

The Alpha *"Universal NOP" – unop* instruction is encoded as a type of load (*ldq_u*) with destination register ($31). These loads were treated as true NOPs in the modified simulator and did not perform any LSQ, WC, or L1 cache accesses.

A brief description of the benchmarks used is given in Table 2.

6 Performance Evaluation

This section presents an experimental evaluation of the proposed Way Cache. Several factors are examined in order to evaluate the potential of the Way Cache: its hit rate and how it varies with Way Cache size and L1 cache associativity, the difference in behavior for data and instruction caches, and the impact of cached LSQ on *WC* performance.

Figure 3 shows the Way Cache hit rates for 32–, 64–, and 128–entry *WCs* and an 8–way set associative L1 data cache. The average hit rates are slightly higher for integer benchmarks (up to 85%) than for floating–point ones (up to 80%).

Table 2. Benchmark description

SpecINT	Description	SpecFP	Description
bzip2	Compression	applu	Partial Differential Equations
crafty	Game Playing: Chess	apsi	Pollutant Distribution
eon	Computer visualization	art	Neural Network
gap	Group theory	equake	Seismic Wave Propagation
gcc	C optimizing compiler	facerec	Face Recognition
gzip	Compression	fma3d	Finite-element Simulation
parser	Word processing	galgel	Computational Fluid Dynamics
perlbmk	PERL interpreter	lucas	Primality Testing
twolf	Place and route simulator	mesa	3-D Graphics
vortex	Object–oriented database	mgrid	Multi-grid Solver
vpr	Circuit place & and route	sixtrack	Nuclear Physics
		swim	Shallow Water Modeling
		wupwise	Quantum Chromodynamics

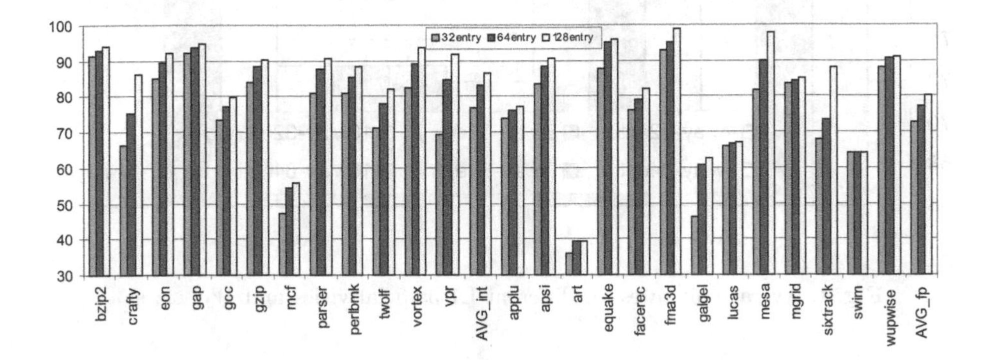

Fig. 3. Way Cache hit rate for an 8 way L1 data cache

mcf has the lowest WC hit rate of all SpecINT benchmarks. This benchmark does a lot of pointer chasing which reduces the DL1 hit rate. More importantly, it has not been correctly ported for LP64 systems (i.e. systems that store C data types long and pointer as 64–bit quantities) further gratuitously increasing the DL1 miss rate. This can be fixed by applying a 6 line change to the benchmark code, but was intentionally not done in this paper. The *art* benchmark would benefit from a high-level compiler transformation that transforms the fundamental data structure used by the benchmark from an array of structures to a structure of arrays. This would greatly improve the DL1 access patterns and performance. The compiler used in this paper does not make the above transformation, hence the low *WC* hit rate for this benchmark. The *galgel* and *swim* benchmarks benefit from another high level compiler transformation that significantly affects the DL1 access pattern: loop interchange. The fact that this optimization was not performed for the binaries used here explains the lower WC hit rates obtained. If the issues enumerated above were to be solved the *WC* average performance would be significantly improved.

It is easy to observe in Figure 3 that the hit rates increase significantly with the *WC* size. However, even for the smallest *WC*, average hit rates of over 76% (SpecINT) and 73% justify adding a *WC* to a processor's D–cache system. A 128–entry *WC* may be a bit large and possibly not fast enough, but only detailed VLSI design can determine this. All of the results below only use 32– and 64–entry *WC* given that the LSQ used in the processor has 64 entries. Also, only the INT and FP averages are presented from here on to avoid making graphs too busy.

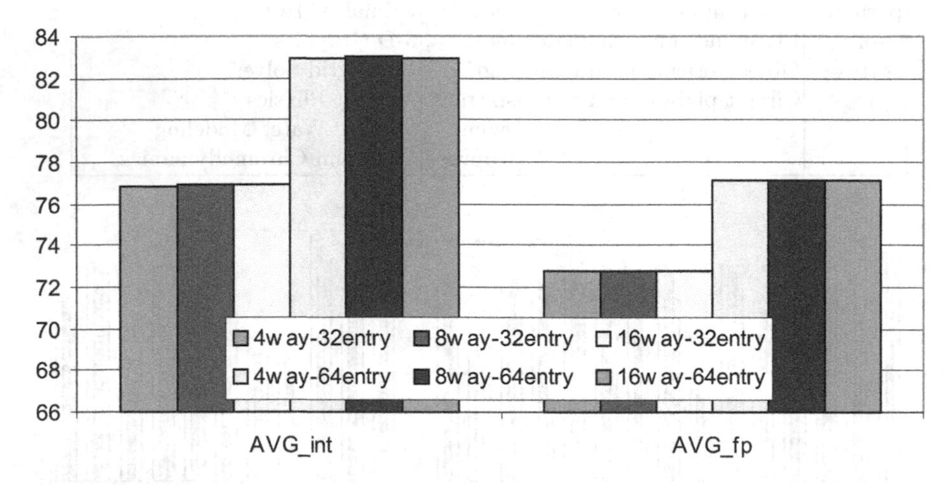

Fig. 4. Average hit rates for different L1 associativities and *WC* sizes

One of the main motivations for introducing the Way Cache is the trend to increase cache associativity in modern processors. Figure 4 shows the average hit rates for 32– and 64–entry *WC* and 4, 8 and 16–way associative caches. It can be observed that the *WC* hit rates are almost constant with associativity (this trend is also observed for individual benchmarks). This is very important since the accuracy of previously proposed way-prediction schemes tends to degrade with higher associativity. The fact that Way Cache works because of spatial and temporal locality, which does not vary very much with cache associativity, explains this trend.

The potential energy savings from using the Way Cache can be estimated as follows. Most of the L1 cache energy consumption is due to accessing the tag and data arrays, the rest of the components consume much less. On a Way Cache hit only the L1 tag and data array for one way in the set are accessed, so the upper bound on the access energy savings is $Access\_Energy * (N-1)/(N)$ (N being the number of ways). The reason why this is an upper bound is that the Way Cache itself consumes energy reducing the potential energy savings. Given that the energy savings are proportional to the *WC* hit rate, and that the hit rate is high the L1 cache energy savings obtained by using the Way Cache will be significant.

As mentioned in the introduction, the Way Cache also reduces the L1 cache hit latency because the data output multiplexor can be enabled before the data access completes. The precise latency reduction can only be determined after a detailed VLSI design, which is beyond the scope of this paper. The results presented in this paper do not assume a change in the cache hit time.

6.1 Comparison to Way Prediction

Recall that a major difference between the way predictor and the Way Cache is that for the former in the case of a mis-prediction the L1 access has to be repeated, accessing all the L1 cache ways. This costs 2 extra cycles in our model, one cycle less than full a cache access (no new TLB and LSQ accesses are needed).

Figure 5 shows the average way prediction hit rates for a 4, 8 and 16–way L1 data cache. The hit rates decrease significantly with the increase in associativity. This puts way prediction at a significant disadvantage when compared to the Way Cache given the trend of increasing L1 associativity. Way Cache also has higher hit rates for the 64–entry configuration.

A way predictor incurs a performance penalty on a way prediction miss. Figure 6 shows the execution time increase for way prediction over the Way Cache. The performance penalty follows the mis–prediction trend: it increases with increased associativity. For most benchmarks the performance penalty is under 5%, but it can be as high as 25.6% for galgel and 18.7% for mgrid. Way Cache, not being a predictor, does not incur performance penalties, even on WC misses.

6.2 Way Caching and Cached Load–Store Queue

[12] proposed caching data in a modified Load/Stored Queue called Cached Load/Store Queue (CLSQ). The CLSQ contains data from both stores still in

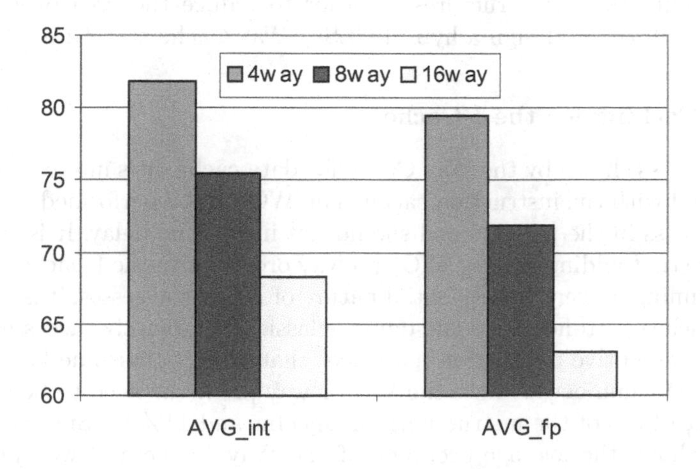

Fig. 5. Average way prediction hit rate for different L1 cache associativities

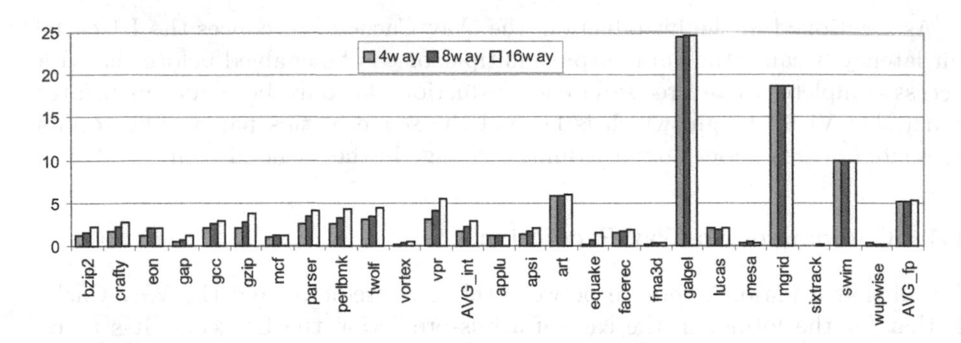

Fig. 6. Average way prediction performance degradation vs associativity

execution (as in a standard LSQ) and also data from completed load and store instructions. The CLSQ address tag is checked prior to an L1 cache access and, on match, the L1 cache access is avoided and the CLSQ data is used. On average, the 64-entry CLSQ reduces the number of DL1 accesses by 39% for SpecINT (up to 60% for *bzip2*) and 23% for SpecFP (up to 65% for *fma3d*).

The Way Cache performance needs to be evaluated in a system containing a CLSQ since CLSQ causes a significant reduction in L1 cache accesses. In addition, CLSQ use causes a significant change in locality which may affect the WC behavior (or that of a way predictor, if it was used).

As it turns out, the WC hit rate actually increases in all situations, especially for the SpecFP benchmarks, with better hit rates for higher associativities. This shows that even though the CLSQ significantly alters the memory reference patterns, the Way Cache still works very well. The WC obtains good performance in conjunction with CLSQ (and perhaps other devices that try to accomplish the same goal).

Both the CLSQ and the Way Cache use addresses as tags, thus it may be possible to unify the two structures in order to reduce the area overhead. A future study will try to design a hybrid CLSQ–Way Cache structure.

6.3 Way Caching for the I-Cache

Given the success shown by the Way Cache for data cache, it is interesting to see if it works well with the instruction cache. The WC can be performed in parallel with BTB access in the pipeline and should not incur time delay. It is therefore assumed here that adding either a WC or a way predictor to the I–cache does not change the timing. Given the sequential nature of I–cache accesses, it is assumed that the I–cache controller can avoid doing "classical" associative accesses to the I–cache for consecutive instruction references that do not cross line boundaries. This reduces the number of times the WC or way predictor need to be accessed to an average of 9% of the instructions for SpecInt and 11% for SpecFP.

Figure 7 shows the average accuracy of the Way Cache and way predictor for an 8–way instruction cache. The 64–entry *WC* performs better that the way

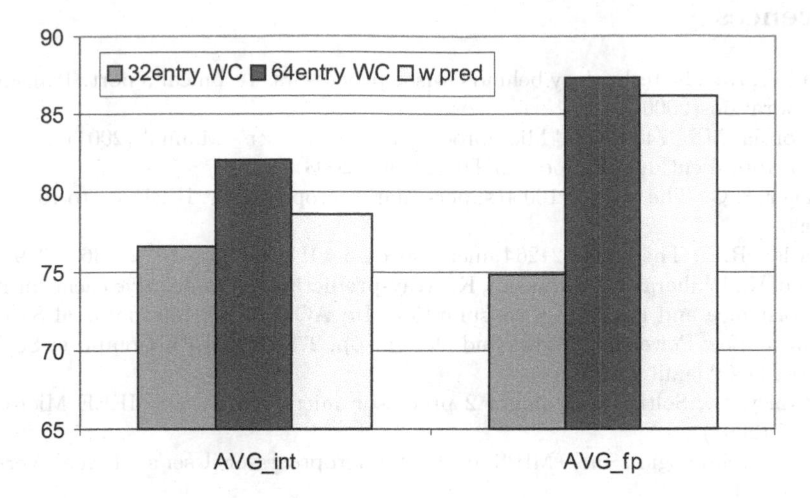

Fig. 7. Way Cache and Way Prediction hit rates for an 8–way I–cache

predictor. It thus has the potential to provide energy savings and performance improvement for instruction cache.

7 Conclusions

This paper presented the design and evaluation of the *Way Cache*, a device that can be used to reduce the associative cache access latency and energy usage. The Way Cache is not a predictor and thus does not suffer from delay penalties in case of mis–prediction. The Way Cache has an important property: its performance improves with the increase in associativity, which continues to increase in current and future processors. Way prediction becomes less effective with increased associativity.

The Way Cache was shown to be highly accurate in supplying way information, especially as its size increases to 64 or even 128 entries, which is still quite small for a cache size even if fully–associative. The WC access latency can be completely overlapped with other CPU activities, such as the LSQ access for data or the BTB access for instructions. It was suggested that the WC can perhaps be integrated in the LSQ. The WC is even effective after the memory reference stream has been significantly altered by using the cached LSQ.

The Way Cache was shown to be effective for the instruction cache as well. Future work will furhter study the integratation of the Way Cache in the I–cache system.

Use of a larger Way Cache has shown that there is further potential for improvement. Finding methods to only put in the Way Cache addresses that are more likely to be referenced in the future and using a better replacement policy are future directions of research.

References

1. Klaiber, A.: The technology behind Crusoe processors. Technical report, Transmeta Corporation (2000)
2. Motorola: MPC7450 RISC Microprocessor Family User's Manual (2001)
3. Intel: Intel Pentium M Processor Datasheet (2003)
4. Yeager, K.C.: The MIPS R10000 superscalar microprocessor. IEEE Micro 16, 28–40 (1996)
5. Kessler, R.E.: The Alpha 21264 microprocessor. IEEE Micro 19, 24–36 (1999)
6. Inoue, K., Ishihara, T., Murakami, K.: Way-predicting set-associative cache for high performance and low energy consumption. In: ACM/IEEE International Symposium on Low Power Electronics and Design, pp. 273–275. IEEE Computer Society Press, Los Alamitos (1999)
7. McNairy, C., Soltis, D.: Itanium 2 processor microarchitecture. IEEE Micro 23, 44–55 (2003)
8. MIPS Technologies, Inc.: MIPS R10000 Microprocessor User's Manual Version 2.0.(1996)
9. Tang, W., Veidenbaum, A., Nicolau, A., Gupta, R.: Simultaneous way-footprint prediction and branch prediction for energy savings in set-associative instruction caches. In: IEEE Workshop on Power Management for Real-Time and Embedded Systems (2001)
10. Witchel, E., Larsen, S., Ananian, C.S., Asanovic, K.: Direct addressed caches for reduced power consumption. In: Proceedings of the 34th Annual International Symposium on Microa rchitecture (MICRO-34) (2001)
11. Burger, D., Austin, T.M.: The SimpleScalar tool set, version 2.0. Technical Report TR-97-1342, University of Wisconsin-Madison (1997)
12. Nicolaescu, D., Veidenbaum, A., Nicolau, A.: Reducing data cache energy consumption via cached load/store queue. In: Proceedings of the 2003 International Symposium on Low Power Electronics and Design, pp. 252–257. ACM Press, New York (2003)

Design of Fast Collective Communication Functions on Clustered Workstations with Ethernet and Myrinet*

Dongyoung Kim and Dongseung Kim

Department of Electrical Engineering
Korea University
Seoul, 136-701, Republic of Korea
Telephone nos.: +82 2 3290 3232(voice); Fax: +82 2 928 8909
dkim@classic.korea.ac.kr

Abstract. In cluster computing, current communication functions under MPI library are not well optimized. Especially, the performance is worse if there are multiple sources and/or destinations involved, which are the cases of collective communication. Our algorithms uses multidimensional factorization and pairwise exchange communication/dissemination methods to improve the performance. They deliver better performance than previous algorithms such as ring, recursive doubling and dissemination algorithms. Experimental results show the improvement of 50% or so over MPICH version 1.2.6 on a Linux cluster.

Keywords: cluster computing, MPI, broadcast, message passing communication.

1 Introduction

Message passing is performed to efficiently exchange data among parallel/ distributed computing processors (or nodes). However, the operation is usually slow compared to recent high-performance CPUs, and the overall execution time of parallel programs is critically dependent on the speed of interprocessor communication[7,8]. A great deal of efforts has been made to improve the communication performance in areas of interconnection network, switching method, routing algorithm, and system communication kernel. For programmers, the execution performance relies on software writing since they write parallel programs using standardized communication functions such as those supported by PVM[4] or MPI[5]. However, the collective communication functions like in MPI_Bcast[11] have shown poor performance because of naive implementation. For this reason, many programmers have often written their own collective communication routines using point-to-point communication functions.

* This research was supported by Korea Science and Engineering Foundation(grant no.: R01-2001-0341-0). Preliminary results of the paper are to appear at Int. Conf. on Parallel and Distributed Systems, July 22, 2005.

J. Labarta, K. Joe, and T. Sato (Eds.): ISHPC 2005 and ALPS 2006, LNCS 4759, pp. 105–116, 2008.

To improve MPI collective communication functions, we develop new MPI functions based on factorization and pairwise-exchange dissemination algorithms. Among a complete set of MPI collective communication functions, we focus only on allgather, and then extend the algorithm to apply to broadcast and allreduce. This is because allgather consists of common procedure of other collective communication.

2 Communication Model and Collective Communication

Instead of general model of parallel computation, we focus on cluster computers for practical purpose. They consist of distributed-memory parallel computing nodes, and each node has one computational processor indexed from to $P - 1$. We assume that the time taken to send a message between any two nodes can be modeled as $\alpha + n\beta$, where α is the start-up time per message, β is the transfer time per bytes, and n is the number of bytes transferred. The time taken is independent of the distance between the communication nodes, and the network is fully connected. When the communication requires arithmetic operations in reduction, the cost is denoted by γ. In parallel computation, the communication is usually performed by send and receive, with one sender and one receiver, which are called *point-to-point communication.* However, collective communication functions are frequently used such as broadcast and reduce that invoke multiple senders, multiple receivers, or both. For simplicity, we consider the case of collective operations where all processors take part in the communication as either senders or receivers. Broadcast gives out a common message from a sender to all other processors. Scatter is to send a private message from a singe sender to the respective receivers. Gather is to collect an individual message from each sender at a single receiver, whereas reduce performs a designated arithmetic operation onto the gathered data to draw one resultant data. Allgather is to send an individual message from each sender to all receiver processors, whereas allreduce performs a designated arithmetic operation on the gathered data to find a set of resultant values, then it is sent to all processors. allgather is important since it includes core communication procedures of other collective communication functions such as scatter, gather, broadcast, all-to-all broadcast, allreduce, etc. Thus, we like to focus on the algorithmic details and the ways to implement allgather. Let P, n, and M denote the number of processors, the integrated message size for *collective communication*, and the message size of one processor. Each processor and each message are distinguished by their indexes. For example, initial messages of P processors of $P_0, P_1, \cdots, P_{P-2}, P_{P-1}$ for gather operations are respectively $M_0, M_1, \cdots, M_{P-2}, M_{P-1}$ where M_i is stored at processor i (P_i). For simplicity we also assume that processors have messages of equal length if applicableWe will use the relationship $M \cdot P = n$. If necessary, we assume $P = 2^k$, where k is an integer. The indexes of the communication partners are usually found by addition or subtraction to the sender/receiver ids with *modular-P* arithmetic to avoid out-of-range value.

3 Previous Implementation of Collective Communication

Three algorithms are reported previously to improve the performance. *Ring, recursive doubling,* and *dissemination algorithms* are introduced below.

In the old MPICH version, messages of each processor travel along a virtual circle of processors. Each processor P_i sends a copy of size-M data received from P_{i-1} in the previous step to P_{i+1} which was originated from P_{i-1}, and receives a size-M message from P_{i-1} which is a copy of M_{i-2}. This action continues like a pipeline operation until all processors have $P \cdot M$ messages, $M_0, M_1, \cdots, M_{P-2}, M_{P-1}$. The total number of steps of this ring algorithm[14] is $P-1$, and the execution time taken is $\sum_{k=1}^{P-1}\left(\alpha + \frac{1}{P}n\beta\right) = (p-1)\,\alpha + \frac{P-1}{P}n\beta$.

Recursive doubling. Algorithm [7] is used in the version of MPICH 1.2.5 to implement allgather. We assume that there are $P = 2^k$ processors for some integer k. In the first step, each two adjacent processors (groups) exchange their messages. Thus, $P_0\&P_1, P_0\&P_1, \cdots, P_{P-2}\&P_{P-1}$ are the pairs to exchange messages. In the following steps, new groups are created to include twice the processors in the previous step, and intergroup message exchange is performed between each two adjacent groups. For example, in step 2, there will be $P/2$ groups $G_0\left(P_0, P_1\right), G_1\left(P_2, P_3\right), G_2\left(P_4, P_5\right), G_3\left(P_6, P_7\right), \cdots$ and $P_0\&P_2, P_1\&P_3, P_4\&P_6, P_5\&P_7, \cdots$ are the pairs to exchange their data of size-$2M$ which include their own data as well as another from its partner. In a similar manner, in the j-th step there are 2^{j-1} processors in each group, and the size of data to exchange is $2^{j-1} \cdot M$. In this way, all processors collect all data with the size $P \cdot M$ of in $\log P$ steps. The total time in this case is $\sum_{k=1}^{\log P}\left(\alpha + \frac{2^{k-1}}{P}n\beta\right) = \log P\alpha + \frac{P-1}{P}n\beta$. If P is not a power of 2, additional communication process (post processing) is needed [1].

Dissemination algorithm. For the allgather is based on dissemination barrier algorithm [6]. In the first step, P_i sends M_i to $P_{(i+1) \bmod P}$ and receives $M_{(i-1) \bmod P}$ from $P_{(i-1) \bmod P}$. Note that this communication is not an exchange between a pair of processors, but is composed of a send to one processor in the right (in circular way), and a receive from one in the left. In the second step, each P_i sends $M_i + M_{(i-1) \bmod P}$ to $P_{(i+2) \bmod P}$, and receives $M_{(i-2) \bmod P} + M_{(i-3) \bmod P}$ from $P_{(i-2) \bmod P}$. In the j-th step, P_i sends a message of size $2^{j-1} \cdot M$ to $P_{(i+2^{j-1}) \bmod P}$. The communication repeats $\log P$ times. If the number of processors is not a power of two, an additional step is needed in which each processor sends a message whose size is $\left(P - 2^{\lfloor \log P \rfloor}\right) \cdot M$ to its partner processor that has not received complete data yet. The total time taken by this method is given by $\sum_{k=1}^{\lceil \log P \rceil}\left(\alpha + \frac{2^{k-1}}{P}n\beta\right) = \lceil \log P \rceil \alpha + \frac{P-1}{P}n\beta$. It has the smallest number of steps for allgather without using pairwise-exchange [1].

4 New Algorithms

A new method called factorization has been reported simultaneously by E.Chan et al.[2] and Kims[8]. After factoring the number of processors, they logically form a multidimensional mesh, and messages are propagated dimension-by-dimension using proper communication algorithms selected according to message length for best performance in the cluster computer.

In factorization algorithm, the number of processors P should be factorized, i.e. $P = \prod_{i=1}^{k} P_i$, where P_is are some integers. When the factorized groups exchange data, various algorithms such as ring, dissemination, or recursive doubling can be applied. For simplicity and regularity, only ring algorithm is used in this analysis. Processors are logically configured such as $c \times b \times a$ for $c \leq b \leq a$. The factor 2(two) is advantageous since more data can be exchanged in one step by pairwise-exchange. In the first step, each group consists of a processors and there are $c \times b$ groups. $a \cdot M$ data in each group are collected(all-gathered) internally. The merge is done by ring algorithm in this step. In the second step, there will be c groups where each group consists of $b \times a$ processors. Like in the previous step, $b \cdot (a \cdot M)$ data are merged within each group. The grouping and merging continues until all processors get the complete merged data of $M_0, M_1, \cdots, M_{P-2}, M_{P-1}$. When the factor is 2, pairwise-exchange is used in merging that gives better performance in TCP/IP protocol using than a send and a receive operations. Figure 1 shows an example of factorization for 12 processors where processors are partitioned into three dimensional $(2 \times 2 \times 3)$ groups. The total time taken is

$$\sum_{i=1}^{P_1-1} (\alpha + \tfrac{1}{P} n\beta) + \sum_{i=1}^{P_2-1} (\alpha + \tfrac{P_1}{P} n\beta) + \sum_{i=1}^{P_3-1} (\alpha + \tfrac{P_1 P_2}{P} n\beta) + \cdots + \sum_{i=1}^{P_k-1} \left(\alpha + \tfrac{P}{P P_k} n\beta\right)$$

$$= \sum_{i=1}^{k} (P_i - 1)\,\alpha + \tfrac{(P_1-1)+(P_1P_2-P_1)+(P_1P_2P_3-P_1P_2)+\cdots+(P-P_1P_2\cdots P_{k-1})}{P} n\beta$$

$$= \sum_{i=1}^{k} (P_i - 1)\,\alpha + \tfrac{P-1}{P} n\beta\ .$$

We may use dissemination algorithm instead of ring algorithm in data merging steps. The time in this case is

$$\sum_{m=1}^{\lceil \log P_1 \rceil} \left(\alpha + \tfrac{2^{m-1}}{P} n\beta\right) + \sum_{m=1}^{\lceil \log P_2 \rceil} \left(\alpha + \tfrac{2^{m-1} P_1}{P} n\beta\right) + \cdots + \sum_{m=1}^{\lceil \log P_k \rceil} \left(\alpha + \tfrac{2^{m-1} P}{P P_k} n\beta\right)$$

$$= \sum_{i=1}^{k} \lceil \log P_i \rceil \alpha + \tfrac{P-1}{P} n\beta\ .$$

If we apply both ring algorithm (k times) and dissemination algorithm (l times) in $P = \prod_{i=1}^{k} P_i \cdot \prod_{j=1}^{l} P_j$ to maximize the performance, the total time becomes $\sum_{i=1}^{k} \lceil \log P_i \rceil \alpha + \sum_{j=1}^{l} (P_j - 1)\,\alpha + \frac{P-1}{P} n\beta$.

The performance of factorization algorithm is dependent on k (the number of factors), where $P = \prod_{i=1}^{k} P_i$. Greater k will reduce the iteration steps, thus the communication overhead shrinks due to reduced α term. When P is not factored or the current factors are not good enough, we may add some dummy processors or subtract a part of P before applying the algorithm. Suppose we add P_α processors to yield the total number of processors Q to be factorized. Factorization algorithm applies as before on Q processors (for example, if the number of processors is 15, Q is 16 P_α with of 1). If we use ring algorithm in every steps, the total time taken by this approach for $Q = \prod_{i=1}^{k} Q_i$ is $\sum_{i=1}^{Q_1-1} (\alpha + \frac{1}{P} n\beta) + \sum_{i=1}^{Q_2-1} \left(\alpha + \frac{Q_1}{P} n\beta\right) + \cdots + \sum_{i=1}^{Q_k-1} \left(\alpha + \prod_{j=1}^{k} Q_j \frac{1}{P} n\beta\right) = \sum_{i=1}^{k} (Q_i - 1)\,\alpha + \frac{Q-1}{P} n\beta$.

Another way to have greater partitions of factorization is to use smaller value Q by removing a few processors from P. For example, if $P = 17, Q$ is 16 and, P_b is 1. The compensation is performed at the end for the removed processors. Factorization algorithm is applied to Q processors in the same way on Q processors as before, then, proper communication follows for the excluded P_b processors. For some collective communication functions like broadcast, however, the compensation can be ignored or reduced.. If we use ring algorithms at every step, the total time taken by this approach for $Q = \prod_{i=1}^{k} Q_i$ is

$$\sum_{i=1}^{Q_1-1} (\alpha + \tfrac{1}{P} n\beta) + \sum_{i=1}^{Q_2-1} \left(\alpha + \tfrac{Q_1}{P} n\beta\right) + \cdots + \sum_{i=1}^{Q_k-1} \left(\alpha + \textstyle\prod_{j=1}^{k} Q_j \tfrac{1}{P} n\beta\right) + T(n, P_b) = \sum_{i=1}^{k} (Q_i - 1)\,\alpha + \alpha + \tfrac{Q-1}{P} n\beta + T(n, P_b)$$

Pairwise-exchange dissemination algorithm

This algorithm follows the core pattern of dissemination algorithm in the implementation of allgather as before. However, the actual data communication is performed by pairwise-exchange instead of previous send-to-the-right and receive-from-the-left pattern. It utilizes efficient communication function that minimizes the overhead in the previous send-to-one and receive-from-another by *exchange* within a pair of processors. The way of partitioning and communication differs whether or not P is even.

Suppose P is an even number. In the first step, P processors are partitioned into two equal sized groups, $G_L\left(P_0 : P_{k/2-1}\right)$ and $G_R\left(P_{k/2} : P_k\right)$. P_i in G_L exchanges M_i with $P_{i+P/2}$ in G_R (Refer to Figure 2 for 12 processors). In the second step, P_i in G_L exchanges $2 \cdot M$ data composed of both M_i of its own and $M_{(i+1) \bmod (P/2)}$ from the previous communication partner with $P_{P/2+(i+1) \bmod (P/2)}$ in G_R. In each step k, P_i exchanges $2^{k-1} \cdot M$ data with processor $\frac{P}{2} + \left(i + 2^{k-1} - 1\right) \bmod \frac{P}{2}$.

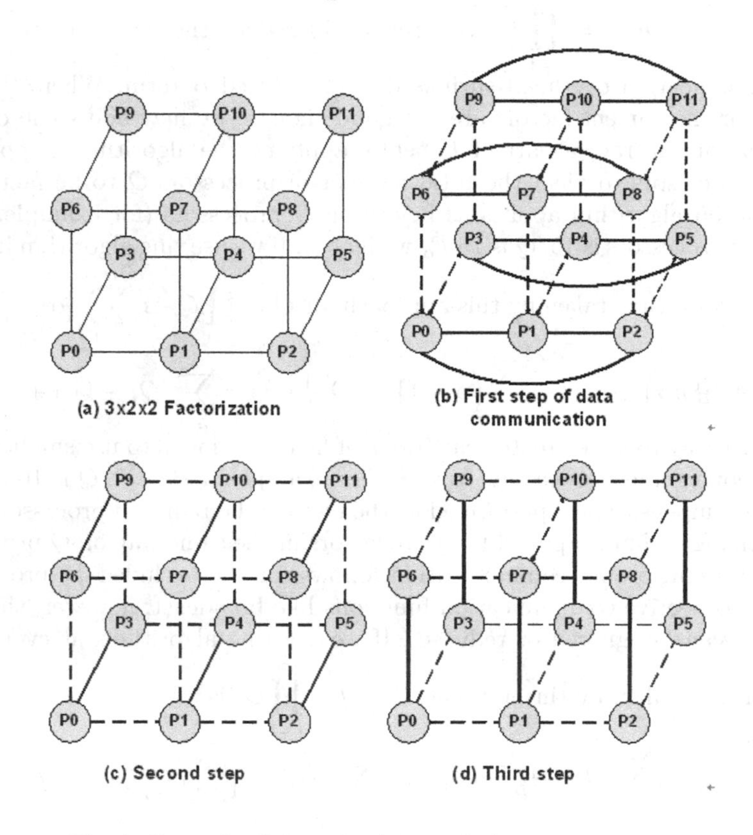

Fig. 1. Example of factorization applied to 12 processors

This pattern of communication repeats $\log P$ times. More formal description is given in Figure 4. The sendrecv here is a combined function of both send and receive between two processors. If P is not a power-of-two, an additional step is needed, in which each processors sends additional $(P - 2\lfloor \log P \rfloor) \cdot M$ data to the corresponding destination and to fill up the yet-to-be-received part. The total time taken by this approach is given by $\sum_{k=1}^{\lceil \log P \rceil} \left(\alpha + \frac{2^{k-1}}{P} n\beta\right) = \lceil \log P \rceil \alpha + \frac{P-1}{P} n\beta$.

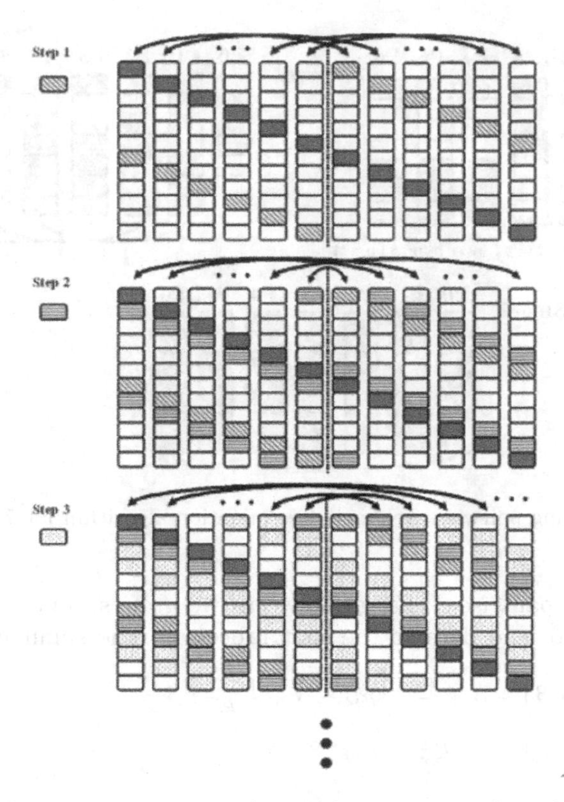

Fig. 2. Allgather using pairwise-exchange dissemination algorithm for 12 processors Shading/color represents the data movement at the corresponding step

Now, if P is odd, we apply both pairwise-exchange dissemination algorithm and general dissemination algorithm in two partitions, respectively. In the first stage, the processors are partitioned into two nearly equal sized groups $G_L\left(P_0 : P_{\lceil P/2 \rceil}\right)$ (with even processors) and $G_R\left(P_{\lceil P/2 \rceil+1} : P_P\right)$ (with odd processors) respectively like shown in Figure 3. Because the execution in the even number of processors is superior to one in the odd number of processors, the group with even processors uses pairwise-exchange dissemination algorithm in $G_L\left(P_0 : P_{\lceil P/2 \rceil}\right)$, and the odd sized group adopts general dissemination algorithm in $G_R\left(P_{\lceil P/2 \rceil+1} : P_P\right)$. In the second stage, each processor in two groups except the root processor exchanges their own data with its partner processor. Finally, a root processor receives the remaining required data of size $\lceil P/2 \rceil \cdot M$ from any other processor (This last stage can be ignored for broadcast because a root processors has already all data). The total time needed by this approach consists of three parts, $\sum_{k=1}^{\lceil \log \lfloor P/2 \rfloor \rceil} \left(\alpha + \frac{2^{k-1}}{P} n\beta\right)$, which is the time to communicate within each group, and $\alpha + \frac{\lceil P/2 \rceil}{P} n\beta$, which is the time

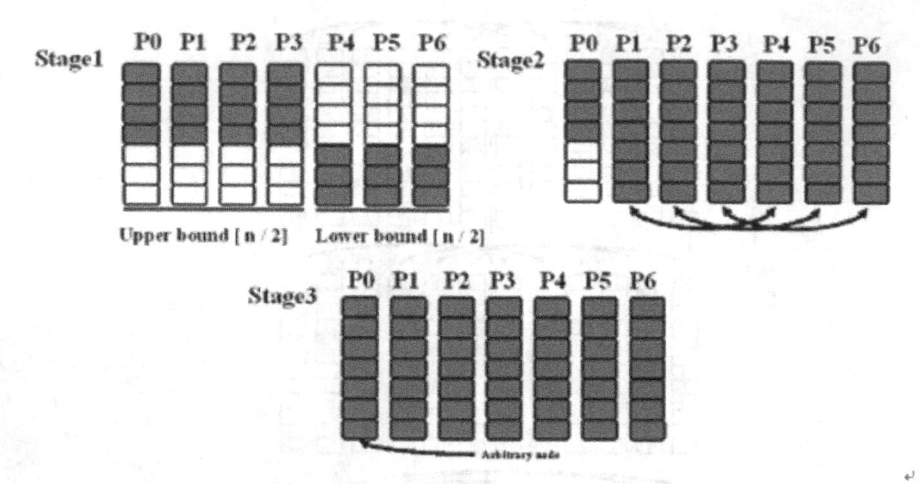

Fig. 3. Allgather using pairwise-exchange dissemination algorithm for 7 processors

to communicate by pairwise-exchange between two groups, and $\alpha + \frac{\lceil P/2 \rceil}{P} n\beta$, which is the time to send data to the root processor. The summation results $\sum_{k=1}^{\lceil \log \lfloor P/2 \rfloor \rceil} \left(\alpha + \frac{2^{k-1}}{P} n\beta \right) + \alpha + \frac{\lceil P/2 \rceil}{P} n\beta + \alpha + \frac{\lceil P/2 \rceil}{P} n\beta =$ $\lceil \log(P-1) \rceil \alpha + \frac{P-1}{P} n\beta + \frac{P-1}{2P} n\beta$.

5 Experiments and Discussion

New algorithms have been implemented and run to compare to old functions in MPICH 1.2.6, ring algorithm, and dissemination algorithm. The parallel system used in the experiments consists of a cluster with 16 nodes where each node has a 1GHz AMD processor, a 30GB local disk, and 256MB main memory. All nodes are logically fully connected by Fast Ethernet and Myrinet switches. With TCP protocol in the ethernet, messages can be sent and received simultaneously if they do not exceed the buffer size (The usual buffer size is 16K Bytes to 64K Bytes, but it depends on various factors like the feature of network card, switch, system kernel, and so on [3,10]). If the message size is greater than the threshold, i.e. the buffer size, it cannot be sent and received simultaneously anymore not to overflow the buffer [12]. Below the threshold, messages can be sent and received simultaneously with full performance. When the communication is performed only on two processors (this is the case of using *pairwise-exchange* communication), the bandwidth shows the peak since the communication overhead is minimized due to optimizing the exchange by piggybacking data on ACK packets [1]. These phenomena may not exist in Myrinet. Thus, to compare the algorithmic performance under different networks, both ethernet and Myrinet switches are employed in the experiments.

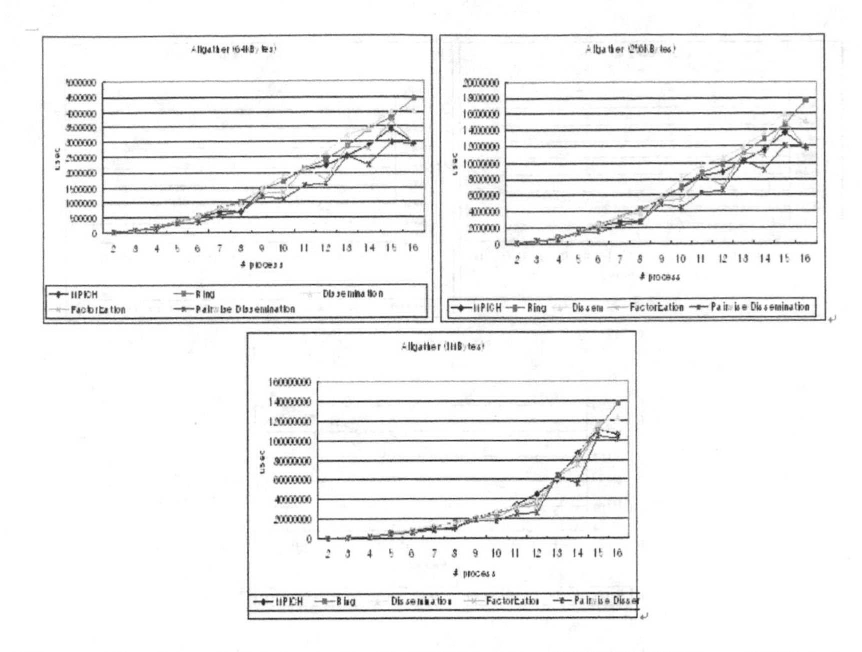

Fig. 4. Allgather performance in Fast Ethernet with 64K, 256K, and 1M bytes

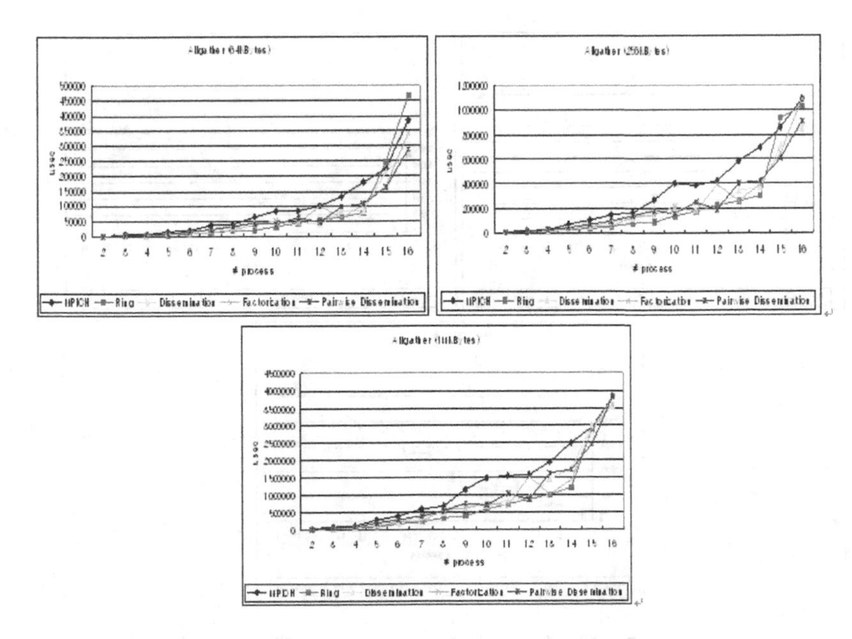

Fig. 5. Allgather performance in Myrinet with 64K, 256K, and 1M bytes

In the experiments, the overall sizes of the messages are measured by those contributed by each processor in a run. Thus, if the message size per processor is 256K bytes and if we use 4 processors, the total data accumulated at the

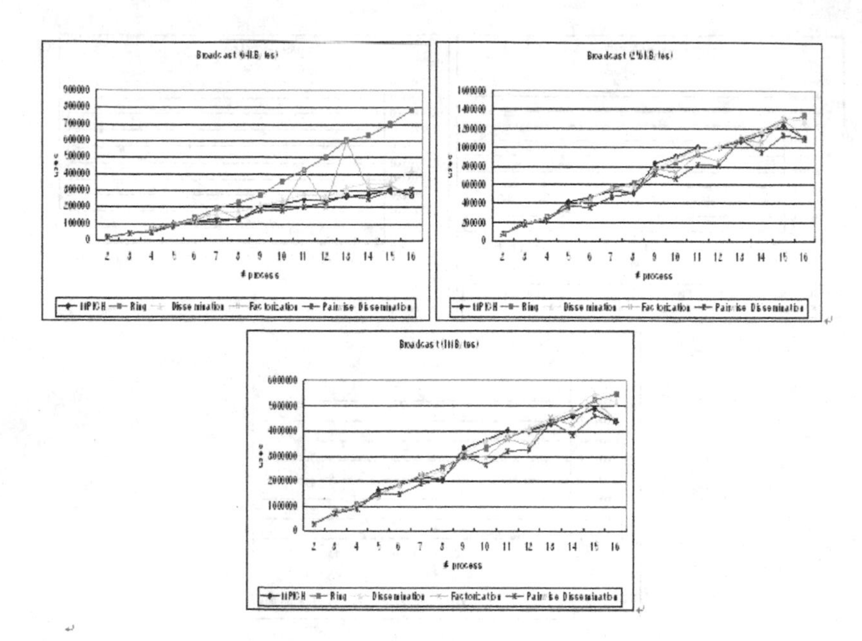

Fig. 6. Broadcast performance under Fast Ethernet

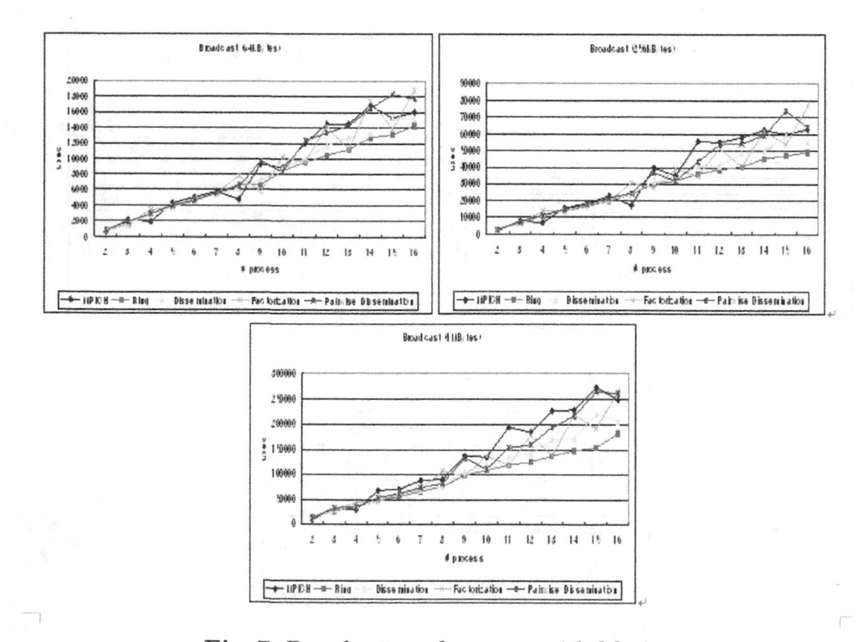

Fig. 7. Broadcast performance with Myrinet

end of allgather is 1M bytes. The data sizes of 64K, 256K, and 1M bytes are selected to represent short, mid-size, and long messages, respectively. In the graphs, "MPICH", "Ring" and "Dissemination" denote the functions in MPICH 1.2.6, ring algorithm, and dissemination algorithm respectively. "Factorization" means factorization algorithm and "Pairwise dissemination" does pairwise-exchange dissemination algorithms.

Figure 4 shows the execution times of allgather on the fast ethernet cluster. Note that the figures are drawn with different scales in the Y axis. The performance of short message exchange improves due to minimizing the overhead in start-up time accumulation. The best algorithm of allgather is pairwise exchange dissemination algorithm that requires only $\log P$ steps, and MPICH is the worst since due to no performance optimization since it was implemented quite a few years ago.

The communication time of allgather for long messages is dependent on the bandwidth term. Pairwise-exchange dissemination algorithm is the best because it employs pairwise-exchange communication in all stages of communication. Factorization algorithm that executes with partitioning of the processors shows the second best performance. Dissemination and ring algorithms, however, give poor performance because they do not lessen the start-up overhead of the point to point communication, or they can not shorten the number of iterations and minimize TCP traffic. Similar performance results can be observed in broadcast.

Figure 5 shows the execution times of allgather under Myrinet. They show quite different results compared to ethernet except for MPI library function. Ring algorithm performs better than the new algorithms since the overhead in each message communication is minimized, and ring-like message propagation is not so bad compared to ethernet communication because Myrinet uses firmware or hardware to enhance those communication. Hence, our algorithms are not the best.

Figures 6 and 7 compare the performance of broadcast using ethernet and Myrinet. When using ethernet cluster, new algorithms are more advantageous for collective communication. However, similar to previous results of allgather, more complex algorithms to reduce the overhead may not improve the performance much under Myrinet. Also in this case, since broadcast employed in MPICH 1.2.6 does not use scatter and gather, it shows good performance.

6 Conclusions and Future Work

This paper reports new collective communication algorithms with comparisons of their performance with previous ones. We verify analytically and experimentally that they consistently perform better. More research is under progress to expand the cluster size and to include other types of networks. Other collective communication functions like all-to-all type functions will be included in the enhancement soon.

References

1. Benson, G., Chu, C., Huang, Q., Caglar, S.: A comparison of MPICH allgather algorithms on switched networks, Recent advances in Parallel Virtual Machine and Message Passing Interface, 10th European PVM/MPI Users' Group Meeting. In: Dongarra, J.J., Laforenza, D., Orlando, S. (eds.) Recent Advances in Parallel Virtual Machine and Message Passing Interface. LNCS, vol. 2840, pp. 335–343. Springer, Heidelberg (2003)

2. Chan, E., Heimlich, M., Purkayastha, A., Geijn, R.: On Optimizing Collective Communication. In: Proceedings of 2004 IEEE International Conference on Cluster Computing, San Diego, USA, pp. 145–155 (September 2004)
3. Farrell, P., Ong, H.: Factors involved in the performance of computations on Beowulf clusters. Electronic Transactions on Numerical Analysis 15 (2003)
4. Geist, A., et al.: Parallel Virtual Machine, A User's Guide and Tutorial for Networked Parallel Computing. MIT Press, Cambridge (1994)
5. Gropp, W., Lusk, E., Dose, N., Skjellum, A.: A High-Performance, Portable Implementation of the MPI Message Passing Interface Standard
6. Hensgen, D., Finkel, R., Manber, U.: Two algorithms for barrier synchronization. International Journal of Parallel Programming 17(1), 1–17 (1988)
7. Hwang, K.: Advanced Computer Architecture: Parallelism, Scalability, Programmability. McGraw-Hill, New York (1993)
8. Kim, D., Kim, D.: Fast Broadcast by the Divide-and-Conquer Algorithm. In: Proceedings of 2004 IEEE International Conference on Cluster Computing, San Diego, USA, pp. 487–488 (September 2004)
9. Lee, K., Yoon, I., Kim, D.: Fast broadcast by message segmentation. In: Proceedings of 1999 Parallel and Distributed Processing Techniques and Applications, Monte Carlo Resort, Las Vegas, Nevada, USA, June 28 - July 1, 1999, pp. 2358–2364 (1999)
10. MPI and Embedded TCP/IP Gigabit Ethernet Cluster Computing, 27th Annual IEEE Conference on Local Computer Networks (LCN 2002), 6 - 8 November 2002, pp.733–734 (2002)
11. MPICH - A protable implementation of MPI. `http://www.mcs.anl.gov/mpi/mpich`
12. Pallas MPI Benchmarks - PMB, Part MPI-1. `http://www.pallas.com`
13. Sistare, S., Varrt, R., Loh, E.: Optimization of MPI collective on clusters of large-scale SMPs. In: Proceedings of SC99: High Performance Networking and Computing (November 1999)
14. Thakur, R., Rabenseifner, R., Gropp, W.: Optimization of Collective Communication Operations in MPICH, Argonne National Laboratory

Dynamic Load Balancing in MPI Jobs

Gladys Utrera, Julita Corbalán, and Jesús Labarta

Departament d'Arquitectura de Computadors (DAC)
Universitat Politècnica de Catalunya (UPC)
{gutrera,juli,jesus}@ac.upc.es

Abstract. There are at least three dimensions of overhead to be considered by any parallel job scheduling algorithm: load balancing, synchronization, and communication overhead. In this work we first study several heuristics to choose the next to run from a global processes queue. After that we present a mechanism to decide at runtime weather to apply Local process queue per processor or Global processes queue per job, depending on the load balancing degree of the job, without any previous knowledge of it.

1 Introduction

Scheduling schemes for multiprogrammed parallel systems can be viewed in two levels. In the first level processors are allocated to a job, in the second, processes from the job are scheduled using this pool of processors. When having more processes than processors allocated to a job, processors must be shared among them.

We will work on message-passing parallel applications, using the MPI [16] library, running on shared-memory multiprocessors (SMMs). This library is worldwide used, even on SMMs, due to its performance portability on other platforms, compared to other programming models such as threads and OpenMP and because it may also fully exploit the underlying SMM architecture without careful consideration of data placement and synchronization. So the question arises whether to map processes to processors and then use a local queue on each one, or to have all the processors share a single global queue. Without shared memory a global queue would be difficult to implement efficiently [8].

When scheduling jobs there are three types of overhead that must be taken into account: load balance, to keep the resources busy, most of the time; synchronization overhead, when scheduling message-passing parallel jobs; and communication overhead, when migrating processes among processors losing locality.

Previous work [24] has studied the synchronization overhead generated when scheduling processes from a job need to synchronize each other frequently. It proposes a scheme based on the combination of the best benefits from Static Space Sharing and Co-Scheduling, the Self-Coscheduling. The communication overhead is also studied in [25] when applying malleability to MPI jobs. In this work we present a mechanism, the Load Balancing Detector (LDB), to classify

J. Labarta, K. Joe, and T. Sato (Eds.): ISHPC 2005 and ALPS 2006, LNCS 4759, pp. 117–129, 2008.

applications depending on their balance degree, to decide at runtime the appropriate process queue type to apply to each job, without any previous knowledge of it.

We evaluate first several heuristics to decide the next process to run from the global queue depending on the number of unconsumed messages, the process timestamp, the sender process of the recently blocked or the process executed before the recently blocked one if it is ready. We obtained the best performance when selecting the sender process of the recently blocked process or if is not possible the one with the greater number of unconsumed messages.

When a process from a parallel job arrives to a synchronization point and cannot continue execution, we do blocking immediately and then context switching to a process in the ready queue. We present a novelty ratio to estimate load balance from a job by observing the coefficients of variation of the number of context switches and the user execution times, among processes. We constructed an algorithm that calculates these at runtime, and after deciding the type of application (balanced or imbalanced), applies immediately the appropriate type of queue. Results show that we obtained better performance when choosing the appropriate queue type at runtime, than applying a predefined queue type to all the applications without taking into account their balance degree.

The rest of the paper is organized as follows. In Section 2 we discuss the related work. Then in Section 3 the execution framework and in Section 4 follows the scheduling strategies evaluated. Section 5 describes the proposal of this paper. Section 6 shows the performance results and Section 7 the conclusions and the future work.

2 Related Work

There is a lot of work in this area; here we mention some work related to implementations of local and global queues from the literature.

Using local queues of processes at each processor is a natural approach for distributed memory machines. It is also suitable to shared memory machines as there is some local memory as well. Provided that only one processor will use each local queue, there won't be no contention and no need to locks. However there must be a decision where to map processes to processors and how to schedule them to satisfy the communication requirements to minimize the synchronization overhead.

There is an interesting and extensive discussion about using local and global queues in [8]. Local queues have been used in Chrysalis [14] and Psyche [19] on the BBN Butterfly. The issue involved in the use of local queues is load balancing, by means of migrating processes after they have started execution. The importance of load balancing depends on the degree of imbalance of the job and hence of the mapping. Then a scheduling algorithm must ensure the synchronization among processes [24][2][18][3][17][9]. There is also an interesting proposal in [5] where they do load balancing by creating threads at loop level.

A global queue is easy to implement in shared memory machines. It is not the case for distributed systems. The main advantage of using such queues is that they provided automatic load balancing or named load sharing as in [8]. However this approach suffers from queue contention [1], lack of memory locality and possible locks overhead, which is eliminated local per-processor, thereby reducing synchronization overhead. Other work in process scheduling has considered overhead associated with reloading the cache on each context switch when a multiprocessor is multiprogrammed. [21] argued that if a process suspends execution for any reason, it should be resumed on the same processor. They showed that ignoring affinity can result in significant performance degradation. This is discussed also in [26].

Global queues are implemented in [6]. Here they implement a priority queue based on the recent CPU usage plus a base priority that reflects the system load. However this issue is not crucial as in most cases the cache is emptied after a number of other applications have been scheduled [12].

A combined approach is implemented in [7], where at each context switch a processor would choose the next thread to run from the local queue or the global queue depending on their priority.

About job classification at runtime there is a work in [10], where they classify each process from a job and varying the working set when applying gang scheduling.

3 Execution Framework: Resource Manager (CPUM)

Here we present the characteristics of our resource manager that is the Cpu Manager (CPUM). It implements the whole mechanism to classify applications depending on their balance degree and takes all the necessary actions to apply the appropriate queue type to each job. It also is in charge of deciding the partition size, the number of folding times depending on the maximum multiprogramming level (MPL) established, the mapping from processes to processors.

The CPUM is a user-level scheduler developed from a preliminary version described in [15]. The communication between the CPUM and the jobs is done through shared memory by control structures.

In order to get control of MPI jobs we use a dynamic interposition mechanism, the DiTools Library [20]. This library allows us to intercept functions like the MPI calls or a system call routine which force the context switch among process and is invoked by the MPI library when it is performing a blocking function.

All the techniques were implemented without modifying the native MPI library and without recompilation of the applications.

The CPUM wakes up periodically at each quantum expiration and examines if new jobs have arrived to the system or have finished execution, updates the control structures and if necessary depending on the scheduling policy, redistributes processors, context switch processes by blocking and unblocking them or change based on the mechanism proposed in this article the process queue type to each job.

4 Processor Sharing Techniques Evaluated

In this section we describe the main characteristics of each scheme used for the evaluation as well as the LDB.

4.1 Local Queues

As the number of processors assigned to a job could be less than its number of processes, there may be a process local queue at each processor. We choose the next process to run in a round robin fashion, and as soon as a process reaches a synchronization point where it cannot proceed execution (e.g. an MPI blocking operation), it blocks immediately freeing the resource.

Notice that processes in local queues are assigned to the processor during the whole execution, avoiding migrations and preserving locality.

4.2 Global Queues

While in Local queues we have a process queue assigned fixed to each processor, in Global queues we have a unique queue for each application assigned fixed to a processor partition. The unassigned processes are kept in the global queue until the scheduler select them to run on a newly freed processor.

We implemented several heuristics to choose the next process to run, which we describe below:

- *Timestamp*: Each process has a timestamp which is incremented when it is waiting in the global queue, in order to do aging. The selected process will be the one with the greater timestamp. This turn out to be the classical round robin.
- *Unconsumed messages*: The selected process will be the one with the greater number of unconsumed messages.
- *Sender*: The selected process will be the sender of the recently blocked process if it has also the number of *unconsumed messages* equal or greater than zero. This last condition is to ensure that it will have useful work to do when unblocked. If it is not possible, the first with a number of *unconsumed msgs* greater than zero is selected.
- *Affinity*: The selected process will be a process that has already ran in the processor that is asking context switching. If it doesn't exist then the first process with unconsumed messages greater or equal than zero is selected.

After applying the heuristic if there isn't any eligible process, then the context switch is not done. Notice that there will be several process migrations because the last cpu where a process was allocated will not be taken into account except for Affinity. However, in this case it is not a strong condition because if there aren't ready processes that satisfies the condition, then it is ignored the affinity.

Notice also that when applying global queues process mapping to processors is not a crucial decision as in Local queues, since they may be migrated during execution. Another interesting aspect of this approach is that the next process to run is chosen from the global queue, so there will be a higher probability of finding a process which can continue executing than in Local queues.

5 Our Proposal: The Load Balancing Detector (LBD)

We propose a mechanism which decides dynamically, without any previous knowledge of the application, whether to apply Local or Global process queues to it by measuring its load balancing degree.

We have observed that usually the applications at the beginning of the execution have an imbalanced behaviour, because processes start creating, the data is distributed among them, and after that, they normally perform a global synchronization function. So at this point all the jobs behave as imbalanced ones, after that they start doing regular work. As a matter of fact the coefficient of variation[1] (CV) of the number of context switches detected among processes during this period is higher than the rest of the execution due to this chaotic behaviour.

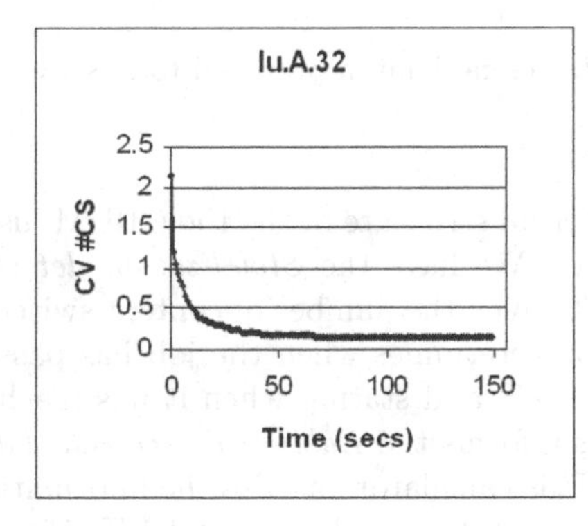

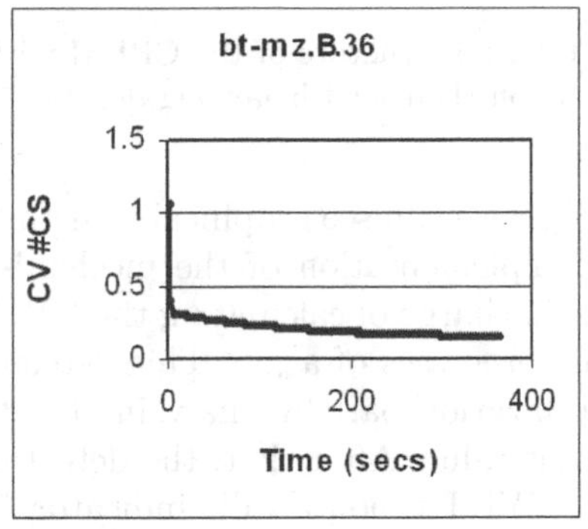

Fig. 1. Coefficient of variation of the number of context switches for a well-balanced application: lu.A and for an imbalanced application: bt-mz.B.36

In Fig. 1 it is possible to see graphically the coefficient of variation of the number of context switches (CVCS) for lu.A as a well-balanced application and for the bt-mz.B.36 as an imbalanced one. Both executed applying Global queue types for the whole running with MPL=4. After a while the CVCS goes down and remains constant until the end of the execution, no matter the load balancing degree of the jobs. We can observe also that the time spent by the job in doing the initializations is variable and depends on it. We base our criterion to measure at runtime the balance degree of an application on two CVs: the number of context switches and the user time. The first one gives us an impression when exactly

[1] Percentage of the standard deviation in the average value.

the job has started executing regularly. The second one quantifies the difference in time spent in calculation by each process in order to measure their balance degree.

The user time per process shows exactly the time spent in calculation. By calculating the coefficient of variation of the user time (CVUT) among processes from a job, it is possible to quantify its load balancing degree. During the calculation of this measure we apply Global queue type to the job, to ensure a fair distribution of the cpu time among processes, until a decision is made. As the beginning of the execution is chaotic, we delay any decision until this phase has finished.

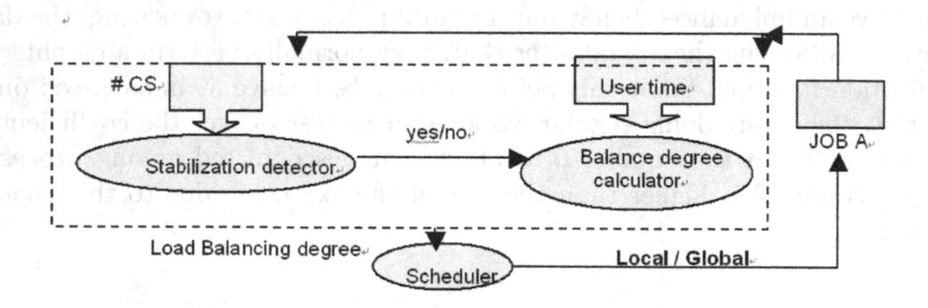

Fig. 2. Internal structure of the CPUM with the mechanism proposed to classify jobs depending on their load balancing degree

In Fig. 2 we can see graphically the internal structure of the the CPUM, used for the implementation of the mechanism. We have the *Stabilization detector* which is in charge of calculating the CVCS given the number of context switches from the processes of a job. This detector determines when the job has passed the initialization part by analyzing the CVCS and stating when it has reached a constant value. After that, the detector informs the *Balance degree calculator* that the CVUT is now a valid indicator. This calculator has also the information about the user time of each process from a job to calculate the CVUT. Finally the scheduler determines with the CVUT if the job must be considered balanced or imbalanced.

Once the job is classified the scheduler decides whether to continue the execution using Global queues or to switch to Local queues.

In Fig. 3, we show graphically the CVUTs for the same executions showed in Fig. 1. It can be viewed clearly that at the beginning of the execution the CVUTs are high, and then they go down depending on the load balancing degree of the job. This means that the current CVUT can give us a true idea of the balance degree of the job.

6 Evaluations

We conducted the experiments in the following way: first we ran isolated all the applications with MPL=2, 4 and 6, applying Global queues described in Section

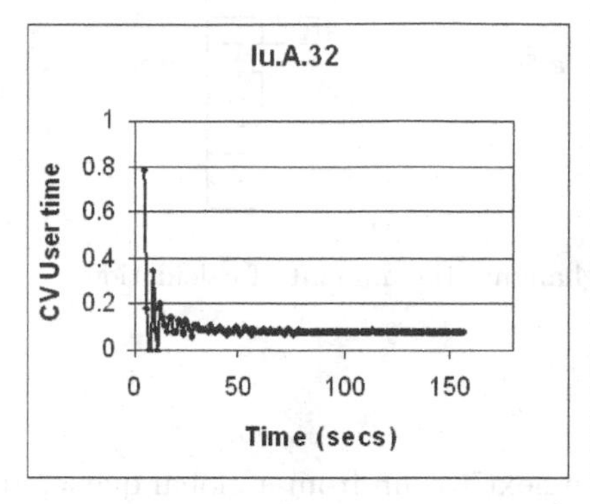

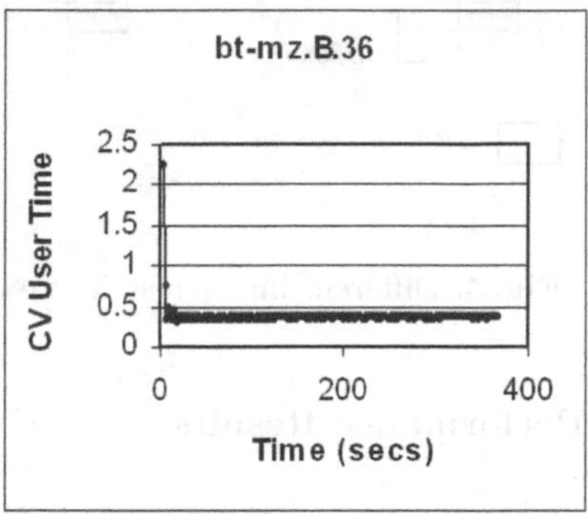

Fig. 3. Coefficient of variation of the user time for two well-balanced applications: cg.B and lu.A and two imbalanced applications: a synthetic u.50x1 and the bt-mz.B.36

4.2 and comparing with the performance under Local queues. From this part we take the heuristic that showed the best performance. In the second part of the evaluation we re-execute all the applications with MPL=4 under Local and Global queues statically and using our LDB described in Section 5.

6.1 Architecture

Our implementation was done on a CC-NUMA shared memory multiprocessor, the SGI Origin 2000 [16]. It has 64 processors, organized in 32 nodes with two 250MHZ MIPS R10000 processors each. The machine has 16 Gb of main memory of nodes (512 Mb per node) with a page size of 16 Kbytes. Each pair of nodes is connected to a network router. The operating system where we have worked is IRIX version is 6.5 and on its native MPI library with no modifications.

6.2 Applications and Workloads

For the evaluations we use as well-balanced applications the NAS Benchmarks: cg, lu, bt, sp, ep, ft and as imbalanced applications the bt-mz multi-zone version [27] and synthetic applications. They consist of a loop with three phases: messages, calculation, and a barrier. The amount of calculation varies in order to generate imbalance. We use two types of imbalance: a) 50% of processes with 2, 3 and 6 times greater in calculation than the other 50%, see Fig. 4; b) a random number of amount of calculation for each process with the biggest 6 times greater than the smallest.

We worked with a number of processes per application of 32, running in isolation on a processor partition sized depending on the MPL. For the first part of the evaluation we used MPLs=2, 4 and 6, so the processor partition sized 16, 8 and 5 respectively. For the second part we executed with MPL=4 so the partition size was 8.

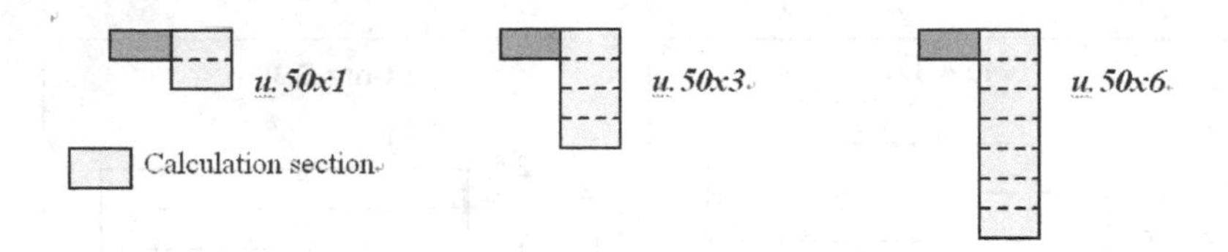

Fig. 4. Different imbalance degree changing the amount of calculation

6.3 Performance Results

First we compare heuristics to choose the next to run from a global queue, and in the second part we evaluate the performance of the LDB.

6.3.1 Global Queues

Here we present the performance results for the applications with different MPLs under Global queues schemes, with all the heuristics described before.

6.3.1.1 Well-balanced Applications. For the well-balanced case we present a high communication degree applications, the cg.B, one with global synchronizations, the ft.A and one low communication degree, the ep.B.

We can see in Fig. 5, the slowdowns with respect to Local queues of well-balanced applications cg.B.32, the ft.A.32 and the ep.B.32 evaluated using global queues with the different heuristics to choose the next to run. For the cg.B.32, the Sender+Msgs heuristic seem to work best. The worst is the Timestamp which does not take into account the synchronization between processes, which is an important issue because the cg is high communication degree.

The main difference between local and global approximations is the number of process migrations. While under Local queues this number is zero, under Global queues it increases significantly, thus generating context switching overhead effect degrading performance.

On the other hand the number of times a process asks context switching diminishes under Global queues because it has a higher chance of choosing the "right" process to run than Local queues, incrementing the number of coscheduled processes.

We conclude that for well-balanced jobs, even Global queues reduce the number of context switches, they don't compensate the locality provided by Local queues.

The ft.A, in spite of being well-balanced performs similar under the Local and Global approaches. This is may be due to it is composed only by global synchronizations, so their processes must executed as much as coscheduled as possible. While locality favours the local approach, the Global approach ensures a more fair distribution of the cpu time among processes.

For the ep.B, the differences are diminished because the processes does not block very often as they rarely call blocking functions.

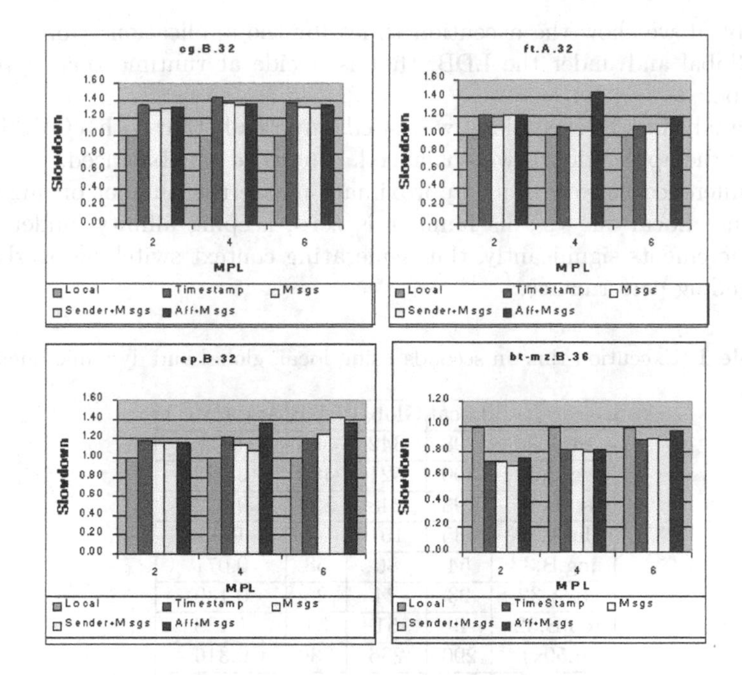

Fig. 5. Slowdown of the execution times under Global queues with respect to Local queues varying the # assigned processes per processor (MPL)

6.3.1.2 Imbalanced Application. For the imbalanced case, we show in Fig. 5 the bt-mz, a multi-zone version from the NAS Benchmarks. Here a Global queue seems a more attractive option than using Local queues as the slowdown is under 1. This is due to the automatic load balancing effect. We can observe that for Sender+Msgs heuristics the coefficient of context switches over the execution time is reduced with respect to Local queues in about 20 % for MPL=4.

In conclusion, analyzing the performance of the jobs showed in the figures above the best option for well-balanced high and low communication degree applications is Local queues, as they have a slowdown greater than 1. On the other hand the imbalanced synthetic application, using Global queues seems a more attractive option. The FT and EP seems to work well under both schemes, global and local. About the heuristics when using Global queues, select the next process to run the one with the greater number of unconsumed messages seems the best option.

For the rest of the evaluation we use MPL=4, as it is the maximum MPL that can be reached before performance degrades [24].

6.3.2 Evaluating Our Proposal: LDB

In this section we present the performance results of the applications using local, global and dynamic queues. Notice that the execution time under the dynamic queue approach is always a little worse than the best of the rest of queue types. This is caused by the initial CV calculation.

In Table 1 we show the execution times for the applications running under Local, Global and under the LDB, that is decide at runtime the appropriate queue type.

After analyzing the executions, we stated empirically that with a CVUT below 0.1 shows the application is well-balanced, otherwise is imbalanced.

The difference between both approximations are the number of migrations, while using Local queues this number is zero, keeping affinity, under Global queues increments significantly, thus generating context switching overhead effect degrading performance.

Table 1. Execution time in seconds using local, global and dynamic queues

	Local	Global	LDB	Avg CVUT
bt.A.36	298	342	306	0.02
cg.B.32	390	521	408	0.025
sp.B.36	193	248	195	0.032
lu.A.32	135	151	140	0.05
mg.B.32	54	80	58	0.071
ft.A.32	23	24	25	0.139
ep.B.32	49	51	51	1.128
u.50x1	290	236	236	0.316
u.50x3	437	315	315	0.482
u.50x6	825	538	538	0.686
u.rand.1	1307	1152	1152	0.14
u.rand.2	1287	1111	1111	0.217
u.rand.3	1279	1046	1046	0.322
bt-mz.B.36	447	368	368	0.386

On the other hand the number of times a process asks for context switching diminishes under Global queues, because a process has a higher chance of choosing the right process to run than in Local queues. In this way the number of coscheduled processes, is increased.

Finally we conclude that in spite of Global queues reduce the number of context switches really taken, they don't compensate the locality provided by the Local queues which obtain the best performance.

7 Conclusions and Future Work

Given the many-to-few relation of processes to processors allocated to a job, the question is whether to map processes to processors and then use a local queue on each one, or to have all the processors share a single global queue. Without shared memory a global queue is difficult to implement efficiently.

In this work we propose a mechanism, the Load Balancing Detector (LDB), to classify applications dynamically, without any previous knowledge of it, depending on their balance degree and apply the appropriate process queue type to each job.

The work consisted on two parts. First we analyze several heuristics to apply in Global queues to choose the next to run. We founded that the sender process of the currently running, performs best. In the second part we evaluate our proposal, the LDB. We use the NAS benchmarks and several synthetic applications.

Our proposal demonstrated to work quite well, especially for the imbalanced jobs. The well-balanced jobs suffer from an overhead which is not crucial; they still work better under the new scheme than under the Global one. However for the FT and EP as they behave as well-balanced jobs, they switched queue type suffering from an overhead, having the worst performance under our proposal. On the other hand, our mechanism applies to each job independently, that means that in a workload there may be jobs executing with different queue types.

For the future we plan to extend our experiments to a wide variety of applications in order to establish a more accurate limit for the coefficient of variation of the user time (CVUT) and context switches (CVCC) among processes, to determine the balance degree of a job.

We are planning also to use the CVUT and the CVCC to map processes to processors in to implement a dynamic load balancing with Local queues.

Acknowledgments

This work was supported by the Ministry of Science and Technology of Spain under contract TIN2004-07739-C02-01 and the HiPEAC European Network of Excellence. And has been developed using the resources of the DAC at the UPC and the European Centre for Parallelism of Barcelona (CEPBA).

References

1. Anderson, T.E., Lazowska, E.D., Levy, H.M.: The performance Implications of Thread Management Alternatives for Shared Memory Multiprocessors. IEEE Trans. on Comp. 38(12), 1631–1644 (1989)
2. Arpaci-Dusseau, A., Culler, D.: Implicit Co-Scheduling: Coordinated Scheduling with Implicit Information in Distributed Systems. ACM Trans. Comp. Sys. 19(3), 283–331 (2001)
3. Bailey, D., Harris, T., Saphir, W., Wijngaart, R., Woo, A., Yarrow, M.: The NAS Parallel Benchmarks 2.0, Technical Report NAS-95-020, NASA (December 1995)
4. Bershad, B.N., Lazowska, E.D., Levy, H.M.: The Performance Implications of Thread Management Alternatives for Shared Memory Multiprocessors. IEEE Trans. on Comp. 38(12), 1631–1644 (1989)
5. Bhandarkar, M., Kale, L.V., de Sturler, E., Hoeflinger, J.: Object-Based Adaptive Load Balancing for MPI Programs. In: Alexandrov, V.N., Dongarra, J.J., Juliano, B.A., Renner, R.S., Tan, C.J.K. (eds.) ICCS 2001. LNCS, vol. 2074, pp. 108–117. Springer, Heidelberg (2001)
6. Black, D.L.: Scheduling support for concurrency and parallelism in the Mach operating system. Computer 23(5), 35–43 (1990), [16] Silicon Graphics, Inc. IRIX Admin: Resource Administration, Document number 007-3700-005 (2000), http://techpubs.sgi.com

7. Bryant, R.M., Chang, H.-Y., Rosenburg, B.: Experience developing the RP3 operating system. Computing Systems 4(3), 183–216 (1991)
8. Feitelson, D.: Job Scheduling in Multiprogrammed Parallel Systems. IBM Research Report RC 19790 (87657) (October 1994), Second Revision (August 1997)
9. Feitelson, D.G., Jette, M.A.: Improved Utilization and Responsiveness with Gang Scheduling. In: Feitelson, D.G., Rudolph, L. (eds.) JSSPP 1997. LNCS, vol. 1291, Springer, Heidelberg (1997)
10. Frachtenberg, E., Feitelson, D., Petrini, F., Fernandez, J.: Flexible CoSheduling: Mitigating Load Imbalance and Improving Utilization of Heterogeneous Resources. In: IPDPS 2003 (2003)
11. Gupta, R.: Synchronization and Comunication Costs of Loop Partitioning on Shared-Memory Multiprocessor Systems. In: Gupta, R. (ed.) ICPP 1999, pp. II:23–30 (1989)
12. Gupta, A., Tucker, A., Urushibara, S.: The impact of operating system scheduling policies and synchronization methods on the performance of parallel applications. In: SIGMETRICS Conf. Measurement & Modeling of Comp. Syst., pp. 120–132 (May 1991)
13. Hofmann, F., Dal Cin, M., Grygier, A., Hessenauer, H., Hildebrand, U., Linster, C., Thiel, T., Turowski, S.: MEMSY: a modular expandable multiprocessor system. In: Dal Cin, M., Bode, A. (eds.) Parallel Computer Architectures. LNCS, vol. 732, pp. 15–30. Springer, Heidelberg (1993)
14. LeBlanc, T., Scott, M., Brown, C.: Largescale parallel programming: experience with the BBN Butterfly parallel processor. In: Proc. ACM/SIGPLAN, pp. 161–172 (July 1988)
15. Martorell, X., Corbalan, J., Nikolopoulos, D., Navarro, J.I., Polychronopoulos, E., Papatheodorou, T., Labarta, J.: A Tool to Schedule Parallel Applications on Multiprocessors: the NANOS CPU Manager. In: Feitelson, D.G., Rudolph, L. (eds.) IPDPS-WS 2000 and JSSPP 2000. LNCS, vol. 1911, pp. 55–69. Springer, Heidelberg (2000)
16. Message Passing Interface Forum. MPI: A Message-Passing Interface standard. Journal of SuperComputer Jobs 8(3/4), 165–414 (1994)
17. Moreira, J.E., Chan, W., Fong, L.L., Franke, H., Jette, M.A.: An Infrastructure for Efficient Parallel Job Execution in Terascale Computing Environments. In: SC 1998(1998)
18. Nagar, S., Banerjee, A., Sivasubramaniam, A., Das, C.R.: A Closer Look at Coscheduling Approaches for a Network of Workstations. In: 11th ACM Symp. on Parallel Algorithms
19. Scott, M., LeBlanc, T., Marsh, B., Becker, T., Dubnicki, C., Markatos, E., Smithline, N.: Implementation issues for the Psyche multiprocessor operating system. Comp. Syst. 3(1), 101–137 (1990)
20. Serra, A., Navarro, N., Cortes, T.: DITools: Applicationlevel Support for oids Dynamic Extension and Flexible Composition. In: Proc. of the USENIX Annual Technical Conference, pp. 225–238 (June 2000)
21. Squillante, M.S., Nelson, R.D.: Analysis of Task Migration in Shared-Memory Multiprocessor Scheduling. In: Proc. of the 1991 ACM SIGMETRICS Conf. on Measurement and Modeling of Comp. Syst., pp. 143–145 (May 1991)
22. Thomas, R., Crowther, W.: The Uniform System: An Approach to Runtime Support for Large Scale Shared Memory Parallel Processors. In: Proc. of the ICPP 1988, pp. 245–254 (August 1998)

23. Tucker, A., Gupta, A.: Process control and scheduling issues for multiprogrammed shared-memory multiprocessors. In: Proc. of the SOSP 1989, pp.159–166 (December 1989)
24. Utrera, G., Corbalan, J., Labarta, J.: Scheduling of MPI applications: Self Co-Scheduling. In: Danelutto, M., Vanneschi, M., Laforenza, D. (eds.) Euro-Par 2004. LNCS, vol. 3149, Springer, Heidelberg (2004)
25. Utrera, G., Corbalan, J., Labarta, J.: Implementing Malleability on MPI Jobs. In: PACT 2004, pp. 215–224 (2004)
26. Vaswani, R., Zahorjan, J.: Implications of Cache Affinity on Processor Scheduling for Multiprogrammed, Shared Memory Multiprocessors. In: Proc. SOSP 1991, pp. 26–40 (October 1991)
27. www.nas.gov/News/Techreports/2003/PDF/nas-03-010.pdf

Workload Characterization of Stateful Networking Applications

Javier Verdú[1], Mario Nemirovsky[2], Jorge García[1], and Mateo Valero[1,3]

[1] Departament d'Arquitectura de Computadors, UPC
Barcelona, Spain
[2] Consentry Networks Inc.
Milpitas, CA, USA
[3] Barcelona Supercomputing Center
Barcelona, Spain
{jverdu,jorge,mateo}@ac.upc.edu
mario@consentry.com

Abstract. The explosive and robust growth of the Internet owes a lot to the "end-to-end principle", which pushes stateful operations to the end-points. The Internet grow both in traffic volume, and in the richness of the applications it supports. A whole new class of applications requires stateful processing.

This paper presents the first workload characterization of stateful networking applications. The analysis emphasizes the study of data cache behaviour. Nevertheless, we also discuss other issues, such as branch prediction, instruction distribution and ILP. Another important contribution is the study of the state categories of the networking applications. The results show an important memory bottleneck that involves new challenges to overcome.

Keywords: Flow, state, non-additional-data, stateless, stateful.

1 Introduction

The explosive and robust growth of the Internet pushes stateful operations to the end-points. The Internet grow both in traffic volume, and in the richness of the applications it supports. The growth also brought along new security issues. The sophistication of attack methods is progressing rapidly.

The attackers take advantage of stateless firewalls that cannot look beyond the single individual packet while inspecting network traffic. The attacks may be spread out in several packets, such as inter-packet signature, or even may be undetectable with signature-based detection systems, such as portscan, unknown attacks, or zero day attacks [7,8]. Additionally, stateless Network Intrusion Detection Systems (NIDSs) may be overwhelmed with the Snot [5] and Stick [2] attacks. These attacks work by generating packets that are expected to trigger alerts by the NIDS. Therefore, more complex firewalls, that keep track of the processed packets, are being developed in order to catch these new attacks.

J. Labarta, K. Joe, and T. Sato (Eds.): ISHPC 2005 and ALPS 2006, LNCS 4759, pp. 130–141, 2008.

Another area of stateful applications is network monitoring, such as the commercial tool, called Cisco NetFlow [3], and the publicly available application, called Argus [4]. They are able to record important information along different granularity levels. From flow level, such as number of packets of a given flow, up to user level, such as where do the users go on the network.

The development of network processors is focused on overcoming the time constraints of both network line rates and networking applications workloads, mainly on the fast-path. Internet traffic continues to grow vigorously close to 100% per year [17] and consequently the traffic aggregation level reflects an incremental trend. On the other hand, concerning the statefulness of the application, there are more flow states maintained and more complex applications generate larger states. Consequently, the memory capacity requirements become a bottleneck.

This paper presents the first workload characterization of stateful networking applications. The analysis emphasizes the study of data cache behaviour. Nevertheless, we also discuss other issues, such as branch prediction, instruction distribution and ILP. Another important contribution is the study of the state categories of the networking applications. Our conclusions show that stateful applications present an important memory bottleneck that involves new challenges to overcome. We demonstrate that the basis of our conclusions can be generalized to other new stateful programs.

The rest of this paper is organized as follows. Section 2 provides related work on current networking benchmark suites and their workload characterizations. In Section 3, we present the benchmark selection, the traffic traces, and the methodology to perform the evaluation. The workload characterization is presented in Section 4. Finally, we conclude in Section 5.

2 Related Work

Benchmarking NPs is complicated by a variety of factors [9], such as emerging applications that do not yet have standard definitions. There is a high interest and an ongoing effort in the NP community to define standard benchmarks [15].

Several benchmarks suites have been published in the NP area: CommBench [25], NetBench [14] and NpBench [12]. Wolf et al. [25] present the CommBench benchmark suite: a set of eight benchmarks classified in Header Processing Applications (HPA) and Payload Processing Applications (PPA). The suite is focused on program kernels typical of traditional routers. The workloads are characterized and compared versus SPEC benchmarks.

Memik et al. [14] present a set of nine benchmarks, called NetBench. The authors categorize the benchmarks into three groups, according to the level of networking application: micro-level, IP-level and application-level. The workloads are compared versus MediaBench programs.

Lastly, Lee et al. [12] propose a new set of ten benchmarks, called NpBench. It is focused on both control and data plane processing. In this case, the benchmarks are categorized according to the functionality: traffic management and quality of service group (TQG), security and media processing group (SMG), and packet processing group (PPG). The study of the workloads is compared with the CommBench workloads.

All the above benchmarks are no stateful applications, since they do not keep track of the previous processed packets. An IDS called Snort [1], which is included in the NetBench suite, although it is not included in the original paper [14], is the single application that presents stateful features. Moreover, depending on Snort's configuration, the statefulness of the processing may vary a lot. There are several publications that present studies about Snort [19,11,20]. However, the workload and the cache behavior of the stateful configuration have not been analyzed yet.

3 Environment and Methodology

3.1 Benchmarks Selection

According to the data management along the packet processing, there are the following application categories: *non-additional-data* applications are those programs that do not need to search any kind of data related to the packet or connection to be able to perform the packet processing. For example, CRC only needs the IP packet header. The *stateless* category includes the applications that generate no record of previous packet processing and each packet processing has to be handled based entirely on its own information. For instance, the packet forwarding do not record information about previous packet forwarding. Unlike non-additional-data category, stateless applications search information related to the packet (e.g. IP lookup data structures). Finally, the third category is the *stateful* applications that keep track of the state of packet processing [13], usually by setting fields of state related to the flows or connections. For example, TCP termination requires to maintain the state of the TCP flows. The main difference between stateful and stateless programs is the former may update a variety of fields within the state. Instead stateless applications only require the value and do not update any information. As our concern is to characterize stateful applications, the results of the other benchmarks are used to explain in a better way the differences among the above categories.

Table 1 shows the selected benchmarks according to the above classification. We select the benchmarks that present a similar performance than the average of the category. Due to the lack of stateful applications within the publicly available benchmark suites, there is only a single application that presents a potentially statefulness feature: Snort 2.3 [1]. We employ three different configurations: *Snort_SLess* is configured to execute stateless preprocessors, and *Snort_Str4* and *Snort_Pscan* are tuned to use the stateful preprocessors called Stream4 and Flow Portscan, respectively. The former is the inspection of establishing TCP connections and their maintenance and prevents attacks such as Snot [5] and Stick [2]. The latter is an engine designed to detect portscans based on flow creation and the goal is to catch one to many hosts and one to many ports scans. Additionally, we select *Argus* [4] (i.e. network Audit Record

Table 1. Selected Benchmarks

App. Category	State Categories	Benchmark	Bench. Suite
Non-Additional-Data	Pkt	AES	NpBench
	Pkt	MD5	NpBench
Stateless	Pkt & Global	Route	NetBench
	Pkt & Global	Nat	NetBench
	Pkt & Global	Snort_SLess	NetBench
Stateful	Pkt & Global & Flow	Snort_Str4	NetBench
	Pkt & Global & Flow	Snort_Pscan	NetBench
	Pkt & Global & Flow & App	Argus	

Generation and Utilization System) even it is not included in any benchmark suite. This application is a fixed-model Real Time Flow Monitor. That is, it can be used to monitor individual end systems or activity on the entire enterprise network.

The stateful granularity level of every application is shown in the column called "State Categories". We discuss and analyze this classification in Section 4.2.

3.2 Traffic Traces

In order to perform a strict comparison among applications from different benchmark suites we cannot use the default traffic traces included in the suites. Obtaining representative network traffic traces always has been an obstacle to overcome. There are several public sites (e.g. NLANR [16], CAIDA [6]) where there are publicly available traffic traces from a wide open range of routers (e.g. MRA, etc.). However, for confidentiality reasons the IP packet addresses of these traces are sanitized [18]. The sanitization of addresses involves the loss of spatial locality of the Internet IP address distribution [10] and it could affect the results of some networking application studies. Verdú et al. [23] shows that the loss of spatial IP address distribution has no significant influence on the evaluation of Snort with a stateful configuration. In fact, our analysis show that stateful applications do not present significant variations. This means that sanitized traffic is representative to do research in stateful applications. The non-additional-data applications are unaffected by the sanitization, because they do not need the IP address to perform any search. Finally, the stateless applications are affected by the use of sanitized traces. However, our studies show that they present no significant variations in the application performance and, moreover, currently there is no better way to perform this analysis.

In order to keep track of TCP connections, the traces have to hold packets in the two directions of the flows (i.e. packets from server to client and vice versa). In other case, the connection does not follow the TCP protocol and the flow state could not be updated. There is a reduced number of public traces with this property. We select traces from an OC12c (622 Mbit/s) PoS link connecting the

Table 2. Bascline Configuration

Processor Configuration	
Fetch Width	4
Queues Entries	64 int, 64 fp, 64 ld/st
Execution Units	6 int, 3 fp, 4 ld/st
Physical Registers	192 int, 192 fp
ROB Size	256 entries
Branch Predictor Configuration	
Branch Predictor Perceptron	256 perceptrons, 4096 x 14 bit local 40 bit global history
Branch Target Buffer	256, 4-way
RAS	32 entries
Memory Configuration	
ICache DCache	64KB, 2-way, 8 banks, 32B lines, 1 cycle access
L2 Cache	2MB, 8-way, 16 banks, 32B lines, 20 cycles access
Main Memory	500 cycles access
TLB	48-entry I + 128-entry D
TLB miss penalty	160 cycles

Merit premises in East Lansing to Internet2/Abilene (i.e. MRA traces within the NLANR site).

Finally, there is no public traces with representative traffic aggregation levels with bidirectional traffic. As we need a trace with a representative traffic aggregation level in order to obtain representative results [24], we synthetically generate traffic traces simulating different bandwidths. From four original traffic traces of the same link we sanitized them using four mechanisms that assure the independence of IP addresses between traces. The traffic trace used simulates to a bandwidth link of roughly 1Gbps, showing 170K active flows on average.

3.3 Evaluation Methodology

We use different analysis tools depending on the target of the study. We instrument the binary code with ATOM [21] and generate statistics for instruction distribution. On the other hand, for studying the performance of selected applications, we use a modified version of the SMTSim simulator [22]. Table 2 shows the configuration employed of a single threaded out-of-order processor. The baseline is an ample configuration in order to understand the actual application behaviour.

We run every benchmark using the selected traffic traces and processing the same number of packets. Before starting to take statistics, we run the applications until the initial stage is finished, such as the creating of IP lookup table. Subsequently, the applications are warmed running enough packets in order to reach the stable behaviour of the program. Our studies indicate that 10K packets are enough. Also, these studies indicate that on average 50K packets is a representative amount of packets for obtaining representative statistics.

4 Characterization of Benchmarks

4.1 Instruction Mix and ILP

Our results show that roughly 50% of instructions are arithmetic, shift and logic operations. The great part of the applications present a similar percentage of branch operations (12% on the average). Only *AES* shows a lower percentage. Finally, an average of 40% of instructions are memory accesses. Moreover, roughly two thirds of the memory accesses are loads.

The instruction distribution shows minor variations according to the application categories. Unlike the non-additional-data applications, the other networking programs require the search of additional data in order to do the packet processing, such as IP lookup or flow state. Due to the search of data structures, stateless and stateful applications are slightly more memory stressed.

On the other hand, we evaluate the instruction level parallelism (ILP) of the applications. The processor configuration presents variations over some of the baseline parameters (see Table 2) towards avoiding any additional performance constraint. There are no limitations on both fetch bandwidth and functional units. The new configuration also presents an oracle branch predictor and a perfect memory system, where every memory access has one cycle latency.

Figure 1 shows the available ILP within the applications. We can observe that the available ILP is independent of the application category, although it is inherent to the application itself. For example, *MD5* presents an ILP of 2,3 against the 4,5 of *AES*, even though both of them belongs to the same category. This difference is emphasized with the average ILP of *Control Plane* and *Data Plane* applications of the NpBench benchmark suite, where the greatest part of the applications belongs to the non-additional-data category.

The evaluated stateful applications present an ILP of 3,7 on average. Actually, the data flow of the standard stateful packet processing present a reduced

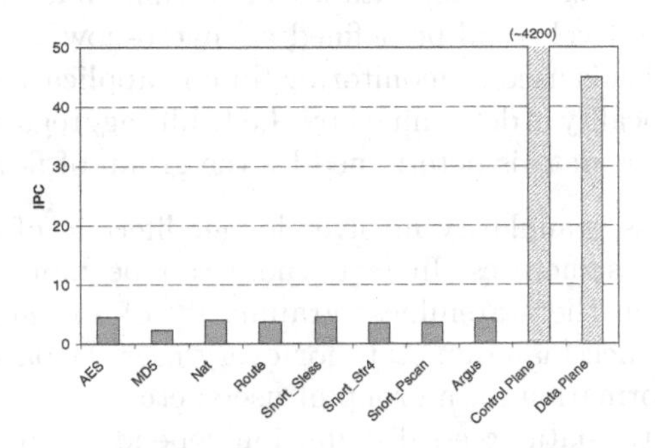

Fig. 1. Available Parallelism

ILP. Since the type of packet processing cannot exploit the ILP like the other applications.

4.2 State Categories

The variables and data accesses of an application are allocated into three segments of the memory model, namely: data segment, heap, and stack. From the application point of view, the data can be classified according to the variable environment, such as global to the application or local to a given function.

In general, the networking applications use to handle a global state. That is, the data structures are shared between packets. Actually, the data can be classified in two subsets. One is the global variables set, such as total number of processed packets. These variables are independent of any network traffic parameter. The other is represented by the global state data structures, such as IP lookup table. This state use to be related to a network feature, such as IP address. Thus, the temporal locality of these data between packets is determined by the network traffic properties. Generally, the lifetime of these data structures is eternal.

Nevertheless, the networking applications shows an additional classification that categorizes the data according to the granularity of the state that they represent:

- **Packet:** The data only cover the current packet processing, such as packet content. The lifetime of these data structures is very short, since they are reset every packet processing. The inter-packet temporal locality of this data set is independent of any network traffic parameter. In fact, depending on the target of the application, the temporal locality within the packet processing use to be very high.
- **Flow:** The data structures are related to the network connection, such as flow state. The temporal locality is determined by the traffic aggregation level of the network link. Although the lifetime is longer than Packet State Data, it is delimited by the lifetime of the flow.
- **Application:** The data structures are associated to the application layer, such as Real Time Protocol (i.e. a transport protocol designed to provide end-to-end delivery services for data with real-time characteristics). A similar granularity state level could be defined as macro-flow, such as counters of a set of flows, which is used in monitoring/billing applications between others. The temporal locality is determined by the traffic aggregation level. However, the lifetime of the data is determined by the group of flows.

Lower statefulness granularity involves longer lifetime of the states and, in general, larger data structures. In fact, there can be more stateful data categories depending on the statefulness granularity of the application, namely: User State Data, which maintain state for every user; Department State Data, which maintain information for a group of users; etc.

Figure 2 depicts the data access distribution depending on the data categories of the evaluated applications. The non-additional-data applications do a very

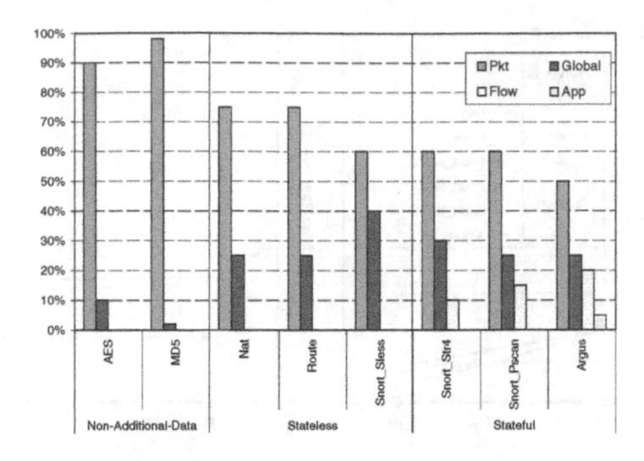

Fig. 2. Data Access Distribution

high rate of data accesses on the Packet State Data. On the other hand, stateless applications use to search data for the packet processing. Due to this, the rate of Packet Data accesses is lower, whereas the Global Data access rate is higher. Depending on the target of the application, this search could be intensive, such as rule matching in a rule-based Intrusion Detection System. Finally, stateful applications present access rates to a wide variety of state data structures. The rate of flow state accesses can vary according to the statefulness granularity level of the application. Due to the properties of every data category, the data access distribution of an application directly affects the data cache behaviour.

4.3 Data Cache Behavior

In this section we discuss the data cache miss rate of the evaluated applications, and we analyze the impact of the data access distribution shown in the above section. We do not include an analysis of instruction cache. As other papers explain [25,14,12], the networking applications present on average near to 100% of instruction cache hit rates. Moreover, our studies show similar results with stateful applications.

Figure 3(a) shows the L1 data cache miss rate using a variety of sizes. The x-axis shows the sizes of L1 data cache. In the legend of the graph we group the benchmarks according to the application category. We can observe that the non-additional-data applications present very reduced data cache miss rate, even with reduced size data caches. The greatest part of the working set is related to the packet state. Thus, the data cache is very useful in order to maintain this type of structures. The working set of stateless applications is larger due to global data structures, such as IP lookup table. As the access rates to these structures is higher, reduced caches present higher miss rates. However, a cache size of 16 KBytes or higher presents a low data cache miss rate of 2% on average. On the other hand, *Snort_SLess* could present higher miss rates if it was a rule-based IDS and there was a large number or rules. Finally, stateful applications present higher miss rates than the rest of applications, due to a larger working set and the variety of data structures that are dependent on networking properties, such as distance between two packets of the same flow.

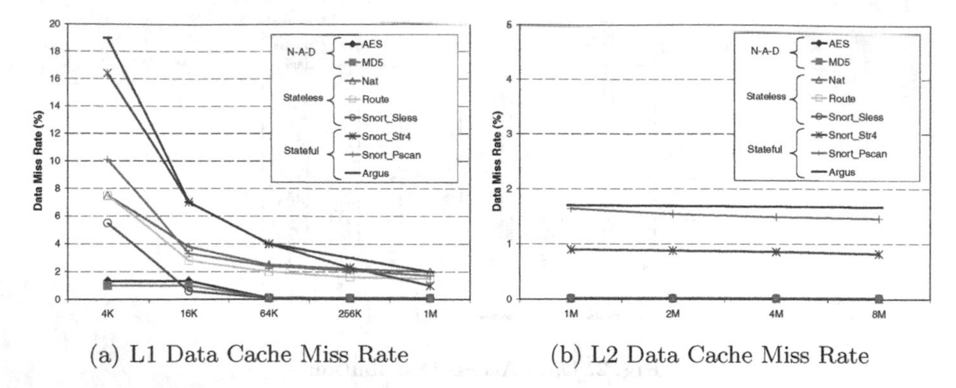

(a) L1 Data Cache Miss Rate (b) L2 Data Cache Miss Rate

Fig. 3. Data Cache behaviour

We analyze the effects of increasing the associativity level, but there are no significant benefits. Because the main problem are not cache conflicts, but the cache inability of maintaining the data of the active flows. However, using a larger cache line size we take advantage of spatial locality within the packet processing itself. For example, if we enlarge from 32B up to 64B cache line size we achieve a 3% reduction on average in the cache miss rate. This fact represents three times the improvement obtain doubling the cache associativity degree.

Nevertheless, the most important impact resides in the L2 data cache miss rate shown in Figure 3(b). Whereas non-additional-data and stateless applications present near to zero miss rate, stateful applications show a saturated miss rate from 1% up to 1,7%. In other words, even using a large L2 data cache, the misses due to stateful data structures cannot be eliminated. Thus, when we use a very large DL1 cache, almost 100% of DL1 misses also are misses in L2.

Depending on the size of flow-states and the traffic aggregation level, we can require several MBytes of data structures for flow-states. In fact, the memory performance of stateful applications is very sensitive to the traffic aggregation level [24], since the memory capacity requirements significantly grows and the temporal locality of flow state is reduced. Obviously, with larger flow-states the sensitivity is higher. Our studies show that *Snort_Str4* and Argus requires roughly 420 Bytes and 1 KByte, respectively. As we mentioned in Section 3.2, we maintain 170K flows on average. Thus, the memory requirements of *Snort_Str4* and Argus are roughly 68 MBytes and 166 MBytes, respectively. Therefore, this requirement will be higher with more statefulness applications that process traffic with higher aggregation level.

4.4 Branch Prediction

In this section we evaluate the behaviour of the branches through the study of branch prediction accuracy. The impact of branches on the IPC will be discussed in Section 4.5. We employ a perceptron predictor [26,27]. Additionally, we also have studied other branch predictors, such as g-share [28]. However, the results show a slightly lower accuracy than the perceptron predictor, and higher sensitivity to the PHT size.

As we can see in Section 4.1, all the benchmarks present between 10% and 20% of branches within the workload, excepting *AES* that shows a lower rate. On average 70% of these branches are conditional branches.

The results show a high branch prediction accuracy for every benchmark. *AES*, *MD5*, and *Snort_SLess* present more than 99% of accuracy. *Snort_Pscan* and *Argus* shows roughly 98% of hit rate. Finally, the lowest accuracy rate is 95% by *Snort_Str4*. Our studies show that, as well as this application presents high branch rate, the branches are more sensitive to the network properties than the other applications. Thus, there is a negative aliasing among independent packets, unlike *Snort_Pscan* and *Argus* that present lower negative aliasign.

The key insight is that certain branches depend on the flow-state itself. Thus, regarding the statefulness of the application, coupled with the traffic aggregation level, the likelihood of obtaining lower branch prediction accuracy could be stronger.

4.5 Impact of Bottlenecks

Figure 4 presents the IPC of every benchmark using four different configurations: the baseline configuration, an oracle branch predictor, a perfect memory system (i.e. every memory access has one cycle latency), and a oracle predictor with a perfect memory system.

The Stateless applications present a baseline IPC of roughly 2,75. The main improvement resides in the branch prediction with a 12% on average. With the perfect configuration of memory and branch prediction we achieve an average of 20%. However, as *Snort_SLess* shows lower miss rates, its preformance improvement with the perfect configuration is only 2%. The main problem is the global data structure properties, such as size and locality. On the other hand, the stateful applications emphasize a lot the properties of global data structures through the accesses to flow and application data structures. Due to this, the memory bottleneck is stressed and L2 data cache is unable to maintain the states of the active flows. A perfect memory system can obtain roughly 3x of speedup on

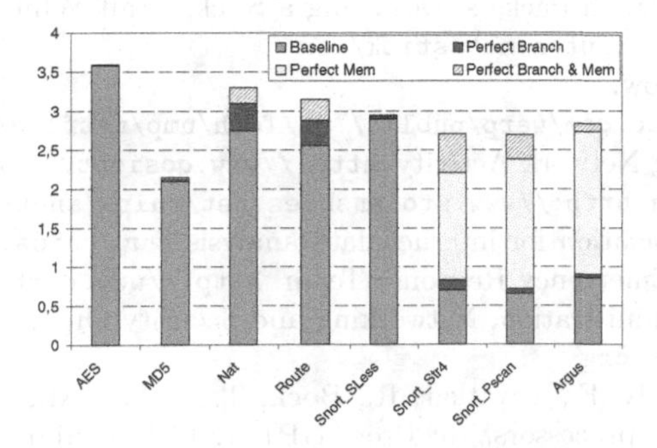

Fig. 4. Impact on IPC

average. The baseline IPC could be lower and the speedup could be emphasized when statefulness of the application is higher. Once the memory bottleneck is overcome, we can achieve an additional improvement of 12% on average.

5 Conclusions

To the best of our knowledge, we present the first workload characterization of stateful networking applications. We analyze the main differences between the networking application categories according to the management of state data. Actually, stateful applications present a variety of statefulness granularity levels.

The main bottleneck of stateful applications resides in the memory system, since even L2 data cache is unable to maintain the state of active flows. The evaluated applications show 3x of speedup on average using a perfect memory system. Nevertheless, this speedup can vary depending on the statefulness level of the application.

As future work we are concerned about to do a deep analysis of the architectural impact of the stateful applications. On the other hand, as the variety of stateful benchmarks is still reduced, we continue looking for other publicly available stateful applications.

Acknowledgements

This work has been supported by the Ministry of Education of Spain under contract TIN–2004–07739–C02–01, and grant BES-2002-2660 (J. Verdú), the HiPEAC European Network of Excellence, and the Barcelona Supercomputing Center. The authors also would like to thank Rodolfo Milito for his review and technical inputs of this work.

References

1. Beale, J., Foster, J.C., Posluns, J., Caswell, B.: Snort 2.0 Intrusion Detection. Syngress Publishing Inc. (2003)
2. Coretez, G.: Fun with Packets: Designing a Stick, Draft White Paper on Stick, `http://www.eurocompton.net/stick/`
3. Cisco IOS NetFlow. `http://www.cisco.com/warp/public/732/Tech/nmp/netflow/index.shtml`
4. Argus - Auditing Network Activity. `http://www.qosient.com/argus`
5. Snot V0.92 alpha. `http://www.stolenshoes.net/sniph/snot-0.92a-README.txt`
6. Cooperative association for internet data analysis. `www.caida.org`
7. The Computer Emergency Response Team. `http://www.cert.org`
8. The System Administration, Networking and Security Organization. `http://www.sans.org`
9. Chandra, P., Hady, F., Yavatkar, R., Bock, T., Cabot, M., Mathew, P.: Benchmarking network processors. In: Proc. NP1, Held in conjunction with HPCA-8, Cambridge, MA, USA (February 2002)

10. Kohler, E., Li, J., Paxson, V., Shenker, S.: Observed structure of addresses in IP traffic. In: Proc. of the 2nd ACM SIGCOMM Workshop on Internet measurment workshop, Pittsburgh, PA, USA (August 2002)
11. Kruegel, C., Valeur, F., Vigna, G., Kemmerer, R.: Stateful intrusion detection for high-speed networks. In: Proc. IEEE Symposium Security and Privacy, IEEE Computer Society Press, CA, USA (2002)
12. Lee, B.K., John, L.K.: Npbench: A benchmark suite for control plane and data plane applications for network processors. In: Proc. of ICCD, San Jose, CA, USA (October 2003)
13. Melvin, S., Nemirovsky, M., Musoll, E., Huynh, J., Milito, R., Urdaneta, H., Saraf, K.: A massively multithreaded packet processor. In: Proc. of NP2, Held in conjunction with HPCA-9, Anaheim, CA, USA (February 2003)
14. Memik, G., Mangione-Smith, W.H., Hu, W.: Netbench: A benchmarking suite for network processors. In: Proc. of ICCAD, San Jose, CA, USA (November 2001)
15. Nemirovsky, A.: Towards characterizing network processors: Needs and challenges. Xstream Logic Inc., white paper (2000)
16. National lab of applied network research. http://pma.nlanr.net/Traces
17. Odlyzko, A.M.: Internet traffic growth: Sources and implications. In: Dingel, B.B., Weiershausen, W., Dutta, A.K., Sato, K.-I. (eds.) Proc. SPIE, Optical Transmission Systems and Equipment for WDM Networking II, vol. 5247 (September 2003)
18. Pang, R., Paxson, V.: A high-level programming environment for packet trace anonymization and transformation. In: Proc. of the ACM SIGCOMM Conference (August 2003)
19. Roesch, M.: Snort- lightweight intrusion detection for networks. In: LISA. Proc. of the 13th Conference on Systems Administration, Seattle, WA, USA (1999)
20. Schaelicke, L., Slabach, T., Moore, B., Freeland, C.: Characterizing the performance of network intrusion detection sensors. In: Proc. of RAID-6, Pittsburgh, PA, USA (September 2003)
21. Srivastava, A., Eustace, A.: ATOM - A system for building customized program analysis tools. In: Proc. ACM SIGPLAN Conf. on Programming Language Design and Implementation, pp. 196–205 (June 1994)
22. Tullsen, D.M.: Simulation and modeling of a simultaneous multithreading processor. In: 22nd Annual Computer Measurement Group Conference (December 1996)
23. Verdú, J., García, J., Nemirovsky, M., Valero, M.: Analysis of traffic traces for stateful applications. In: Proc. of NP3, Held in conjunction with HPCA-10, Madrid, Spain (February 2004)
24. Verdú, J., García, J., Nemirovsky, M., Valero, M.: The Impact of Traffic Aggregation on the Memory Performance of Networking Applications. In: Proc. of MEDEA Workshop, Held in conjunction with PACT-2004, France (September 2004)
25. Wolf, T., Franklin, M.A.: Commbench - a telecommunications benchmark for network processors. In: Proc. of ISPASS, Austin, TX, USA (April 2000)
26. Jimenez, D., Lin, C.: Neural methods for dynamic branch prediction. ACM Transactions on Computer Systems 20(4), 369–397 (2002)
27. Vintan, L., Iridon, M.: Towards a high performance neural branch predictor. In: Proc. of IJCNN, vol. 2, pp. 868–873 (July 1999)
28. McFarling, S.: Combining Branch Predictors. Technical Report TN-36, Compaq Western Research Lab (June 1993)

Using Recursion to Boost ATLAS's Performance

Paolo D'Alberto[1] and Alexandru Nicolau[2]

[1] Department of Electrical and Computer Engineering - Carnegie Mellon University
pdalbert@andrew.cmu.edu

[2] Department of Computer Science - University of California at Irvine*
nicolau@ics.uci.edu

Abstract. We investigate the performance benefits of a novel recursive formulation of Strassen's algorithm over highly tuned matrix-multiply (MM) routines, such as the widely used ATLAS for high-performance systems.

We combine Strassen's recursion with high-tuned version of ATLAS MM and we present a family of recursive algorithms achieving up to 15% speed-up over ATLAS alone. We show experimental results for 7 different systems.

Keywords: dense kernels, matrix-matrix product, performance optimizations.

1 Introduction

In this paper, we turn our attention to a single but fundamental basic kernel in dense and parallel linear algebra such as **matrix multiply** (MM) for matrices stored in double precision.

In practice, software packages such as LAPACK [1] or ScaLAPACK are based on a basic set of routines such as the basic linear algebra subroutines BLAS [2,3]. Moreover, The BLAS is based on an efficient implementations of the MM kernel.

In the literature, we find an abundant collection of algorithms, implementations and software packages (e.g., [4,5,6,7,8,9,10,11]), that aim at the efficient solution of this basic kernel. However, among all ATLAS [11] is one of the most widely recognized and used.

In today's high performance computing, the system performance is the result of a fine and complicated relation between the constituent parts of a processor –i.e., the hardware component, and the sequence of instructions of an application –i.e., the software component. For example, ATLAS [11] is an adaptive software package implementing BLAS that addresses the system-performance problem by careful adaptation of the software component. In practice, ATLAS generates an optimized version of MM tailored to the specific characteristics of the architecture and ATLAS does this custom installation by an combination of micro-benchmarking and an empirical search of the code solution space. In

* This work has been supported in part by NSF Contract Number ACI 0204028.

J. Labarta, K. Joe, and T. Sato (Eds.): ISHPC 2005 and ALPS 2006, LNCS 4759, pp. 142–151, 2008.

this work, we show how an implementation of Strassen's algorithm can further improve the performance of even highly-tuned MM such as ATLAS.

In the literature, other approaches have been proposed to improve the classic formulation of MM by using Strassen's strategy [12] (or Winograd's variant). In fact, Strassen's algorithm has noticeably fewer operations $O(n^{\log_2 7}) = O(n^{2.86})$ than the classic MM algorithm $O(n^3)$ and, thus, potential performance benefits. However, the execution time of data accesses dominates the MM performance and this is due to the increasing complexity of the memory hierarchy realty.

In fact, experimentally, Strassen's algorithm has found validation by several authors [13,14,5] for *simple* architectures, showing the advantages of this new algorithm starting from very small matrices or **recursion truncation point** (RP) [15]. The recursion point is the matrix size n_1 for which Strassen's algorithm yields to the original MM. Thus, for a problem of size $n = n_1$, Strassen's algorithm has the same performance of the original algorithm, and, for every matrix size $n \geq n_1$, Strassen's algorithm is faster than the original algorithm. With the evolution of the architectures and the increase of the problem sizes, the researcher community witnessed the RP increasing [16]. We now find projects and libraries implementing different version of Strassen's algorithm and considering its practical benefits [15,17,18], however with larger and larger RP, mining the practical use of Strassen's algorithm.

In this paper, we investigate recursive algorithms for an empirical RP determination and we embody our ideas so as to combine the high performance of tuned dense kernels –at the low level– with Strassen's recursive division process –at the high level– into a family of recursive algorithms. We present our experimental results for 7 systems where we tested our codes.

Our approach has the following advantages over previous approaches. First, we do not pad the original matrices so as to have even-size or, worse, power-of-two matrices [12]. Second, our codes have no requirements on the matrix layout, thus, they can be used instead of other MM routines (ATLAS) with no modifications or extra overhead to change the data layout before and after the basic computation (unlike the method proposed in [18]). In fact, we assume that the matrices are stored in row-major format and, at any time, we can yield control to a highly tuned MM such as ATLAS's *dgemm*(). Third, we propose a balanced recursive division into subproblems, thus, the codes exploit predictable performance; unlike the division process proposed by Huss-Lederma et al. [15] where for odd-matrix sizes, they divide the problem into a large even-size problem, on which Strassen can be applied, and a small, and extremely irregular, computation. Fourth, we investigate recursive algorithms that can unfold the division process more than once so to achieve further performance (in contrast to [15,18] where the unfolding is limited to one level).

The paper is organized as follows. In Section 2, we present a generalization of Strassen's algorithm better suited for recursion. In section 3, we present our techniques to determine the RP for our codes. In Section 4, we present our experimental results. Finally, Section 5, we present our concluding remarks.

2 Strassen's Algorithm for Any Square-Matrix Sizes

In this section, we show that Strassen's MM algorithm can be generalized quite naturally and more efficiently than previous implementations available in the literature [12,18,15] so that it can be applied to any square-matrix size.

From here on, we identify the **size** of a matrix $\mathbf{A} \in \mathbb{M}^{m\times n}$ as $\sigma(\mathbf{A}) = m \times n$. We assume that an operand matrix $\mathbf{A}$ of size $\sigma(\mathbf{A}) = n \times n$ is logically composed by four **near square** matrices; that is, every submatrix has number of rows r and number of columns c that differ by at most one, i.e., $|r - c| \leq 1$, [9].

The classical MM of $\mathbf{C} = \mathbf{AB}$ can be expressed as the multiplication of the submatrices as follows: $\mathbf{C}_0 = \mathbf{A}_0\mathbf{B}_0 + \mathbf{A}_1\mathbf{B}_2$, $\mathbf{C}_1 = \mathbf{A}_0\mathbf{B}_1 + \mathbf{A}_1\mathbf{B}_3$, $\mathbf{C}_2 = \mathbf{A}_2\mathbf{B}_0 + \mathbf{A}_3\mathbf{B}_2$ and $\mathbf{C}_3 = \mathbf{A}_2\mathbf{B}_1 + \mathbf{A}_3\mathbf{B}_3$. The computation is divided in four basic computations, one for each submatrix composing $\mathbf{C}$. Thus, for every matrix $\mathbf{C}_i$ $(0 \leq i \leq 3)$, the classical approach computes two products, for a total of 8 MMs and 4 **matrix additions** (MA).

Notice that every product is the MM of near square matrices and it computes a result that has the same size and shape of the submatrix destination $\mathbf{C}_i$. Furthermore, if we compute the products recursively, each product is divided in further four subproblems on near square matrices [9].

Strassen proposed to divide the problem into only 7 MMs and to introduce 18 matrix additions/subtractions. When the matrices have power-of-two sizes, $n = 2^k$, all multiplications and additions are among square matrices of the same sizes even if the computation is recursively carried on. We adapt Strassen's algorithm so as to compute the MM for every square matrix size as follows: $\mathbf{C}_0 = \mathbf{M}_1 + \mathbf{M}_4 - \mathbf{M}_5 + \mathbf{M}_7$, $\mathbf{C}_1 = \mathbf{M}_2 + \mathbf{M}_4$, $\mathbf{C}_2 = \mathbf{M}_3 + \mathbf{M}_5$ and $\mathbf{C}_3 = \mathbf{M}_1 + \mathbf{M}_3 - \mathbf{M}_2 + \mathbf{M}_6$ where every $\mathbf{M}_i$ is defined as follow:

$$
\begin{aligned}
\mathbf{M}_1 &= \mathbf{T}_0\mathbf{T}_1 \text{ with } \mathbf{T}_0 = \mathbf{A}_0 + \mathbf{A}_3 \text{ of size } \sigma(\mathbf{T}_0) = \lceil n \rceil \times \lceil n \rceil, \\
&\qquad \text{and with } \mathbf{T}_1 = \mathbf{B}_0 + \mathbf{B}_3 \text{ of size } \sigma(\mathbf{T}_1) = \lceil n \rceil \times \lceil n \rceil \\
&\qquad \text{thus } \sigma(\mathbf{M}_1) = \lceil n \rceil \times \lceil n \rceil \\
\mathbf{M}_2 &= \mathbf{T}_2\mathbf{B}_0 \text{ with } \mathbf{T}_2 = \mathbf{A}_2 + \mathbf{A}_3 \text{ of size } \sigma(\mathbf{T}_2) = \sigma(\mathbf{M}_2) = \lfloor n \rfloor \times \lceil n \rceil \\
\mathbf{M}_3 &= \mathbf{A}_0\mathbf{T}_3 \text{ with } \mathbf{T}_3 = \mathbf{B}_1 + \mathbf{B}_3 \text{ of size } \sigma(\mathbf{T}_3) = \sigma(\mathbf{M}_3) = \lceil n \rceil \times \lfloor n \rfloor \\
\mathbf{M}_4 &= \mathbf{A}_3\mathbf{T}_4 \text{ with } \mathbf{T}_4 = \mathbf{B}_2 - \mathbf{B}_0 \text{ of size } \sigma(\mathbf{T}_4) = \sigma(\mathbf{M}_4) = \lfloor n \rfloor \times \lceil n \rceil \\
\mathbf{M}_5 &= \mathbf{T}_5\mathbf{B}_3 \text{ with } \mathbf{T}_5 = \mathbf{A}_0 + \mathbf{A}_1 \text{ of size } \sigma(\mathbf{M}_5) = \sigma(\mathbf{T}_5) = \lceil n \rceil \times \lfloor n \rfloor \\
\mathbf{M}_6 &= \mathbf{T}_6\mathbf{T}_7 \text{ with } \mathbf{T}_6 = \mathbf{A}_2 - \mathbf{A}_0 \text{ and } \mathbf{T}_7 = \mathbf{B}_0 + \mathbf{B}_1 \\
&\qquad \text{of size } \sigma(\mathbf{M}_6) = \sigma(\mathbf{T}_6) = \sigma(\mathbf{T}_7) = \lceil n \rceil \times \lceil n \rceil \\
\mathbf{M}_7 &= \mathbf{T}_8\mathbf{T}_9 \text{ with } \mathbf{T}_8 = \mathbf{A}_1 - \mathbf{A}_3 \text{ of size } \sigma(\mathbf{T}_8) = \lceil n \rceil \times \lfloor n \rfloor \\
&\qquad \text{and } \mathbf{T}_9 = \mathbf{B}_2 + \mathbf{B}_3 \text{ of size } \sigma(\mathbf{T}_8) = \lfloor n \rfloor \times \lceil n \rceil \\
&\qquad \text{and } \sigma(\mathbf{M}_7) = \lceil n \rceil \times \lceil n \rceil
\end{aligned}
$$

As result of the division process, the matrices $\mathbf{A}_i$, $\mathbf{B}_i$ and $\mathbf{C}_i$ are near square matrices as in the classic algorithm but MA and MMs must be re-defined.

First, we generalize **matrix addition**. Intuitively, when the resulting matrix $\mathbf{X}$ is larger than $\mathbf{Y}$ or $\mathbf{Z}$, the computation is performed as if the matrix operands are extended and padded with zeros. Otherwise, if the result matrix is smaller than the operands, the computation is performed as the matrix operands are cropped to fit the result matrix. See a simple implementation for the addition of two generic matrices in Figure 1.

```
/* C = A+B */
void Add(Mtype *c, int McolC, int mC, int pC,
         Mtype *a, int McolA, int mA, int pA,
         Mtype *b, int McolB, int mB, int pB) {

    int i,j,x,y;
    /* minimum sizes */
    x = min(mA,mB); y = min(pA,pB);

    for (i=0; i<x; i++) {
      /* core of the computation */
      for (j=0;j<y;j++)  c[i*McolC+j] = a[i*McolA+j] + b[i*McolB+j];

      if (y<pA)        c[i*McolC+y] = a[i*McolA+y]; /* A is larger than B */
      else if (y<pB)   c[i*McolC+y] = b[i*McolB+y]; /* B is larger than A */
    }

    /* last row */
    if (x<mA) { /* A is taller than B */
      for (j=0;j<y;j++) c[x*McolC+j] = a[x*McolA+j];

      if (y<pA)         c[x*McolC+y] = a[x*McolA+y];
      else if (y<pB)    c[x*McolC+y] = b[x*McolB+y];
    }
    else if (x<mB) { /* B is taller than A */
      for (j=0;j<y;j++) c[x*McolC+j] = b[x*McolB+j];

      if (y<pA)        c[x*McolC+y] = a[x*McolA+y];
      else if (y<pB)   c[x*McolC+y] = b[x*McolB+y];
    }
}
```

Fig. 1. Addition C-code

Second, we generalize **matrix multiplication** as follows: $\mathbf{X} = \mathbf{Y} * \mathbf{Z}$ where $\sigma(\mathbf{X}) = m \times n$, $\sigma(\mathbf{Y}) = m \times q$ and $\sigma(\mathbf{Z}) = r \times n$ so as $c_{i,j} = \sum_{k=0}^{\min(q,r)} y(i,k) * z(k,j)$.

Notice that the product $\mathbf{A}_0\mathbf{B}_0$, which is a term of $\mathbf{M}_1$, is a **necessary product** and it is required for the computation of $\mathbf{C}_0$; in contrast, $\mathbf{A}_0\mathbf{B}_3$ is an **artificial product**, computed in the same expression, and it must be reduced by MAs (e.g., $\mathbf{M}_1 + \mathbf{M}_4$). The algorithm previously defined computes correctly all necessary products and it annihilates all artificial products.

Both MA and MM, as previously defined, introduce negligible overheads. In fact, the matrices involved in the computations are always near square matrices (i.e., their sizes may differ by at most one) and, thus, the extra control is negligible for the matrix sizes tested in this work.[1] We explain how the two approaches, that is, our version of Strassen's and tuned ATLAS routines are combined in Section 3.

In our codes, the matrix are stored in row-major format and we do not apply any recursive layout strategy as in [18], for the following three reasons. First,

[1] Furthermore, we use the highly tuned ATLAS *dgemm*() to reduce further the effects on the overall performance.

modern memory hierarchics use (4+ way) associative caches for which the effects of cache interferences, due to the matrix layout, is relatively minimal. Second, the MAs in the Strassen's algorithm create a smaller working space where the operands are stored dynamically, so the effect of interference can be reduced further. Third and last, non-standard layout complicates the development of correct and efficient leaf-computation routines for any square matrices; in fact, these leaf routines must be tailored to the type of layout.

The simplicity of our code in conjunction with the performance improvements achievable make our approach a good strategy addition to the already widely used software packages such as ATLAS, especially for large problems. Our pseudo code is presented in Figure 2. We also reorganized the original Strassen's computation so as to use only three temporary matrices, as already proposed in the literature [15].

```
/*
 * | C0 C1 |   | A0 A1 |   | B0 B1 |
 * | C2 C3 | = | A2 A3 | * | B2 B3 |
 */
C mul(A, B) {

  if (Problem_Size < leaf_strassen)
    CC = AA atlas_dgemm  BB;
  else {
    Allocate_workspace(T1,T2,M1);

    T1 = A0 add A3;        T2 = B0 add B3;
    M1 = T1 mul T2;
    C0 = M1;               C3 = M1;

    T1 = A2 add A3;
    M2 = T1 mul B0;
    C2 = M2;               C3 = C3 sub M2;

    T1 = B1 sub B3;
    M3 = A0 mul T1;
    C1 = M3;               C3 = C3 add M3;

    T1 = B2 sub B0;
    M4 = A3 mul T1;
    C0 = C0 add M4;        C2 = C2 add M4;

    T1 = A0 add A1;
    M5 = T1 mul B3;
    C0 = C0 sub M5;        C1 = C1 add M5;

    T1 = A2 sub A0;        T2 = B0 add B1;
    M6 = T1 mul T2;
    C3 = C3 add M6;

    T1 = A1 sub A3;        T2 = B2 add B3;
    M7 = T1 mul T2;
    C0 = C0 add M7;

    Deallocate_workspace();
  }
}
```

Fig. 2. Pseudo Strassen's Algorithm

3 Empirical Considerations on the Recursion Truncation Point

In this section, we propose a technique for determining when the algorithm's strategy must change so as to stop Strassen's and to yield control to the regular MM, the recursion truncation point (RP). In other words, we consider the problem of when to have a recursive call (to Strassen's MM) or a call to an highly tuned *dgemm* (e.g., such as the one offered by ATLAS). We show in Section 4 that the optimal strategy is a function of the problem size and of the underlying system.

Strassen's algorithm embodies different locality properties because its two basic computations exploit different data locality: MM has spatial and temporal locality, and MA has only spatial locality. In fact, consider that the matrix operands fit a cache level, for example L_2, but do not fit the lower cache, such as L_1. Note that the MA does not exploit data locality at the lower levels of cache and, actually, data accesses to/from the CPU during the MA will flush previous contents. In fact, MAs have little data reuse and, thus, data-access latency time cannot be circumvented or hidden; for these applications a memory hierarchy actually slows down the overall performance. In contrast, highly tuned MMs exploit temporal and spatial locality at every level of cache, thus, having fast memory accesses and fast computations. In a hierarchical memory system, the two computations may have drastically different performance. Thus, Strassen's algorithm has a performance edge versus the regular MM only when the savings in MMs, is higher (in execution time) than the cost of the extra additions.

In the literature, we find different and, often contradicting, experimental results about the RP. In fact, a few authors have found that for any problem size Strassen's (or Winograd's variation) is always faster; a few authors have found that the RP is about 500 for some systems and implementations; and a few others, citing private communications, claim that the RP is larger than 1000 [14,16,5,15,18].

Even though the RP is machine and problem-size dependent, however it is straightforward to determine, even if tedious and time consuming. We propose to determine the RP empirically by direct measure of Strassen's MM execution and we do this for recursive Strassen's algorithm with different unfolding levels. This idea is very similar to the one applied for the solution search in ATLAS.

4 Experimental Results

We installed our codes and the software package ATLAS on 7 different architectures, Table 1. Once the installation is finished, we then determined experimentally the RP n_1 based on a simple linear search. Note that for the Fosa system, we could find no problem size for which Strassen's is faster than ATLAS's.

In the following, we present the experimental results for five systems. We use the following terminology: **S-k-unfold** is the Strassen algorithm for which k is the number of times the recursion unfolds before yielding to ATLAS *dgemm*.

Table 1. Systems

System	Processors	n_1	Figure
Fujitsu HAL 300	SPARC64 100MHz	400	Fig. 3
Ultra 5	UltraSparc2 300MHz	1225	Fig. 4
Ultra-250	UltraSparc2 2 @ 300MHz	1300	No
Sun-Fire-V210	UltrasparcIII 1GHz	1150	Fig. 5
ASUS	AthlonXP 2800+ 2GHz	1300	Fig. 6
Unknown server	Itanium 2 @ 700MHz	2150	Fig. 7
Fosa	Pentium III 800MHz	N/A	No

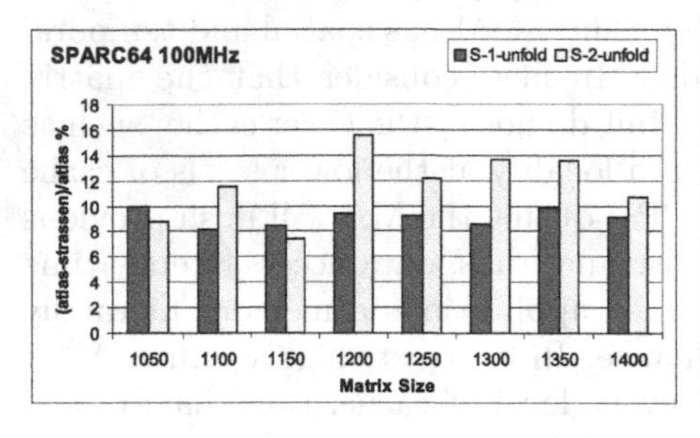

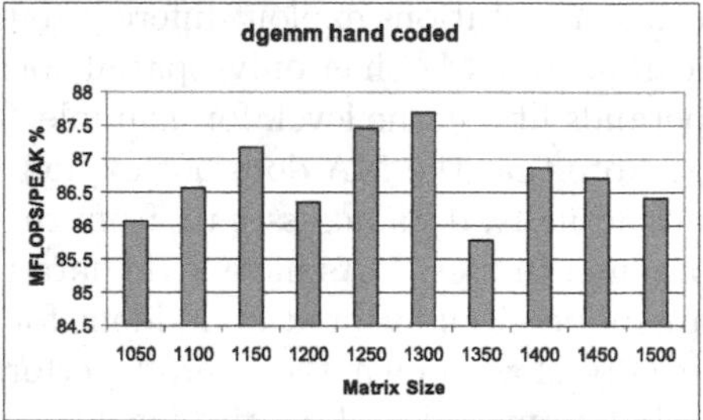

Fig. 3. Fujitsu HAL 300

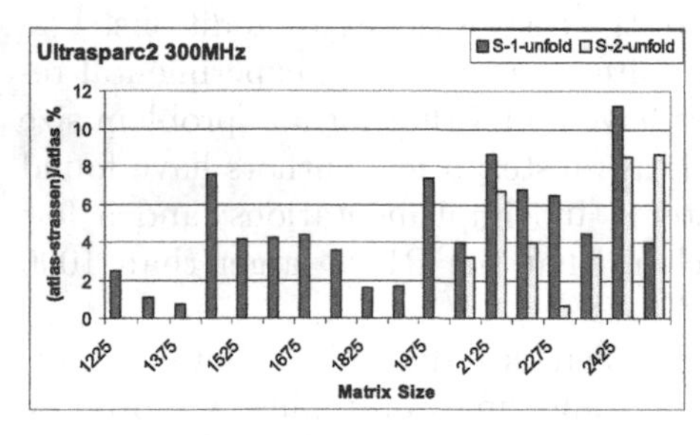

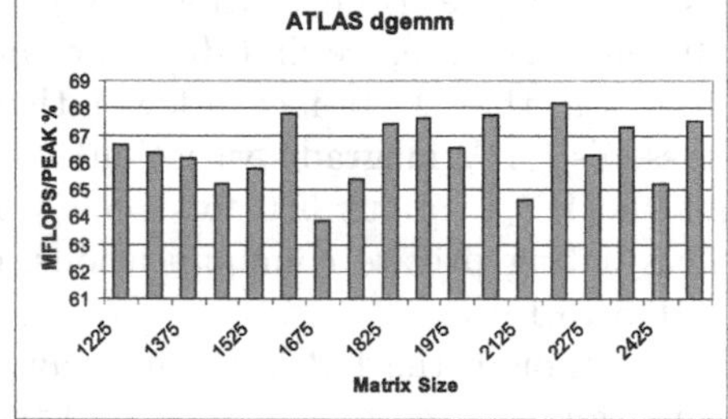

Fig. 4. Ultra 5

(Note we opted to omit negative relative performance and no bar is presented in the charts instead.) The performance obtained by the systems in Table 1, and presented from Figure 3 to Figure 7, are obtained by the collection of the best performance among several trials.

Note that the S-2-unfold algorithm is beneficial for very large problems and for specific systems. However, for the systems in Table 1, the performance improvements are some how limited. We have performance measures of the S-3-unfold algorithm but for the current set of systems, the algorithm has no performance advantage over ATLAS and, thus, we do not report them.

From Figure 3 to Figure 7, we present two measures of performance: relative execution time over ATLAS, and relative MFLOPS for ATLAS *dgemm* over peak

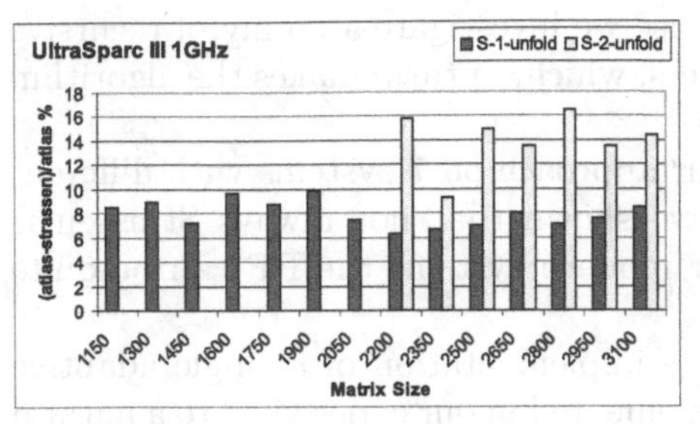

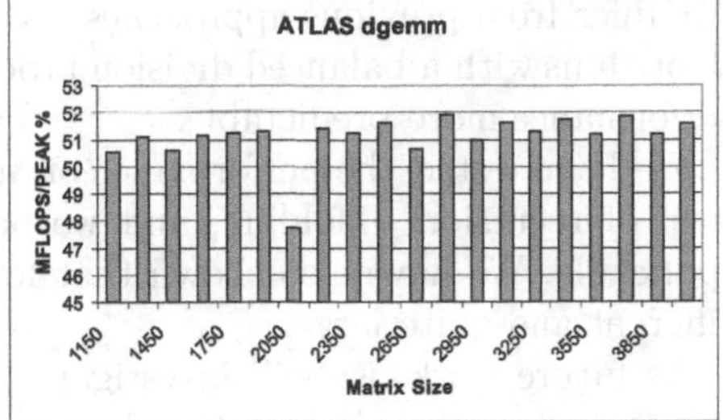

Fig. 5. Sun-Fire-V210

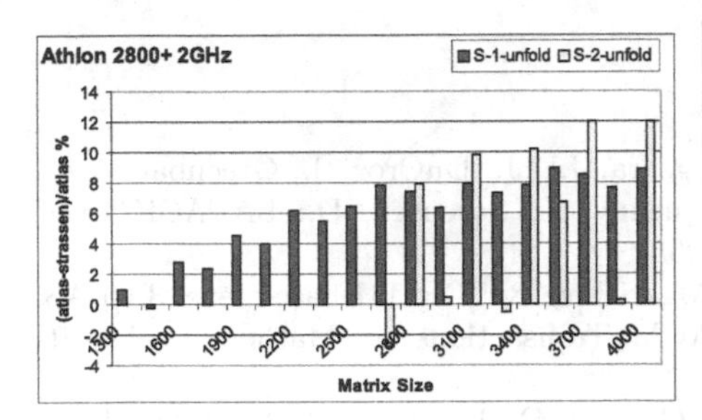

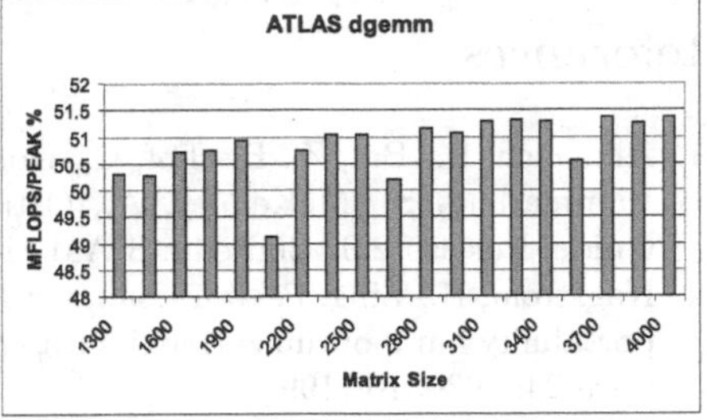

Fig. 6. ASUS A7N8X

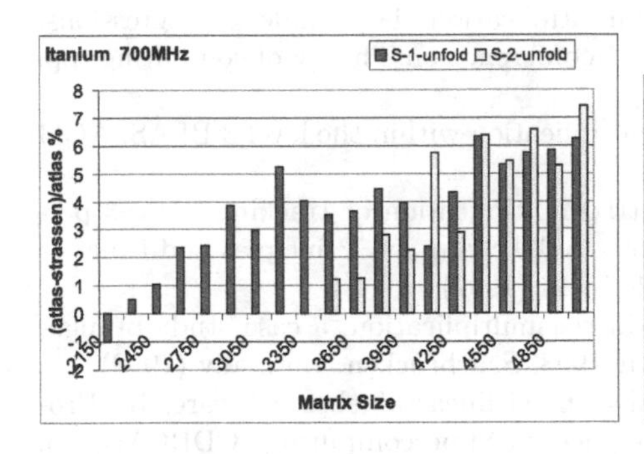

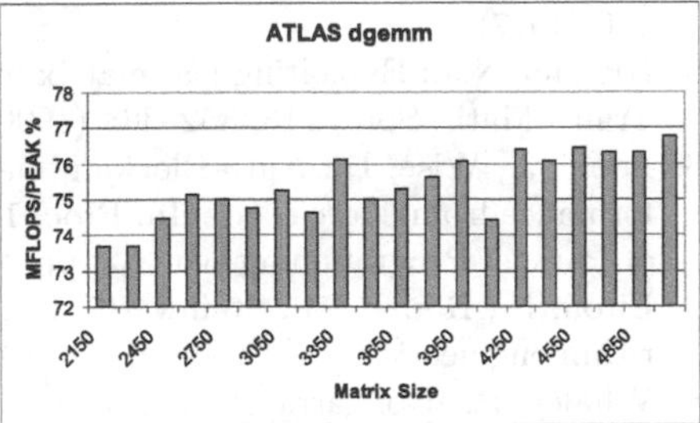

Fig. 7. Linux Itanium 2 700 MHz

performance. In fact, the execution time is what any final user cares comparing two different algorithms. However a measure of performance for ATLAS shows whether or not Strassen's algorithms improve the performance of a MM kernel which is either efficiently or poorly designed.

5 Conclusions

We have presented a practical implementation of Strassen's algorithm, which applies a recursive algorithm to exploit highly tuned MMs, such as ATLAS's.

We differ from previous approaches because we investigate a family of recursive algorithms with a balanced division process, which, in turn, makes the algorithm performance more predictable.

We have tested the performance of our approach on 7 systems with different level of recursion unfolding, and we have shown that not always Strassen is applicable. We have also shown that for modern systems the RP can be quite different and quite large.

As future work, we will investigate the implementation of a single adaptive recursive algorithm. In fact, the ideas implemented in our codes yield to a natural approach for the automatic determination of the RP for a recursive Strassen's algorithm for different systems.

References

1. Anderson, E., Bai, Z., Bischof, C., Dongarra, J.D.J., DuCroz, J., Greenbaum, A., Hammarling, S., McKenney, A., Ostrouchov, S., Sorensen, D.: LAPACK User' Guide, Release 2.0., 2nd edn. SIAM (1995)
2. Kagstrom, B., Ling, P., van Loan, C.: Algorithm 784: GEMM-based level 3 BLAS: portability and optimization issues. ACM Transactions on Mathematical Software 24, 303–316 (1998)
3. Kagstrom, B., Ling, P., van Loan, C.: GEMM-based level 3 BLAS: high-performance model implementations and performance evaluation benchmark. ACM Transactions on Mathematical Software 24, 268–302 (1998)
4. Coppersmith, D., Winograd, S.: Matrix multiplication via arithmetic progressions. In: Proceedings of the 19-th annual ACM conference on Theory of computing, pp. 1–6 (1987)
5. Higham, N.J.: Exploiting fast matrix multiplication within the level 3 BLAS. ACM Trans. Math. Softw. 16, 352–368 (1990)
6. Frens, J., Wise, D.: Auto-Blocking matrix-multiplication or tracking BLAS3 performance from source code. In: Proc. 1997 ACM Symp. on Principles and Practice of Parallel Programming, vol. 32, pp. 206–216 (1997)
7. Eiron, N., Rodeh, M., Steinwarts, I.: Matrix multiplication: a case study of algorithm engineering. In: Proceedings WAE 1998, Saarbrücken, Germany (1998)
8. Whaley, R., Dongarra, J.: Automatically tuned linear algebra software. In: Proceedings of the 1998 ACM/IEEE conference on Supercomputing (CDROM), pp. 1–27. IEEE Computer Society Press, Los Alamitos (1998)
9. Bilardi, G., D'Alberto, P., Nicolau, A.: Fractal matrix multiplication: a case study on portability of cache performance. In: Workshop on Algorithm Engineering 2001, Aarhus, Denmark (2001)
10. Goto, K., van de Geijn, R.: On reducing tlb misses in matrix multiplication. Technical Report Technical Report TR-2002-55, The University of Texas at Austin, Department of Computer Sciences (2002)
11. Demmel, J., Dongarra, J., Eijkhout, E., Fuentes, E., Petitet, E., Vuduc, V., Whaley, R., Yelick, K.: Self-Adapting linear algebra algorithms and software. In: Proceedings of the IEEE, special issue on Program Generation, Optimization, and Adaptation, vol. 93 (2005)
12. Strassen, V.: Gaussian elimination is not optimal. Numerische Mathematik 14, 354–356 (1969)

13. Brent, R.P.: Error analysis of algorithms for matrix multiplication and triangular decomposition using Winograd's identity. Numerische Mathematik 16, 145–156 (1970)
14. Brent, R.P.: Algorithms for matrix multiplication. Technical Report TR-CS-70-157, Stanford University (1970)
15. Huss-Lederman, S., Jacobson, E., Tsao, A., Turnbull, T., Johnson, J.: Implementation of Strassen's algorithm for matrix multiplication. In: Supercomputing 1996. Proceedings of the 1996 ACM/IEEE conference on Supercomputing (CDROM), p. 32. ACM Press, New York (1996)
16. Bailey, D.H., Gerguson, H.R.P.: A Strassen-Newton algorithm for high-speed parallelizable matrix inversion. In: Supercomputing 1988. Proceedings of the 1988 ACM/IEEE conference on Supercomputing, pp. 419–424. IEEE Computer Society Press, Los Alamitos (1988)
17. Bilmes, J., Asanovic, K., Chin, C., Demmel, J.: Optimizing matrix multiply using PHiPAC: a portable, high-performance, Ansi C coding methodology. In: International Conference on Supercomputing (1997)
18. Thottethodi, M., Chatterjee, S., Lebeck, A.: Tuning Strassen's matrix multiplication for memory efficiency. In: Proc. Supercomputing, Orlando, FL (1998)

Towards Generic Solver of Combinatorial Optimization Problems with Autonomous Agents in P2P Networks*

Shigeaki Tagashira, Masaya Mito, and Satoshi Fujita

Department of Information Engineering
Graduate School of Engineering, Hiroshima University
{shigeaki,mito,fujita}@se.hiroshima-u.ac.jp

Abstract. This paper proposes a new class of parallel branch-and-bound (B&B) schemes. The main idea of the scheme is to focus on the functional parallelism instead of conventional data parallelism, and to support such a heterogeneous and irregular parallelism by using a collection of autonomous agents distributed over the network. After examining several implementation issues, we describe a detail of the prototype system implemented over eight PC's connected by a network. The result of experiments conducted over the prototype system indicates that the proposed parallel processing scheme significantly improves the performance of the underlying B&B scheme by adaptively switching exploring policies adopted by each agent participating to the problem solving.

Keywords: P2P system, combinatorial optimization problem, parallel branch-and-bound, autonomous agents, winner determination problem.

1 Introduction

According to the recent advancement of network technologies, it emerges an increasingly strong requirement for high performance computing over the large-scale interconnection networks. In general, a high complexity of server procedures will limit the scalability of distributed systems, and it motivates the study of *fully distributed systems* such as grid computers and pure peer-to-peer (P2P) systems. A P2P system consists of a collection of host computers called nodes or peers [1,5], and those nodes are connected with each other by an interconnection network such as the Internet. In recent years, a lot of important services such as shared file systems and Domain Name Systems (DNS) are constructed over the P2P model, and they have been used in many application fields, such as electronic bulletin board, network auction systems, and so on.

In this paper, we propose a new application field for such fully distributed systems, and discuss several implementation issues to realize it in actual distributed environments. As the concrete target of our research, we will focus our attention to a distributed execution of parallel branch-and-bound (B&B) schemes [2,6],

* This research was partially supported by the Grant-in-Aid for Scientific Research.

J. Labarta, K. Joe, and T. Sato (Eds.): ISHPC 2005 and ALPS 2006, LNCS 4759, pp. 152–163, 2008.

which have been applied to many important fields as a generic solver to generate an optimum solution to computationally hard optimization problems in a relatively short computation time. In addition, as the concrete problem to be solved, we will focus on the Winner Determination Problem (WDP, for short) in combinatorial auctions, which has also been studied extensively in recent years to realize a fair match-making among individual customers participating to e-Markets and e-Auctions (a formal definition of WDP will be given in the next section). It should be worth noting that in most of previous work, parallel B&B schemes are designed by merely focusing on the *data parallelism* that naturally exists in exhaustive tree search schemes. Although it would be slightly complicated compared with a simple OR parallelism, such a small difference is mainly due to the mutual dependency between the upper and the lower bounds, which could be efficiently handled by adopting an appropriate broadcast mechanism within the framework of data parallelism.

In our recent paper [4], we proposed a new class of parallel B&B schemes that could naturally be applied to fully distributed systems such as P2P systems. We examined several design issues toward the implementation of a prototype of the distributed B&B system, and conducted preliminary experiments. In the current paper, we report a detail of our first prototype system. The prototype system is implemented on eight nodes, embedded with three schemes to select appropriate policies (i.e., functions) for exploring a given portion of the search tree. Several experiments were conducted to evaluate the goodness of the proposed system. The result of experiments shows that among three schemes, a dynamic one based on the feedback from participating agents exhibits a good performance compared with the other schemes including those with fixed and uniform policies.

The remainder of this paper is organized as follows. Section 2 describes necessary definitions and concepts. Section 3 describes an overview of the proposed system, and in Section 4, we propose three schemes for selecting an appropriate policy. The result of experiments is given in Section 5, and finally in Section 6, we conclude the paper with future problems.

2 Preliminaries

2.1 Problem

Let $S = \{x_1, x_2, \ldots, x_m\}$ be a set of **goods** sold by the auctioneer. In combinatorial auctions, buyers submit a set of bids to the auctioneer, where a bidding is made on a subset of goods instead of a single good as in classical auctions, and the auctioneer selects a subset of those bids in such a way to maximize the revenue of the auctioneer. A bidder of a selected bid is called a "winner" of the auction. In this paper, we assume that each bidder can submit any number of bids, and can be a winner of several bids, without loss of generality. Note that this assumption enables us to separate bids from bidders.

Let $\mathcal{B} = \{B_1, B_2, \ldots, B_n\}$ be a set of bids submitted by the bidders. Each bid $B_i \in \mathcal{B}$ is an ordered pair $\langle S_i, v_i \rangle$, where S_i is a nonempty subset of S called **bidset** (or simply "bid") and v_i is an integer referred to as the **bid value** (or simply "value"). A subset $\mathcal{B}'$ of $\mathcal{B}$ is said to be **feasible** if any two bids in the subset do not intersect with each other. In addition, a bid $B_i(\in \mathcal{B})$ is said to be feasible with respect to $\mathcal{B}'(\subseteq \mathcal{B})$ if set $\mathcal{B}' \cup \{B_i\}$ is feasible. The **revenue** of subset $\mathcal{B}'(\subseteq \mathcal{B})$, denoted by $r(\mathcal{B}')$, is the sum of bid values contained in $\mathcal{B}'$. The winner determination problem (WDP) is the problem of, given a finite set of bids $\mathcal{B}$, finding a feasible subset $\mathcal{B}'$ of $\mathcal{B}$ with a maximum revenue.

2.2 Branch-and-Bound Method

The basic idea of the branch-and-bound (B&B) method for solving WDP is described as follows. In what follows, a feasible subset of $\mathcal{B}$ is referred to as a **partial solution**. Let $\mathcal{B}'$ be a partial solution. In a **list scheduling** (LS, for short) method, all bids in $\mathcal{B}$ are first given a total ordering, and those bids are sequentially selected to be contained in the partial solution, in such a way that any two selected bids do not intersect with each other. It is known that LS is a "complete" scheme in the sense that for any instance, there exists a total ordering of bids in $\mathcal{B}$ to generate an optimum solution under the scheme. In other words, by attempting LS for all of the $n!$ permutations, we can always find an optimum solution to WDP. This idea can be realized by conducting an exhaustive search in a tree structure satisfying the following three properties: 1) each vertex of the tree corresponds to a partial solution, 2) the root of the tree corresponds to a partial solution with respect to an empty set of bids, and 3) if a vertex x corresponds to a partial solution, then a child of x corresponds to a partial solution that is obtained by greedily appending a single bid to x, which is not contained in x and does not intersect with any bid in x. Note that in such trees, any path from the root to a leaf corresponds to a permutation over a feasible subset of $\mathcal{B}$.

The B&B method performs a depth-first (or best-first) search over the above tree structure in an exhaustive manner. A trick to reduce the execution time is to "prune" subtrees if it is guaranteed that there can exist no better solutions than the currently best one on the subtrees. Such a guarantee is generally realized by evaluating an *upper bound* for each partial solution, which implies that any solution generated from the partial solution can not be better than that bound.

3 Proposed Scheme

3.1 Design Concept

In this paper, we consider a distributed execution of parallel B&B schemes. The main issues for realizing efficient B&B schemes are: 1) how to find a better partial solution quickly, and 2) how to calculate a sharp upper bound quickly. It should be worth noting that those two issues are closely related with each other. That is, the time before finding a better partial solution could be reduced by pruning as many meaningless branches as possible, and the possibility of pruning a branch at a given upper bound could generally be increased by providing a better partial solution.

In our proposed scheme, the function of each agent is designed by focusing on the following two points [4]. The first point is concerned with the upper bound; i.e., *there is a trade-off between the cost and the accuracy of calculating an upper bound.* That is, in general, we could obtain a sharper upper bound by spending more calculation time. However, since the objective of calculating a sharp upper bound is to prune meaningless branches as much as possible, in this context, this problem could be regarded as a simple YES/NO problem (i.e., the result is whether we could prune a subtree or not). Hence in order to realize a pruning with a low calculation cost, we should prepare several procedures for calculating upper bounds, and should apply them sequentially in the order of lower calculation cost. The next point we have to consider is about the lower bound; i.e., *there is a dilemma in determining the expansion order of partial solutions.* In general, a bid order that quickly derives a better lower bound could not derive partial solutions that are unlikely to be pruned by upper bounds. Such a dilemma could particularly be observed when we could determine the bid selecting order for each partial solution independently. More concretely, a subtree that could not be efficiently pruned is a branch whose upper bound could not be accurately calculated, which is generally different from a branch that is likely to derive a better lower bound.

The above problems are due to the fact that *we have to make a selection from several candidates*, and thus, could be relaxed by introducing the notion of parallel execution. First, as for the trade-off on the upper bound, we could resolve it by preparing (at least) two kinds of agents, i.e., basic agent and advanced agent, and by executing those agents concurrently, in such a way that: 1) basic agents calculate the initial upper bound for each partial solution, and 2) advanced agents try to improve the initial upper bound for several selected partial solutions. In the selection of partial solutions, for example, we could take into account the success rate of previously executed pruning operation, the level of partial solution in the search tree, and the expected calculation time for the improvement. In realizing such a mechanism in distributed environments with no centralized control, we have to design each agent in such a way that those selections are conducted in a heuristic and autonomous manner.

On the other hand, as for the dilemma on the way of expansion, we could resolve the problem by expanding several branches simultaneously, while we have to introduce a kind of strategies since the amount of available resources is finite. One possible strategy is to use the following two phase control; i.e., initially, a quick improvement of the lower bound is given a higher priority, and after observing the saturation of the improvement speed, it switches to another heuristic in which a branch that is unlikely to be pruned is given a higher priority.

3.2 Upper Bound Agents

In the prototype system that will be described in the next section, the following two types of upper bound (UB) policies are prepared, and each policy continuously tries to improve the upper bound on partial solutions.

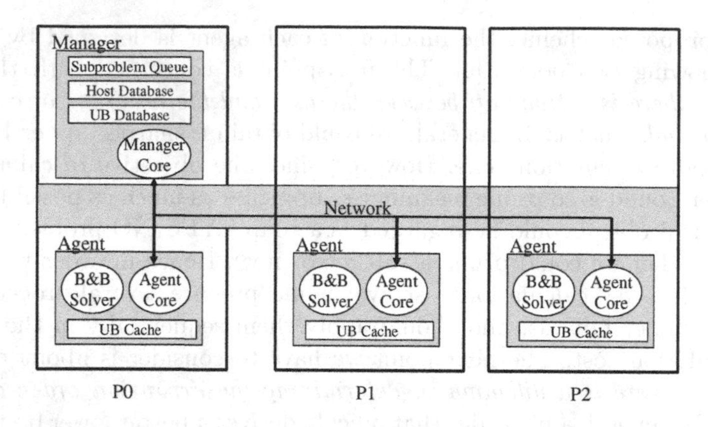

Fig. 1. System configuration of the first prototype system

Type TRV. An agent of this type calculates an upper bound on the revenue that could be derived from the partial solution, based on a heuristic estimation of expected revenue [7]. Given feasible set of bids $\mathcal{B}'(\subseteq \mathcal{B})$, let us define an *estimated revenue* with respect to $\mathcal{B}'$ as $h(\mathcal{B}') \stackrel{\text{def}}{=} \sum_{x \in S'} \left\{ \max_{S_j \ni x, S_j \cap (S-S')=\emptyset} \left(\frac{v_i}{|S_i|} \right) \right\}$ where S' is the set of goods that are not contained in bids in $\mathcal{B}'$. By using the calculated value, an upper bound on the partial solution $\mathcal{B}'$ is calculated as $r(\mathcal{B}') + h(\mathcal{B}')$ since it has already selected bids with total revenue $r(\mathcal{B}')$.

Type LP. An agent of this type calculates an upper bound for each partial solution $\mathcal{B}'$ by solving a linear programming (LP) defined as follows:

$$\begin{aligned} \text{maximize} \quad & \sum_{B_i \notin \mathcal{B}', S_i \cap (S-S')=\emptyset} v_i p_i \\ \text{subject to} \quad & \sum_{B_i \notin \mathcal{B}', S_i \cap (S-S_i)=\emptyset} a_{ji} p_i \leq 1 \text{ for all } j \in S \end{aligned}$$

where S' is the set of goods that are not contained in bids in $\mathcal{B}'$, $0 \leq p_i \leq 1$ and $a_{ji} = 1$ if $x_j \in S_i$ and 0 otherwise. In the above formulation, several bids containing the same good in common can be selected in a fractional manner with fraction p_i, as long as the sum of such fractions does not exceed one. Note that an optimum solution to the above LP is not smaller than an optimum solution to the original problem.

3.3 System Configuration

Figure 1 illustrates the configuration of our first prototype system. In the system, each node is associated with its own **agent**, and a **manager** is associated to the node who owns a problem to be solved (i.e., we invoke one manager for each instance to be solved). A manager consists of a host database, an upper bound database, a subproblem queue (s-queue, for short), and the manager core, which realizes the communication with agents. Although the given instance is handled by the manage in a centralized manner, we are planning to modify it in such a way that the information on the given instance is shared by the participants in a distributed manner (as in Distributed Hash Table, for example). On the other hand, each agent consists of an upper bound cache, B&B solver, and an agent core, which realizes the management of the agent and the communication to the manager.

The basic procedure for solving a given problem over the system is as follows.

- (Initialization) After receiving a problem to be solved, the manager partitions it into several subproblems, and puts them into the s-queue in an appropriate order. In the default setting of our prototype system, the number of subproblems is fixed to 64, and they are sorted in a non-increasing order of a trivial upper bound of the root vertex. On the other hand, each agent who wants to participates to the problem solving registers itself to the host database by sending a message to the manager, and receives the specification of the problem with a set of possible policies from the manager.
- To acquire a subproblem to be solved, each agent sends a request message to the manager. Upon receiving the message, the manager sends back a subproblem contained in the head of the s-queue, in such a way that no two agents with the same policy receive the same subproblem.
- Each agent tries to solve the received subproblem by using the B&B solver with its own policy, and after obtaining a solution to the subproblem, it immediately replies it to the manager. Upon receiving the solution, the manager removes the corresponding subproblem from the s-queue, and notifies the fact to all nodes that have been assigned the same subproblem to interrupt their execution. The interrupted node discards the corresponding subproblem, and tries to acquire the next subproblem from the manager.
- Whenever it finds a better lower bound, the agent informs the fact to the manager, which will be broadcast to all agents participating to the system.
- The agent stops the execution when the s-queue contains no subproblem corresponding to its policy. In addition, when the s-queue becomes empty, the manager terminates its operation after returning the solution to the user.

As an option, we could set up the system such that the upper bounds on subproblems are shared by all participants in the following manner. In the prototype system, upper bounds calculated by each node is locally stored in the upper bound cache with a bit string representing unexplored set of bids. When the option is selected, each node periodically uploads the (differential) contents of this cache to the manager, which will be downloaded by the other agents via a periodical reference to the manager.

4 Switching of Policy

In this section, we propose three schemes to select an appropriate policy for solving a given subproblem in each node. The first two schemes are static ones and the last scheme is a dynamic one. In the static schemes, we adopt LP as the "background" policy, and selectively apply TRV to the instances that could be efficiently solved with it. This approach is based on an observation on the result of our preliminary experiments, in which we compared two policies in terms of the number of solved instances within 1000 seconds and the average computation time for those solved instances (a concrete description of the examined 108 instances will be given in Section 5.1). The result of the preliminary experiments is summarized as follows: 1) the number of solved instances is 76 for LP and 55 for TRV, and 2) the average computation time is 52.2 sec for LP and 49.0 sec for TRV. Thus, we can conclude that LP could solve more instances than TRV, whereas it takes a slightly longer time than TRV. In other words, LP is a good selection for general instances, but for several specific instances, TRV beats the performance of LP. In fact, in the experiment, we discovered an instance that could be solved by TRV in 110 times faster than LP.

4.1 First Static Scheme

The first static scheme is based on an evaluation of the trade-off between TRV and LP. In general, TRV should explore a larger space than LP due to the inaccuracy of the derived upper bound. Thus, if the time required for the additional exploration is shorter than the time required for the calculation of an upper bound in LP, then TRV should be selected instead of LP. The size of the additional space and the time required for the calculation of an upper bound could be approximated by measuring the accuracy of the upper bound at the root vertex of the search tree and its concrete calculation time. Let $UB(p)$ denote the upper bound calculated at the root vertex with policy p, and $T(p)$ denote the calculation time. Then, the first static selection scheme selects policy TRV if and only if $UB(\mathsf{LP})/UB(\mathsf{TRV}) > \theta_{upper}$ and $T(\mathsf{LP})/T(\mathsf{TRV}) > \theta_{time}$ for some thresholds θ_{upper} and θ_{time}.

4.2 Second Static Scheme

In the preliminary experiment, we found that an instance could efficiently be solved by TRV if it has an (optimum) solution consisting of small number of bids. More concretely, TRV is better than LP if the solution contains less than ten bids, and the superiority of the policy will be decreased as increasing the number of bids contained in the solution. Let $\tilde{m}$ be an estimated number of bids contained in an optimum solution, a formal definition of which will be given later. According to the above observations, we designed the second static scheme as follows: The scheme selects TRV with probability 1 if $\tilde{m} \leq \theta_1$, and selects TRV with probability 0.5 if $\theta_1 < \tilde{m} \leq \theta_2$, where θ_1 and θ_2 are predetermined thresholds. The estimation of value $\tilde{m}$ could be conducted as follows. Let d be an average number of bids conflicting with a bid. For each $i \geq 1$, let a_i be an integer defined as follows:

$$a_i \stackrel{\text{def}}{=} \begin{cases} n & \text{if } i = 1 \\ \left(1 - \frac{d}{n}\right) a_{i-1} - 1 & \text{otherwise} \end{cases}$$

Note that this formula provides an estimation of the number of selectable bids after selecting the first i bids, in the following sense: By selecting the $(i-1)$st bid, among a_{i-1} remaining candidates, $d \times a_{i-1}/n$ bids become unselectable in expectation, which reduces the number of candidate bids from a_{i-1} to $a_{i-1}-(d\times a_{i-1}/n+1)$. Note that this estimation assumes no locality on the selection of bids. By solving the above recurrence, we have $a_i = \left(1-\frac{d}{n}\right)^i \left(n+\frac{n}{d}\right) - \frac{n}{d}$, and since $\tilde{m}$ is equal to the smallest integer k such that $a_k = 0$, we have $\tilde{m} = \frac{\log 1/(d+1)}{\log(n-d)/n}$.

4.3 Dynamic Scheme

Next, we propose a dynamic scheme which adaptively selects an appropriate policy according to the characteristics of the subproblems having been solved by the agents. A concrete procedure for the proposed scheme is described as follows:

- The manager initializes local counters c_L and c_T to one.
- When a subproblem is sent out to an agent, the manager determines the policy of the agent concerned with the subproblem to LP with probability $c_L/(c_L+c_T)$ and to TRV with probability $c_T/(c_L+c_T)$.
- If it receives a solution from an agent with policy LP (resp. TRV), the manager increments its local counter c_L (resp. c_T) by one.

It should be worth noting that the dynamic scheme could adapt itself to various kinds of instances and could add new policies relatively easily (i.e., by simply preparing a local counter corresponding to the new policy), although it would take a relatively long time to converge to an appropriate policy.

5 Evaluation

5.1 Environment

To evaluate the goodness of the proposed scheme, we conducted several experiments. The experiments was conducted over eight PCs with the following specifications: CPU: Pentium4 3.2G, Memory: 2G, Network: 1GbE, Operating System: FreeBSD 5.3. Two thresholds in the first static scheme are fixed as $\theta_{upper} = 0.7$ and $\theta_{time} = 400$; and those in the second static scheme are fixed as $\theta_1 = 10$ and $\theta_2 = 30$. The option on the sharing of upper bounds is not selected, and in all experiments, we fixed the timeout of each run to 1000 seconds.

As the benchmark set, we adopted three benchmark suites, Random, Uniform, and Locality, a brief description of which could be stated as follows [8]:

Random: Each bid B_i is constructed by selecting k_i goods from S without replacement and by assigning a bid value v_i to it, where k_i is a random value drawn from $\{1, 2, \ldots, m'\}$, where $m' \leq m$, and v_i is a random value drawn from $\{1, 2, \ldots, Max\}$.

Uniform: Modify Random in such a way that the size of each bid is fixed to a constant k.

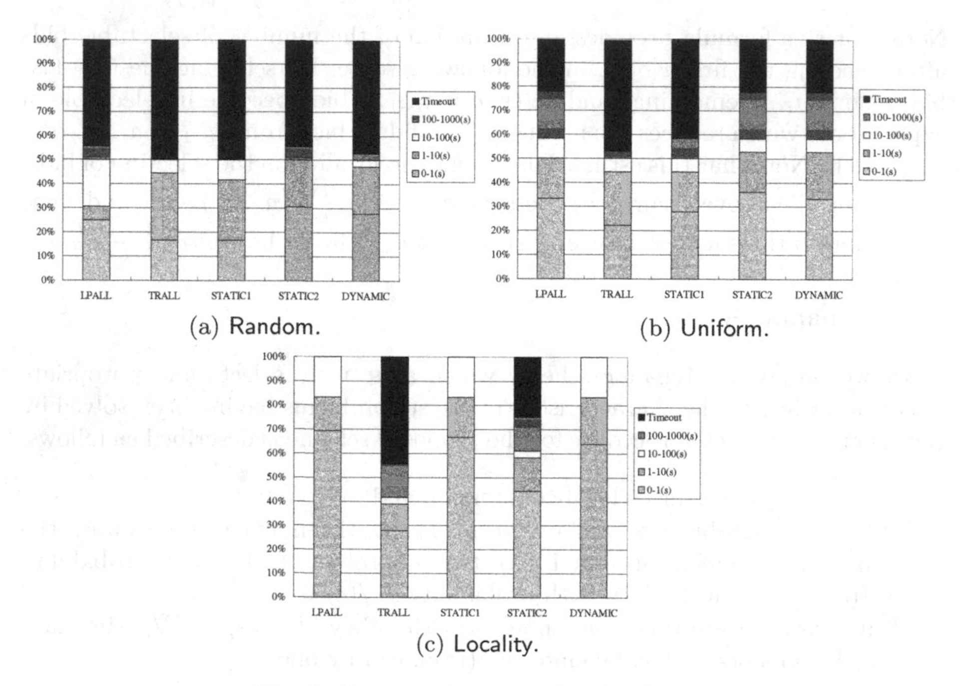

Fig. 2. Computation time of each scheme

Locality: Modify Random in such a way that the goods selected by the bids follow a locality according to the Zipf's first law [9][1].

For each suite, we varied the average number of goods in a bid as 3, 6, and 9; the total number of goods as 50, 200, and 350; and the number of bids is 100, 300, 500, and 700; i.e., we prepared $3 \times 3 \times 4 = 36$ instances for each suite.

As an initial assessment, we evaluate the performance of the system by assigning LP to all agents (LPALL) or by assigning TRV to all agents (TRVALL). In the experiment, we compare the number of instances (among 36 instances for each suite) for which a given scheme exhibits the better performance than the other scheme. The result is summarized as follows: for LPALL, such number of instances for Random, Uniform, and Locality are 19, 9, and 30, respectively, and for TRVALL, they are 9, 11, and 6, respectively. Thus, we can conclude that LPALL is better than TRVALL for Uniform or Locality, and TRVALL is better than LPALL for Random.

5.2 Results

Figure 2 compares the distribution of the computation time for each scheme. From the figure, we could observe that:

[1] In the law, the ith element w_i in S is associated with a probability $p_i = 1/(i \times Q)$, where $Q \stackrel{\text{def}}{=} \sum_{i=1}^{|S|}(1/i)$. Note that $\sum_{i=1}^{|S|} p_i = 1$ holds by definition.

Table 1. The number of instances for which an optimum policy was selected

	static 1		static 2	
	success	failed	success	failed
Random	10	10	19	1
Uniform	13	15	27	1
Locality	31	5	14	22

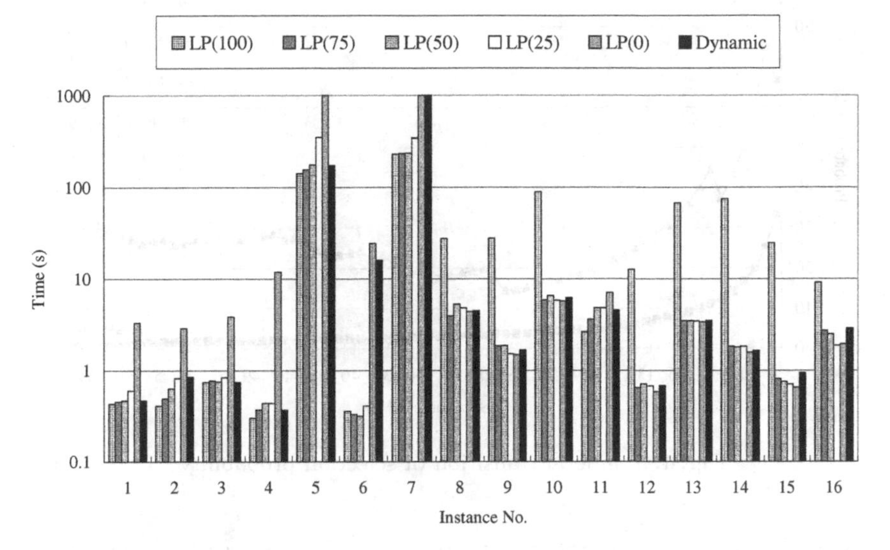

Fig. 3. Result for dynamic scheme

- The goodness of two static schemes depends on the class of given instances; e.g., the first scheme is not good for Uniform and Random, and the second scheme is not good for Locality.
- The performance of the dynamic scheme is relatively stable independent of the class of instances; e.g., it is as good as LPALL for Uniform and Locality, and it is as good as TRVALL for Random.

In order to examine the goodness of the static schemes in more detail, we verified the appropriateness of the policy selected by the schemes. Table 1 shows the number of instances for which an appropriate policy is successfully selected by the schemes, where the selection with probability 0.5 in the second scheme is considered to be successful. As is shown in the table, the first static scheme could not select an appropriate policy except for Locality, which is due to the inaccuracy of estimation at the root of the search tree. Conversely, the goodness of the scheme for Locality is due to the tightness of the estimation at the root, which makes the scheme to always select LP to generate a good solution.

On the other hand, the second static scheme could not select an appropriate policy for Locality, although it could make an appropriate selection for Uniform and Random. The badness for Locality is due to the inaccuracy of the number

Table 2. Probability of selection in Dynamic

Instance No.	1	2	3	4	5	6	7	8	9	10	11	12	13	14	15	16
Probability [%]	99	13	93	97	97	4	31	96	1	1	39	6	6	1	60	1

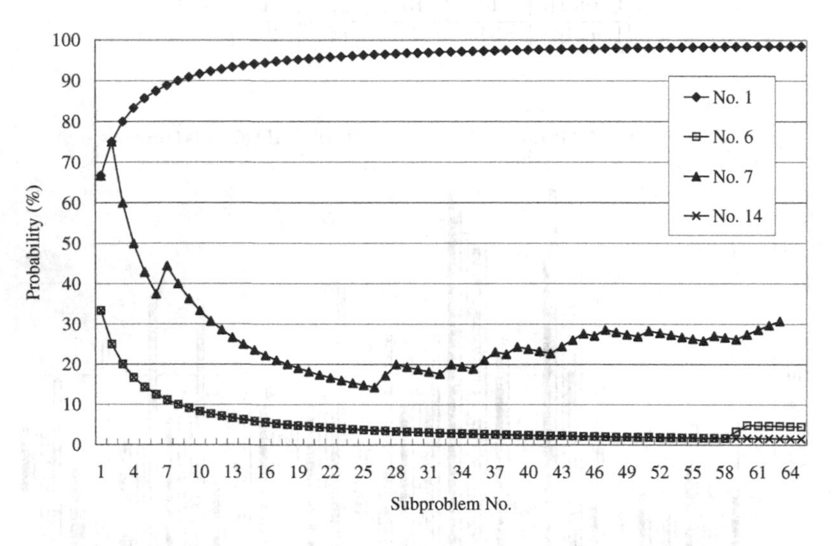

Fig. 4. Temporal transition of selection probability

of bids in an optimum solution for the instances contained in Locality. In fact, the actual number of bids contained in an optimum solution is 26,65, 28.54, and 25.83 for Random, Uniform, and Locality, respectively, whereas the estimated number of bids are 30.68, 21.56, and 7.28, respectively.

In contrast to the static schemes, the dynamic scheme exhibits a stable performance for all classes of the instances. An advantage of the dynamic scheme is that it allows each agent to have its own policy. In order to examine the impact of this advantage to the performance, we compare the performance of the dynamic scheme with schemes with a fixed percentage of LP policy, where the percentage is varied as 100%, 75%, 50%, 25%, and 0%. Figure 3 illustrates the result for Random (the case of LP with x% is denoted as LP(x) in the figure), and the probability of selecting LP is summarized for each instance in Table 2. Note that this figure omits instances that could be solved in one second, and of course, omits instances that could not be solved within 1000 seconds. From the figure, we could observe the superiority of the dynamic scheme. Although there are several instances for which the other schemes outperform the dynamic scheme (e.g., instances No. 6 and 7), we could conclude that the dynamic scheme could select an appropriate policy.

To see this in more detail, we evaluated how the probability of selecting LP transits during the execution of the scheme. Figure 4 summarizes the result for bad instances (No. 6 and 7) and good instances (No. 1 and 14). The horizontal axis of the figure represents the sequence number of subproblems sent out to the agents, and the vertical axis represents the probability of selecting LP for the subproblem with a given sequence number. As is shown in the figure, for good instances, it converges to an appropriate policy, although it takes relatively long time before convergence. On the other hand, for bad instances, it often makes a wrong decision, which causes unnecessary vibration of the probability. By a detailed analysis of such instances, we found that such instances could be effectively solved by appropriately solving a specific subproblem, and to efficiently solve them, we have to select a specific policy, whereas the most of the remaining subproblems do not strongly rely on the selection of the policy. An improvement of the proposed dynamic scheme in such a way to in corporate with such a situation is left as a future problem.

6 Concluding Remarks

In this paper, we proposed a new class of parallel B&B schemes, and described a detail of our first prototype system implemented over eight PC's. The result of experiments conducted over the prototype system indicates that the dynamic selection of exploring policies could improve the overall performance of the underlying B&B scheme, and it could efficiently support the functional parallelism residing in the original B&B scheme. We are now extending the prototype system in such a way that the search space submitted by each node is shared by all nodes participating to the problem solving.

References

1. Clarke, I., Sandberg, O., Wiley, B., Hong, T.W.: Freenet: A Distributed Anonymous Information Storage and Retrieval System. In: ICSI workshop on Design Issues in Anonymity and Unobsevability, pp. 46–66 (July 2000)
2. Clausen, J., Perregaard, M.: On the best search strategy in parallel branch-and-bound: Best-First Search versus Lazy Depth-First Search. Annals of Operations Research 90(1), 1–17 (1999)
3. Fujishima, Y., Leyton-Brown, K., Shoham, Y.: Taming the computational complexity of combinatorial auctions: Optimal and approximate approaches. In: Proc. IJCAI 1999, pp. 548–553 (1999)
4. Fujita, S., Tagashira, S., Qiao, C., Mito, M.: Distributed Branch-and-Bound Scheme for Solving the Winner Determination Problem in Combinatorial Auctions. In: Proc. AINA 2005, March 28–30, 2005, Tamkang University, Taiwan (2005)
5. Gnutella, `http://gnutella.wego.com/`
6. Portable Parallel Branch-and-Bound Library, `http://wwwcs.upb.de/fachbereich/AG/monien/SOFTWARE/PPBB/ppbblib.html`
7. Sakurai, Y., Yokoo, M., Kamei, K.: An efficient approximate algorithm for winner determination in combinatorial auctions. In: ACM Conf. on Electronic Commerce, pp. 30–37 (2000)
8. Sandholm, T.: Algorithm for optimal winner determination in combinatorial auctions. Artificial Intelligence 135(1-2), 1–54 (2002)
9. Zipf, G.K.: Human Behavior and Principle of Least Effort. Addison-Wesley, Boston (1949)

New Evaluation Index of Incomplete Cholesky Preconditioning Effect

Takeshi Iwashita[1] and Masaaki Shimasaki[2]

[1] Academic Center for Computing and Media Studies, Kyoto University, Yoshida-Honmachi Sakyo-ku, Kyoto, Japan
iwashita@media.kyoto-u.ac.jp
[2] Graduate School of Engineering, Kyoto University, Kyoto Daigaku Katsura, Nishikyo-ku, Kyoto, Japan
simasaki@kuee.kyoto-u.ac.jp

Abstract. In Incomplete LU (ILU) preconditioning, orderings often affect the effect of preconditioning. The authors recently proposed a simple evaluation way for orderings in the ILU preconditioning technique. The present paper introduces the evaluation method in unstructured analyses in which the effect of preconditioning is not easily estimated. The evaluation index, which has a simple relationship with the matrix norm of the remainder matrix, is easily computed without additional memory requirement. The computational cost of the index is trivial in the total iterative solution process. The effectiveness of the method is examined by numerical tests using coefficient matrix data from the Matrix Market, a finite-difference analysis of Poisson equation, and a 3-d electromagnetic field analysis.

Keywords: Iterative method, ILU preconditioning, Convergence rate, Convergence evaluation index, Ordering.

1 Introduction

The ILU factorization preconditioning is one of the most popular preconditioning techniques for Krylov subspace iterative methods [1]. In this preconditioning method, it is well-known that the preconditioning effect is significantly affected by the ordering of the unknowns. Furthermore, since reordering technique has been a well-known parallelization way of ILU preconditioning [2], the relationship between ordering and convergence has been intensively investigated by several researchers [3,4].

Most of previous investigations on orderings were mainly performed in finite difference analyses. In these studies, the early and important work was done by Duff and Meurant [3]. They indicated the significant effect of orderings on convergence in ILU-preconditioned iterative solvers, and then proposed a use of the norm of a remainder matrix for evaluation of orderings. The remainder matrix $\boldsymbol{R}$ is given by $\boldsymbol{R} = \boldsymbol{M} - \boldsymbol{A}$, where $\boldsymbol{M}$ is the preconditioning matrix and $\boldsymbol{A}$ is the coefficient matrix. While the evaluation method has been confirmed

J. Labarta, K. Joe, and T. Sato (Eds.): ISHPC 2005 and ALPS 2006, LNCS 4759, pp. 164–175, 2008.

by various numerical tests, the remainder matrix has been commonly used as a tool for examining convergence in subsequent research works [5,6]. Following Duff and Meurant's work, Doi, Lichnewsky and Washio performed a series of works paying a special attention to "incompatible nodes" [5,7]. In this research, they proposed an evaluation index for orderings, which is called "incompatibility ratio". The incompatibility ratio, which has a unique value for a fixed ordering, is easily calculated. Next, the authors recently proposed a new evaluation index for orderings, which is called "S.R.I. (Simple Remainder Index)" [8]. This evaluation index estimates the effects of all nodes including non-incompatible nodes, which are not evaluated in incompatibility ratio.

In contrast to studies on ordering of nodes in finite difference analyses, the effect of orderings in unstructured analyses has been rarely discussed. In the unstructured analysis, the effect of individual property of a problem is not trivial. Therefore, it is not easy to evaluate orderings in a simple way. But, we have tried to propose an evaluation method for orderings in unstructured analyses by permitting a small range of errors. Based on the results of Duff and Meurant's research, we finally proposed a new evaluation index "P.R.I. (Precise Remainder Index)" [8]. This index has a simple relationship with the remainder matrix norm in a special case. Both of computational cost and memory requirement for computing the index are trivial in the total iterative solution process. In this paper, we examine our evaluation method in four numerical tests.

2 ILU Preconditioning

This paper deals with a following n-dimensional linear system of equations:

$$\boldsymbol{A}\boldsymbol{u} = \boldsymbol{f}. \tag{1}$$

While the coefficient matrix is symmetric and positive or semi-positive definite, our evaluation method is explained in a general format including a non-symmetric coefficient matrix case.

When the linear system of equations is solved by means of iterative methods, preconditioning techniques are often used [4]. In this technique, the linear system is transformed into the preconditioned system

$$(\boldsymbol{K}_1^{-1}\boldsymbol{A}\boldsymbol{K}_2^{-1})(\boldsymbol{K}_2\boldsymbol{u}) = \boldsymbol{K}_1^{-1}\boldsymbol{f}, \tag{2}$$

which accelerates the convergence of a basic iterative method. The matrix $\boldsymbol{M} = \boldsymbol{K}_1\boldsymbol{K}_2$ is called a preconditioner matrix. In ILU(0) preconditioning, the preconditioning matrix is given by

$$\boldsymbol{M} = \boldsymbol{L}\boldsymbol{D}^{-1}\boldsymbol{U}, \tag{3}$$

where $\boldsymbol{L}$, $\boldsymbol{D}$ and $\boldsymbol{U}$ are a lower triangular matrix, a diagonal matrix and a upper triangular matrix, respectively. These matrices $\boldsymbol{L}$, $\boldsymbol{D}$ and $\boldsymbol{U}$ are derived from ILU factorization of the coefficient matrix $\boldsymbol{A}$ as follows:

$$\boldsymbol{A} = (\boldsymbol{L}\boldsymbol{D}^{-1}\boldsymbol{U}) - \boldsymbol{R}_{ilu}, \tag{4}$$

where R_{ilu} is the matrix of the elements that are dropped during the incomplete factorization.

In the ILU preconditioned iterative method algorithm, the transformation (2) is not performed explicitly, and the preconditioning step is given by the solution of a linear system:

$$(LD^{-1}U)z = x. \tag{5}$$

Since the preconditioner matrices depend on ordering of the unknowns, the preconditioning effect is also affected by the ordering. Moreover, the degree of parallelism in the solution (5) depends on the ordering.

3 New Evaluation Index for Orderings

3.1 Remainder Matrix

A typical way to evaluate the preconditioning effect is through checking the condition number or the eigenvalues of the preconditioned coefficient matrix. But, the computational cost of computing eigenvalues is generally high. Thus, the following method proposed by Duff and Meurant is widely used for evaluating preconditioning effect.

Duff and Meurant's method [3]
The effect of ILU preconditioning is evaluated by the norm of the *remainder matrix* R,

$$R = M - A. \tag{6}$$

A smaller norm of R results in better convergence.

In the ILU preconditioning case, it holds

$$R = R_{ilu}. \tag{7}$$

Duff and Meurant used the Frobenius norm of the remainder matrix in their research [3]. Their numerical tests of finite difference analyses confirmed the effectiveness of their method. However, when the ILU preconditioning technique is applied to a general sparse coefficient matrix, the additional memory requirement and computational cost for computing the Frobenius norm of the remainder matrix are not small compared with the iterative solution process. Therefore, another practical method is required for evaluating preconditioning effects in unstructured analyses.

3.2 P.R.I.

In this subsection, we describe a new evaluation index, which is called P.R.I. (Precise Remainder Index) [8]. In order to construct a new evaluation index, we consider the remainder matrix in ILU preconditioning. The remainder matrix R

```
R = O
I_rp = 0
for I = 1 to n − 1
for J = I + 1 to n
for K = I + 1 to n
if ã_{J,I} ≠ 0 & ã_{I,K} ≠ 0 & ã_{J,K} ≠ 0 then
ã_{J,K} = ã_{J,K} − ã_{J,I} * ã_{I,K}/ã_{I,I}
// (ILU factorization)
endif
if ã_{J,I} ≠ 0 & ã_{I,K} ≠ 0 & ã_{J,K} = 0 then
r_{J,K} = r_{J,K} + ã_{J,I} * ã_{I,K}/ã_{I,I}
// (Computation of R)
I_rp = I_rp + |ã_{J,I} * ã_{I,K}/ã_{I,I}|
// (Computation of P.R.I.)
endif
end for
end for
end for
```

Fig. 1. Algorithm of ILU factorization with computing remainder matrix and P.R.I.

can be computed by using the algorithm associated with ILU factorization as is shown in Fig. 1. When we focus on an arbitrary element $r_{J,K}$ in the algorithm, the element is updated several times depending on the non-zero element pattern of the coefficient matrix. Since these update quantities are different to each other, storing all dropped fill-ins is necessary for computing the exact remainder matrix entries. Thus, in the P.R.I. evaluation, we use a summation of the absolute values of the updates of the remainder matrix. The algorithm of calculating the P.R.I. value I_{rp} is shown in Fig. 1. The additional memory requirement for the calculation is for only one variable. Moreover, its computational cost is generally much smaller than the iteration process. When the coefficient matrix $\boldsymbol{A}$ has the same signs in all diagonal entries and also has the same signs in all non-diagonal entries, a simple relationship between the remainder matrix norm and the P.R.I. is given as follows:

$$||\boldsymbol{R}||_A = I_{rp}, \tag{8}$$

where $||\boldsymbol{R}||_A$ is defined as a sum of the absolute values of all entries of $\boldsymbol{R}$, and is given by

$$||\boldsymbol{R}||_A = \sum_{IJ} |r_{IJ}|. \tag{9}$$

The operator $||\cdot||_A$ satisfies the definition of the matrix norm shown in the reference 9.

3.3 P.R.I. for Variants of ILU Preconditioning

In practical analyses, ILU factorization can fail due to pivot breakdown. One of remedies for the breakdown is modification of the coefficient matrix before the factorization, for example, diagonal sifts [10] [11]. Let $\boldsymbol{A}_{\Delta m}$ be a modification term, then the modified factorization is given by

$$\boldsymbol{A} + \boldsymbol{A}_{\Delta m} = (\boldsymbol{L}_m \boldsymbol{D}_m^{-1} \boldsymbol{U}_m) - \boldsymbol{R}_{ilum}. \tag{10}$$

For this factorization, the remainder matrix is written as follows:

$$\boldsymbol{R} = \boldsymbol{R}_{ilum} + \boldsymbol{A}_{\Delta m}. \tag{11}$$

Here, we define the P.R.I. value for the factorization (10) as

$$I_{rpm} = I_{rp0} + ||\boldsymbol{A}_{\Delta m}||_A. \tag{12}$$

In (12), I_{rp0} is the I_{rp} value calculated in the algorithm shown in Fig. 1.

Next, we consider ILU(l) preconditioning, in which some fill-ins are allowed. Although some fill-ins are admitted in the factorization, the preconditioning matrix is given by (4) as well as in ILU(0) preconditioning case. Therefore, the P.R.I. value for ILU(l) preconditioning can be defined by a summation of the absolute values of the dropped fill-ins, and it is given by the algorithm in Fig. 1.

4 Numerical Results

In this paper, we present four numerical results: 1) Two coefficient matrices data downloaded from the Matrix Market [12], 2) Finite difference analysis of Poisson equation, 3) Electromagnetic field analysis (finite edge-element method). Since the coefficient matrices are symmetric, the ICCG (Incomplete Cholesky Conjugate Gradient) method is used. The convergence criterion of the iterative method is given by $||\boldsymbol{r}||_2/||\boldsymbol{b}||_2 < 10^{-7}$ where $\boldsymbol{r}$ and $\boldsymbol{b}$ are the residual vector and the right-hand side vector, respectively. The correlation between the P.R.I. and convergence is examined by using 51 random orderings.

4.1 Matrix Market Data (1)

Fig. 2 shows the relationship between the P.R.I. and the number of iterations in the numerical test for CYLSHELL S1RMQ4M1 data from the Matrix Market. Fig. 3 depicts the convergence behavior. The coefficient matrix arises from a finite element structure analysis with shell type elements. The information about the coefficient matrix is shown in Table 1. Since strong correlation between the iteration count and the P.R.I. value is shown in Fig. 2, our evaluation index gives good estimates of convergence. The correlation coefficient reaches 0.86.

4.2 Matrix Market Data (2)

This subsection describes the numerical result using CYLSHELL S3RMT3M1 data from the Matrix Market. This data set is derived from a finite element analysis with an unstructured triangular mesh. Fig. 4 depicts the relationship

Table 1. Coefficient matrix of S1RMQ4M1 from Matrix Market

Number of unknowns	5489
Number of entries	143300*2
Number of nonzero entries	262411
Band width	192
Max eigenvalue	6.874×10^{5}
Min eigenvalue	3.80×10^{-1}
Condition number	1.81×10^{6}

Table 2. Coefficient matrix of S3RMT3M1 from Matrix Market

Number of unknowns	5489
Number of entries	112505*2
Number of nonzero entries	217669
Band width	192
Max eigenvalue	9.67×10^{3}
Min eigenvalue	3.90×10^{-7}
Condition number	2.48×10^{10}

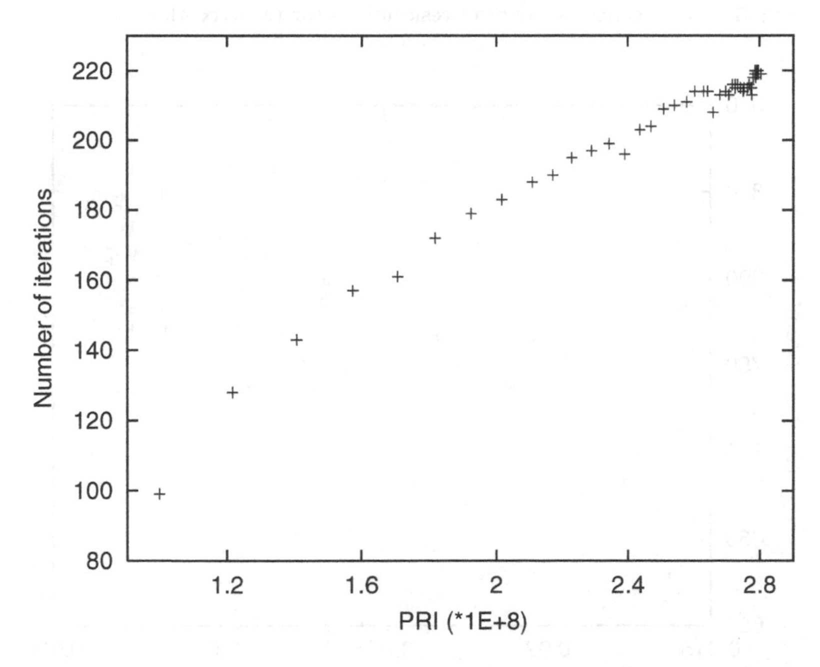

Fig. 2. Relationship between number of iterations and P.R.I. (Matrix Market data 1)

between the number of iterations and the P.R.I. values. In this numerical test, many spikes are observed in the convergence behavior as is shown in Fig. 5. Therefore, the convergence estimates by the P.R.I. include some errors especially

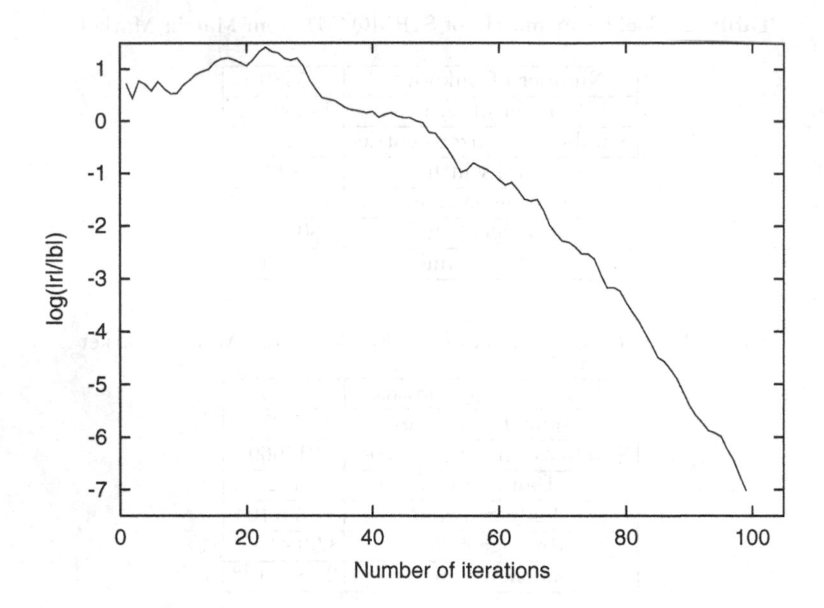

Fig. 3. Convergence behavior of residual vector (Matrix Market data 1)

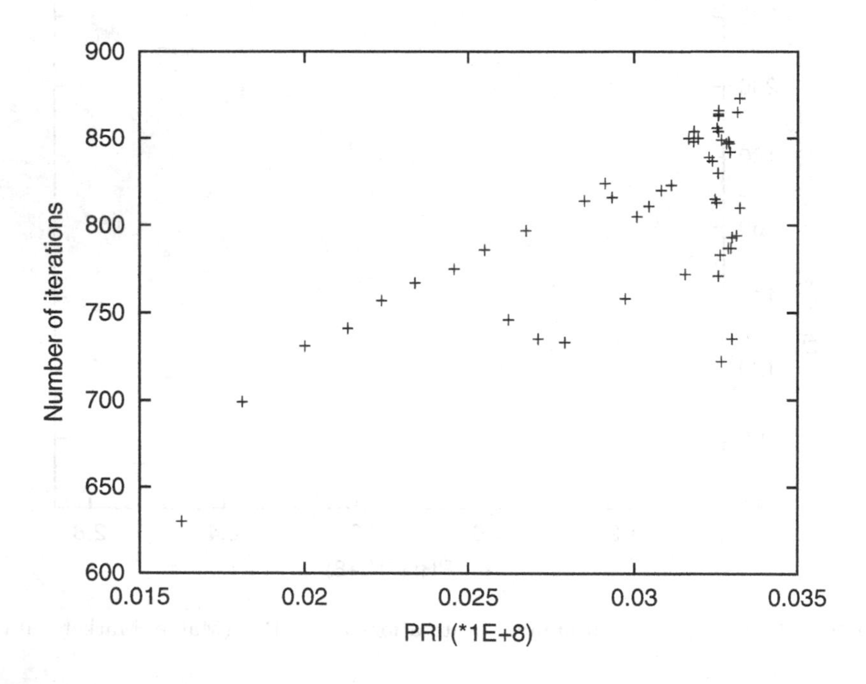

Fig. 4. Relationship between number of iterations and P.R.I. (Matrix Market data 2)

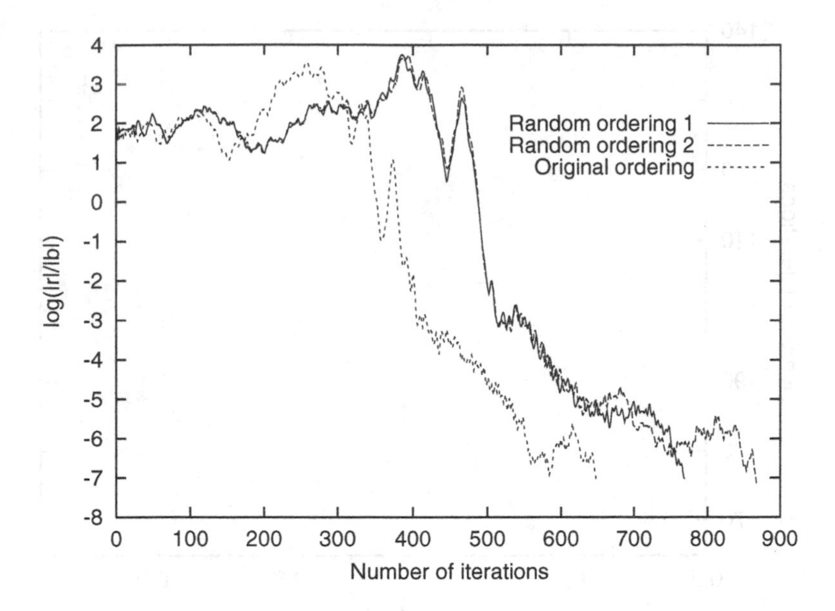

Fig. 5. Convergence behavior of residual vector (Matrix Market data 2)

when orderings have high P.R.I. values. However, a high correlation coefficient value, which is over 0.8, is obtained in the whole random ordering test.

4.3 Finite Difference Analysis of Poisson Equation

This subsection deals with a following two-dimensional Poisson equation with the Dirichlet boundary condition:

$$
\begin{gathered}
-\nabla \cdot (\kappa \nabla u) = f \\
in\ \Omega(0,1) \times (0,1) \\
u(x,y) = 0\ on\ \delta\Omega \\
if\ (\tfrac{1}{4} \le x \le \tfrac{3}{4} \& \tfrac{1}{4} \le y \le \tfrac{3}{4})\ then \\
\kappa = 100.0 \\
else \quad \kappa = 1.0.
\end{gathered} \tag{13}
$$

The equation is solved by means of 5-point finite difference scheme. The grid size is 100×100 and the original ordering is lexicographical ordering. The right-hand side f is given by 0.5 sin $(i_d + 1)$ where i_d is the node number.

Fig. 6 shows the relationship between the number of iterations and the P.R.I. values. Fig. 7 plots the convergence behavior when the original ordering is used. Fig. 6 indicates that the preconditioning effect can be estimated by the P.R.I. value.

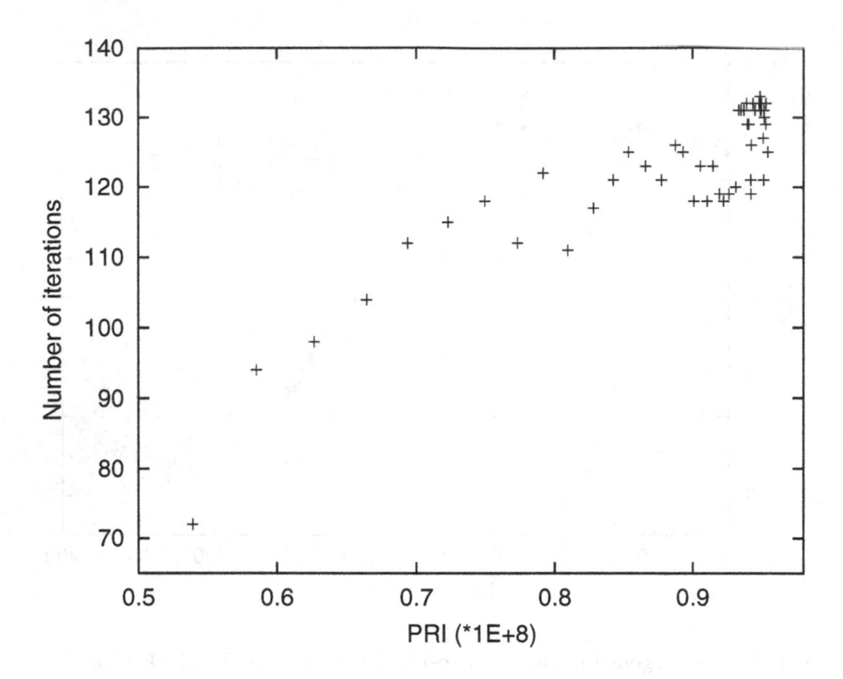

Fig. 6. Relationship between number of iterations and P.R.I. (Finite difference analysis of Poisson equation)

4.4 Three-Dimensional Eddy-Current Analysis (Finite Edge-Element Analysis)

In this subsection, we use test data of a three-dimensional eddy-current analysis. The basic equation, which is derived from quasi static form of Maxwell equations, is given as follows:

$$\nabla \times (\nu \nabla \times \boldsymbol{A}_m) = -\sigma \frac{\partial \boldsymbol{A}_m}{\partial t} + \boldsymbol{J}_0, \tag{14}$$

where $\boldsymbol{A}_m$, ν, σ, and J_0 are the magnetic vector potential, the magnetic reluctivity, the electrical conductivity, and the exiting current, respectively. The basic equation is solved by using the Galerkin method with $A-$formulation and the backward time difference method [13].

In the present analysis, we use the IEEJ standard benchmark model of 3-D eddy current analyses [14]. The analyzed model is discretized by first-order brick-type edge elements. Table 3 lists the discretization data. The original linear system is assembled with the lexicographical ordering. Since the analyzed model includes non-conductive region, the coefficient matrix results in a semi-positive definite matrix. The relationship between the P.R.I. and the preconditioning effect is examined in one time step of the time-dependent calculation. Fig. 8 shows the relationship between the number of iterations and the P.R.I. values

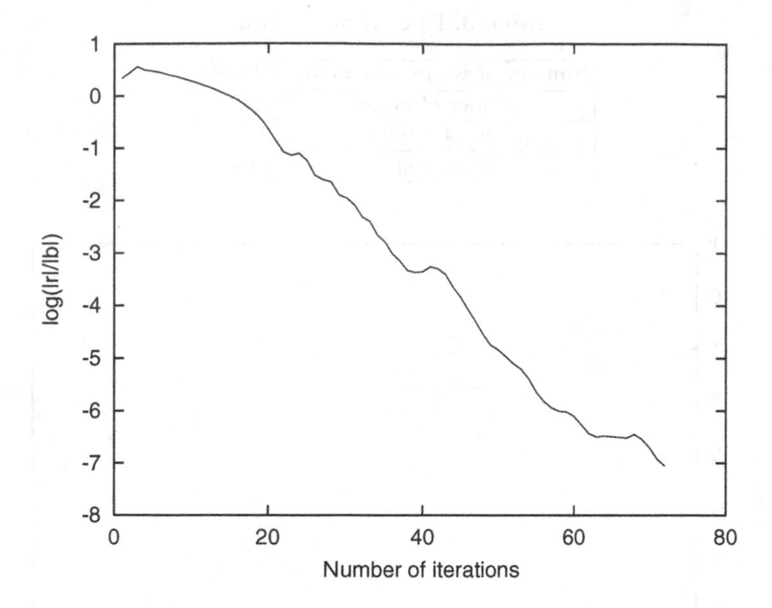

Fig. 7. Convergence behavior of residual vector (Finite difference analysis of Poisson equation)

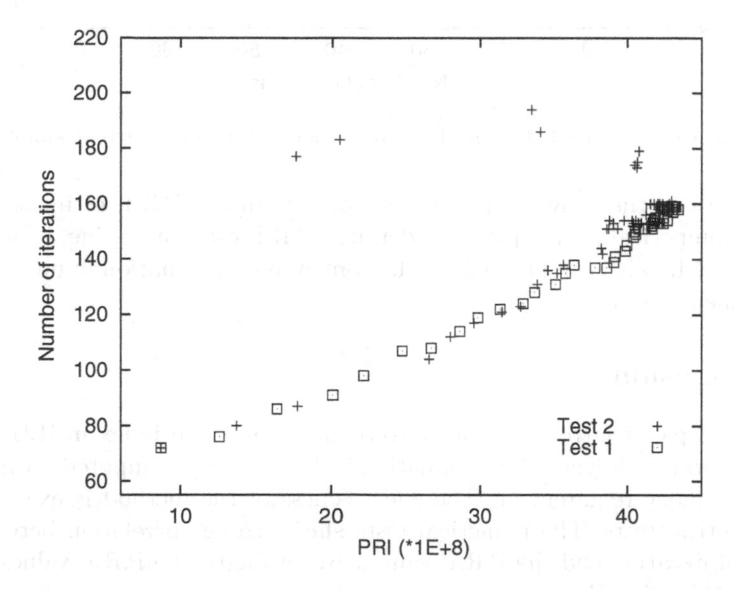

Fig. 8. Relationship between number of iterations and P.R.I. (3-d eddy-current analysis)

in two tests. Two different sets of random orderings are used in the numerical test. Fig. 9 depicts the convergence behavior. In Test 1, we can observe strong correlation between the P.R.I. and the convergence. On the other hand, in some

Table 3. Discretization data

Number of volume elements	327680
Number of nodes	342225
Number of unknowns	1011920
Time step	1 msec

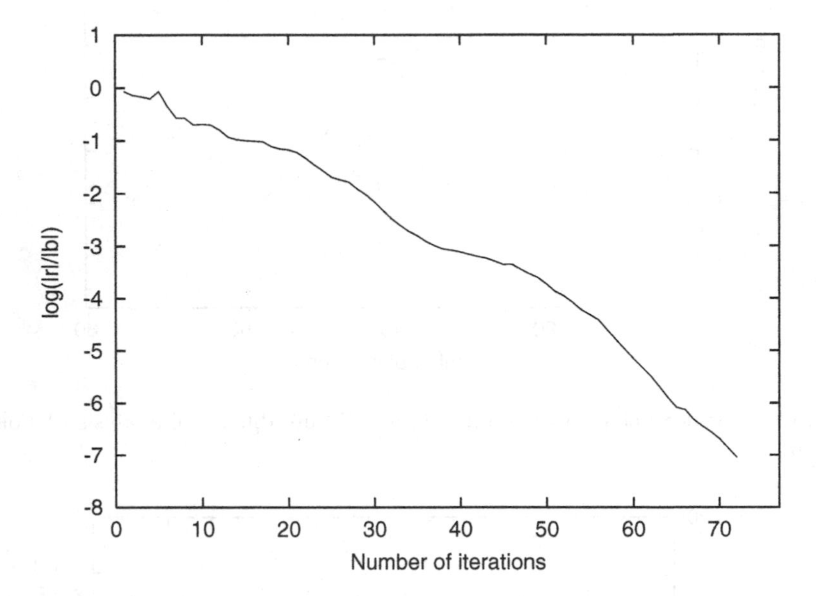

Fig. 9. Convergence behavior of residual vector (3-d eddy-current analysis)

cases of Test 2, the convergence rate is worse than the P.R.I. estimation. But, since two numerical results plot the identical P.R.I. estimation line, it is implied that the P.R.I. value can be used for the convergence estimation in unstructured finite element analyses.

5 Conclusion

The present paper introduces the convergence evaluation index in ILU preconditioned iterative solvers. The evaluation index is easily computed without additional memory requirement. The effectiveness of the method is examined by four numerical tests. The numerical tests show strong correlation between the number of iteration and the P.R.I. values. Accordingly, the P.R.I. values can be used for estimating ILU preconditioning effects.

References

1. Meijerink, J., van der Vorst, H.A.: An Iterative Solution Method for Linear Systems of Which the Coefficient Matrix Is a Symmetric M-matrix. Mathematics of Computation 31, 148–162 (1977)

2. van der Vorst, H.A., Chan, T.F.: Parallel Preconditioning for Sparse Linear Equations. ZAMM. Z. angew. Math. Mech. 76, 167–170 (1996)
3. Duff, I.S., Meurant, G.A.: The Effect of Ordering on Preconditioned Conjugate Gradients. BIT 29, 635–657 (1989)
4. Saad, Y.: Iterative Methods for Sparse Linear Systems, 2nd edn. SIAM, Philadelphia, PA (2003)
5. Doi, S., Washio, T.: Ordering Strategies and Related Techniques to Overcome the Trade-off Between Parallelism and Convergence in Incomplete Factorization. Parallel Computing 25, 1995–2014 (1999)
6. Eijkhout, V.: Analysis of parallel incomplete point factorizations. Linear Algebra Appl. 154–156, 723–740 (1991)
7. Doi, S., Lichnewsky, A.: A Graph-Theory Approach for Analyzing the Effects of Ordering on ILU Preconditioning. INRIA report 1452 (1991)
8. Iwashita, T., Nakanishi, Y., Shimasaki, M.: Comparison criteria for parallel orderings in ILU preconditioning. SIAM J. Sci. Comput. 26(4), 1234–1260 (2005)
9. Demmel, J.: Applied Numerical Linear Algebra. SIAM, Philadelphia, PA (1997)
10. Fujiwara, K., Nakata, T., Ohashi, H.: Improvement of Convergence Characteristic of ICCG Method for the $A - \phi$ Method Using Edge Elements. IEEE Trans. Magn. 32(3), 804–807 (1996)
11. Benzi, M., Tuma, M.: A robust incomplete factorization preconditioner for positive definite matrices. Numer. Linear Algebra Appl. 10, 385–400 (2003)
12. `http://math.nist.gov/MatrixMarket/`
13. Iwashita, T., Sokabe, R., Mifune, T., Shimasaki, M.: Three-dimensional Finite Brick-Type Edge-Element Eddy Current Analysis Using Parallelized Linear-System Solvers. In: PARELEC 2000. Proc. Int. Conf. Parallel Computing in Electrical Engineering, pp. 203–207 (2000)
14. Nakata, T., Takahashi, N., Imai, T., Muramatsu, K.: Comparison of Various Methods of Analysis and Finite Elements in 3-D Magnetic Field Analysis. IEEE Trans. Magn. 27, 4073–4076 (1991)

T-Map: A Topological Approach to Visual Exploration of Time-Varying Volume Data

Issei Fujishiro[1], Rieko Otsuka[2], Shigeo Takahashi[3], and Yuriko Takeshima[1]

[1] Tohoku University, 2-1-1 Katahira, Aoba-ku, Sendai 980-8577, Japan
[2] Hitachi, Ltd., 1-280 Higashi-koigakubo, Kokubunji-shi, Tokyo 185-8601, Japan
[3] The University of Tokyo, 5-1-5 Kashiwanoha, Kashiwa-shi, Chiba 277-8561, Japan

Abstract. The rapid advance in high performance computing and measurement technologies has recently made it possible to produce a stupendous amount of time-varying volume datasets in a variety of disciplines. However, there exist a few known visual exploration tools that allow us to investigate the core of their complex dynamics effectively. In this paper, our previous approach to topological volume skeletonization is extended to capture the topological features of large-scale time-varying volume datasets. A visual exploration tool, termed T-map, is presented, where pixel-oriented information visualization techniques are deployed so that the user can identify partial 4D spatiotemporal domains with characteristic changes in a topological sense, prior to detailed and comprehensible volume visualization. A case study with datasets from atomic collision research is performed to illustrate the feasibility of the proposed tool.

1 Introduction

At this moment, it is routinely performed in a large variety of disciplines to visualize a single regular-grid volume dataset with 256^3–512^3 voxels interactively. This is due to intensive *volume visualization* R&D efforts over the past two decades, including software optimization [1]; the advent of special hardware devices [2]; and the development of commodity PC cluster systems with enhanced volume graphics cards [3] and GPUs [4].

The recent rapid increase in the performance of computing and measurement environments as well as in opportunities to use the global information infrastructure has made it possible to investigate the correlations among model parameters in complex scientific/engineering problems. A tremendous number of runs/observations produce a multi-dimensional array of large-scale *time-varying (4D)* volume datasets, which may not fit into even the virtual memory space of modern computing facilities [5]. This background motivates us to develop a novel scheme which can provide the user with the "serendipity" [6] by managing and visualizing those datasets effectively.

As an initial step for establishing a promising scheme, this paper extends our previous concept of *volume data mining (VDM)* [7] to propose a scheme, termed *T-map (Topology-map)*, which takes full advantage of knowledge in the field of *differential topology* to capture the topological structure of a 4D volume. This

J. Labarta, K. Joe, and T. Sato (Eds.): ISHPC 2005 and ALPS 2006, LNCS 4759, pp. 176–190, 2008.

is one of the main characteristics to differentiate the T-map from existing 4D volume analysis tools, which rely primarily on geometric tracking of isosurface components [8,9].

In the T-map, analyzed results are represented compressively into artificially-designed spaces, termed *index spaces*, where the user is allowed to carry out drill-down manipulations interactively, along with comprehensible rendering techniques [7,10] to effectively visualize a limited number of volumes chosen from regions of interest in the index spaces.

The remainder of this paper is organized as follows. As preliminaries, Sect. 2 describes two kinds of topological analysis methods for isosurfaces and snapshot volumes. Section 3 presents the T-map with a special focus on how to design its index spaces and related drill-down manipulations. In Sect. 4, the feasibility of the T-map will be illustrated with an application to a practical problem from atomic collision research. Lastly Sect. 5 concludes the paper with several remarks on related future extensions.

2 Preliminaries

This section is devoted to an overview of topological tools that allow the user to capture both local and global features of isosurfaces and snapshot volumes. Also a basic idea is shown in which a topological index is employed to quantify the resultant graph structures.

Before proceeding to the overview, let us make clear the basic structural relationships among several related concepts. In general, an n-dimensional (n-D) scalar field can be viewed as an $(n+1)$-D hyper-surface ($n \geq 3$). For example, consider a time-varying 3D field $F(x, y, z, t)$, called *4D volume* interchangeably, where (x, y, z) denotes an ordinary 3D spatial coordinate, and t time. The behavior of the field F can be delineated through an analysis of the 5D hyper-surface:

$$w = F(x, y, z, t), \tag{1}$$

where F is a single-valued function of (x, y, z) and t. Fixing a time $t = T$ yields a *snapshot volume* from the 4D volume:

$$V(T) \equiv \{ (x, y, z, w) \in \mathbb{R}^4 | w = f(x, y, z) = F(x, y, z, T)\}. \tag{2}$$

Similarly, if we fix a field value $w = W$ as well as time $t = T$, an isosurface $IS(W, T)$ can be implicitly specified:

$$IS(W, T) \equiv \{ (x, y, z) \in \mathbb{R}^3 | W = f(x, y, z) = F(x, y, z, T)\}.$$

2.1 Reeb Graph (RG)

The *Reeb graph* (*RG*) representation was originally imported into the computer graphics community by Shinagawa and Kunii [11] to describe surface topology. Consider a kind of RG, which represents the topological skeleton of a surface in 3D space along a predefined axis for the *height function*. In this setting, the node indicates a critical point, such as a *peak*, a *pass*, or a *pit*, and its link a set of connected cross sectional contours. Figure 1(a) shows a depressed sphere on the left, and its corresponding RG on the right.

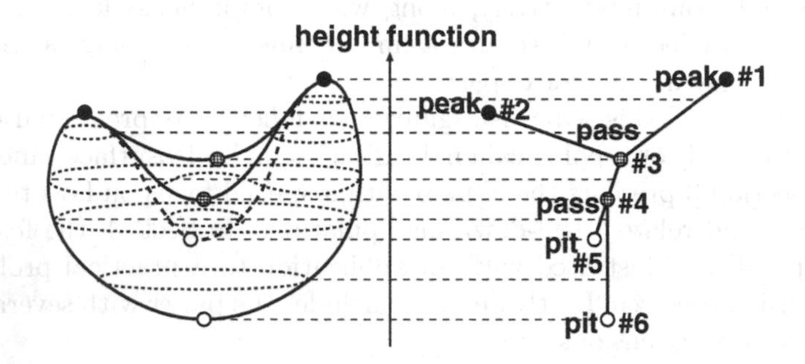

Fig. 1. An example of Reeb graph construction. A depressed sphere (left) and its Reeb graph (right).

Takahashi, et al. [12] proposed a robust geometric algorithm to extract an RG from a surface described in the form of 3D mesh. While the algorithm effectively constructs the RG from a polygonal surface, applicable surface types are limited to topological spheres. The algorithm was extended in [13] so as to characterize isosurfaces of arbitrary topological type, which may arise in real world datasets. In this paper, we are going to use the extended RG extraction algorithm.

In general, changing the directions of the height axis may give different RGs. This implies that the RG is an unambiguous, but non-unique representation of the surface topology. Recently, a geometric transformation-invariant RG representation using a *geodesic* distance was proposed in [14]. Such an RG is tailored for topological matching of 3D shapes, but seems to be rather insensitive to the difference in local shape of isosurfaces.

2.2 Volume Skeleton Tree (VST)

In order to capture the topological skeleton of a volumetric field, we extended the RG representation described in Sect. 2.1 to what we call *Volume Skeleton Tree* (*VST*) [10].

Cutting a volume dataset at different field values will produce the topological changes of isosurfaces, where isosurfaces are split or merged. A volumetric critical point will appear at a contact point between such splitting or merging isosurfaces. More specifically, critical points of a volume are classified into four types: a maximum (index 3), a saddle (index 2), a saddle (index 1), and a minimum (index 0). In this paper, the symbols C_3, C_2, C_1, and C_0 will be used to represent the above critical points, where each subscript denotes the index of the corresponding critical point.

Figure 2 depicts isosurface behaviors around the critical points. At a maximum, a new topological sphere appears in 3D space. Conversely, an existing sphere disappears at a minimum. At a saddle of index 2, two isosurfaces are merged while an existing isosurface is split at a saddle of index 1. For such saddles, in particular, the topological changes become more complicated when we take into account embeddings of isosurfaces in 3D space. Figure 2 classifies such changes in isosurface shapes depending on their embeddings in 3D space.

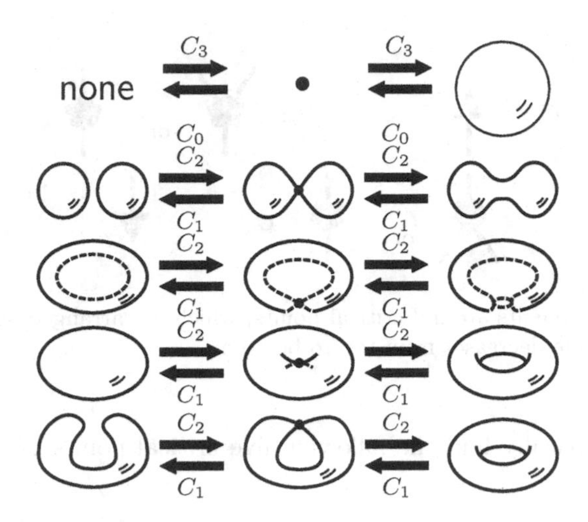

Fig. 2. Classification of isosurface changes at volumetric critical points

By adding a virtual minimum to the volume function (2), a volume dataset becomes a topological 3D sphere S^3. The Euler–Poincaré formula states that the critical points of a 3D sphere S^3 must satisfy the following condition:

$$\#\{C_3\} - \#\{C_2\} + \#\{C_1\} - \#\{C_0\} = 0, \tag{3}$$

where $\#\{C_i\}$ represents the number of critical points of index i.

VST is a level-set graph that represents the splitting and merging of isosurfaces with respect to the field value, and effectively delineates the transition of such evolving isosurfaces. The node of the graph represents a critical point, and its link one connected component of a varying isosurface in between. The VST is constructed by assembling components as shown in Fig. 3, where each component contains one critical point.

For constructing a VST, we have developed a *topological volume skeletonization* algorithm [10] as a 3D extension of the RG extraction algorithm [12]. In this paper, we employed a robust and efficient version of the algorithm [15].

Here, we take as an example, the following analytic volume function, which is hereafter referred to as *Bread Basket* (*BB*):

$$
\begin{array}{rcl}
w = f(x,y,z) & = & 4c^2\left((x-R)^2+(z-R)^2\right) \\
& & -\left((x-R)^2+y^2+(z-R)^2+c^2-d^2\right)^2 \\
& & +4c^2\left((x+R)^2+(z+R)^2\right) \\
& & -\left((x+R)^2+y^2+(z+R)^2+c^2-d^2\right)^2,
\end{array} \tag{4}
$$

where $c > d > 0,\ c^2 + d^2 \geq 6R^2$.

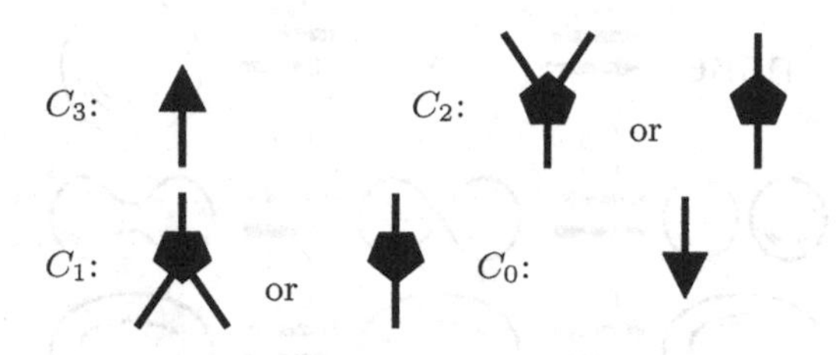

Fig. 3. VST components around critical points, which are arranged so that the corresponding field value decreases from top to bottom

We can analytically find the following five critical points of the BB volume (4):

$$
\begin{array}{rl}
p_{1,2}: & \left(\mp\sqrt{\frac{c^2+d^2-2R^2}{2}}, 0, \pm\sqrt{\frac{c^2+d^2-2R^2}{2}}\right), \\
p_{3,4}: & \left(\mp\sqrt{\frac{c^2+d^2-6R^2}{2}}, 0, \mp\sqrt{\frac{c^2+d^2-6R^2}{2}}\right), \\
p_5: & (0,0,0),
\end{array} \tag{5}
$$

and their corresponding *critical field values*:

$$
\begin{array}{rl}
f(p_{1,2}) &= 8c^2d^2, \\
f(p_{3,4}) &= 8c^2d^2 - 16R^2(c^2+d^2-4R^2), \\
f(p_5) &= -2(c^2-d^2)^2 + 8R^2(c^2+d^2-R^2).
\end{array} \tag{6}
$$

Isosurfaces of the BB volume function (4) grow as shown in Fig. 4(a), and the resultant VST is shown in Fig. 4(b). Here, two *croissant*-shaped isosurfaces appear at p_1 and p_2 first, and then they meet at p_3 and p_4 to become a torus (*donut*). The hole of the torus is filled at p_5, which makes the isosurface a sphere (*loaf*) again. This isosurface evolution suggests that the critical points p_1 and p_2 are of index 3 (i.e. C_3), p_3 and p_4 are of index 2 (i.e. C_2), and p_5 is of index 1 (i.e. C_1). Furthermore, p_6 represents the virtual minimum. Clearly, these critical points satisfies the Euler–Poincaré formula:

$$2 - 2 + 1 - 1 = 0.$$

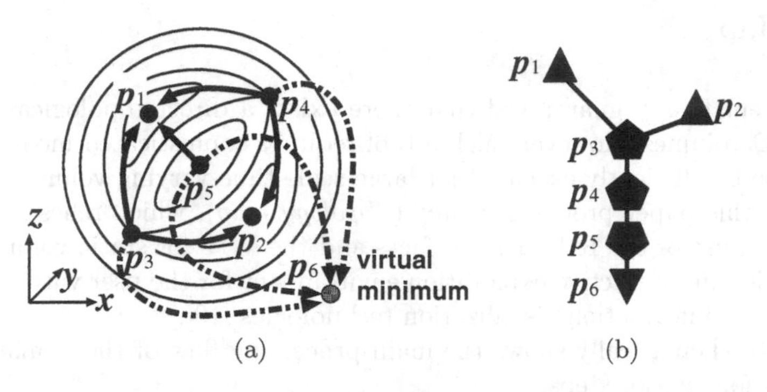

Fig. 4. Bread Basket Volume. (a) Critical points of the function (4); (b) Corresponding VST.

2.3 Quantizing RG and VST

In order to compare the topological equivalence between a pair of isosurfaces extracted from a 3D volume and a pair of snapshot volumes extracted from a 4D volume, we need to examine the homogeneity of their RGs and VSTs. To this end, we borrowed a *topological index* (*T-index*) from molecular similarity research [16]. Indeed, we can derive a pair of characteristic quantities from the adjacent and distance matrices for an RG or a VST. Suppose we have the characteristic polynomial $\phi_C(g)$ and the distance polynomial $\phi_D(g)$ of the graph g:

$$\phi_C(g) = \sum_{i=0}^{n} a_i x^i$$
$$\phi_D(g) = \sum_{i=0}^{n} a'_i x^i$$

Then, the two characteristic quantities $I_C(g)$ and $I_D(g)$ are defined as follows:

$$I_C(g) = \sum_{i=0}^{n} |a_i|$$
$$I_D(g) = \sum_{i=0}^{n} |a'_i|.$$

The pair $(I_C(g), I_D(g))$ has been commonly used in computational chemistry to identify a molecule structure qualitatively, with few exceptions [17].

Recall herein the Reeb graph RG_{DS} for the depressed sphere shown in Fig. 1(a). A simple eigensystem analysis shows that the characteristic polynomial and distance polynomial are $\phi_C(x) = x^6 - 5x^4 + 4x^2$ and $\phi_D(x) = x^6 - 65x^4 - 296x^3 - 504x^2 - 352x - 80$. Therefore, $(I_C(RG_{DS}), I_D(RG_{DS}))$ equals $(10, 1298)$. The eigensystem analysis tool we use here is the Householder method [18].

It should be noted here that an isosurface extracted from a volume consists of more than one connected component. In this case, we must compute an index pair for each of the connected components, and adopt the list of index pairs to represent the RGs for the isosurface. The VST always has a single connected component.

3 T-Map

It is mathematically guaranteed that there exists a direct topological analysis tool for 4D volumes. However, such a tool requires sophisticated memory management to be efficiently executed for large-scale time-varying volume datasets. Therefore, this paper proposes *T-map* (*Topology map*), which relies heavily on an extended use of the RG for isosurfaces and the VST for static volume fields, and provides an interactive exploration environment for the user with the aid of pixel-oriented information visualization technologies [19].

Figure 5 schematically shows the main processing flow of the T-map, which consists of four major steps:

Step 1: Partition 4D volume dataset
A given large-scale 4D volume dataset is partitioned into a series of snapshot volumes. If needed, each of snapshot volumes can be downsized.

Step 2: Extract topological structures
For each of the snapshot volumes generated in Step 1, the corresponding topological features are analyzed in terms of RGs and VSTs.

Step 3: Visualize topological changes
The results of topological feature analysis performed in Step 2 are abstracted onto a pixel map form, termed *index space*, through a pixel-oriented information visualization [19]. The design of pixel glyphs strongly owes the topological graph quantification described in Sect. 2.3. The user is allowed to modify the space interactively by using a set of predefined manipulations.

Step 4: Locate significant spatiotemporal regions
With the aid of the 4D volume drill-down mechanisms, the user can visually identify a small set of snapshot volumes within a spatiotemporal subregion on the index space, and visualize the subregion with judiciously designed visualization techniques. In this paper, opacity/color transfer functions are accentuated semi-automatically for comprehensible volume visualization [7,10].

As in Sect. 2.2, the following 4D analytic volume function, referred to as *Time-series Bread Basket* (*TBB*) volume, will be used throughout this section:

$$\begin{aligned} w &= g(x, y, z, t) \\ &= 16c^2t(1-t)((x-R)^2 + (z-R)^2) \\ &\quad -((x-R)^2 + y^2 + (z-R)^2 + c^2 - d^2)^2 \\ &\quad +16c^2t(1-t)((x+R)^2 + (z+R)^2) \\ &\quad -((x+R)^2 + y^2 + (z+R)^2 + c^2 - d^2)^2, \end{aligned} \tag{7}$$

$$\text{where } c > d > 0,\ c^2 + d^2 \geq 6R^2, t \in [0, 1].$$

Note that $g(x, y, z, 0.5)$ reduces to $f(x, y, z)$, which has already been analyzed in (4). In the following experiments, the parameter values were set as follows: $c = 0.60, d = 0.50, R = 0.20$. We sampled the TBB function into a series of 65 snapshot volumes commonly with $64 \times 64 \times 64$ voxels.

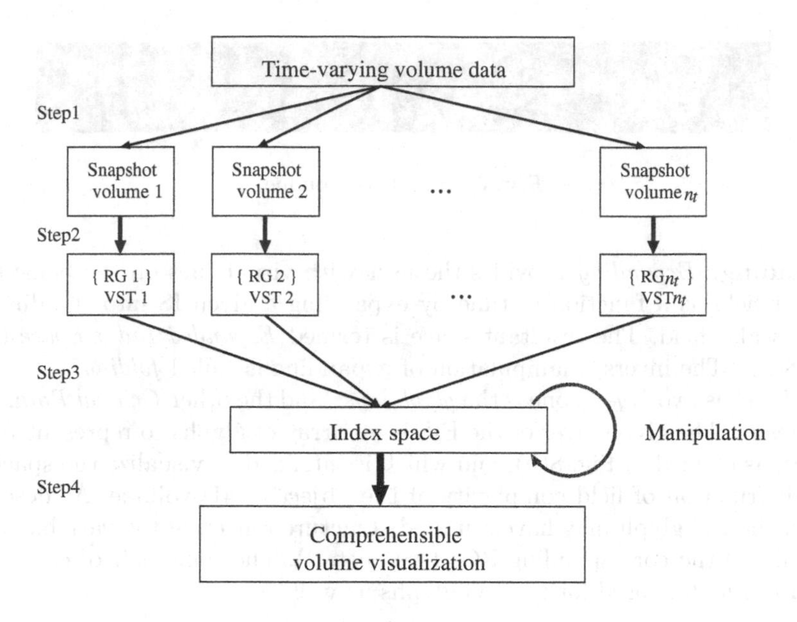

Fig. 5. Overview of processing flow in T-map.

3.1 Index Space

Index space (*IS*) is defined as a time-series of pixel glyphs, each of which represents the T-index of the VST extracted from the corresponding snapshot volume. Figure 6 shows the VST glyph design adopted here.

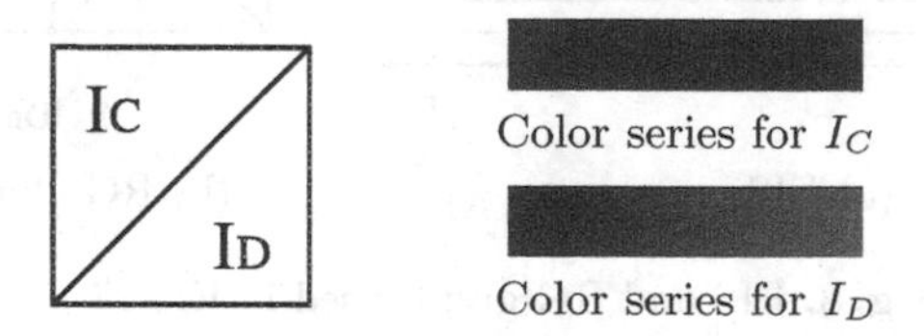

Fig. 6. Definition of VST glyph in T-map IS.

Figure 7 visualizes the IS of the TBB volume. A glance at the IS leads to a fact that the TBB volume is symmetric about the time $t = 0.5$. More detailed analysis can be realized through interactive use of progressive drill-down manipulations, which will be presented in the next subsection.

3.2 Progressive Drill-Down in T-Map

A set of T-map data manipulations is proposed herein to realize an interactive environment for drilling down volumetric data progressively.

Fig. 7. IS of TBB volume

Expanding. *Expanding* provides the user with visual cues on the behavior of a scalar field as a function of time by expanding a given IS into the direction of the scalar field. The resultant space is termed *Expanded Index Space* (*EIS*) (Fig. 8(a)). The inverse manipulation of expanding is called *folding*.

An EIS has two *layers*: one is the *glyph layer* and the other *Critical Point layer* (*CP layer*). The glyph layer of the EIS is an array of glyphs to represent RG T-indices, as defined in Fig. 8(b), and which is intended to visualize the space v.s. time distribution of field complexity of the objective 4D volume. As described before, the RG glyph may have a nested structure reflecting the variable-length index list of the corresponding RGs (see Sect. 2). The same pair of color series shown in Fig. 6 is used for the RG glyphs as well.

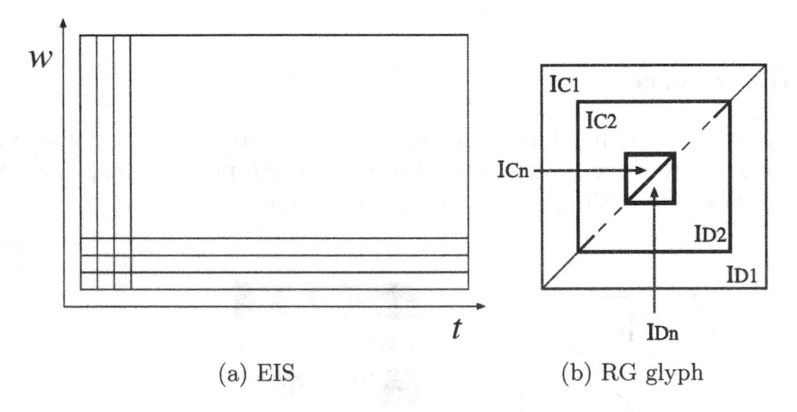

(a) EIS (b) RG glyph

Fig. 8. EIS and RG glyph used in EIS

On the other hand, a CP layer of EIS plots the location of critical field values with X-shaped marks on each vertical line of snapshot volumes. A CP layer is expected to track each of the critical points in VST as a function of time. To this end, different mark colors are used to distinguish the type of critical points (C_3: red; C_2: yellow; C_1: green; C_0: blue).

Figure 9 visualizes the EIS of the TBB volume, where the symmetrical structure of the field is visualized more precisely. In addition, we can understand that all the critical field values are concentrated in the region with relatively high field values. Note that the unimportant regions with very low field values have been truncated. Furthermore, tracing VST critical points on the CP layer allows the user to visually confirm the time-evolution of the volumetric field (see Fig. 4(a)):

– t=0.0 (1.0): Two spheres meets at a single CP p_5 of the BB volume;
– t=0.071 (0.929): Two CPs p_1 and p_2 of the BB volume appears/disappears;
– t=0.142 (0.858): Two CPs p_3 and p_4 of the BB volume appears/disappears;
– t=0.5: The case reduces to (4) (Sect. 2.2), as mentioned above.

Note that these critical timings are found analytically by substituting $c = 0.60$, $d = 0.50$, $R = 0.20$ into (7), and solving two quadratic equations with respect to t.

The well-known *fisheyes* and *magic lenses* metaphors [20] can be introduced to realize the *focus+context* display and probing of ISs and EISs in the T-map.

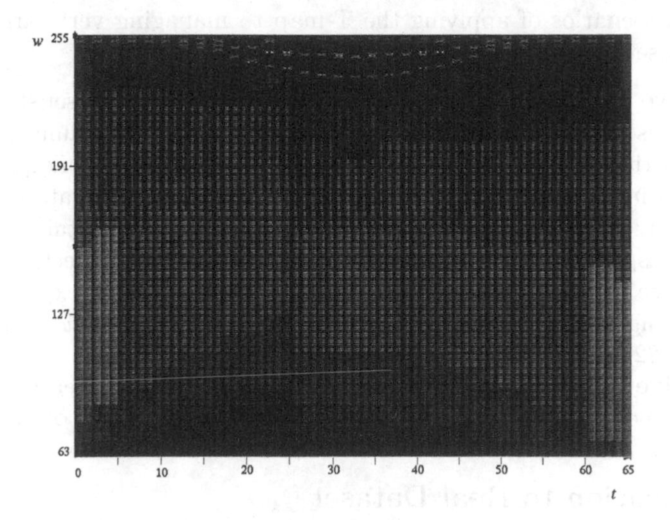

Fig. 9. EIS of TBB volume

Displaying Details on Demand. Once a set of target timings has been found, we take advantage of accentuated volume rendering for the limited number of snapshot volumes. The key to comprehensible volume rendering is the design of *transfer functions* (*TFs*), which map physical field values of given volume samples to optical properties such as color and opacity. In the T-map, informative images of a single volume can be visualized with topologically-accentuated TFs around the critical field values specified in the corresponding VST [7,10]. If a restricted sequence of snapshot volumes are selected, TFs used commonly for visualizing time-varying volume datasets [21] can also be designed more effectively.

3.3 Potentials of T-Map

Let us recall the concept of RG described in Sect. 2.1. If we take time as the axis for the height function, and examine the sequence of snapshot volumes in terms of VST, we can find a particular time $T_{critical}$, which is termed *critical timing* (*CT*), exactly when the topological equivalence of consecutive snapshot volumes is not maintained. The homogeneity of VSTs can be checked using the pair of VST index quantities in Sect. 2.3. Since we do not have any information between adjacent snapshot volumes, a true CT can not always be found on the sampling time step. We may possibly pass over minor CTs in a case with sparse time steps. However, we can rely on human visual perception to approximate the location of CTs with a combination of the above-described functionalities.

In general, since a partial series of snapshot volumes around CPs are of particular interest for the user, it is desirable to provide the user with a semi-automatic mechanism to shift his or her focus to the partial series, and to visually explore the details with topologically-accentuated volume rendering schemes [10,13].

Potential scenarios of applying the T-map to managing very large-scale 4D volume datasets include:

- **Selective data migration:** The objective 4D volume dataset stored in tape libraries is examined entirely by the topological surface/volume skeletonization algorithms in an offline manner to make the corresponding T-map IS and EIS in advance. Then the user is allowed to investigate those spaces to visually identify partial temporal regions of interest containing CTs, and the corresponding portion of the dataset is migrated selectively into the disk space of a computing environment for further visual exploration. This challenging issue is in marked contrast to the concept of *out-of-core visualization* [22].
- **Adaptive computational steering:** If necessary, more detailed recalculation can be requested within the selected temporal regions of interest.

4 Application to Real Dataset

In order to perform the feasibility study, we applied the T-map scheme to a large-scale time-varying volume dataset, called H^+-H here, which consists of data for 61^3 volumes over 10^4 time steps for simulated intermediate-energy collision of a proton and a hydrogen atom [7]. Prior to the analysis of the volume field topology, each snapshot volume has been downsized into 30^3 voxels.

The simulation deals with a fundamental atomic collision problem, and is very important in that the problem has a wide spectrum of applications such as nuclear fusion, material sciences, and radiology. The purpose here is to investigate

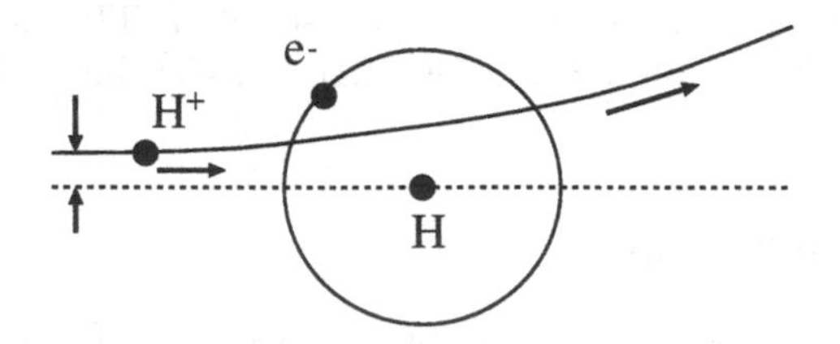

Fig. 10. Proton–hydrogen atom collision process

Fig. 11. IS of H^+–H dataset

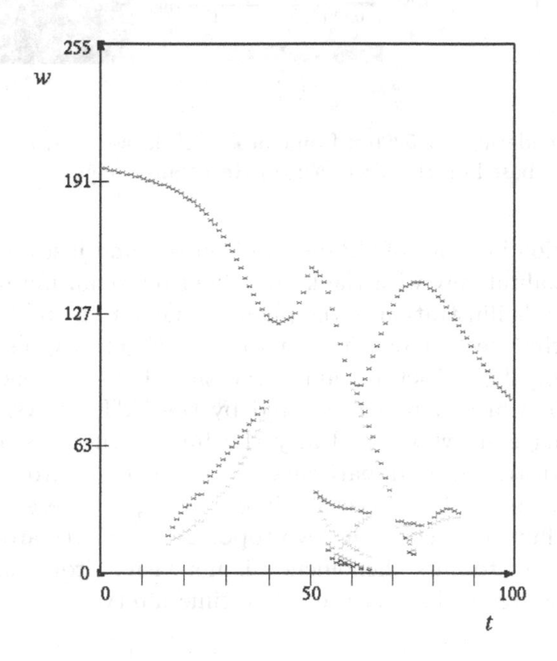

Fig. 12. CP layer of EIS of H^+–H dataset

how the positive charge of an incident proton affects the behavior of an electron around the target hydrogen atom (Fig. 10).

Figure 11 shows the IS of the H^+–H dataset, where 100 VST glyphs are visualized at 100 step intervals between 1st frame through 10,000th frame.

It is widely known that the stationary electron density distribution around a hydrogen atom constitutes a completely layered structure of spherical isosurfaces. Therefore, without producing a volume-rendered animation of the entire time sequence, we can identify easily from the IS space, an approximate timing of the collision as the portion with a least complexity of VST glyphs. Indeed, the collision occurs around 5,000th frame, as indicated with a triangle icon in Fig. 11. The fact was verified from other simulated information of the minimum simulated distance between the incident proton and the target hydrogen atom.

Figure 12 shows the CP layer of the EIS space expanded from the IS in Fig. 11. When taking a look at the expanded space, it turns out that the most

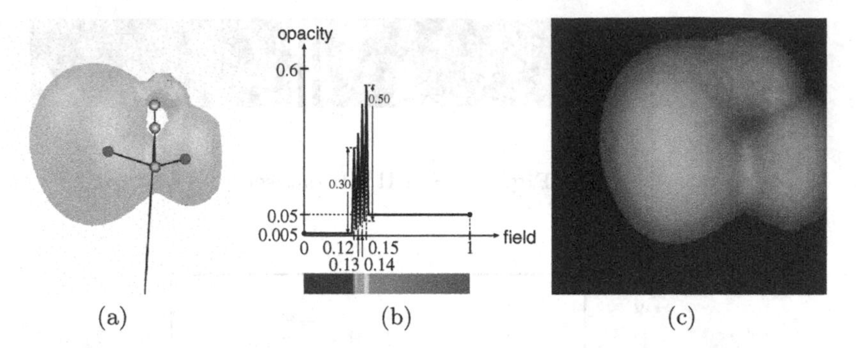

Fig. 13. Volume rendering the 5300th frame of H^+–H dataset. (a) A critical isosurface with VST; (b) TFs based on the VST (a); (c) Resultant image.

complicated portion in terms of CP distribution is found just after the collision. This coincides qualitatively with the knowledge from quantum physics.

A comprehensible illustration of the phenomena can be obtained by visualizing the 3D distorted electron density distribution. Figure 13(c) shows a volume-rendered snapshot of the 5300th frame. The snapshot was generated with the TFs in Fig. 13(b) which were accentuated by the VST analysis in Fig. 13(a). The resultant image shows more clearly the inner structures of the distorted electron density distribution. In particular, it can be found from the image that a topologically enhanced TFs succeeded in capturing an interesting arm-shaped electron density flux connecting the two topological spheres around the hydrogen nuclei. This easy-to-miss phenomenon is not visible from animations which commonly use the same TFs over the entire time interval.

5 Conclusions and Future Work

This paper proposed T-map as a topologically-based visual data mining environment for large-scale time-varying volume datasets. The T-map relies on a mathematically rigorous theory, and allows the user to grasp the topological features of a given 4D scalar field. The present scheme can be extended to handle a higher dimensional array of datasets for the purpose of parameter study. For example, the IS and EIS can serve as a 2D-scrollable strip chart for visual data mining spreadsheet called *HyperCells* [23].

Many related issues still remain open for future research, including:

- Introducing the *level of detail* control of RGs and VSTs for the multiresolutional analysis to ISs and EISs.
- Providing a mechanism of similarity search for isosurfaces, snapshot volumes, and partial sequences of 4D volumes based on the topological *proximity* of RGs and VSTs.
- Extending VST to directly characterize a given 4D volume dataset represented as the samples of the 5D hyper-surface (1) to locate 4D critical points.

– Enriching T-map manipulations so that they cover the basic visual-information-seeking functions stated in the *Shneiderman Mantra* [24].

Acknowledgements

This work has been partially supported by Japan Society of the Promotion of Science under Grants-in-Aid for Scientific Research (C) 13680401 and (B) 18300026 and Young Scientists (B) 17700092, and the Ohkawa Foundation for Information and Telecommunications.

References

1. Meissner, M., Huang, J., Bartz, D., Mueller, K., Crawfis, R.: A practical evaluation of popular volume rendering algorithms. In: Crawfis, R., Cohen-Or, D. (eds.) Proc. IEEE Volume Visualization and Graphics Symposium 2000, vol. 151, pp. 81–90 (2000)
2. Pfister, H., Hardenbergh, J., Knittel, J., Lauer, H., Seiler, L.: The VolumePro real-time ray-casting system. In: Rockwood, A. (ed.) Proc. ACM SIGGRAPH 1999, pp. 251–260 (1999)
3. Muraki, S., Ogata, M., Ma, K.L., Koshizuka, K., Kajihara, K., Liu, X., Nagano, Y., Shimokawa, K.: Next-generation visual supercomputing using PC clusters with volume graphics hardware devices. In: CD-ROM Proc. ACM/IEEE SuperComputing 2001 (2001)
4. Muraki, S., Lum, E.B., Ma, K.L., Ogata, M., Liu, X.: A PC cluster system for simultaneous interactive volumetric modeling and visualization. In: Koning, A., Machiraju, R., Silva, C.T. (eds.) Proc. IEEE Symposium on Parallel and Large-Data Visualization and Graphics 2003, pp. 95–102 (2003)
5. Chen, L., Fujishiro, I., Nakajima, K.: Optimizing parallel performance of unstructured volume rendering for the Earth Simulator. Parallel Computing 29, 355–371 (2003)
6. Ramakrishnan, N., Grama, A.Y.: Data mining: From serendipity to science. IEEE Computer 32, 34–37 (1999)
7. Fujishiro, I., Azuma, T., Takeshima, Y., Takahashi, S.: Volume data mining using 3D field topology analysis. IEEE Computer Graphics and Applications 20, 46–51 (2000)
8. Ji, G., Shen, H.W., Wenger, R.: Volume tracking using higher dimensional isosurfacing. In: Turk, G., van Wijk, J.J., Moorhead, R. (eds.) Proc. IEEE Visualization 1991, pp. 209–216 (1991)
9. Chen, J., Silver, D., Jiang, L.: The Feature Tree: Visulaizing feature tracking in distributed amr datasets. In: Koning, A., Machiraju, R., Silva, C.T. (eds.) Proc. IEEE Symposium on Parallel and Large-Data Visualization and Graphics 2003, pp. 103–110 (2003)
10. Takahashi, S., Takeshima, Y., Fujishiro, I.: Topological volume skeletonization and its application to transfer function design. Graphical Models 66, 24–49 (2004)
11. Shinagawa, Y., Kunii, T.L.: Constructing a Reeb graph automatically from cross sections. IEEE Computer Graphics and Applications 11, 44–51 (1991)
12. Takahashi, S., Ikeda, T., Shinagawa, Y., Kunii, T.L., Ueda, M.: Algorithms for extracting correct critical points and constructing topological graphs from discrete geographical elevation data. Computer Graphics Forum 14, 181–192 (1995)

13. Fujishiro, I., Takeshima, Y., Takahashi, S., Yamaguchi, Y.: Topologically-accentuated volume rendering. In: Post, F.H., Nielson, G.M., Bonneau, G.P. (eds.) Data Visualization: The State of The Art, pp. 95–108. Kluwer Academic Publishers, Dordrecht (2003)
14. Hilaga, M., Shinagawa, Y., Kohmura, T., Kunii, T.L.: Topology matching for fully automatic similarity estimation of 3D shapes. In: Fiume, E. (ed.) Proc. ACM SIGGRAPH 2001, pp. 203–212 (2001)
15. Takahashi, S., Nielson, G.M., Takeshima, Y., Fujishiro, I.: Topological volume skeletonization using adaptive tetrahedralization. In: Hu, S.M., Pottmann, H. (eds.) Proc. Geometric Modeling and Processing 2004, pp. 227–236. IEEE Computer Society Press, Los Alamitos (2004)
16. Johnson, M.A., Maggiora, G.M. (eds.): Concepts and Applications of Molecular Similarity. John Wiley & Sons, Chichester (1990)
17. Kier, L.B., Hall, L.H.: Molecular Connectivity in Chemistry and Drug Research. Academic Press, London (1976)
18. Press, W.H., Flannery, B.P., Teukolsky, S.A., Vetterling, W.T.: Numerical Recipes in C. Cambridge University Press, Cambridge (1988)
19. Keim, D.A.: Designing pixel-oriented visualization techniques: Theory and applications. IEEE Transactions on Visualization and Computer Graphics 6, 59–77 (2000)
20. Card, S.K., Mackinlay, J.D., Shneiderman, B.: Readings in Information Visualization: Using Vision to Think. Morgan Kaufmann, San Francisco (1998)
21. Jankun-Kelly, T.J., Ma, K.L.: A study of transfer function generation for time-varying volume data. In: Mueller, K., Kaufman, A. (eds.) Volume Graphics 2001, pp. 33–43. Springer, Heidelberg (2001)
22. Farias, R., Silva, C.: Out-of-core rendering of large, unstructured grids. IEEE Computer Graphics and Applications 21, 42–50 (2001)
23. DeCoste, D.: Visualizing massive multivariate time-series data. In: Fayyad, U., Grimstein, G.G., Wierse, A. (eds.) Information Visualization in Data Mining and Knowledge Discovery, pp. 95–97. Morgan Kaufmann, San Francisco (2001)
24. Shneiderman, B.: Designing the User Interface Strategies for Effective Human-Computer Interaction, 3rd edn., ch. 15. Addison-Wesley, Reading (1998)

Cross-Line — A Globally Adaptive Control Method of Interconnection Network

Takashi Yokota[1], Masashi Nishitani[2],
Kanemitsu Ootsu[1], Fumihito Furukawa[3], and Takanobu Baba[1]

[1] Department of Information Science, Utsunomiya University,
7–1–2 Yoto, Utsunomiya-shi, Tochigi, 321–8585 Japan
{yokota, kim, baba}@is.utsunomiya-u.ac.jp
[2] Technology Division, The Japan Research Institute, Ltd.
[3] Learning Technology Laboratory, Teikyo University

Abstract. An ordinal interconnection network is composed of many independent routers that can cooperate as a communication subsystem in a massively parallel system. Many routing algorithms are proposed in the past, however, they do scarcely utilize global network information. In this paper, we propose a new adaptive routing method, Cross-Line, that makes efficient use of global information over the network. The algorithm achieves global routing control efficiently by collecting just one-bit information of each virtual channel in the x- and y-directions. Analytical and simulation results reveal the effectiveness of the algorithm.

1 Introduction

An economical but effective interconnection network is strongly required for massively parallel systems. After a communication message is generated at a source node, it is relayed via multiple routers until its destination. Each router determines the proper direction for the message to go through. Routing function can only be acquired as a result of co-operative work of individual routers.

Many routing algorithms are presented in the past[1,2]. Many of them improve the communication performance, although, no one succeeds in globally optimal routing. This is an essential problem of interconnection networks. Routers are connected to each other by limited number of links. Thus a router can acquire quite localized information by itself and it can perform only locally optimal routing. We can expect performance improvement by introducing globally optimal control mechanisms.

This paper presents a practical and efficient routing method, called *Cross-Line*[3,4]. The method includes an efficient function for collecting global information of network status and a routing algorithm that uses the collected information. The method can be implemented as a simple extension of the ordinal adaptive routing method. Evaluation results reveal the effectiveness of the proposed method.

J. Labarta, K. Joe, and T. Sato (Eds.): ISHPC 2005 and ALPS 2006, LNCS 4759, pp. 191–198, 2008.

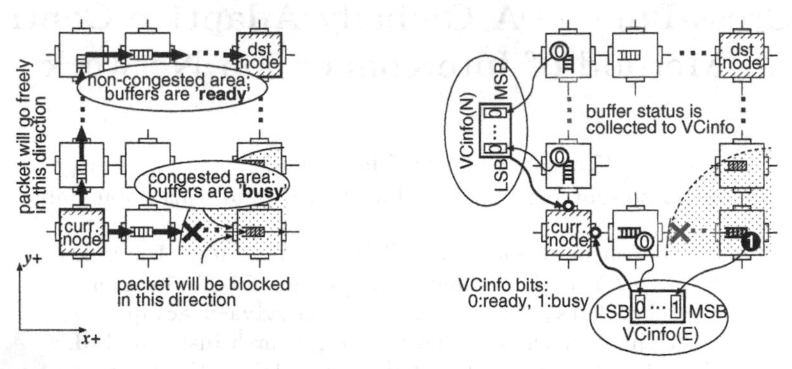

(a) Congested area and buffer status. (b) VCinfo represents congested area.

Fig. 1. Basic idea of global congestion information

2 Basic Design for Global Adaptability

Network performance is extremely degraded in a badly congested situation[1,2]. The source of the problem is a chain reaction of packet blocking. As a blocked packet is stopped at a packet buffer, the filled buffer blocks other packets. Thus the chain reaction spreads over the network. Our essential idea is to prevent chain reactions by introducing a globally optimal routing method. The method should be not only effective but also practical. We discuss the basic method in this section.

Here, we assume 2-dimensional torus networks for simplification. We define **congestion** as a situation in which packets are blocked among multiple neighboring routers. And we define a **congested area** as a congested portion of the network. If a router has global information of congestion, the router can properly guide packets in a less congested area.

Our basic idea is illustrated in Figure 1. A packet is routed in x- and y-directions. A router determines whether the packet should go straight or turn. We use a one-bit information per packet buffer, which represents ready/busy state of the buffer. The bit is transferred to the *straight* direction and forms a global congestion information. As the one-bit information is equivalent to virtual channel status, we call the collected information **VCinfo**.

Neighboring router's status is directly reflected as a flow-control signal in the corresponding output port. The LSB of VCinfo represents the state. A router transfers its own VCinfo to the reverse direction. The neighboring router receives the information and stores it into its own VCinfo after shifting left one bit. A ready/busy signal in the corresponding output port is set to the LSB of the VCinfo. Thus i-th bit in VCinfo represents the corresponding buffer status. A router can determine appropriate routing direction by its own VCinfo. For example in Figure 1, `curr.node` has a packet destined for `dst.node`. VCinfo in $x+$ direction shows a congested area, while $y+$ VCinfo does not include congested area information. Thus the router can properly select $y+$ direction.

```
route( int cx, int cy, int dx, int dy ){
  // (cx, cy) ... address of current router
  // (dx, dy) ... destination address
  int  wx = cx;
  int  wy = cy;
  for( i=0 ; i<MaxHops ; i++ ){
    if( wy==dy || wx==dx )
      break;
    // compare VCinfo values from LSB
    if( VCinfo_x[i]==READY && VCinfo_y[i]==BUSY  ) Goto_X;
    if( VCinfo_x[i]==BUSY  && VCinfo_y[i]==READY ) Goto_Y;
    wx = next_x(wx);
    wy = next_y(wy);
  }
  // not determined
  if( wx==dx && wy!=dy )  Goto_Y;
  else                    Goto_X;
}
```

Fig. 2. Routing algorithm of Cross-Line

3 Cross-Line Routing

3.1 Routing Algorithm

Our proposed Cross-Line method is composed of two items; VCinfo mechanism and routing algorithm. The former offers collection, distribution and sharing mechanism for global information of congested area. We describe the routing algorithm here. A router can detect congested area by consulting VCinfo values. Basically the router determines the smoother direction by comparing two VCinfo values according to possible directions and virtual channels. The router compares VCinfo values bit-by-bit from LSB as far as both bits are within the packet's traveling distance. When the bit-wise comparison encounters different bit values, the router detects imbalanced situation and selects '0' bit direction[1].

Figure 2 shows the Cross-Line routing algorithm in C-like code. `(cx, cy)` is a two-dimensional address expression of the current router, and `(dx, dy)` means destination address. `Goto_X` and `Goto_Y` mean selected directions. Function `next_x()` (`next_y()`) calculates x-axis (y-axis) address of the neighboring router.

3.2 Appropriateness of the Design

Here, we discuss appropriateness of the Cross-Line method by introducing a simple mathematical model that represents imbalanced congestion situation. We introduce an abstract congestion measure C that represents degree of congestion

[1] We suppose that 0 means ready and 1 busy in VCinfo.

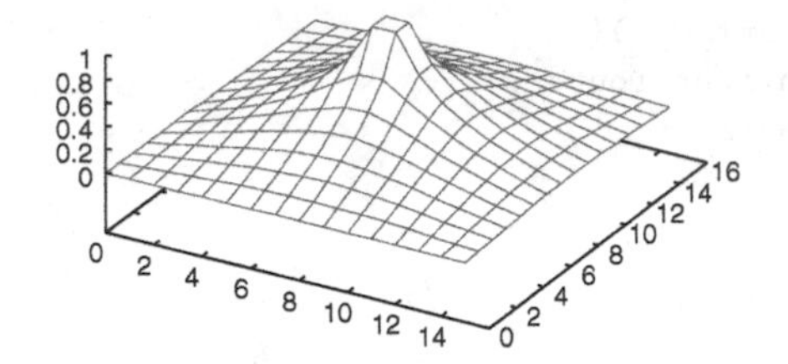

Fig. 3. Example congestion measure

dimension order	5.505e4
random walk	5.628e4
adaptive	5.245e4
cross-line	4.767e4
optimal	3.801e4

Fig. 4. Calculation results of path costs

at each router. A packet buffer is ready when C is under a certain threshold. Roughly speaking, VCinfo represents gradient of C at the corresponding position. Thus Cross-Line method is roughly equivalent to choosing the direction with smaller gradient out of $\frac{\partial C}{\partial x}$ and $\frac{\partial C}{\partial y}$.

We define a **path cost** as a summation of C values along a packet's trail. If a packet goes through congested area, the path cost becomes high. An optimal algorithm minimizes the packet cost. For simplification, we assume that only one congested area exists at the center of the system as shown in Figure 3. We calculated sum of path costs for possible combinations of source and destination nodes. Figure 4 shows results of some routing algorithms. All of them route shortest path. In 'random walk,' a packet goes into randomly selected direction and the listed value is an average of 100 trials. In 'adaptive' routing, a packet changes its direction only when its nearest-neighbor is in congested area. 'Optimal' value is the minimum value among possible routes. Cross-Line achieves about 14% better path cost against dimension order routing, whereas adaptive routing achieves only 5% improvements. These results reveal the appropriateness of Cross-Line algorithm — Cross-Line has high potential in avoiding congested area.

4 Evaluation

4.1 Evaluation Model

We used the following algorithms for comparison. **(a) dimension order routing** (dim.order) $\cdots$ A packet goes through x-direction first and then y-direction. **(b) deterministic routing** (det.) $\cdots$ This routing algorithm is equivalent to *Cross-Line algorithm without VCinfo reference*, i.e., `MaxHops` in Figure 2 is zero. This algorithm is used for comparison purpose to clarify the effectiveness of Cross-Line (`MaxHops` > 0). **(c) ideal** (ideal) $\cdots$ This routing algorithm directly refers packet buffer states, not VCinfo. VCinfo has essential delay by transferring state information of packet buffers.

We use the following channel selection function for evaluation purpose. At packet generation, packet direction is determined out of four combinations of x and y directions $(x\pm, y\pm)$. We assign initial virtual channel 0 for $(x+, y+)$ and $(x-, y-)$ packets and 1 for $(x-, y+)$ and $(x+, y-)$ packets. After that, when a packet goes across 'date-line,' its virtual channel number is incremented by two. Date-lines are placed at $x = 0$ and $y = 0$. This channel selection algorithm is applied for any algorithm in the evaluation.

We built an interconnection network simulator. Cross-Line method and comparison algorithms are implemented. We use four-flit length packets and each packet buffer has three flit capacity. VCinfo is transfered only when no message data goes through the port. Thus collection and distribution of VCinfo do not interfere in ordinal message communication[5].

In our simulation, the system is 32×32 two-dimensional torus. Uniform random traffic and 5% hot-spot traffic patterns are used. In the former pattern, packet destination is randomly selected. In the latter pattern, 95% of generated packets go to randomly selected destinations, and remaining 5% of packets go to a particular destination (a center node $(16, 16)$). Each node generates packets at a given clock interval. Latency is measured from the first flit of a packet is injected into the network to the last flit is transfered to the destination processor. Simulation time was 2,000,000 cycles, but initial 1,000,000 cycles are ignored for obtaining stable results.

4.2 Performance Comparison

Performance results are shown in Figure 5. In these graphs, normalized accepted throughput is used in x-axis, which is calculated by $P/(T/l * N)$ where P is the total number of received packets, T is simulation time (1,000,000 cycles), l is packet size in flit (4), and N is the number of nodes (1,024). In these graphs, CrossLn, (ideal), det., and dim.order mean Cross-Line, ideal case, deterministic routing, and dimension order routing, respectively. Figure 5(b) includes enlarged plots at critical saturation points.

Comparison between det. and CrossLn clarifies the effect of adaptive routing in Cross-Line. As we can find in Figure 5(a), Cross-Line achieves about 1.79 times improvement in peak throughput (0.068 (det.) and 0.122 (CrossLn)). Cross-Line achieves about 93% peak performance of its ideal case (ideal). This result shows that VCinfo is effective even it has essential delay. Cross-Line also achieves better peak performance than that of dimension order routing, but average latency is larger. The long average latency originates from the baseline routing algorithm.

Hot-spot performance results (in Figure 5(b)) show different natures. In non-congested situations, average latency curves overlap each other. We can find, in the enlarged graph, CrossLn and ideal improve peak throughput.

4.3 Sufficient VCinfo Bits

We have also investigated Cross-Line performance under limitation of maximum VCinfo references. We actually limit the `MaxHops` value in Figure 2. Figure 6 shows the results. In these graphs, normalized offered traffic is a simple index of l/v where l is packet size in flits and v is packet generation interval in clocks. In these graphs, `MaxHops` values are specified in parentheses. CrossLn(full) shows no limitation on `MaxHops`. Especially, `MaxHops` $= 1$ is equivalent to ordinal adaptive

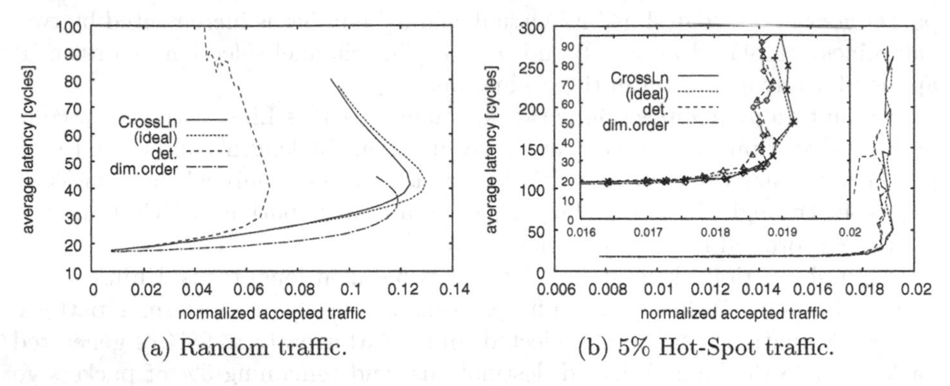

(a) Random traffic.

(b) 5% Hot-Spot traffic.

Fig. 5. 32×32 traffic performance

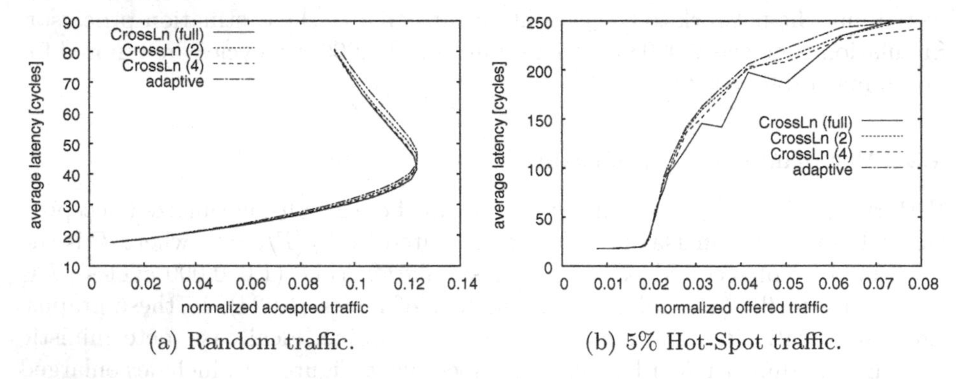

(a) Random traffic.

(b) 5% Hot-Spot traffic.

Fig. 6. Performances under VCinfo reference limitation

routing method that refers only ready/busy information of output port. Thus we use adaptive as a comparison algorithm. of the Cross-Line method. From the graphs, we can find that at least four bits of VCinfo is required.

4.4 Cross-Line as an Adaptive Routing

We can compare Cross-Line algorithm with ordinal adaptive one (adaptive) in the Figure 6 graphs. In uniform random traffic pattern, average latency of Cross-Line is lower than that of adaptive. For example, at the peak accepted throughput (around 0.124), Cross-Line achieves 41.72 cycles of average latency and adaptive achieves 43.65 cycles, improvement factor of Cross-Line is about 4.4%. But no significant difference is found in peak throughput.

In hot-spot traffic pattern, especially heavily congested case, Cross-Line performs better (shorter latency) than adaptive routing. Cross-Line shows a zigzag curve in Figure 6 (c), this represents that actual effect of Cross-Line is sensitive to the network situation. But, in many cases, Cross-Line can route packets appropriately and achieves better average latency than adaptive routing.

5 Related Work

Thottethodi et al. proposed a throttling algorithm with a self-tuning function[6]. Their method uses the number of transient packets in the network. While the count exceeds a self-tuned threshold, packet injection is suppressed. The method uses a kind of global information, however, the information only shows average status of the network and thus it cannot represent locality in congestion. Cross-Line handles global information as a collection of 1-bit local states. Note that Cross-Line offers routing function whereas Thottethodi's method provides throttling function. These methods do not conflict with each other.

Speculative Selection Routing (SSR [7]) counts the number of transient packets in each router, and the counter value is delivered to neighborhood routers. Each router determines packet's routing by the counter values. The method is disadvantageous in distributing the values. Because of multiple bits of counters, distribution area is limited. Furthermore, SSR's routing function has high complexity. Since VCinfo represents multiple states of routers at a time, complexity in routing function of Cross-Line is much less than that of SSR.

Singh et al. claim globally adaptive routing in their proposed algorithms, GOAL [8] and GAL [9]. They introduced the quadrant idea defined by source node (s_x, s_y) and destination (d_x, d_y) so that global load balancing is enabled. Their algorithms work well in some classes of problems, however, the algorithms do not use any global information. Thus their approach is completely different from ours.

Some routing methods employ adaptive functions for avoiding congested situations. For example, [10] employs special queues to release blocked packets from congestion. As another example, [11] prepares dedicated buffers, into which packets escape from congestion. After congestion situation disappears, escaped packets are re-injected into the network. These methods are symptomatic, whereas Cross-Line takes a preventive way by using global information.

6 Conclusions

A large-scale interconnection network, composed of huge number of routers, has an essential problem, i.e., the lack of globally optimal control. Our basic idea in this paper is introducing global but decentralized control to avoid routing toward congested area.

We introduced VCinfo (virtual channel information) as a collection of one-bit information of a packet buffer in each virtual channel. Each router has VCinfo for each virtual channel in each output port. Contents of VCinfo are forwarded in the *straight* direction. By using VCinfo values, a router can select a proper direction for a packet to avoid congested area.

By introducing an abstract congestion measure, we show theoretically high potential of Cross-Line in avoiding congested area. Furthermore, simulation results reveal Cross-Line's high efficiency. Cross-Line improves peak throughput in both uniform random and 5% hot-spot traffic communication patterns. It relaxes long latency in a heavily congested situation caused by the hot-spot traffic.

Through deep considerations of simulation results, we can guess temporal behavior of congestion area in various congestion situations. However, we still need understanding accurate congestion behavior to improve the proposed Cross-Line mechanism. Congestion behaviors are very complex, so clarifying accurate behavior of congested area is our future work.

Acknowledgments. This research was supported in part by Grant-in-Aid for Scientific Research ((B) 14380135 and (C)16500023) and Young Scientists ((B) 17700047) of Japan Society for the Promotion of Science (JSPS).

References

1. Duato, J., Yalamanchili, S., Ni, L.: Interconnection Networks: An Engineering Approach. Morgan Kaufmann Pub., San Francisco (2003)
2. Dally, W.J., Towles, B.: Principles and Practices of Interconnection Networks. Morgan Kaufmann Pub., San Francisco (2004)
3. Nishitani, M., Ezura, S., Yokota, T., Furukawa, F., Ootsu, K., Baba, T.: Preliminary Research of a Novel Routing Algorithm Cross-Line Using Dynamic Information. IPSJ SIG-Note 2004(20), 7–12 (2004)
4. Nishitani, M., Ezura, S., Yokota, T., Ootsu, K., Baba, T.: Preliminary Research of a Novel Routing Algorithm Cross-Line Using Dynamic Information. In: PDCS 2004. Proc. 16th IASTED International Conference on Parallel and Distributed Computing and Systems, pp. 107–112 (2004)
5. Yokota, T., Matsuoka, H., Okamoto, K., Hirono, H., Sakai, S.: Virtual Control Channel and its Application to the Massively Parallel Computer RWC-1. In: Proc. HiPC 1997, pp. 443–448 (1997)
6. Thottethodi, M., Lebeck, A.R., Mukherjee, S.S.: Self-Tuned Congestion Control for Multiprocessor Networks. In: Proc. HPCA-7, pp. 107–118 (2001)
7. So, T.C., Oyanagi, S., Yamazaki, K.: Speculative Selection in Adaptive Routing on Interconnection Networks. IPSJ Transactions on Advanced Computing Systems 44(Sig 11(ACS 3)), 147–156 (2003)
8. Singh, A., Dally, W.J., Gupta, A.K., Towles, B.: Towles: GOAL: A Load-Balanced Adaptive Routing Algorithm for Torus Networks. In: Proc. ISCA 2003, pp. 194–205 (2003)
9. Singh, A., Dally, W.J., Towles, B., Gupta, A.K.: Globally Adaptive Load-Balanced Routing on Tori. Computer Architecture Letters 3(1), 6–9 (2004)
10. Duato, J., Johnson, I., Flich, J., Naven, F., García, P., Nachiondo, T.: A New Scalable and Cost-Effective Congestion Management Strategy for Lossless Multistage Interconnection Networks. In: Proc. HPCA-11, pp. 108–119 (2005)
11. Flich, J., López, P., Malumbres, M.P., Duato, J., Rokicki, T.: Improving Network Performance by Reducing Network Contention in Source-Based COWs with a Low Path-Computation Overhead. In: Proc. IPDPS 2001 (2001)

The Bandwidth Expansion Effectiveness of Cache Levels Block Prefetch

Youngkwan Ju, Bongyong Uh, and Sukil Kim

Dept. of Computer Science, Chungbuk National University
Gashindong 12, Cheongju, Chungbuk, Republic of Korea
{rainbow,uby,ksi}@chungbuk.ac.kr

Abstract. Most cache architectures exploit only a second level cache prefetch. In this paper, we propose the hierarchical prefetch cache architecture which allows prefetch between all levels of caches. We discovered that this architecture has a virtual effect of expanding memory bus bandwidth. According to an experimental analysis using 10 benchmark programs, the proposed architecture that employs all level cache prefetcher obtained a maximum 11% increased performance when compared to both architecture with expanded bus bandwidth and architecture with employment only a level 2 cache prefetcher. This shows our proposed architecture has an effectiveness of memory-bus bandwidth expansion.

Keywords: Prefetch Cache, Memory Hierarchy Architecture, Memory Bandwidth.

1 Introduction

In the modern computer architecture, the speed gap between central process unit and main memory is getting greater. Furthermore, the memory reference ratio of the application whose size is getting bigger occupies about 43% to whole instruction set [1]. It means whole execution time of an application will be extended as the CPU stalling time extends due to frequent memory references. In order to reduce the CPU stalling time, the latency time of memory reference should be shortened. A number of techniques have been proposed: hierarchical cache architecture [2, 3, 4] reducing the latency time of average memory reference with low cost, instruction reordering [5] executing other instructions not related to memory reference during fetching instruction related to it, and cache prefetching technique [3] fetching data and loading to cache or register before the time that data is actually needed by front-end processor. Especially, the cache prefetching technique is able to reduce cache miss rate greatly by adding simple prefetch cache controller at the existing hierarchical cache architecture, thus many researches have been conducted upon this area [2, 3, 6, 7].

The One Block Lookahead (OBL) technique [8] fetches single consecutive memory block based on memory block address that causes a cache miss, similarly the Multi Block Lookahead (MBL) technique [9] does multiple ones. These

J. Labarta, K. Joe, and T. Sato (Eds.): ISHPC 2005 and ALPS 2006, LNCS 4759, pp. 199–210, 2008.

techniques have a simple algorithm to determine prefetch block address, however, if memory block address to be prefetched does not exist in prefetched blocks, the efficiency of prefetched memory blocks in the cache is decreased or becomes zero. Consequently, they have a drawback of cache pollution which useful memory blocks are replaced frequently by loading prefetch block to cache.

In order to reduce the drawback like cache pollution, more accurate prefetch algorithms were studied. We can raise predicting memory block technique [7,10, 11,12] as an example that predicts memory block based on previously-referred memory address blocks. This technique can reduce cache pollution, since it predicts memory address to be loaded more accurately than OBL and MBL.

Regarding to cache architecture, a prefetch buffer architecture [9] stores prefetch blocks onto a buffer and moves only the referred block to the cache to reduce cache pollution. Also a dual cache architecture [10,13] is to store prefetch blocks onto a prefetch buffer which is added to an ordinary cache architecture.

So far, these techniques and architectures were mainly considered only for a simple cache architecture, wherein prefetch blocks are stored onto prefetch buffer located between level 1 cache and main memory. Considering 2 level cache architecture, no architecture fully exploit effectiveness of prefetch in each level of caches. In this paper, we propose hierarchical prefetch cache architecture to fully exploit all levels of cache prefetch.

The rest of this paper is organized as follows. Section 2 analyzes existing hierarchical cache architecture and describes motivation of this research. Section 3 proposes a hierarchical cache architecture employing prefetch technique suggested by this paper. Section 4 describes the cache architecture with expanded bus bandwidth based on such hierarchical cache architecture. Section 5 performs a simulation to compare cache architectures and compare their results too. Section 6 concludes the paper.

2 Related Work

There has been numerous researches in order to reduce memory access latency time. The most fundamental and easiest way is to expand bus bandwidth such as primary cache replication [18], cache compression [18] and multi-port architectures [19]. However the method of expanding bus bandwidth has a drawback of increasing the hardware complexity. Therefore, a number of alternative methods were suggested such as hierarchical memory architecture. It reduces memory access latency time by employing multi-level cache between main memory and a processor. Recently, the hierarchical memory architecture that consists of level 2 cache is being applied with wide range from a personal computer to a super computer.

The size of level 1 (L_1) cache is very small, however, latency time of reference is very low. For instance, the Pentium [14] and the Pentium III [15] of Intel Cooperation have built-in L_1 cache of 16KB size and 8KB size respectively. The PowerPC 60X [16] of Motorola Cooperation has built-in L_1 cache of 32KB size. Also reference latency time of these L_1 caches have 2˜3 cycles.

The level 2 (L_2) cache has more capacity than L_1 cache but the access latency of L_2 is 12˜45 cycles [7,14,15,16,17]. It is noticeable that the typical latency time ratio of L_2 cache to main memory are 10˜20, while the latency time ratio of L_2 cache to L_1 cache are 12˜30. The previous researches have focused on reduction of level 2 cache miss number in order to overcome the speed gap between L_1 cache and main memory.

2.1 Prefetch Buffer Architecture

The Prefetch Buffer Architecture [9] stores prefetched block to high-speed buffer (B_p) located between memory and L_1 cache as depicted as Figure 1(a). In this architecture, the CPU refers L_1 cache in parallel with B_p. if a memory block stored at B_p is referred, then this block is moved from B_p to L_1 cache in the preparation of its reuse. However, it has a shortcoming of cache pollution, since storing prefetch block into L_1 cache still replaces L_1 cache block.

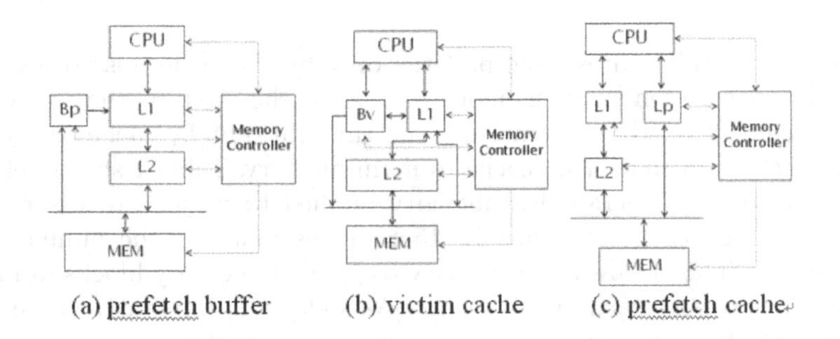

Fig. 1. Cache Architectures for Prefetch

2.2 Victim Cache Architecture

The victim cache architecture [20] employs a buffer which stores memory blocks replaced compulsorily from L_1 cache temporarily in order to reduce the number of cache pollution. This architecture depicted at Figure 1(b) employed victim buffer(B_v) located between L_1 cache and L_2 cache. If a block of victim buffer is referred, this block is transferred to L_1 cache from victim buffer via L_2 cache. By employing relatively small buffer-B_v, this architecture has an advantage of reducing the number of cache pollution caused by prefetching.

2.3 Prefetch Cache Architecture

The prefetch cache architecture [6] employs another level 1 cache only storing prefetch block. In times of L_2 cache miss, demand request blocks are stored to L_1 cache while prefetch memory block is loaded to prefetch cache. Therefore, this architecture has an advantage of preventing cache pollution, since prefetch block do not give any influence on L_1 cache. Figure 1(c) depicts this prefetch architecture, where L_p means prefetch cache. Because processor refers L_1 cache in parallel with L_p cache unlikely 1(a) and 1(b), this architecture has a merit that any referred block in L_p cache do not need to be moved to L_1 cache.

Table 1. Reference Latency Delay Ratio of cache to main memory

Processor	Ratio	
	L_2/L_1	L_2 Cache Size
Alpha 21164	16 / 2	96KB
PowerPC 604	29 / 2	512KB
MPC7400	30 / 2	1024KB
UltraSPARC	38 / 2	512KB
Pentium	12 / 2	256KB
Pentium II	35 / 3	256KB
Pentium III	45 / 3	512KB
Pentium IV	40 / 3	512KB

These three architectures issue prefetch only by L_2 cache miss to overcome the speed gap between main memory and L_1 cache. But in the hierarchical memory architecture, the reference latency time ratio of L_1 cache to L_2 cache is much bigger than that of L_2 cache to main memory. Table 1 shows reference latency time ratio of hierarchical memory applied to major processors. Also, the number of L_1 cache miss has 10˜15 times as much as the number of L_2 cache miss [17]. This means it is necessary to prefetch memory blocks from main memory whenever L_2 cache miss occurs. As well, it is imperative to prefetch memory blocks from L_2 cache when L_1 cache miss occurs.

On the other hand, architectures that employ prefetch algorithm have an effect of logical expanding bus bandwidth in terms of issuing prefetch order only when bus bandwidth is not used by demand fetch order. In this paper, based on these observations, we propose a cache architecture adding on level 1 cache prefetcher upon traditional prefetch cache architecture and will prove that our suggested one has an effect of virtually expanding bus bandwidth.

3 Hierarchical Prefetch Memory Architecture

3.1 Target Architecture

Figure 2 depicts the hierarchical memory architecture proposed in this paper. This architecture employs L_1 cache and L_p cache as level 1 cache and stores demand block and prefetch block to L_1 cache and L_p cache respectively. Therefore, the basic architecture is very similar with prefetch cache architecture shown at Figure 1(c).

In the Figure 3, cache management procedure during referring instructions is shown. The gray area indicates prefetching function of L_1 cache added on existing cache architecture. It means that L_1 cache miss firstly request demand block from L_2 cache, and then determine the block to be prefetched and store it to L_p cache. This eventually moves two consecutive memory blocks from L_2 cache. Thus, the target architecture avoids L_1 cache pollution by preventing inaccurate prediction of prefetcher from replacing existing L_1 cache block.

In case that prefetch block does not exist at L_2 cache, this block is not requested from main memory. In other words, if prefetch block directed by L_1 cache miss does not exist at L_2 cache, no prefetch occurrs. The prefetch algorithms using by L_1 cache and L_2 cache will be discussed at Section 3.2.

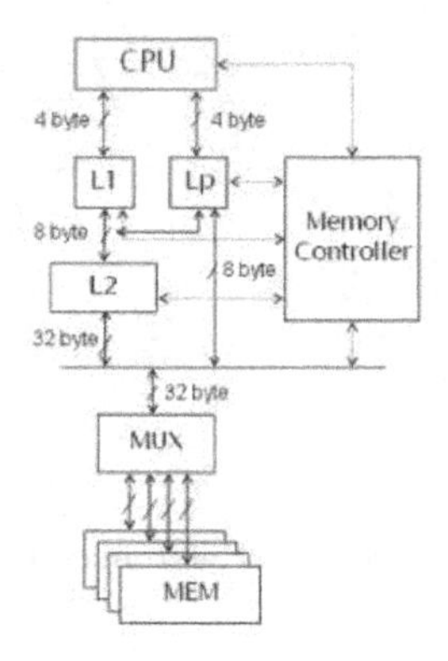

Fig. 2. Target Architecture

We suppose that the bandwidth between processor and L_1 cache is 1 word (4 byte) and bandwidth between L_2 cache and L_p cache as well as L_1 cache is 2 word. But the main memory uses 4-way interleaving access, thus 4 words are referred from every 4 independent memory module and are transferred to L_2 cache. All non-memory reference instruction takes only a single cycle. The L_1 cache employs write-through replacement policy. The L_2 cache is 512KB size with direct-mapping replacement policy and operates with multi-bank for 4-way interleaving access. And its line size is 128 byte. This L_2 cache uses write-back replacement policy. The main memory is composed of four-memory banks for memory accesses of demand on fetch and prefetch. Also the address bus and the data bus have 4 access latency cycles with external memory and the length of prefetch request queue has 64 entries.

3.2 Prefetch Algorithm

The proposed architecture requires prefetchers at each cache level. To begin with, the L_1 cache prefetcher directed by L_1 cache miss loads demand block from L_2 cache and stores it to L_1 cache, and then determines next block to be fetched and stores it to L_p cache. The L_2 cache prefetcher has the same architecture as that of the existing prefetcher, which requests memory block from L_2 cache and stores it to L_2 cache and stores corresponding block to L_1 cache. Subsequently, this L_2

cache prefetcher determines prefetch block from main memory and then loads it. In contrast, the procedure of level 1 cache prefetching should be terminated within two cycles. Therefore the algorithm used for level 1 cache would be faster. Prefetch algorithm used in this paper employs the OBL(One Block Lookahead) algorithm, which is able to determine prefetch block with the least latency.

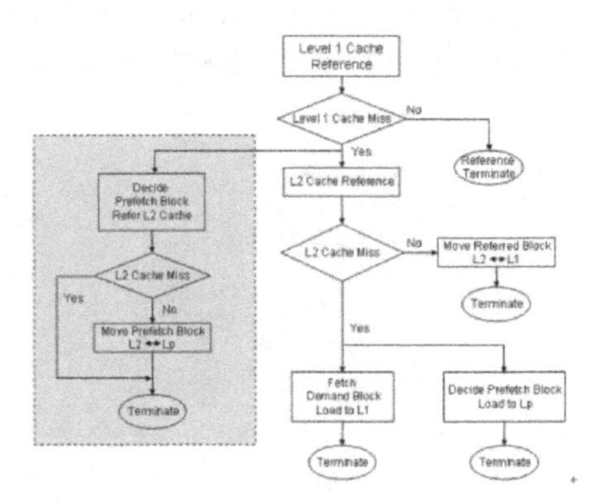

Fig. 3. Flowchart of Memory Reference of Target Architecture

Numerous researches regarding level 2 cache prefetcher technique has been conducted. For example, the OBL and MBL which prefetch single or multiple consecutive memory blocks have an advantage of being consisted of simple algorithm, however, if memory address to be prefetched does not exist at prefetched multiple blocks, the efficiency of prefetch memory block stored to the cache becomes very low, thus cache pollution occurs. The Correlation Prediction Table Prefetching technique [21] predicts next memory block address based on already-referred memory address blocks. Therefore the efficiency of this technique is very high in case that the pattern of referring memory block address is constant. So it can predict next memory block address more accurately, however it has a drawback of requiring more memory capacity than the OBL. The Filtering Table Technique is more complex and advanced than techniques mentioned above. It keeps the information for all prefetch block whether referred by main processor or not, and then permit any prefetching issue to be conducted only if the block address has been referred by main processor. This paper adopted these techniques of Correlation Prediction Table and Filtering Table Technique as the second level prefetcher of the proposed architecture.

If OBL is applied as L_1 cache prefetcher algorithm, it has the same efficiency as that of L_1 cache when loading two consecutive blocks from L_2 cache in pipe-line manner. This scheme would provide the same efficiency as the architecture with 2 times of bandwidth between L_1 cache and L_2 cache. This implies that using two prefetchers in memory controller would virtually improve bus bandwidth of the memory system. Efficiency of the architecture will be studied in detail in the next section.

4 Hierarchical Memory Architecture with Expanded Bandwidth

The systemic design of the cache architecture with expanded bandwidth twice between L_1 cache and L_2 cache is shown at Figure 4. L_1 cache miss incurs both demand cache block and next consecutive block from L_2 cache and store them to L_1 cache using expanded bandwidth. This cache architecture will be identical with that of Section 3 unless it has L_1 cache prefetcher. In times of L_2 cache miss, the operation with external memory has no difference with architecture of Section 3 in order to analyze the efficiency of bandwidth expansion.

When it comes to hardware complexity, the hierarchical prefetch architecture described at Section 3 requires only an ADDER, since L_1 cache prefetcher employs OBL algorithm that selects a next block based on missed address. Apparently adding a single ADDER is much more inexpensive methodology than physically expanding the bandwidth between L_1 cache and L_2 cache as shown Figure 4.

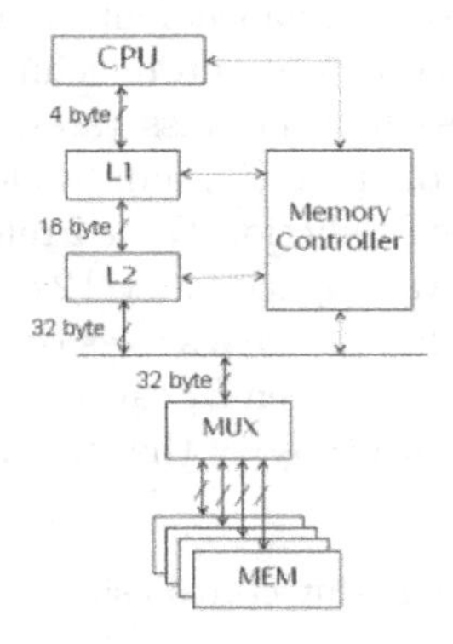

Fig. 4. Architecture with Expanded Bandwidth

5 Experiments and Performance Analysis

5.1 Simulation Environment

In order to analyze the efficiency of the proposed hierarchical cache architecture, we generated traces of instruction of an application running on ALPHA machine. Then the traces were analyzed by ATOM simulator [22]. We designed a cache simulator running on multi-thread manner.

We measured cache miss rate of each cache level, coverage and total execution cycles of 10 different benchmark programs. The benchmarks include SPEC2000, MediaBench. Table 2 shows the characteristics of each benchmark program used at this benchmark. We chose 7 programs from MediaBench and 3 programs from SPEC2000 INT. In the table, f_1 and f_2 denotes L_1 cache miss rate and L_2 cache miss rate in an architecture without prefetch logic, respectively.

Table 2. Benchmark Applications

Benchmark	Description	f_1	f_2
Cjpeg	JPEG image Compression	.0202	.0121
Djpeg	JPEG image Decompression	.0173	.0084
Mpegenc	MPEG2 Compression	.0031	.0014
Mpegdec	MPEG2 Decompression	.0009	.0006
Epic	EPIC image Compression	.0201	.0148
Unepic	EPIC image Decompression	.0345	.0122
Ghostscript	An interpreter for the PostScript language	.0407	.0232
Gcc	C Programming Language Compiler	.0160	.0066
Gzip	GNU Compression	.0367	.0319
Perl	PERL Programming Language	.0420	.0257
TimberWolfSC	Place and Route Simulator	.0886	.0555

5.2 Simulation Result and Analysis

In order to compare performance enhancement of the hierarchical cache architecture, we employ various metrics about benchmarks described at Table 2: instruction per cycle (IPC), each cache miss rates, coverage. The based cache architectures are architecture of employing non-prefetch.

We suppose that Correlation Prediction Table Prefetching technique [6,21] and Filtering Table prefetching [7] was applied to level 2 cache prefetcher and OBL was applied to level 1 cache prefetcher. Table 3 shows the experimentation results for 7 architectures. The bandwidths between L_1 cache and L_2 cache on both NC2 and NF2 are two times compared with those of the existing architectures.

Table 3. Cache Architectures using at Experiment

Legend	L1 Prefetcher	L2 Prefetcher
TR	None	None
NC	None	Correlation Table Algorithm
NF	None	Filtering Table Algorithm
NC2	None (bandwidth*2)	Correlation Table Algorithm
NF2	None (bandwidth*2)	Filtering Table Algorithm
OC	OBL	Correlation Table Algorithm
OF	OBL	Filtering Table Algorithm

5.2.1 The Number of Instruction Per Cycle

We compare performances of cache architectures by analyzing total numbers of cycle upon 5 cache architectures shown in Table 3 for the same benchmarks. We employ IPC as a test basis, since the length of executed instructions of every benchmark is different.

Figure 5 depicts IPC of all benchmarks. As we can see in Figure 5, OC and OF architecture that employ both L_1 cache prefetcher and L_2 cache prefetcher have much higher IPC than NC and NF that employ only L_2 cache prefetcher. This means that performances of OC and OF are much improved by employing level 1 cache prefetcher consisted of simple algorithm than existing architectures employing only L_2 cache prefetcher. In addition, it showed that these architectures had an efficiency of expanding bandwidth, for resulting in the higher IPC than both NC2 and NF2 with expanded bandwidth. Also, regarding level 2 prefetcher, the architecture of Filtering Table technique resulted in better performance than that of Correlation Prediction Table technique.

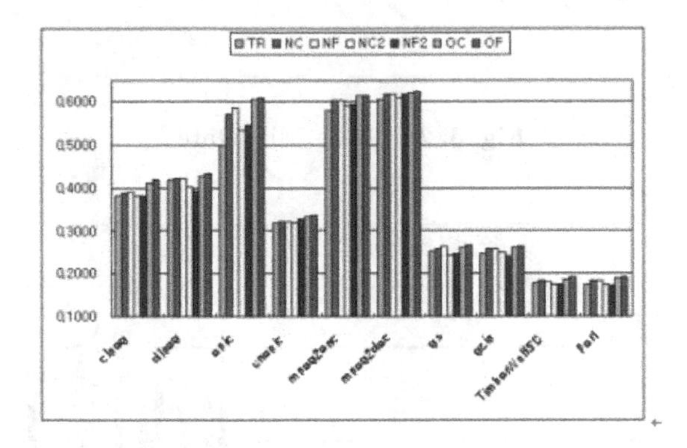

Fig. 5. The number of IPC

5.2.2 Cache Miss Rate Comparison

As the miss rate of L_1 cache and L_2 cache grows, the average time of memory access latency increases. Similarly, as the miss rate of L_1 cache becomes large, the number of prefetch from L_2 cache increases. Also as the miss rate of L_2 cache becomes large, the number of prefetch from main memory increases. So we test and analyze each miss rate of L_1 cache and L_2 cache in order to compare 7 cache architectures. Figure 6 and Figure 7 show the results of L_1 cache miss rate and L_2 cache miss rate respectively.

Figure 6 depicts L_1 cache miss rate of 7 cache architectures. As shown at Figure 6, OC and OF which employing L_1 cache prefetcher proposed in this paper has lower miss rate than other architecture which employs only L_2 cache prefecher. On the architectures, such as NC2 and NF2, loading two consecutive cache blocks from L_2 cache does not reduce miss rate, rather increase that rate unless the memory reference pattern is sequential. In Figure 6, NC shows better results than NF in the case of unepic and TimberWolfSC benchmarks, since the filtering table size of 16 KB is relatively smaller than the Correlation Prediction Table size of 2 MB. Thus, these tables are not able to accommodate the whole working set range, and as a result, it shows lower performance.

Figure 7 depicts L_2 cache miss rate of 7 cache architectures. The L_2 cache miss rate of OC and OF is lower than that of NC and NF. However, the number of L_2 cache miss between OC and OF and between NC and NF showed almost equal,

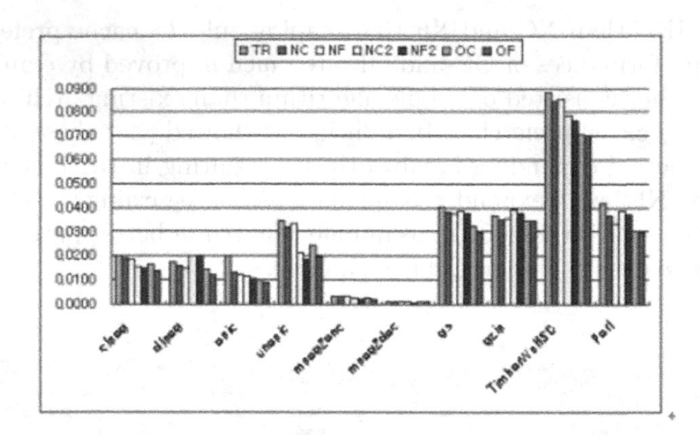

Fig. 6. L_1 Cache Miss Rate

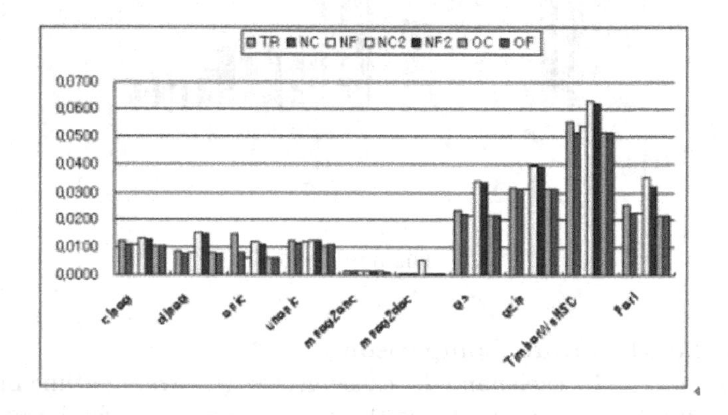

Fig. 7. L_2 Cache Miss Rate

since efficiency of L_2 cache prefetcher is dependent of benchmark program's characteristic. On the other hand, referring to the architectures of NC2 and NF2, the very rare use of second block loaded from L_2 cache in times of L_1 cache miss resulted in increasing L_1 cache miss rate and only to increase overall total latency.

6 Conclusions

In this paper, we proposed a cache architecture that employed L_1 cache prefetcher onto the existing hierarchical cache architecture that has only L_2 cache prefetcher. We compared the performances of the proposed architecture with a physically expanded bus bandwidth architecture. An experiments upon various benchmarks showed that the architecture employing both L_1 cache prefetcher and L_2 cache prefetcher resulted in the fastest execution time on all benchmarks. Also this simulation proves that employing a simple prefetch algorithm such as OBL draw the same efficiency as extending bandwidth between L_1 cache and L_2 cache. The experimentation among 10 benchmark programs showed that the efficiency of the hierarchical architecture employing L_1 cache prefetcher and L_2 cache prefetcher was achieved up to 11% compared with that of architecture employing only L_2 cache prefetcher.

Consequently, the employment a cache prefetcher at each cache level in the hierarchical memory architecture is expected to improve overall architecture performance.

References

1. Grama, A., Gupta, A., Karapis, G., Kumar, V.: Introduction to Parallel Computing, 2nd edn. Addison Wesley, Reading (2003)
2. Fritts, J.: Multi-Level Memory Prefetching for Media and Streaming Processing. In: Proceedings International Conference on Multimedia and Expo (2002)
3. Bear, J.L., Wang, W.H.: Architectural Choices for Multi-level Cache Hierarchies. In: Proceedings 16th international Conference on Parallel Processing, pp. 258–256 (1987)
4. Moon, H.J., Jeon, J.N., Kim, S.: Design of A Media Processor Equipped with Dual Cache. Journal Korean Institution Science Society 29(9), 573–581 (2002)
5. Gaddis, N.B., Butler, J.R., Kumar, A., Queen, W.J.: A 56-entry instruction reorder buffer, Solid-State Circuits Conference. In: IEEE International Digest of Technical Papers. 43rd ISSCC, pp. 212–213 (February 1996)
6. Joseph, D., Grunwald, D.: Prefetching Using Markov Predictors. In: Proceedings 24th Inl. Symp. Computer Architecture, pp. 252–263 (June 1997)
7. Zhang, X., Lee, H.S.: A hardware-based cache pollution filtering mechanism for aggressive prefetches. In: Proceedings 2003 International Conference on Parallel Processing, pp. 286–293 (October 6-9, 2003)
8. Smith, A.: Sequential Program Prefetching in Memory Hierarchies. IEEE Computer 11(2), 7–21 (1997)
9. Jouppi, N.P.: Improving Direct-mapped Cache Performance by the Addition of a Small Fully associative Cache and Prefetch Buffers. In: Proceedings of the 17th Annual International Symposium on Computer Architecture, pp. 364–373 (May 1990)
10. Horel, T., Lauterbach, G.: UltraSPARC-III: Designing Third-generation 64-bit Performance. IEEE Micro 19(3), 73–85 (1999)
11. Chen, T.F., Baer, J.L.: Effective Hardware-Based Data Prefetching for High Performance Processors. IEEE Transactions on Computers 44(5), 609–623 (1995)
12. Jeon, Y.S., Moon, H.J., Jeon, J.N., Kim, S.: A Hardware Cache Prefetching Scheme for Multimedia Data with Intermittently Irregular Strides. KIPS Architecture 31(11), 658–672 (2004)
13. Chan, K.K., Hay, C.C., Keller, J.R., Kurpanek, G.P., Schumacher, F.X., Zheng, J.: Design of the HP PA 7200 CPU. Hewlett-Packard Journal 47(1), 25–33 (1996)
14. Pentium Processor User's Manual, Vol.1, Pentium Processor Databook, Intel (1993)
15. IA-32 Intel Architecture Software Developer s Manual, Vol.1, Basic Architecture, Intel (2004)
16. Denamn, M.: PowerPC 604. Hot Chips VI, 193–200 (1994)

17. Mutlu, O., Kim, H.S., Armstrong, D.N., Patt, Y.N.: Cache Filtering Techniques to Reduce the Negative Impact of Useless Speculative Memory References on Processor Performance. In: SBAC-PAD 2004. 16th Symposium Computer Architecture and High Performance Computing, October 27-29, 2004, pp. 2–9 (2004)
18. Lee, J.S., Hong, W.K., Kim, S.D.: Design and Evaluation of On-Chip Cache Compression Technology. In: Proceedings the 17th IEEE International Conference on Computer Design, pp. 184–191 (1999)
19. Rivers, J.A., Tyson, G.S., Davidson, E.S., Austin, T.M.: On High-Bandwidth Data Cache Design for Multi-Issue Processors. In: Proceedings of the 30th Annual International Symposium on Micro architecture, pp. 46–56 (December 1997)
20. Lee, J.H., et al.: An Intelligent Cache System with Hardware Prefetching for High Performance. IEEE Transactions on Computers 5(5), 607–617 (2003)
21. Solihin, Y., Lee, J., Torrellas, J.: Correlation prefetching with a user-level memory thread. IEEE Transactions on Parallel and Distributed Systems 14, 563–580 (2003)
22. Srivastava, A., Eustace, A.: ATOM: A System for Building Customized Program Analysis Tools. In: Proceedings ACM SIGPLAN 1994, pp. 196–205 (1994)

Implementation and Evaluation of the Mechanisms for Low Latency Communication on DIMMnet-2

Yasuo Miyabe[1], Akira Kitamura[1], Yoshihiro Hamada[2], Tomotaka Miyasiro[1], Tetsu Izawa[1], Noboru Tanabe[3], Hironori Nakajo[2], and Hideharu Amano[1]

[1] Faculty of Science and Technology Keio University, 3-14-1 Hiyoshi, Kohoku-ku, Yokohama, 223-8522 Japan
pdarch@am.ics.keio.ac.jp
[2] Tokyo University of Agriculture and Technology
[3] Corporate Research and Development Center, Toshiba

Abstract. DIMMnet-2 is a network interface for PC cluster, plugged into a DIMM slot. Connecting network interface into commonly used memory bus reduces the cost of building PC cluster compared with using expensive machines with recent high performance I/O bus like PCI-X. Moreover, low latency communication from the host CPU can be achieved. In this paper, implementation of the mechanisms for low latency communication on the DIMMnet-2 prototype board by making the best use of the memory slot is shown. Its latency for 4 Bytes data transfer is only 1.4 μs which is lower than those of InfiniBand and QsNET II on condition those host processes are Intel Xeon.

Keywords: DIMMnet-2, Network interface, PC cluster.

1 Introduction

PC clusters have been widely used because of its high degree of performance per cost. Since both high bandwidth and low latency are required in the networks for such PC clusters, not only a general purpose network such as Gigabit Ethernet but also a special low latency network such as Myrinet[1], Quadrics Network (QsNET)[2] and InfiniBand[3] are used.

The network interfaces of such networks are usually connected to a high performance I/O bus like PCI-X or PCI-Express. However, using such high performance I/O buses tends to increase the price of each node of PC cluster systems. The motherboard with PCI-X slot is mostly for high end servers and more expensive than that used in commodity PCs. PCI-Express 16X (its bandwidth is 8GBytes/s) slots have been used for graphic devices in commodity PCs. But if non-graphic devices such as network interfaces are plugged into the slot, the slot works as PCI-Express 1X (0.5GBytes/s)[4][5] whose bandwidth is not enough for the networks of PC clusters.

J. Labarta, K. Joe, and T. Sato (Eds.): ISHPC 2005 and ALPS 2006, LNCS 4759, pp. 211–218, 2008.

Moreover, the latency of a host chipset occupies a large part of the total latency of communication. For example, the latency on host chipset for 8 Bytes data transfer on QsNET II[6] is 75% of the whole latency.

We have proposed the network interfaces plugged into memory slots, especially DIMM slots, to solve these problems. The bandwidth of a memory bus has improved along with Moore's law, and will be enough for the future networks for PC clusters. Additionally, the memory slots can be accessed from the host with low latency.

In this paper, mechanisms for low latency communication of DIMMnet-2 plugged into DIMM slots are introduced. Our evaluation shows that the prototype of DIMMnet-2 can transfer 4 Bytes data in 1.49 μs.

2 DIMMnet-2

DIMMnet-2 is the second generation of the network interface plugged into a DIMM slot that we have proposed. The first generation of them are called DIMMnet-1[7]. DIMMnet-1 was designed for plugged into an SDR SDRAM slot and connected each other by experimental switches. In contrast, DIMMnet-2 is designed for plugged into a DDR or DDR2 SDRAM slot and connected using commercial InfiniBand switches.

Most commodity PCs are not equipped with a high performance I/O bus like a PCI-X or a PCI-Express X8 that can cope with 10Gbps class network. Although servers provide such I/O buses, they are expensive as a node of PC clusters.

In contrast, most commodity PCs have memory slots which are equipped a high bandwidth, and they evolved rapidly to pursuit the performance improvement of CPUs. In addition, memory bus has the important merit that it can be accessed from host CPUs with low latency. Therefore, by using a memory slot for connecting a network interface, PC clusters with a high performance per cost can be built.

2.1 Prototype Board

The prototype board of DIMMnet-2 using an FPGA as a controller is now available. The aim of this prototype is validation of logic functions before making an ASIC.

Fig. 1 shows the photograph of the prototype board. The network controller is implemented on the Xilinx Virtex-II Pro (XC2VP70) which has high speed serial interface called RocketIO. This prototype board is connected to InfiniBand Switches (4X: 10Gbps) by using the RocketIO.

This board has two DDR SO-DIMMs. They are used not only for communication buffers but also for data memory space for the host CPU. We plan to provide globally shared memory with larger capacity than that in usual memory board by installing multiple memory modules on the board.

In this version, with the limitation of FPGA performance, the main operating frequency is set to be 100MHz, and only PC-1600 DDR SDRAM slot can be used. By replacing the FPGA with the ASIC version controller or a new generation FPGA such as Xilinx Virtex-4, connection with common memory slots will be available.

Fig. 1. Photograph of DIMMnet-2 prototype board

2.2 Network Interface Controller

Fig. 2 shows the block diagram of DIMMnet-2 network interface controller.

Since it is difficult to handle 64bit DDR signals used with DDR SDRAM, the controller converts 64bit DDR signals into 128bit SDR signals with *DDR Host Interface* and *DDR SO-DIMM Interface*. These modules also generate and interpret DDR SDRAM commands.

In DIMMnet-1 using an SDR SDRAM slot, the CPU in the host PC can access directly the SO-DIMMs on the DIMMnet-1 board. However, we have known that such direct accessing is difficult when a high speed DDR SDRAM slot is used from experience of DIMMnet-1.

Thus, the indirect accessing method is adopted in DIMMnet-2. In this method, the CPU accesses the SO-DIMMs through buffers in the network interface controller. The buffers for the indirect access are *Prefetch Window* and *Write Window* in the *CoreLogic* which is the main block of the controller. Fig. 3 shows the block diagram of the CoreLogic.

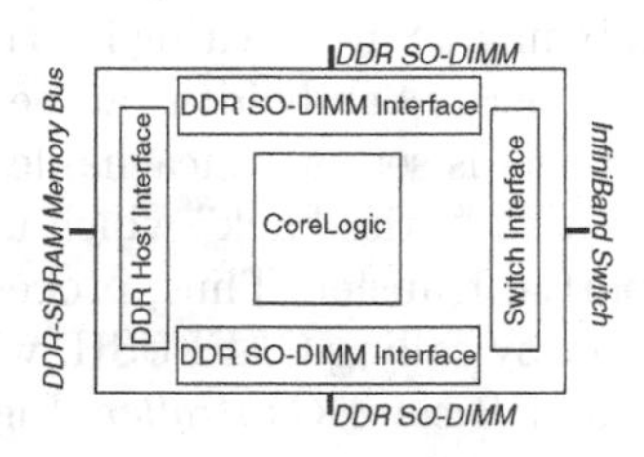

Fig. 2. Block diagram of network interface

For writing data, the CPU writes data into Write Window, and then writes "write" command into the *Command Register* with its address and data size to write transaction. *Window Controller* interprets the command and starts *Write Unit*. Then, the written data is transferred from Write Window to SO-DIMM by Write Unit.

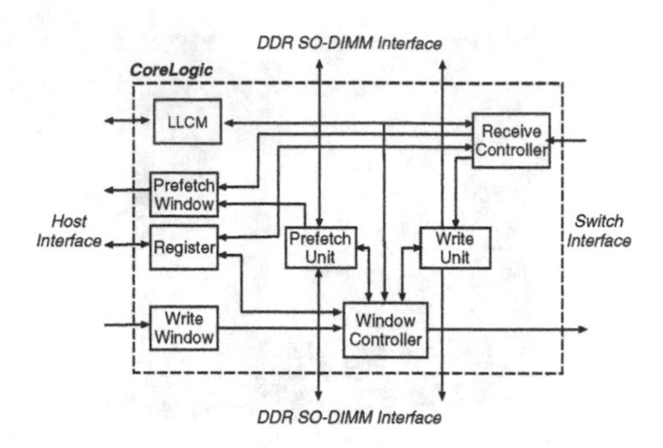

Fig. 3. Block diagram of CoreLogic

For reading data, the CPU writes "prefetch" command into the command register with its address, and then Window Controller starts *Prefetch Unit* for transferring data from SO-DIMM to Prefetch Window. After a fixed delay or changing the value of the *Status Register*, the CPU can read data from Prefetch Window. If the issue timing of "prefetch" is well scheduled, the latency for data transfer between SO-DIMM and Prefetch Window is almost hidden.

Sophisticated prefetch access commands including stride vector access are available. That is, the interface controller collects the only required data words from two SO-DIMMs, and writes them into Prefetch Window. Since the CPU can access required words with the sequential access manner, the latency can be drastically reduced. In such a sense, the controller equips a memory control facilities.

CoreLogic has another buffer called by *LLCM (Low Latency Common Memory)*. This is used for general purpose, for example, for receiving small data or flags for the communication completion.

These buffers and registers are mapped on individual physical address, and MTRR (Memory Type Range Register) which can be used on Intel Pentium Pro or later IA32 processors is suitable for increasing the access speed to them. Write Window is set as "write combining". Since writing into the write combining memory does not pollute cache, the write bandwidth can be enhanced compared with other types. LLCM and registers is set as "uncachable". To increase read bandwidth, Prefetch Window is set as "write back" which uses burst transfer instead of uncachable which uses partial transfer. Thus, processes must maintain cache coherence of Prefetch Window by calling CLFLUSH, which flushes a cache line.

Switch Interface (Fig. 2) and *Receive Controller* (Fig. 3) are used for communication and described later.

3 The Mechanisms for Low Latency Communication

3.1 Communication of DIMMnet-2 Overview

In DIMMnet-2 system, a process is identified by a pair of Process Group ID (PGID) and Process ID (PID). PGID identifies the groups of processes executing the same parallel processing. PID identifies the processes in the same PGID. In contrast, in the network of InfiniBand, a DIMMnet-2 board is identified by Local ID (LID) defined by the InfiniBand Architecture, and processes sharing the same board are identified by Window ID (WID). Sender processes convert receiver PID to receiver LID and WID using the table made by the privilege process, and inform the controller of the LID and WID.

PGID is used to avoid interferences from irrelevance processes. When the controller constructs the packet, the controller finds the sender PGID using sender WID as a key in the table set into the DIMMnet-2 by the privilege process, and then the sender PGID is written into the header of the packet. When the packet is received, the controller compares receiver PGID and the PGID of the header. If they are different, the packet is disposed.

The typical communication of DIMMnet-2 involves the following steps.

1. Sender process writes receiver LID and WID to the control register with other information such as data size.
2. Window Controller reads data from SO-DIMM or Write Window, and then generates a corresponding packet.
3. Window Controller transfers the packet to Switch Interface. Switch Interface encapsulates the packet into InfiniBand packets and sends them to the network.
4. When Receiver Controller detects the incoming packet to Switch Interface, it reads the header of the packet.
5. If the PGID of the packet is correct, the data from Switch Interface are written into Prefetch Window or LLCM by Receive Controller or into SO-DIMM by Write Unit.
6. If the specific flag of the header is valid, the statuses such as sender LID etc. are written to LLCM, and then the status register is updated.

3.2 Block on the Fly (BOTF)

For the low latency communication, DIMMnet-2 supports Block On The Fly (BOTF)[8] data transfer. In BOTF, user process writes a whole packet data into Write Window, and then writes the command into the control register to start packet transfer. Window Controller can send a BOTF packet with a small latency because it can generate a packet easily from the data in the Write Window.

This method can also reduce the overhead of building request for the controller. All information for sending packet is written into Write Window as the packet header. When we reuse Write Window, we don't have to write the same information as the previous transfer, such as the receiver LID.

User process can send any packet by BOTF. However, the controller always rewrites the fields of PGID and packet size in the correct value for protection.

4 Evaluation

In this section, we show the latency of the BOTF implemented on the prototype board of DIMMnet-2.

4.1 Method for Measuring the Latency

The latency was evaluated with a ping-pong message transfer between two DIMMnet-2 boards connected with a two meters InfiniBand cable. Other specifications of the environment are shown in Table 1.

Table 1. Measurement environment

CPU	Pentium 4 2.6GHz
Chipset	VIA VT8751A
Memory	PC-1600 DDR-SDRAM 512MBytes $\times 1$
	DIMMnet-2 $\times 1$
OS	RedHat8 (Kernel 2.4.27)

We evaluated the following times as measures of the latency. The measures are started when a user process starts to write data into Write Window.

(a) When a receive process detects that the data have been written in Prefetch Window by polling the status register.
(b) When a receive process detects that the data have been written in SO-DIMM by polling the status register.
(c) After (a), when a receive process finishes copying received data from Prefetch Window to a buffer allocated by the receive process.
(d) When a receive process detects that the data has been written in LLCM by polling LLCM, and reads out it.

4.2 Result and Discussion

Fig. 4 shows the latency of (a), (b), and (c). (The size that can be received to LLCM is limited to 8 Bytes in the current implementation.) When 4 Bytes data were transfered, the latencies of (a) (b) (c) and (d) were 1.74 μs, 1.81 μs, 1.98 μs, and 1.49 μs, respectively.

This shows that the latency of receiving to LLCM was the lowest. Incoming data to LLCM can be detected by polling where the next data are received, so detecting and reading data can be done simultaneously. However, this mechanism doesn't work well in case of incoming data to Prefetch Window. Since the memory type of Prefetch Window is "write back", polling Prefetch Window needs to call CLFLUSH at each reading to flush cache lines to increase the overhead. Thus, detecting incoming data to Prefetch Window is done by polling the status register whose value changes after completion the reception processing.

The difference between (a) and (b) is caused by the latency for writing data to SO-DIMM.

Table 2 compares the latency of DIMMnet-2 and other networks. The value of the table contains the latency of a switch. If DIMMnet-2 connects to an InfiniBand switch and retransmission is supported, the latency increases by 0.4 μs and becomes 1.89 μs.

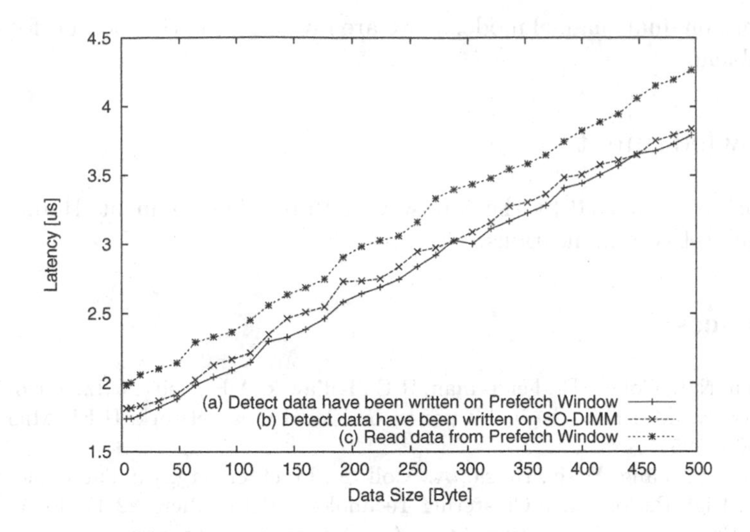

Fig. 4. Result of the latency evaluation

Table 2. Comparison of latency with other networks

DIMMnet-2 BOTF	1.49 μs (1.89 μs)
InfiniHost RDMA (PCI-Express, Xeon)	3.8 μs [9]
QsNET II RDMA (PCI-X, Opteron, AMD)	0.97 μs [10]
QsNET II RDMA (PCI-X, Xeon, Serverworks GC-LE)	2.68 μs [10]

Table 2 shows the latency of DIMMnet-2 is small as half of the latency of InfiniHost, which is an interface of InfiniBand provided by Mellanox, though both DIMMnet-2 and InfiniHost use InfiniBand X4. In addition, the latency of DIMMnet-2 is lower than that of QsNET II with Xeon processors, but is not lower than that of QsNET II with Opteron processors. Considering that the current board is a prototype and motherboards supporting PC-1600 is outdated and comparatively low performance, the latency of DIMMnet-2 is quite well compared with QsNET II.

5 Conclusion and Future Work

In this paper, we introduced DIMMnet-2, which is the network interface plugged into a memory slot, and the latency of BOTF is evaluated. Although the current board is a prototype, the latency for 4 Bytes data transfer is only 1.48 μs which is lower than those of InfiniBand and QsNET II when Intel Xeon is used as a host CPU.

Now, we are connecting multiple DIMMnet-2 nodes with an InfiniBand switch. Further measurement and performance evaluation are needed with such a parallel system before the ASIC migration. And, in this implementation, the memory slot can't work on dual channel mode, so we are investigating the method for solving this problem.

Acknowledgment

This work is supported by the Ministry of Public Management, Home Affairs, Posts and Telecommunications.

References

1. Boden, N.J., Cohen, D., Felderman, R.E., Kulawik, A.E., Seitz, C.L., Seizovic, J.N., Su, W.-K.: Myrinet - A gigabit per second local area network. IEEE Micro 15(1), 29–36 (1995)
2. Petrini, F., Fang, W.-c., Hoisie, A., Coll, S., Frachtenberg, E.: The Quadrics Network: High-Performance Clustering Technology. IEEE Micro 22(1), 46–57 (2002)
3. InfiniBand Trade Association. http://www.infinibandta.org/
4. Intel 915G/915GV/910GL/915P Express Chipset Datasheet (Sepember 2004), http://www.intel.co.jp/
5. Intel 925X/925XE Express Chipset Datasheet (November 2004), http://www.intel.co.jp/
6. Hewson, D., Beecroft, J., Hewson, D., McLaren, M., Roweth, D.: QsNet II: Performance Evaluation (2003), http://www.quadrics.com/
7. Tanabe, N., Hamada, Y., Nakajo, H., Imashiro, H., Yamamoto, J., Kudoh, T., Amano, H.: Low latency communication on DIMMnet-1 network interface plugged into a DIMM slot. In: Parallel Computing in Electrical Engineering 2002, pp. 9–14 (2002)
8. Tanabe, N., Yamamoto, J., Hamada, Y., Nakajo, H., Kudoh, T., Amano, H.: BOTF:A High Bandwidth Communication Mechanism of DIMMnet-1 Network Interface Plugged into a DIMM slot. IPSJ Journal (2002)(In Japanese)
9. Liu, J., Mamidala, A., Vishnu, A., Panda, D.K.: Evaluating InfiniBand Performance with PCI Express. IEEE Micro 25(1), 20–29 (2005)
10. Beecroft, J., Addison, D., Hewson, D., McLaren, M., Petrini, F., Roweth, D.: Quadrics QsNet^{II}: Pushing the Limit of the Design of High-Performance Networks for Supercomputers. IEEE Micro, (accepted for publication), Available from http://www.c3.lanl.gov/fabrizio/papers/ieeemicro-elan4.pdf

Computationally Efficient Parallel Matrix-Matrix Multiplication on the Torus

Ahmed S. Zekri and Stanislav G. Sedukhin

Graduate School of Computer Science and Engineering, The University of Aizu,
Aizu-Wakamatsu City, Fukushima 965-8580, Japan
{d8062103, sedukhin}@u-aizu.ac.jp

Abstract. In this paper, we represent the computation space of the $(n \times n)$-matrix multiplication problem $C=C+A \cdot B$ as a 3D torus. All possible time-minimal scheduling vectors needed to activate the computations inside the corresponding 3D index points at each step of computing are determined. Using the projection method to allocate the scheduled computations to the processing elements, the resulting array processor that minimizes the computing time is a 2D torus with $n \times n$ processing elements. For each optimal time scheduling function, three optimal array allocations are obtained from projection. All the resulting allocations of all the optimal scheduling vectors can be classified into three groups. In one group, matrix C remains and both matrices A and B are shifted between neighbor processors. The well-known Cannon's algorithm belongs to this group. In another group, matrix A remains and both matrices B and C are shifted. In the third group, matrix B remains while both matrices A and C are shifted. The obtained array processor allocations need n compute-shift steps to multiply $n \times n$ dense matrices.

Keywords: matrix-matrix multiplication, 3D torus, array processor, space-time mapping.

1 Introduction

High performance computing (HPC) is constrained mainly by the processor-memory gap. Managing the movement of data between memory and processor is very crucial to enhance the performance. A very naive strategy is to reuse the loaded data as much as possible by means of blocked algorithms expressed in terms of level-3 BLAS (Basic Linear Algebra Subroutines). Meanwhile, most of linear algebra algorithms depend heavily on matrix operations [1]. Hence, the motivation of this research is to investigate efficient methods to perform parallel matrix-matrix multiplication on the best array processor with minimum number of computing steps.

Consider the $(n \times n)$-matrix multiplication problem $C=C+A \cdot B$. The sequential way to compute matrix C is through the following process: for each element $c(i,j)$ where $i,j \in \Omega = \{0,1,...,n-1\}$, on each iteration $k \in \Omega$ we should compute $c(i,j)=c(i,j)+a(i,k) \cdot b(k,j)$. This process takes n^3 multiply-add operations

J. Labarta, K. Joe, and T. Sato (Eds.): ISHPC 2005 and ALPS 2006, LNCS 4759, pp. 219–226, 2008.

to complete matrix-matrix multiplication. The index space of the underlying problem can be defined by $\Im = \{(i,j,k)^T : i,j,k \in \Omega\}$ where for each point $p=(i,j,k)^T \in \Im$, we should compute $c(i,j)$ such that $\Im$ is a partial ordered set of computations. Note that because the summation is associative and commutative, a notation of $k \in \Omega$ means that we can correctly accumulate a result in any order, resulting in $n!$ possible ways of pairing. We repeatedly apply this process for each element of matrix C. Ultimately we will have $n^2 \times n!$ possible ways of computing the matrix-matrix multiplication.

Generally, for sequential computing there is almost no difference in what way we select k under the assumption that all initial and intermediate data are equally fast accessed from memory. So, for some ordered set of pairs $i,j \in \Omega$, the classical algorithm to compute matrix-matrix multiplication starts with k=0 and proceeds with an incrementing step until k=n-1. For parallel processing, however, it is not trivial to choose one way from the exponential solution space. Obviously, we have to introduce some heuristics to guide the selection (or scheduling) procedure. One such heuristic is to maximize reusing data in each iteration step k. To do so we can compute matrix C by using an outer product of two n-vectors on each iteration step k, i.e., *for all* elements of matrix C on *each* k-th iteration $k \in \Omega$, we update matrix C by; $C=C+a_k \cdot b_k^T$ where, a_k and b_k are the k-th column and k-th row of matrices A and B, respectively [2], [3], [4]. In this case, we can introduce a linear time scheduling function *Step*(p)=k, $p \in \Im$, such that $n \times n$ array of processing elements (PEs), where the (i, j)-th PE holds the initial and intermediate values of $c(i,j)$, can compute the matrix-matrix multiplication in n steps. The alignment of the two vectors a_k and b_k can be done inside or outside the array processor. Unfortunately, this approach is based on broadcasting of two n-vectors among an array of PEs to form $(n \times n)$-matrix of outer products and therefore is not scalable. The data broadcasting can be transformed to data pipelining (like in systolic array processors) by using a linear time scheduling function *Step*(p)=i+j+k which introduces additional n-1 steps to fill the pipeline and n-1 steps to flush the pipeline resulting totally in 3n-2 steps for matrix-matrix multiplication on $n \times n$ array processors [5], [3]. Another efficient way to reduce broadcasting is proposed by Fox, Otto and Hey [6]. In this method, data broadcasting through columns is replaced by local circular upward shifts of the elements of matrix B. The elements of matrix A are still broadcast along the array rows. The time scheduling in this case can be described by the modular function *Step*(p)=[-i+k] mod n [7]. This algorithm calculates matrix C in n steps on an $(n \times n)$ array processor. It is clear that this solution is scalable in one direction only.

As a natural extension to the above discussion it seems that removing broadcasts completely is the best way. In this case, it is better to reuse all elements of matrices A and B to update matrix C on each computational step such that each step includes n different iteration steps $k \in \Omega$ for n^2 different (i,j) in a predefined order. The reuse in this case is performed by cyclical rotation of the elements of matrices A, B and C in three orthogonal directions such that one direction is assigned to each matrix. What is required is that the elements $a(i,k)$, $b(k,j)$, and $c(i,j)$ should meet with each other in correct place and time. It is clear that three orthogonal directions for all elements of matrices A, B and C form a 3D torus where the orbital processing is based on the modular time scheduling function $Step(p)=\pi^T \cdot p \bmod n$. The optimal values of the scheduling vector π will be discussed later. If we consider all possible directions of rotations inside the 3D torus, all optimal time scheduling vectors that lead to equally efficient ways to solve the matrix-matrix multiplication problem can be determined. To allocate the computation points to the minimum number of PEs, the projection method [2]is applied. The resulting number of PEs is n^2 (the lower bound) organized in a 2D torus which is a natural projection of the 3D torus. The number of steps needed to complete computations in all array allocations is n.

To compute the resulting allocations on the 2D toroidal array processor, some communication overhead is needed to align data into PEs. For example, in Cannon's algorithm [8], [9], which is one of the resulting array allocations, input matrices are skewed at first. Then, at each step of computing, each PE multiplies the current elements of A and B and accumulates the resulting C element then shifts input data to neighbor PEs. After n steps of compute-shift operations, C is calculated. The data alignment overhead will not be addressed in this paper.

In Section 2 we present time-space mapping, modular scheduling and the projection method. The optimal scheduling vectors of the 3D torus computation space and the minimum number of PEs to execute the scheduled computations are presented in Section 3. Section 4 shows all optimal scheduling functions for the 3D torus computation space and the optimal allocations of the 2D toroidal array processor. Section 5 concludes the paper.

2 Time-Space Mapping

Given a set of computations inside an index space, time-space transformation is defined by finding i) a suitable time scheduling function to arrange the computations at the index points into groups and ii) assigning these groups to PEs in some logical sequence or simply time-steps assuming that the parallel execution and the accompanied data movement of each group takes one unit time. In linear time-space mapping, time scheduling and space allocation are combined into one mapping $T \cdot p$, where $p \in \Im$ and T is a 3×3 non-singular transformation matrix whose first row is the time scheduling vector and the second and third rows compose a 2×3 space transformation matrix which maps the 3D index space into the 2D array processor space [10], [1]. Our interest in this work is in modular time scheduling. So, we will use two separate functions for the time-space mapping, namely *Step*(p) as a modular scheduling function and *Allocation*(p) as a linear allocation function. The latter function determines the coordinates of PEs in the array processor where the scheduled computations and data movement are executed. The definitions of the two functions are given below. Some conditions exist to preserve data dependencies and to prevent allocating more than one computation to the same PE at the same time (see [11], [12], [10] for details).

2.1 Time Scheduling

A scheduling function specifies the sequence of computing steps in the array processor. It is a mapping from the 3D index space to a 1D schedule space. For details on different scheduling functions proposed in the literature, see for example [2], [13], [11]. The type of scheduling determines the topology and the interconnection pattern of the array processor and vice versa. For example, linear scheduling can be used with broadcast communications. In square mesh array processors with wrap around connections, modular scheduling is the best alternative. Modular mapping have been used in [7] to derive data distribution independent programs for matrix-matrix multiplication. In our paper, we use modular time scheduling functions, which are characterized by a linear scheduling followed by a modulo operation, to schedule computations inside the index space $\Im$,

$$Step(p) = [\alpha\, i + \beta\, j + \gamma\, k] \bmod n,$$

where $Step(p) : Z^3 \rightarrow Z$ and $\pi = (\alpha, \beta, \gamma)^T$ is a time scheduling vector with integer components. The mod operation is used to guarantee the generation of the required sequence of steps and to manipulate the wrap around connections.

2.2 Space Transformation

Different allocation functions exist to map the computation points onto the physical array processor [11], [2], [10]. In the well-known projection method, the 3D index space is mapped to 2D space to form the array processor. Because the computation space of the matrix-matrix multiplication problem is a 3D torus, the array processor resulting from the projection will be a 2D torus. We can define the space transformation by the following linear map:

$$Allocation(p) = S \cdot p,$$

where $Allocation(p){:}Z^3{\rightarrow}Z^2$, $p \in \Im$ and S is a $2{\times}3$ space transformation matrix.

Generally, the shape and size of the index space determine the directions of projection. We will call the direction that allows the allocation of exactly one computation point at a time step to exactly one PE an admissible direction of projection. Said alternatively, admissibility holds if and only if the scalar product $\pi^T{\cdot}v \neq 0$, where v is the direction of projection [11]. We need to construct the matrix S such that its rows are orthogonal to the direction of projection which means that $S{\cdot}v{=}0$. It is also assumed that the rows of S are linearly independent. The admissible directions of projection that result in the minimal number of PEs will be discussed next.

3 Optimal Mapping

Optimality here means that the time-space mapping has the minimum computing time on the minimum number of PEs. The objective now is to determine the values of the coefficients α, β, and γ of the function *Step*(p) which define all possible optimal time scheduling vectors used to schedule computations inside the 3D torus. We also determine the admissible directions of projections to assign the scheduled computations to the minimal number of PEs in the array processor.

The optimal scheduling vectors of the 3D torus computation space can be easily determined by a simple enumeration. Let $r_0, r_1, ..., r_{m-1}$ be ordered objects connected in a ring. If we start from r_l, $0 \leq l < m$ then we have two paths to traverse the ring: $r_l, r_{l+1}, ..., r_{m-1}, r_0, ..., r_{l-1}$ and $r_l, r_{l-1}, ..., r_0, r_{m-1}, ..., r_{l+1}$. The subscript of the next object to traverse in the two paths is obtained by the modular operations $[l+1] \bmod m$, and $[l-1] \bmod m$, respectively, where l is the subscript of the current object. We will say that the traverse order of index l in the first path is increasing while in the second is decreasing. The 3D torus space comprises from three rings along i, j and k axes where each ring has two degrees of freedom to rotate. Therefore, eight different combinations of simultaneous rotations along i, j and k axes are possible. An increasing/decreasing order of an index will indicate that the corresponding component of the time schedule vector is assigned a positive/negative sign. The magnitudes of the schedule vector components must be chosen so that data dependencies are preserved while the communications are local. In addition, at any step of computing the data is moved in three directions which mean that none of the components equals to zero. Therefore, we found that each of the three components α, β, and γ equals either +1 or -1. By finding all combinations we get eight possible optimal time scheduling vectors. In fact, they are four vectors with their negatives. This means

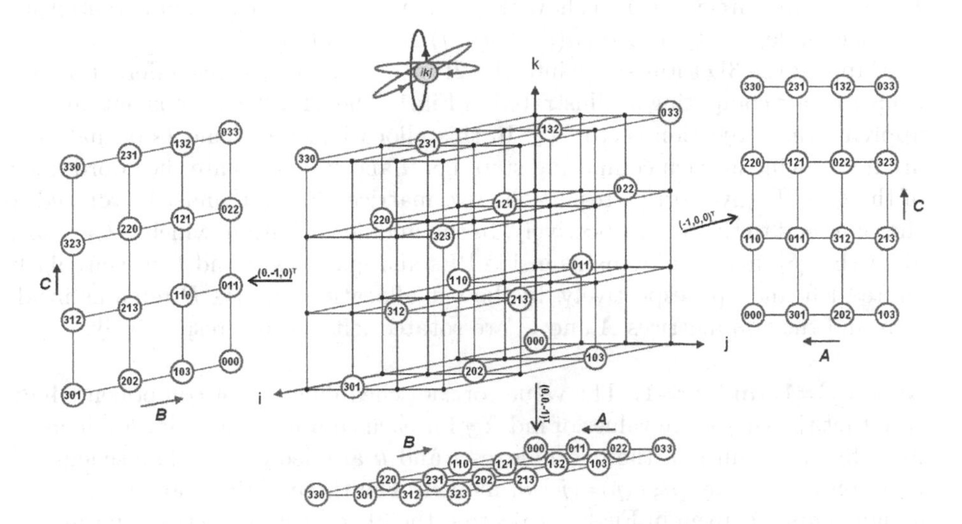

Fig. 1. The distributions of A, B and C at *Step*(p)=0, π=$(-1,-1,1)^T$. In the three data allocations resulted from projection, matrix A, C, and B remains in the left, middle, and right allocations, respectively.

that we have only four scheduling functions described by four pairs of vectors, as will be seen in the next section.

Now, we determine the minimum number of PEs to use with the optimally scheduled computations inside the index space $\Im$. In fact, there are many admissible projection directions to allocate the 3D computation points to PEs. For example, the direction $v = (1,1,1)^T$ is admissible and there are many valid constructions of the matrix S so that $S \cdot v = 0$. The resulting array processor in this case has a hexagonal shape with $3n^2$-$3n$+1 PEs. But, we seek directions that result in the minimum number of PEs in the processor array. It is obvious that the lower bound of the number of PEs (in planar arrays) associated with the cubical $n \times n \times n$ index space $\Im$ is n^2. This means that the best admissible directions of projections that should be used are the unit vectors $e_1 = (1,0,0)^T$, $e_2 = (0,1,0)^T$, and $e_3 = (0,0,1)^T$ and their negatives. In the next section, we will show that using the resulting $n \times n$ toroidal array all the array allocations need n steps of compute-shift operations to solve the ($n \times n$)-matrix multiplication problem.

4 Optimal Time-Space Allocations

α=-1, β=-1, and γ=1. In this case, the time scheduling function is given by *Step*(p) =[k-i-j] mod n. The rings along i and j axes are rotated in reverse (decreasing) order. The ring along k-axis is rotated in normal (increasing) order. The distributions of A, B and C elements inside the 3D torus are shown in Fig. 1. We use the circles (ikj) to show the active points at each step of computing. At each circle, $c(i,j) = c(i,j) + a(i,k) \cdot b(k,j)$ is computed.

Mapping the 3D torus space into the 2D space, we have three different allocations to start computing as illustrated in Fig.1. The left allocation is obtained by applying the projection vector -e_2. In this allocation, the elements of matrix A are not moving at each computing step, i.e., i and k indexes are the coordinates of the PEs. To give correct results the two matrices B and C must be rotated to the right and to the top, respectively. In the middle allocation, which is Cannon's algorithm [8], matrix C remains inside PEs and matrices A and B are circularly shifted left and up, respectively. In the left allocation, matrix B remains inside PEs and the two matrices A and C are rotated left and up, respectively.

α=-1, β=1, and γ=-1. The values of the scheduling vector components indicate that the successive values of index j for each computation point are increasing while the values of the two indexes i and k are decreasing. The scheduling function is given by *Step*(p)=[j-i-k] mod n. The three optimal array processor assignments are given in Fig.2. Projecting the 3D computation space parallel to the j, k and i axes gives the three allocations (a), (b) and (c), respectively.

α=1, β=-1, and γ=-1. The values of α, β and γ indicate that i-index is increasing from step to the next while j and k indexes are decreasing. The scheduling function is given by *Step*(p) = [i-j-k] mod n.

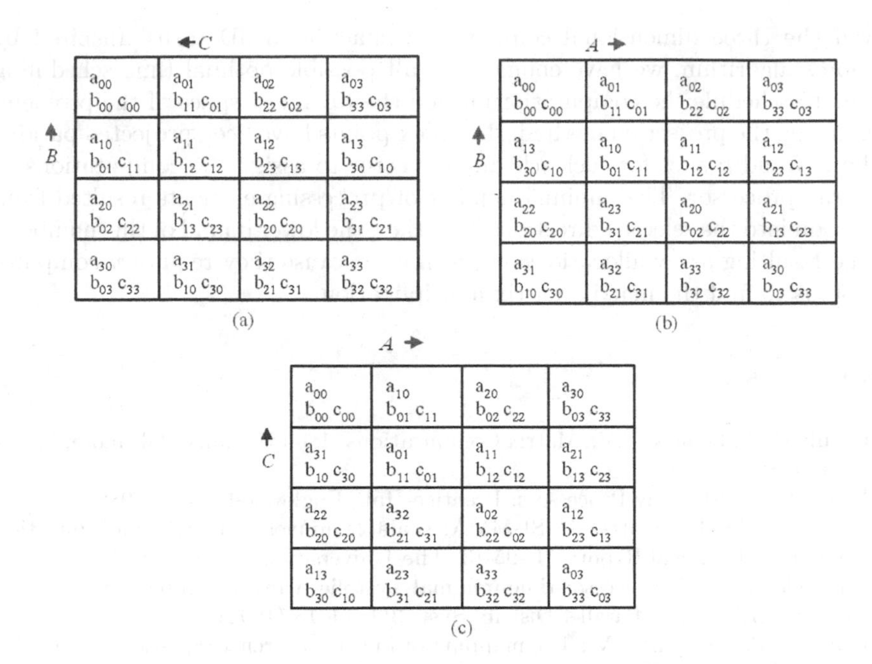

Fig. 2. The optimal array processor allocations at *Step*$(p) = 0$ for $\pi = (-1, 1, -1)^T$

α=-1, β=-1, and γ=-1. The scheduling function is given by *Step*$(p)=[-i-j-k] \bmod n$, where all indexes are rotated in decreasing order.

The remaining four scheduling vectors are the negative of the previously discussed ones. According to the properties of the modulo operation, for each scheduling vector π_i and its negative $-\pi_i$, the distribution of A, B and C elements will be the same at the steps $Step_{\pi_i}(p)=q \bmod n$ and $Step_{-\pi_i}(p)=[n-q] \bmod n$ where $q = 0, 1, ..., n-1$.

When solving the (2×2)-matrix multiplication problem, all the four described scheduling functions will match resulting in only one 3D scheduling. The reason behind this special case is that although there are two directions of rotation for each ring in the 3D torus, they have the same traverse path r_0, r_1. This means that the direction of rotation has no effect (in this case) on the distribution of the data inside the computation space. Therefore, there are only three optimal 2D data allocations to parallelize the (2×2)-matrix multiplication problem on the 2×2 toroidal array processor.

5 Conclusions

A torus provides high-bandwidth nearest-neighbor connectivity while using local communications. It is scalable and can be directly applied to many scientific and data-intensive applications. In this paper, we investigated efficient ways to multiply two $n \times n$ dense matrices on $n \times n$ toroidal array processor. We have viewed the three dimensional computation space as a 3D torus. Inspired by Cannon's algorithm, we have enumerated all possible optimal time scheduling vectors to schedule the computations inside the 3D index space of the problem. Then, using the projection method, the index points have been projected parallel to the i, j and k axes for each scheduling vector to assign the computations to the array processor. The minimal number of processing elements resulted from projection into the selected directions is n^2 (i.e., the lower bound of this number). All the resulting array allocations are optimal because they require n compute-shift steps to find the matrix-matrix multiplication.

References

1. Golub, G.H., Loan, C.F.V.: Matrix Computations. John Hopkins, Baltimore, Maryland (1989)
2. Kung, S.: VLSI Array Processors. Prentice-Hall, Englewood Cliffs (1988)
3. van de Geijn, R., Watts, J.: SUMMA: scalable universal matrix multiplication algorithm. Technical Report TR-95-13, The University of Texas (April 1995)
4. Cappello, P.R.: A processor-time-minimal systolic array for cubical mesh algorithms. IEEE Trans. Parallel Distrib. Syst. 3(1), 4–13 (1992)
5. Quinton, P., Dongen, V.V.: The mapping of linear recurrence equations on regular arrays. Journal of VLSI. Signal Processing 1
6. Fox, G., Otto, S., Hey, A.: Matrix algorithms on a hypercube I: Matrix multiplication. Parallel Computing 4, 17–31 (1987)
7. Lee, H.J., Fortes, J.A.: Modular mappings and data distribution independent computations. Parallel Processing Letters 7(2), 169–180 (1997)
8. Cannon, L.: A Cellular Computer to Implement the Kalman Filter Algorithm. PhD thesis, Montana State University (1969)
9. Lee, H.J., Robertson, J.P., Fortes, J.A.B.: Generalized Cannon's algorithm for parallel matrix multiplication. In: ICS 1997. Proceedings of the 11th international conference on Supercomputing, pp. 44–51. ACM Press, New York (1997)
10. Weston, J.H., Zhang, C.N., Y., Y.F.: Some space considerations of space-time mappings into systolic arrays. In: Int. Conf. on Parallel and Distributed Processing Techniques and Apps., Sunnyvale, California (August 1996)
11. Lavenier, D., Quinton, P., Rajopadhye, S.: Advanced systolic design. In: Digital Signal Processing for Multimedia systems. Signal Processing Series, pp. 657–692. Marcel Dekker (1999)
12. Alain Darte, M.D., Robert, Y.: A characterization of one-to-one modular mappings. Parallel Processing Letters 5(1), 145–157 (1996)
13. Darte, A., Robert, Y.: Constructive methods for scheduling uniform loop nests. IEEE Trans. Parallel Distrib. Syst. 5(8), 814–822 (1994)

A New Dynamic Load Balancing Technique for Parallel Modified PrefixSpan with Distributed Worker Paradigm and Its Performance Evaluation

Makoto Takaki, Keiichi Tamura, Toshihide Sutou, and Hajime Kitakami

Graduate School of Information Sciences, Hiroshima City University,
3-4-1 Ozukahigashi, Asaminami-ku, Hiroshima, 731-3194, Japan
{makoto,ktamura,toshihide,kitakami}@db.its.hiroshima-cu.ac.jp

Abstract. In order to extract the frequent patterns that can become motif at high speed from amino acid sequences, we are developing the parallel Modified PrefixSpan with the distributed worker paradigm. This paper presents a new dynamic load balancing technique for the parallel Modified PrefixSpan with the distributed worker paradigm and its performance evaluation. The characteristics of the dynamic load balancing are the small-grain task and the Cache-based Random Steal schema. This paper explains these characteristics and presents performance evaluations with the PC cluster of 100 nodes.

Keywords: data mining, parallel computing.

1 Introduction

In the field of molecular biology, there has been an increased interest in using the data mining techniques to discover motifs from amino acid sequences. A motif is a featured sequential pattern in amino acid sequences. It is regarded as a function that has been conserved in the process of molecular evolution.

In order to extract the motifs effectively, the algorithms which can extract the frequent sequential patterns with fixed-length wildcard regions and variable-length wildcard regions are required. A Modified PrefixSpan [7] can extract the frequent sequential patterns which include fixed-length wildcard regions and variable-length wildcard regions.

In our previous studies [10] [11], we have developed a parallel Modified PrefixSpan with a master-worker paradigm [13] [4] and its dynamic load balancing technique. The Modified PrefixSpan has a high parallelism. However, the load imbalance is very large. We have developed a new dynamic load balancing technique, a master-task-steal schema, for the master-worker paradigm.

The master-worker paradigm has a performance bottleneck of the master process on a large-scale PC cluster. Therefore, we have been developing a parallel Modified PrefixSpan with a distributed worker paradigm [13]. In the distributed

J. Labarta, K. Joe, and T. Sato (Eds.): ISHPC 2005 and ALPS 2006, LNCS 4759, pp. 227–237, 2008.

worker paradigm, the performance bottleneck of the worker proccss hardly occurs. However, the most efficient dynamic load balancing technique for the distributed worker paradigm is not able to be applied to the parallel Modified PrefixSpan because the load imbalance is very large.

This paper proposes a new dynamic load balancing technique for the parallel Modified PrefixSpan with the distributed worker paradigm. There are two characteristics of the proposed dynamic load balancing technique.

(1) Cache-based Random Steal Schema
Applications with a distributed worker paradigm always use a Random Steal (RS) [5] schema as a dynamic load balancing technique. However, there is a performance bottleneck, if the load is concentrated on a part of the worker processes. In the Modified PrefixSpan, the load is concentrated on a part of the worker processes. To address this problem, we propose a Cache-based RS schema. In the Cache-based RS schema, although the load is concentrated on a part of the worker processes, the performance bottleneck can be avoided.

(2) Small-Grain Task
A process that extracts frequent $(k + 1)$-length sequential patterns from a frequent k-length sequential pattern is called an small-grain task. The small-grain task is effective for dynamic load balancing. The small-grain task is adapted as a task in the parallel processing of the Modified PrefixSpan.

We evaluated the Cache-based RS schema on an actual large-scale PC cluster. In the experiments, three types of datasets that include motifs named Zinc Finger, Kringle and Leucine were used. Experimental results show that the speed-up ratios of the Cache-based RS schema are superior to those of the RS schema.

The rest of this paper is organized as follows: Section 2 explains the algorithm of the Modified PrefixSpan. Section 3 describes the related work. Section 4 shows the parallelism of the Modified PrefixSpan with the distributed worker paradigm. Section 5 proposes the Cache-based RS schema. Section 6 shows the experimental results. Section 7 is the conclusions.

2 Modified PrefixSpan

This section explains problem definition and the basic algorithm of the Modified PrefixSpan.

2.1 Problem Definition

Let $\Sigma = \{$A, C, D, E, F, G, H, I, K, L, M, N, P, Q, R, S, T, V, W, Y $\}$ be a set of all letter alphabet in amino acid sequences. A sequence s is denoted as $(a_1 a_2 \cdots a_m)$, where a_l is a letter, i.e., $a_l \in \Sigma$, $a_l = s[l]$, for $1 \leq l \leq m$. A sequence database S is a set of tuples $\langle sid, s_{sid} \rangle$, where sid is a sequence id and s_{sid} is a sequence. Table 1 shows an example of amino acid sequences.

Table 1. Example of Amino Acid Sequences

sequence id	sequence
1	FKYAKWL
2	SFVKTA
3	ALR
4	MSKPL
5	FSKFLMAW

A k-length sequential pattern is denoted as:

$$\langle pat^k \rangle = \langle A_1 - x(i_1, j_1) - A_2 - x(i_2, j_2) - \cdots - x(i_{k-1}, j_{k-1}) - A_k \rangle.$$

A symbol A_j is called a character element. A symbol "-" means that next element is continued. A symbol $x(i_n, j_n)$ represents wildcard regions. If $0 \leq i_n \leq j_n$, this region is called a variable-length wildcard region. If $i_n = j_n$, this region is called a fixed-length wildcard region, and this region can be represented by $x(i_n)$.

For example, the sequence s_1 includes a sequential pattern "F**A" which is denoted as $\langle F - x(2) - A \rangle$. The sequence s_1, s_2, and s_5 include the sequential pattern $\langle F - x(2,5) - A \rangle$. The sequence s_1 includes $\langle F - x(2) - A \rangle$, the sequence s_2 includes $\langle F - x(3) - A \rangle$, the sequence s_5 includes $\langle F - x(2) - A \rangle$ and $\langle F - x(5) - A \rangle$.

The support count of a k-length sequential pattern $\langle pat^k \rangle$ is the number of tuples containing $\langle pat^k \rangle$ in S. Given a positive integer, min_sup as the support count threshold, the k-length sequential pattern is called a frequent k-length sequential pattern if $\langle pat^k \rangle$ is contained by at least min_sup tuples.

Each $\langle pat^k \rangle$ has a projected database that keeps the subsequent positions of the rightmost characters of $\langle pat^k \rangle$. In order to extract the $\langle pat^{k+1} \rangle$ from a $\langle pat^k \rangle$, the projected databases of the $\langle pat^k \rangle$ are needed. The projected database of $\langle pat^k \rangle$ is denoted as $PDB(\langle pat^k \rangle)$.

There are three user-specified parameters, min_sup, the maximum number of wildcards, the maximum number of errors. The maximum number of wildcards is represented as wc_max. The maximum number of errors is represented as ε_{max}.

The value of wc_max is the maximum number of wildcards of which you may appear continuously in a frequent sequential pattern. The value of $\varepsilon_n = j_n - i_n$ $(1 \leq n \leq k_{max})$ is called as a error. The value of ε_{max} indicates the maximum number of ε_n.

2.2 Basic Philosophy

The basic philosophy of the Modified PrefixSpan is as follows:

- **phase 1:**
  The algorithm scans a sequence database once to find all frequent 1-length sequence patterns in the sequence database. The algorithm creates $PDB(\langle pat^1 \rangle)$ for each $\langle pat^1 \rangle$.

- **phase 2:**
  For each $\langle pat^k \rangle$, the algorithm constructs $(k+1)$-length sequential patterns by using the $PDB(\langle pat^k \rangle)$. If the support count of a $(k+1)$-length sequential pattern is more than min_sup, the $(k+1)$-length sequential pattern is a $\langle pat^{k+1} \rangle$. If no $\langle pat^{k+1} \rangle$ is extracted, the algorithm terminates the frequent sequential pattern extraction. Otherwise, $k := k+1$ and go to "phase2";

3 Related Work

There are many studies on frequent sequential pattern extraction of sequential database. Most of these studies adopt, an *apriori like* [3], a candidate generation-and-test approach. The *apriori like* approach may still be expensive, especially when long and numerous patterns are encountered. To overcome this problem, a new methodology, called *tree projection*, was developed by Jiawei Han[8]. This approach mines the frequent sequential patterns without candidate generation.

To the best of our knowledge, there is little work on the parallel processing of *tree-projection-based* frequent sequential pattern extraction algorithms. There are many studies on the parallel processing of the *apriori like* approach [2] [9] [12]. However, these parallelisms are not adapted to the *tree-projection-based* approach. The main cost of the *apriori like* approach is candidate generation and scanning sequence databases. The main cost of the *tree-projection-based* approach is the construction of frequent sequential patterns from the postfix databases and the generation of the projected database of frequent sequential patterns.

The type of work most related with the present research is the parallel tree-projection-based sequence mining algorithm proposed by Valerie Guralnik and George Karypis [6]. Comparisons of our work and Guralnik's work are as follows:

Guralnik's work focuses on transaction database. While our work focuses on amino acid sequences. The number of amino acid sequences is several thousand at most, however the length of a sequence is over 100 or 1000. The maximum length of pattern is very large.

Guralnik defines a sub-tree search as a task. The processing time of a task can be estimated beforehand.We define the small-grain task as a task, because

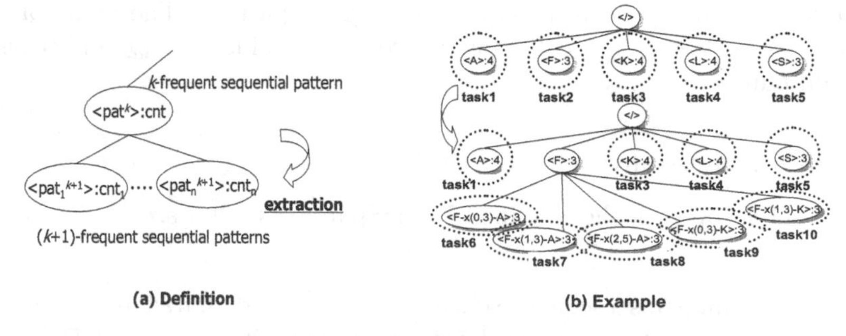

Fig. 1. Small-grain task

the load imbalance of Modified PrefixSpan is very large and the processing time of a task cannot be estimated beforehand.

In Guralnik's work, the RS schema is adapted. The effective results were obtained, because the bias of work load is small. Moreover, the number of machines are comparatively few(16 machines, 32CPUs). We evaluated the proposed technique on the PC cluster of 100 nodes. The RS schema is effective when the number of sites is comparatively few. However, the performance is decremented when the number of nodes is about one hundred. Thus, the modification of the RS schema was needed.

4 Parallel Modified PrefixSpan with Distributed Worker Paradigm

4.1 Task Definition

A process that extracts $(k+1)$-frequent sequential patterns from a k-frequent sequential pattern is called a small-grain task (Figure 1 (a)). The small-grain task is adapted as a task in the parallel Modified PrefixSpan, because the load imbalance of the Modified PrefixSpan is very large. Moreover, the extraction processing time cannot be estimated beforehand.

Figure 1 (b) shows an example of the small-grain task. There are five 1-frequent sequential patterns. For each 1-frequent sequential pattern, the pattern is considered to be one small-grain task. Five 2-frequent sequential patterns are extracted when *task2* is executed, and five small-grain tasks, *task6, task7, task8, task9, task10* are generated newly.

A small-grain task can be executed independently. When $(k+1)$-frequent sequential patterns are extracted from the k-frequent sequential pattern, the contents of processing that has been needed to extract a k-frequent sequential pattern, and the information of another frequent sequential patterns are not needed. This is because the Modified PrefixSpan must examine only the subsequence which follows the rightmost characters of $\langle pat^k \rangle$ when $(k+1)$-frequent sequential pattern is extracted.

4.2 Algorithms

This section shows the processing steps of the parallel Modified PrefixSpan with the distributed worker paradigm which has the RS schema. One worker process begins to extract 1-frequent sequential patterns. This worker process is called a leader worker process. Each worker process has a *local task pool* that stores small-grain tasks and a P_local_i that stores frequent sequential patterns. Each worker process executes concurrently the **Main Function.**

Main Function :

(1) If the worker process is the leader worker process, the worker process sends the parameters to all worker processes. Otherwise, the worker process receives the parameters from the leader worker process.

(2) If the worker process is the leader worker process, go to the **Initialize Subroutine.**
(3) If the *local task pool* is not empty, go to the next step. Otherwise, go to the step (6).
(4) The worker process extracts all $(k+1)$-frequent sequential patterns from the k-frequent sequential pattern. The extracted $(k+1)$-frequent sequential patterns are inserted into P_local_i. Each process, that extracts $(k+2)$-frequent sequential patterns from a $(k+1)$-frequent sequential pattern is inserted into *local task pool.*
(5) Go to the step (3).
(6) Go to the **Random Steal Subroutine.**
(7) If the *local task pool* is not empty, go to the step(3). Otherwise, go to the **Finalize Subroutine.**

Random Steal Subroutine :

(1) The worker process selects a donor worker process (*DWP*) randomly. Then the worker process sends a *task request message* to the *DWP.*
(2) The following processes are selected by the reply of the *task request message.*
   (a) If a small-grain task is returned as the reply of the *task request message*, the received small-grain task is inserted into the *local task pool.* Return to the **Main Function.**
   (b) Otherwise, the worker process selects a donor worker process (*DWP*) randomly. However, the worker process that has already sent the *task request message* is not selected. Go to the step (1).

Initialize Subroutine :
This subroutine is executed only on the leader worker process.

(1) First of all, the leader worker process extracts 1-frequent sequential patterns from the sequence databases S. The 1-frequent patters are inserted into P_local_l.
(2) The leader worker process inserts the small-grain tasks into the *local task pool.* The data structure of a small-grain task consists of a pair of $\langle pat^1 \rangle$ and PDB($\langle pat^1 \rangle$).

Task Steal Trap Subroutine:
This subroutine is executed on the donor worker process which receives the *task steal message* from another worker process.

(1) If the *local task pool* is not empty, go to the step (2). If the *local task pool* is empty and the donor worker process is not executing a small-grain task, go to the step (3). If the *local task pool* is empty and the donor worker process is executing a small-grain task, go to the step (4).
(2) The donor worker process pops a small-grain task from the *local task pool.* The donor worker process sends the small-grain task to the requested worker process as a reply.

(3) The donor worker process sends the *null message* to the requested worker process as a reply.
(4) Wait for executing the small-grain task by the donor worker process. If the execution of small-grain task has finished, return to the step (1).

Finalize Subroutine :

The worker process sends the P_local_i to the leader worker process. The leader worker process receives the P_local_i by all worker processes, and insert the frequent sequential patterns into the P_local_l.

5 Cache-Based Random Steal Schema

5.1 Key Idea

In the RS schema, if the load is concentrated on a part of the worker processes, many messages of the task request are generated until a worker process that holds a task is reached. The load concentrates on a part of the worker processes in the parallel processing of the Modified PrefixSpan.

To address the performance bottleneck of the RS schema, we propose the Cache-based RS schema. The key idea of the Cache-based RS schema is to cache the ID of the donor worker process. The cached ID is denoted as CID.

In the Cache-based RS schema, as a worker process becomes idle, it randomly selects a donor worker process first. The worker process caches the ID of the donor worker process (DWP). As a worker process becomes idle again, the worker process looks up the *CID*. If the value of *CID* is valid, the worker process selects the DWP whose ID is the *CID* as a donor worker process and sends DWP a task requests. If the value of *CID* is invalid, the worker process, selects a donor worker process randomly.

5.2 Algorithm

The **Random Steal Subroutine** that is modified by the Cache-based RS schema is as follows. The worker process has the value CID which stores the ID of the donor worker process.

Random Steal Subroutine:

(1) If the value of CID is invalid, the worker process randomly selects a donor worker process (DWP). Otherwise, the worker process selects the donor worker process (DWP) whose ID is CID.
(2) The worker process sends a *task request message* to the DWP.
(3) The following processes are selected by the reply of the *task request message*.
   (a) If a small-grain task is returned as a reply of the *task request message*, the received small-grain task is inserted into the *local task pool*. The ID of the DWP sets on the CID. Return to the **Main Function**.
   (b) Otherwise, the worker process randomly selects a donor worker process (DWP). However, the worker process that has already sent the *task request message* is not selected. Go to the step (1).

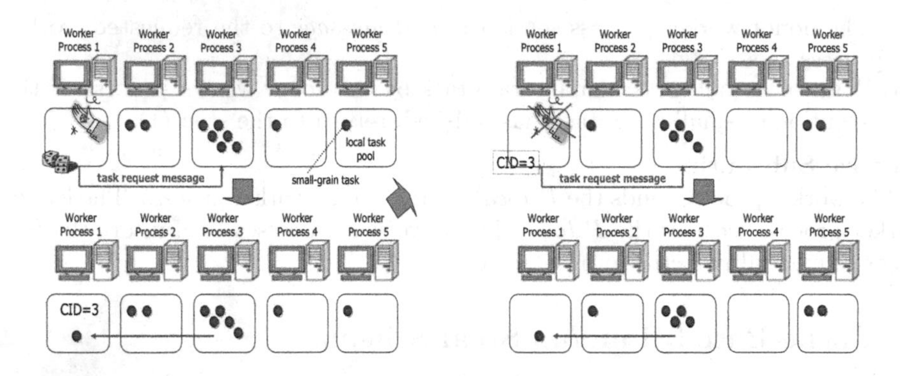

Fig. 2. Example of Cache-based RS Schema

5.3 Example

Figure 2 shows an example of the Cache-based RS schema. There are five worker processes.

First, as the *worker process 1* becomes idle, the *worker process 1* selects randomly the donor worker process. Supposing that the donor worker process is *worker process 3*, the *worker process* sends *task request message* to *worker process 3*.

Second, as the *local task pool* of *worker process 3* is not empty, the *worker process 3* sends a small-grain task to the *worker process 1*. The *worker process 1* receives the small-grain task from *worker process 3* and the *worker process 1* substitutes the ID of *worker process 3* for *CID* in *worker process 1*.

Third, as the *worker process 1* becomes idle again, the *worker process 1* does not select the donor worker process randomly. The *worker process 1* selects the worker process whose ID is equal to *CID*. Therefore *worker process 3* is selected as the donor worker process. The *worker process 1* sends the *task request message* to *worker process 3*. In the RS schema, as the *worker process 1* becomes idle again, the *worker process 1* select the donor worker process randomly.

6 Experimental Results

6.1 Environments

There are one hundred personal computers, each configured with a 2.8GHz Pentium4 processor, 1.0GB memory, and 40GB disk. The personal computers were connected to a 1,000 Mbit/sec Ethernet. Fedora Core 2.0 was used as the operating system. MPICH version 1.2.6 was used as the MPI library.

6.2 Speed-Up

The datasets used in this evaluation were provided by PROSITE [1]. This experiment, each parameter was as follows: Kringle dataset, *min_sup*=59 (85%),

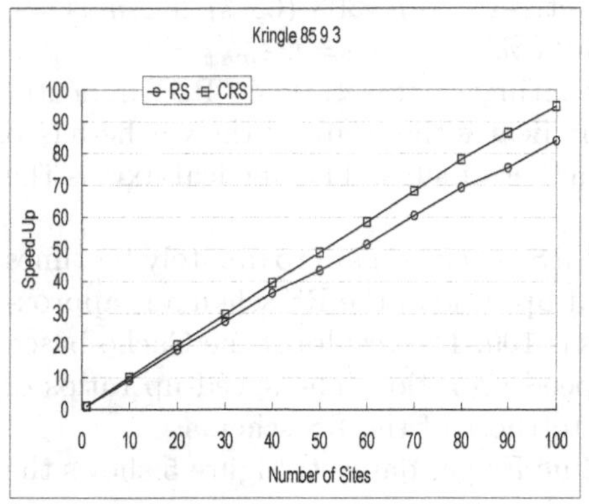

Fig. 3. Speed-Up (Kringle)

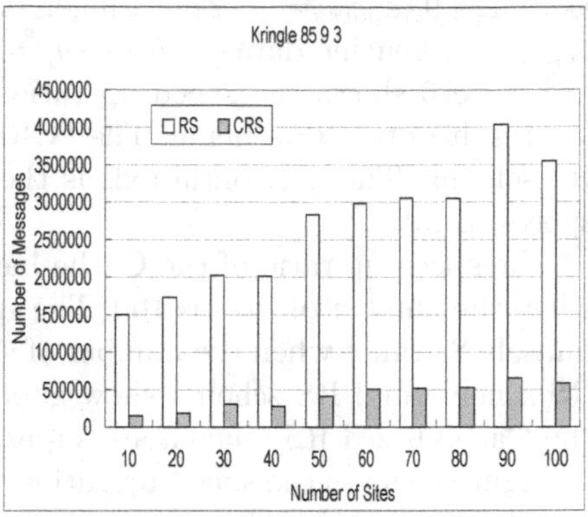

Fig. 6. # of Task Request (Kringle)

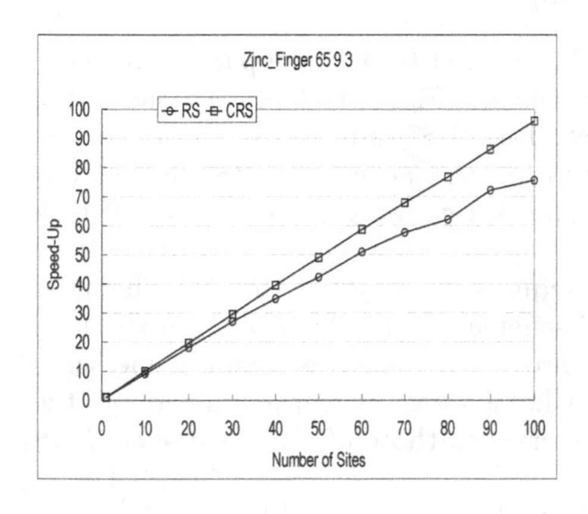

Fig. 4. Speed-Up (Zinc Finger)

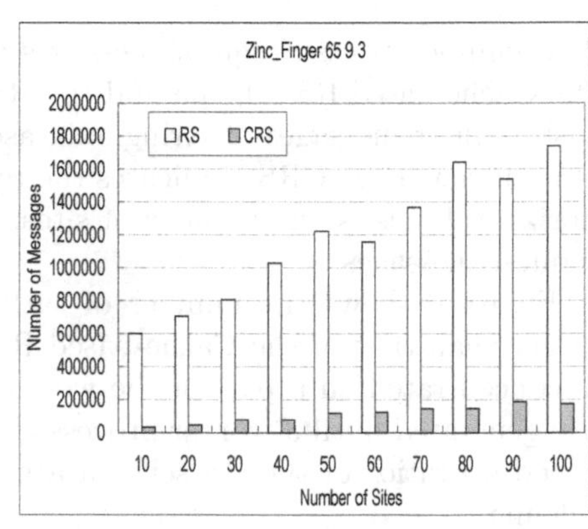

Fig. 7. # of Task Request (Zinc Finger)

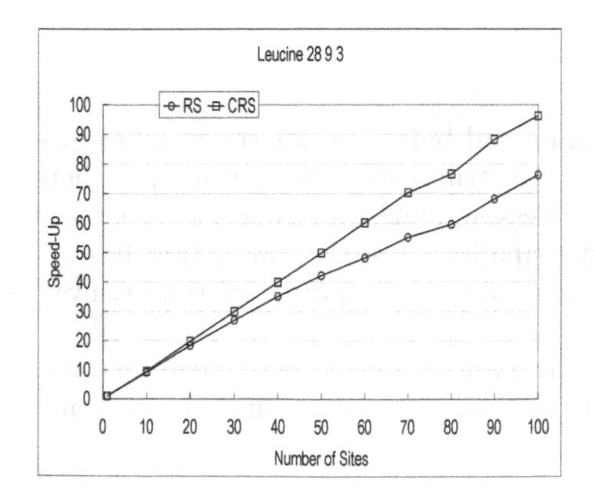

Fig. 5. Speed-Up (Leucine)

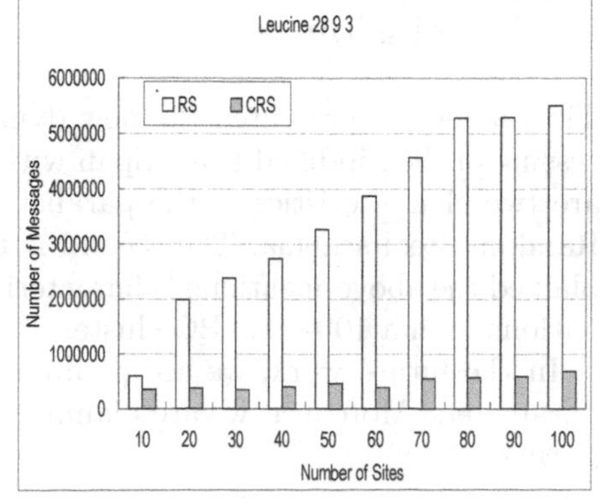

Fig. 8. # of Task Request (Leucine)

wc_max=9, $\varepsilon_{max} = 3$; Zinc Finger dataset, *min_sup*=303 (65%), *wc_max*=9, $\varepsilon_{max} = 3$; Leucine dataset, *min_sup*=103 (28%), *wc_max*=9, $\varepsilon_{max} = 3$.

Figure 3 shows the speed-up ratios of Kringle dataset. The "RS" indicates the result of the RS schema. The "CRS" indicates the result of the Cache-based RS schema. The horizontal axis is the number of sites. The vertical axis is the speed-up ratio.

The speed-up ratio of the Cache-based RS schema is approximately 95 times when the number of sites is 100. The speed-up ratio of the RS schema is approximately 83 times when the number of sites is 100. The results of the Cache-based RS schema and RS schema show good speed-up ratios. The speed-up ratios of the Cache-based RS schema are superior to those of the RS schema.

Figure 4 shows the speed-up ratios of Zinc Finger dataset. Figure 5 shows the speed-up ratios of Leucine dataset.

6.3 Number of Task Request Messages

The number of task request messages is measured for the comparison between the Cache-based RS schema and the RS schema. Figure 6 shows the number of task request messages of Kringle dataset. The "RS" indicates the result of the RS schema. The "CRS" indicates the result of the Cache-based RS schema. The horizontal axis is the number of sites. The vertical axis is the number of task request messages.

Figure 6 shows the number of task request messages of the RS schema is larger than that of the Cache-based RS schema. In the RS schema, if the load is concentrated on a part of the worker processes, many task request messages are generated until a worker process that holds a task is reached. Therefore the speed-up ratios of the RS schema are inferior to those of the Cache-based RS schema.

Figure 7 shows the speed-up ratios of Zinc Finger dataset. Figure 8 shows the speed-up ratios of Leucine dataset.

7 Conclusion

This paper has presented the new dynamic load balancing for the parallel processing of the Modified PrefixSpan with the distributed worker paradigm. There are two characteristics of the parallel processing. The first is the Cache-based Random Steal schema. The second is the small-grain task. This paper has explained the above-mentioned characteristics and presented the performance evaluations with a 100-scale PC cluster.

In the future work, we are planning the performance evaluation by various parameters. Moreover, we are planning to develop this paradigm with grid computing.

Acknowledgements

This work was supported in part by a Hiroshima City University Grant for Special Academic Research (General Studies, No.3106), Grand-in-Aid for Yong Research (B) (No.16700114) from the Ministry of Education, Culture, Sports, Science and Technology in Japan and a Grant-in-Aid for Scientific Research (C) (No.17500097) from the Japanese Society for the Promotion of Science.

References

1. http://kr.expasy.org/prosite/
2. Agrawal, R., Shafer, J.C.: Parallel mining of association rules. IEEE Trans. Knowl. Data Eng. 8(6), 962–969 (1996)
3. Agrawal, R., Srikant, R.: Mining sequential patterns. In: Proceedings of the Eleventh International Conference on Data Engineering, pp. 3–14. IEEE Computer Society Press, Los Alamitos (1995)
4. Carriero, N., Gelernter, D.: How to write parallel programs: a guide to the perplexed. ACM Computing Surveys 21(3), 323–357 (1989)
5. Eager, D.L., Lazowska, E.D., Zahorjan, J.: A comparison of receiver-initiated and sender-initiated adaptive load sharing (extended abstract). In: Proceedings of the 1985 ACM SIGMETRICS conference on Measurement and modeling of computer systems, pp. 1–3. ACM Press, New York (1985)
6. Guralnik, V., Karypis, G.: Parallel tree-projection-based sequence mining algorithms. Parallel Comput. 30(4), 443–472 (2004)
7. Kitakami, H., Kanbara, T., Mori, Y., Kuroki, S., Yamazaki, Y.: Modified prefixspan method for motif discovery in sequence databases. In: Ishizuka, M., Sattar, A. (eds.) PRICAI 2002. LNCS (LNAI), vol. 2417, pp. 482–491. Springer, Heidelberg (2002)
8. Pei, J., Han, J., Mortazavi-Asl, B., Wang, J., Pinto, H., Chen, Q., Dayal, U., Hsu, M.-C.: Mining sequential patterns by pattern-growth: The prefixspan approach. IEEE Transactions on Knowledge and Data Engineering 16(11), 1424–1440 (2004)
9. Shintani, T., Kitsuregawa, M.: Parallel mining algorithms for generalized association rules with classification hierarchy. In: SIGMOD 1998. Proceedings ACM SIGMOD International Conference on Management of Data, pp. 25–36. ACM Press, New York (1998)
10. Sutou, T., Tamura, K., Mori, Y., Kitakami, H.: Design and implementation of parallel modified prefixspan method. In: Veidenbaum, A., Joe, K., Amano, H., Aiso, H. (eds.) ISHPC 2003. LNCS, vol. 2858, pp. 412–422. Springer, Heidelberg (2003)
11. Takaki, M., Tamura, K., Sutou, T., Kitakami, H.: Dynamic load balancing for parallel modified prefixspan. In: PDPTA 2004. Proceedings of The 2004 International Conference on Parallel and Distributed Processing Techniques and Applications, pp. 352–358. CSREA Press (2004)
12. Tamura, M., Kitsuregawa, M.: Dynamic load balancing for parallel association rule mining on heterogenous pc cluster systems. In: Proceedings of 25th International Conference on Very Large Data Bases, pp. 162–173. Morgan Kaufmann, San Francisco (1999)
13. Wilkinson, B., Allen, M.: Parallel Programming Techniques and Applications Using Networked Workstations and Parallel Computers. Prentice-Hall, Englewood Cliffs (1999)

Performance-Based Loop Scheduling on Grid Environments

Wen-Chung Shih[1], Chao-Tung Yang[2], and Shian-Shyong Tseng[1,3]

[1] Department of Computer and Information Science
National Chiao Tung University
Hsinchu 300, Taiwan, R.O.C.
{gis90805,sstseng}@cis.nctu.edu.tw
[2] High-Performance Computing Laboratory
Department of Computer Science and Information Engineering
Tunghai University
Taichung 407, Taiwan, R.O.C.
ctyang@thu.edu.tw
[3] Department of Information Science and Applications
Asia University
Taichung 413, Taiwan, R.O.C.
sstseng@asia.edu.tw

Abstract. Loop scheduling and load balancing on parallel and distributed systems are critical problems, but it is difficult to cope with these ones, especially on the emerging grid environments. Previous researchers proposed some useful self-scheduling schemes, which were applicable to PC-based cluster and grid computing environments. In this paper, we generalized this concept and proposed a general approach, named PLS (Performance-Based Loop Scheduling). To verify our approach, a grid platform was built, and two application programs, matrix multiplication and Mandelbrot, were implemented with MPI to be executed in this testbed. Experimental results showed that our approach was efficient and robust, in terms of the range of α value.

Keywords: Parallel loops, Self-scheduling, Grid computing, Heterogeneous, Performance estimation.

1 Introduction

One objective of grid computing is to virtualize various computing and data resources dispersed in the world, and the resulting single system image enables users to easily access to every type of resources [6].

Recently, the emerging standardization of sharing resources and the availability of higher network bandwidth result in the realization of grid computing [8].

Loop scheduling and load balancing on parallel and distributed systems are critical problems, but it is difficult to cope with these problems, especially on the

J. Labarta, K. Joe, and T. Sato (Eds.): ISHPC 2005 and ALPS 2006, LNCS 4759, pp. 238–245, 2008.

emerging grid environments. Previous researchers have proposed some useful self-scheduling schemes, which are applicable to PC-based cluster and grid computing environments [8, 19, 20]. These two-phased schemes collect system configuration information, and distribute some portion of the work load among slave nodes according to their CPU clock speed ratio. After that, the remaining work load is scheduled by some well-known self-scheduling scheme.

Traditionally, self-scheduling schemes partition the size of loop iterations according to a formula instead of node performance, so additional slave nodes may not get good performance. Intuitively, we may want to partition problem sizes according to CPU clock speed. However, the CPU clock is not the only factor which affects node performance. Many other factors also have dramatic influences in this respect, such as the amount of memory available, the cost of memory accesses, and the communication medium between nodes, etc.

In this paper, we generalize this concept and propose a general approach, named PLS (Performance-based Loop Scheduling). This approach estimates the performance ratio of the clusters, and determines the performance ratio of each node accordingly. To verify our approach, a grid platform is built, and two application programs, matrix multiplication and Mandelbrot, have been implemented with MPI to be executed in this testbed.

The rest of this paper is organized as follows. In section 2, the background about parallel loop self-scheduling and grid computing is reviewed. In section 3, we define our model and describe our approach. Next, our system configuration is specified and experimental results on two application programs are also presented in section 4. Finally, the concluding remarks are given in the last section.

2 Background

In this section, a prerequisite for our research is described. First, we review previous work about self-scheduling schemes. Next, the evolution of grid computing is outlined.

2.1 Self-scheduling Schemes

In general, parallelizing compilers make parallel loop scheduling decisions either statically at compile time or dynamically at run time [15]. Self-scheduling is a large class of dynamic loop scheduling schemes, and [8] provides a good related review about them. In this subsection, we restate self-scheduling schemes briefly. Basically, a self-scheduling scheme uses a function to calculate the chunk-size at each step. At each scheduling step, the master computes the chunk-size and assigns this amount of workload to an idle slave. Different methods to compute the chunk-size have resulted in different scheduling schemes. The well-known schemes include Pure Self-Scheduling (PSS), Chunk Self-Scheduling (CSS), Guided Self-Scheduling (GSS), Factoring Self-Scheduling (FSS) and Trapezoid Self-Scheduling (TSS). The different chunk sizes for a problem with the number of iterations $N = 1024$ and the number of processors $p = 4$ are shown in Table 1.

Table 1. Sample partition size

Scheme	Sample partition size
PSS	1, 1, 1, 1, 1, 1, 1, 1, 1,···
CSS(128)	128, 128, 128, 128, 128, 128, 128, 128, 128, ···
FSS	128, 128, 128, 128, 64, 64, 64, 64, 32, ···
GSS	256, 192, 144, 108, 81, 61, 46, 34, 26, ···
TSS	128, 120, 112, 104, 96, 88, 80, 72, 64, ···

In [19], the authors have revised known loop self-scheduling schemes to fit extremely heterogeneous PC cluster environments. An approach was proposed to partition loop iterations by two phases and it achieved good performance in any heterogeneous environment.

2.2 Grid Computing

Since the term "metacomputing" was presented by Larry Smarr [17], this concept has evolved from a "Seamless Web" to a global grid environment. Basically, grid computing is intended to partition a job and distribute the sub-jobs to computers all over the world, maybe several thousand computers. Grid computing can be seen as one sort of specialization of traditional parallel processing. However, several driving forces make grid computing a promising trend.

Grid computing has a lot of various applications, and can be implemented in different ways. Nevertheless, the building, administration, and operation of a global grid environment require new technology, which involves grid architectures, software protocols and middleware [1, 3, 4, 5, 7, 8, 10, 11, 12, 13, 14].

3 Our Approach: PLS (Performance-Based Loop Scheduling)

In this section, the concept of performance ratio is described first. Then, we present the algorithm of our scheme.

3.1 Performance Ratio

We propose to partition $\alpha\%$ of workload according to the performance ratio of all nodes, and the remaining workload is dispatched by some well-known self-scheduling scheme. Using this approach, we do not need to know the real computer performance. However, a good performance ratio is desired to estimate performance of nodes accurately.

To estimate the performance of each slave node, we define a performance function (PF) for a slave node S as follows:

$$PF_S(V_1, V_2, \cdots, V_M) \tag{1}$$

where
- S represents a slave node S.
- V_i $(1 < i < M)$ is a parameter of the performance function.

In this paper, we do not formulate our performance function explicitly. Instead, execution time is utilized to estimate the value of PF's for all nodes. The PF obtained by simulation execution can estimate performance of computing nodes rather accurately. The calculation of performance ratio (PR) is presented as follows. First, the PF of all nodes are estimated by experimental simulation. Execution time of the target program on all computing nodes is recorded, and their reciprocals are taken to form the performance function values. Performance ratio (PR) is defined as the ratio of these PF's. For instance, assume the PF of 2 nodes are $1/2$ and $1/3$. Then, the PR is $1/2 : 1/3$; i.e., the PR of the two nodes are 3:2. In other words, if there are 5 loop iterations, 3 will be assigned to the former and 2 will be assigned to the latter.

3.2 Our Algorithm

Our algorithm is also a two-phased scheme. In phase one, the performance ratio of slave nodes is estimated by experimental simulation. Then, we partition $\alpha\%$ of workload according to the performance ratio of all slave nodes, and the remaining workload is dispatched by some well-known self-scheduling scheme. The algorithm of our approach is modified from [19], and is described as follows.

Algorithms MASTER and SLAVE in pseudo code:
Module MASTER
/* perform task scheduling and load balancing */

```
Initialization
Gather performance ratio of all slave nodes
r = 0;
for (i = 1; i < number_of_slaves; i++) {
   partition % of loop iterations according to the
   performance ratio;
   send data to slave nodes;
   r++;
}
Partition (100-)% of loop iterations into the task
queue using some known self-scheduling scheme
Probe for returned results
Do {
   Distinguish source and receive returned data
   If the task queue is not empty then
      Send another data to the idle slave
      r -- ;
   else
      send TAG = 0 to the idle slave
```

```
} while (r > 0)
Finalization
END MASTER
```

Module SLAVE /* worker */

```
Initialization
Probe if some data in
While (TAG > 0) {
        Receive initial solution and size of subtask work
        and compute to fine solution
     Send the result to the master
     Probe if some data in
}
Finalization
END SLAVE
```

4 Experimental Results

The grid environment includes three clusters which are located in three universities respectively. Cluster 1, located in Providence University, has five nodes. Cluster 2, located in Hsiuping Institute of Technology, has four nodes. Cluster 3, located in Tunghai University, has four nodes. We use the following tools to build the grid.

- Globus Toolkit 3.0.2
- MPICH library 1.2.6

We have implemented two application programs in the C language, with message passing interface (MPI) directives for parallelizing code segments to be processed by multiple CPU's. For readability of experimental results, the description of our implementation for all programs is listed in Table 2.

Table 2. Description of our implementation for all programs

AP	Name	Description	Reference
Matrix	GSS	Dynamic scheduling(GSS)	[16]
Multiplication	NGSS	Fixed α scheduling + GSS	[19]
Mandelbrot	PGSS	Our scheduling + GSS	

4.1 Application 1: Matrix Multiplication

The matrix multiplication is a fundamental linear algebra operation in many numerical applications. Therefore, its efficient implementation on parallel processing systems is an important issue. In this section, we study implementations for matrix multiplication on a grid environment. The performance ratio is

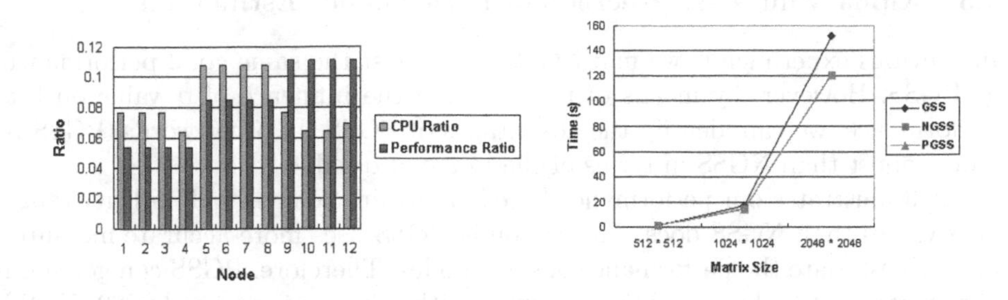

Fig. 1. Matrix multiplication with GSS (a) performance ratio (b) execution time

determined by simulation of input size 512×512. The resulting simulated performance of our grid nodes are illustrated in Fig.1(a).

This experiment consists of three scenarios: GSS, NGSS and our PGSS. With respect to each algorithm, three types of input sizes are taken: 512, 1024 and 2048. Empirical results are illustrated in Fig.1(b), showing NGSS and PGSS get better performance than GSS does.

4.2 Application 2: Mandelbrot

The Mandelbrot set is a problem which does the same computing job on different data points [2]. However, these same jobs on different points might take different time to converge. In this section we study implementations for the Mandelbrot set on a grid environment and develop a performance prediction model for these implementations. The performance ratio is determined by simulation of input size 128*128. The resulting simulated performance of our grid nodes are illustrated in Fig.2(a).

This experiment consists of three scenarios: GSS, NGSS and our PGSS. With respect to each algorithm, three types of input sizes are taken: 128, 512, and 1024. Empirical results are illustrated in Fig.2(b), showing NGSS and PGSS get better performance than GSS does.

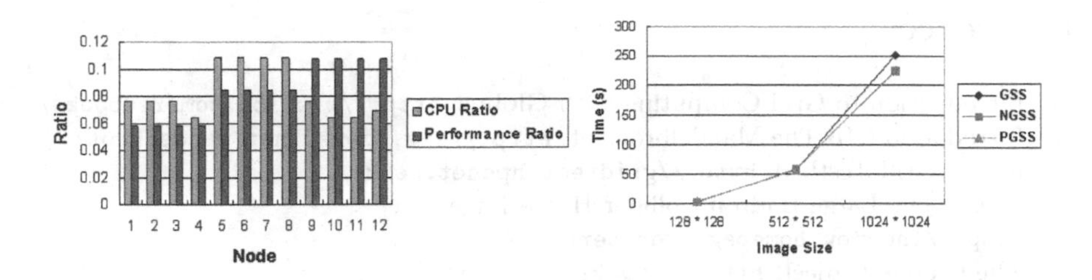

Fig. 2. Mandelbrot with GSS (a) performance ratio (b) execution time

4.3 Alpha Values vs. Precision of Performance Estimation

In previous experiments, we find NGSS has almost the same good performance as PGSS. However, by means of investigating the influence of α value on the performance, we can identify the advantages of PGSS. In other words, PGSS is more robust than NGSS in terms of the range of α value.

Fig.3 illustrates our performance-based algorithm behaves well in lager range of α values than NGSS does. The reason is PGSS uses more accurate measurement to estimate the performance of slave nodes. Therefore, PGSS can get good performance at higher α values. However, with a heuristic of α value 80, NGSS only gets good performance at smaller α range.

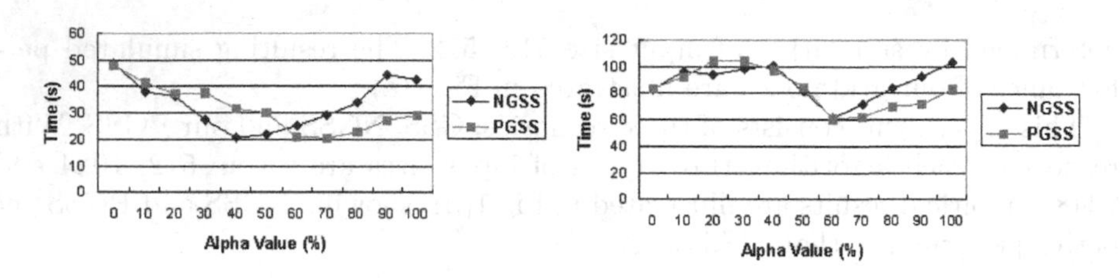

Fig. 3. Execution time vs. α values (a) Matrix Multiplication (b) the Mandelbrot set

5 Conclusions and Future Work

On grid environments, performance of known self-scheduling schemes has not been investigated well. In this paper, we have proposed a performance-based approach, and compared it with previous algorithms by experiments on two application programs. In both case, our approach could obtain obvious performance improvement over GSS. Besides, our approach could work with larger range of α values. In our future work, we will implement more types of application programs to verify our approach. Furthermore, we hope to find better ways of modeling the performance function, not just by empirical simulation.

References

1. Introduction to Grid Computing with Globus, `http://www.ibm.com/redbooks/`
2. Introduction To The Mandelbrot Set, `http://www.ddewey.net/mandelbrot/`
3. KISTI Grid Testbed, `http://gridtest.hpcnet.ne.kr/`
4. LHC - The Large Hadron Collider Home Page, `http://lhc-new-homepage.web.cern.ch/`
5. The Globus Project, `http://www.globus.org/`
6. What Is Grid Computing, `http://www-1.ibm.com/grid/about_grid/what_is.shtml/`
7. Baker, M.A., Fox, G.C.: Metacomputing: Harnessing Informal Supercomputers. In: High Performance Cluster Computing, Prentice-Hall, Englewood Cliffs (1999)

8. Cheng, K.-W., Yang, C.-T., Lai, C.-L., Chang, S.-C.: A Parallel Loop Self-Scheduling on Grid Computing Environments. In: Proceedings of the 2004 IEEE International Symposium on Parallel Architectures, Algorithms and Networks, KH, China, pp. 409–414 (May 2004)
9. Chronopoulos, A.T., Andonie, R., Benche, M., Grosu, D.: A Class of Loop Self-Scheduling for Heterogeneous Clusters. In: Proceedings of the 2001 IEEE International Conference on Cluster Computing, pp. 282–291 (2001)
10. Czajkowski, K., Foster, I., Kesselman, C.: Resource Co-Allocation in Computational Grids. In: HPDC-8. Proceedings of the 8th IEEE International Symposium on High Performance Distributed Computing, pp. 219–228 (1999)
11. Czajkowski, K., Fitzgerald, S., Foster, I., Kesselman, C.: Grid Information Services for Distributed Resource Sharing. In: HPDC-10. Proceedings of the 10th IEEE International Symposium on High-Performance Distributed Computing, pp. 181–194 (August 2001)
12. Foster, I., Kesselman, C.: Globus: A Metacomputing Infrastructure Toolkit. International Journal of Supercomputer Applications and High Performance Computing 11(2), 115–128 (1997)
13. Foster, I., Kesselman, C., Tuecke, S.: The Anatomy of the Grid: Enabling Scalable Virtual Organizations. International Journal of Supercomputer Applications and High Performance Computing 15(3), 200–222 (2001)
14. Foster, I.: The Grid: A New Infrastructure for 21st Century Science. Physics Today 55(2), 42–47 (2002)
15. Li, H., Tandri, S., Stumm, M., Sevcik, K.C.: Locality and Loop Scheduling on NUMA Multiprocessors. In: Proceedings of the 1993 International Conference on Parallel Processing, vol. II, pp. 140–147 (1993)
16. Polychronopoulos, C.D., Kuck, D.: Guided Self-Scheduling: a Practical Scheduling Scheme for Parallel Supercomputers. IEEE Trans. on Computers 36(12), 1425–1439 (1987)
17. Smarr, L., Catlett, C.: Metacomputing. Communications of the ACM 35(6), 44–52 (1992)
18. Tang, P., Yew, P.C.: Processor self-scheduling for multiple-nested parallel loops. In: Proceedings of the 1986 International Conference on Parallel Processing, pp. 528–535 (1986)
19. Yang, C.-T., Chang, S.-C.: A Parallel Loop Self-Scheduling on Extremely Heterogeneous PC Clusters. Journal of Information Science and Engineering 20(2), 263–273 (2004)
20. Yang, C.-T., Cheng, K.-W., Li, K.-C.: An Efficient Parallel Loop Self-Scheduling on Grid Environments. In: Jin, H., Gao, G.R., Xu, Z., Chen, H. (eds.) NPC 2004. LNCS, vol. 3222, pp. 92–100. Springer, Heidelberg (2004)

Reconfigurable Middleware for Grid Environment

Sungju Kwon[1], Jaeyoung Choi[1], and Jysoo Lee[2]

[1] School of Computing, Soongsil University,
1-1 Sangdo-5Dong, Dongjak-Gu, Seoul, 156-743, Korea
{lithlife,choi}@ssu.ac.kr
[2] Supercomputing center, KISTI,
Daejon, Korea

Abstract. A component in application is a functional unit with well-defined interfaces. It encapsulates its internal states and provides services to other components or applications. By modularizing required functions into components, a component-based system can easily reuse those components and provide a flexible application structure with dynamic reconfiguration. In this paper, we propose a component-based middleware, called MAGE, which uses a service-oriented interface to provide transparency of platform, implementation language, and location. MAGE can dynamically reconfigure its architecture to adapt to Grid environments.

1 Introduction

A component is a unit of composition with well-defined interfaces [1]. Like an object, a component encapsulates its internal states and provides services to other components or applications. Components are distinguished from objects in that they are explicitly required interfaces and conformance to a binary standard. All interfaces are managed by a component framework. A component framework is defined by Szyperski as "collections of rules and interfaces that govern the interaction of a set of components plugged into them" [1]. A component framework is a concept of reusable architecture, which provides means of enforcing architectural properties and a component interface for communicating with each other in the binary format. Also, the framework manages the lifecycle of the component. A component hides its internal states and implementation details from clients and allows itself to be accessed only through published interfaces. Therefore, it is possible to hide implementation dependency among components. A component-based system has the ability to change its internal structure dynamically without disturbing users. An application based on component-architecture can be dynamically configure itself by adding, removing, and modifying components.

Dynamic reconfiguration [2,3,4] is a mechanism that can replace a functional unit of an application with another without stopping the application. Dynamic reconfiguration is very useful for adaptable and highly available systems [5]. An adaptable system actively deals with environmental changes. A highly available

J. Labarta, K. Joe, and T. Sato (Eds.): ISHPC 2005 and ALPS 2006, LNCS 4759, pp. 246–253, 2008.

system provides continuous services without interruption of the application even though there is a fault. Adaptive systems and high availability systems are useful for Grid [6], ubiquitous [7], and mobile environments. Currently this dynamic reconfiguration technology is an active research topic in several areas with their own specific formats. These applications have their own reconfiguration technologies, which have difficulties in adapting to a new application. In order to provide a more generalized technology and platform-independent format, a framework is required which manage its components and provide service interfaces among components. Therefore, a component-based architecture is more useful and flexible.

In this paper, we propose a dynamically reconfigurable middleware for Grid environments called MAGE (Modular and Adaptive Grid Environment). MAGE is a reconfigurable framework, which is a base architecture to provide adaptability, high availability, and scalability. MAGE can replace components easily and minimize physical dependency among components. MAGE is a service-oriented architecture to support transparency for platform, language, and communication protocol. Based on this framework, application developers can build Grid applications much simple way.

This paper is organized as follows: Section 2 describes related works including a service-oriented architecture and a component-based middleware. We explain MAGE architecture in Section 3, and a service-oriented flow of MAGE in Section 4. Finally, Section 5 includes the conclusion and future works.

2 Related Works

Grid systems should be adaptable and highly available. Grid middleware systems such as Globus toolkit and GridLab have been studied for more efficient resource usage.

Globus toolkit [8] is an open source software toolkit which provides resource sharing on Grid environments. The Globus toolkit provides services and libraries for resource monitoring, resource management, security, and file management. GGF (Global Grid Forum) defines OGSA (Open Grid Services Architecture) as a service-oriented architecture for distributed computing which provides Grid services to support distributed interaction of services and computing infrastructure. OGSA manages a lifecycle of services through service naming, creation, discovery, and destruction. OGSA extends Web services to manage not only permanent services but also temporary services. However, the OGSA has three problems. First, it is difficult to achieve high speed network performance. The current Web services mainly use SOAP protocol which is not suitable for scientific applications that generate and handle massive data to compute complicated problems. Second, compared with the component-based middleware, OGSA has more constraints on fault tolerance, continuous state management, automated logging, load balancing, and event distribution. Third, it does not support system reconfiguration according to the change of its state.

GridLab [9] project, which was developed in the EU, provides a unified access to Grid middleware through service-oriented interfaces. GridLab consists of user space and capability space for hierarchical structure. These two spaces are integrated by GAT (Grid Application Toolkit). GAT abstracts services that are required by applications. Therefore, GAT can act as an interface between application programs and Grid middleware environments. The GAT provides GAT API for application developers to access different Grid environments in a unified way. The service-oriented architecture of GridLab provides scalability and adaptability for environmental changes. On the other hand, services in an application-level are provided on a large scale and have problems of reusability and adaptability, compared with services in a component-level. Also, GridLab's service interface is statically linked at compile time. This means that it cannot changes the configuration of middleware at the runtime.

GRIDKIT[10] is a component-based middleware framework for configurable and reconfigurable Grid computing. It particularly focuses on lightweight component-based technology to construct an extensible family of open and programmable overlay networks. GRIDKIT provides four main domains: Service Binding, Resource Discovery, Resource Management, and Grid Security. These four domains of middleware functionality are implemented in GRIDKIT as independently and horizontally. Each domain implemented as a framework, which is highly reconfigurable. MAGE provides functionality similar to GRIDKIT. MAGE, however, is more focused on a network transparency with easily customizable components. APIs in MAGE provide a rich set of information for component control and installation. MAGE automatically install dependent components based on component dependency and version information.

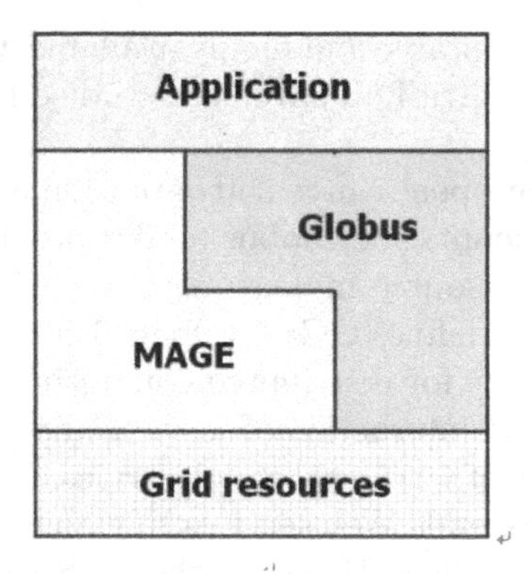

Fig. 1. Relationship between MAGE and Globus

3 MAGE Architecture

MAGE acts as a component management framework. MAGE does not provide application specific functions. MAGE can be executed as a standalone application or with a legacy grid middleware through interface components. Fig. 1 shows the relationship between MAGE and legacy Grid middleware.

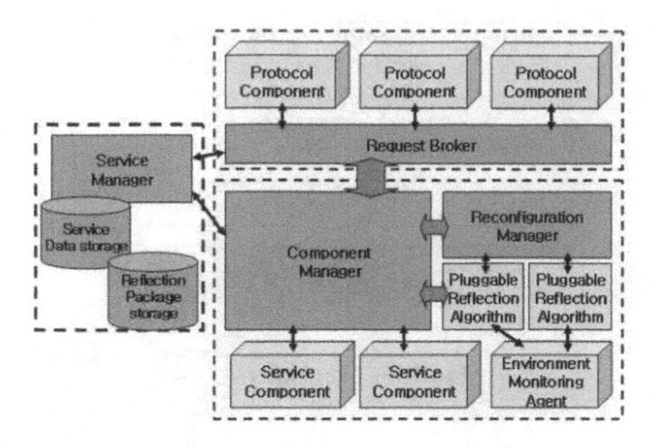

Fig. 2. Reflective MAGE Architecture

MAGE architecture consists of four parts as shown in Fig. 2: Request-Broker, Component-Manager, Service-Manager, and Reconfiguration-Manager. By using this architecture, MAGE provides transparency of the running platform, programming language, and communication protocol. Each element of the architecture has the ability to be located at any location, implemented with any language, and executed on any platform. It is because each element is executed independently from others and communicates with others using internal protocols. Each element can not only exist as a 1:1 relationship but also as M:N relationship. This relationship can be arranged by the system administrator. These separations of elements provide flexibility for the application architecture.

3.1 Request-Broker

The Request-Broker accepts requests from clients and translates into an internal message format. The actual communication is performed by a protocol-component inside the Request-Broker. MAGE provides message translation APIs for protocol-components. A system administrator can dynamically install protocol-components. Change of a client protocol only replaces a correspondent protocol-component. Request-Broker searches for matching Component-Manager and delivers user requests to Component-Manager. Fig. 3 shows a request flow inside Request-Broker.

3.2 Service-Manager

Service-Manager manages registered service information such as service location, dependency information, or supported operations. Request-Broker and Component-Manager use this information to find matching services. Service-Manager provides dependency information of a service to Component-Manager for automatic installation of components.

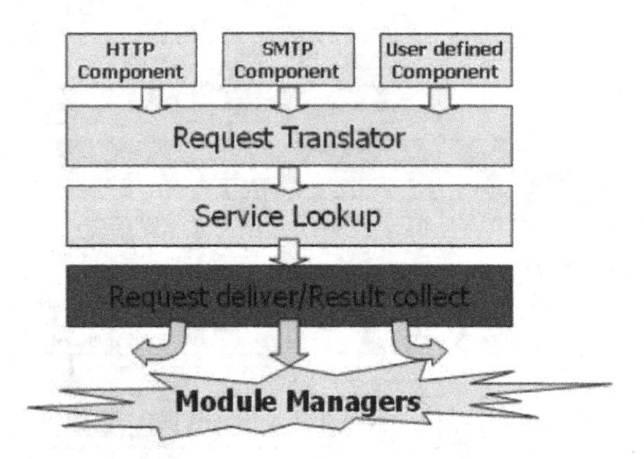

Fig. 3. Request Flow inside Request-Broker

3.3 Component-Manager

Component-Manager maintains the lifecycle of components. A component can be installed using descriptor file. The descriptor file contains the service list and dependent information of component. Component-Manager registers component information with Service-Manager after completing the installation process. Clients and other services use the information when they search appropriate services. A component developer does not need to know about the communication method with clients. A component developer only needs to implement predefined interfaces of the MAGE. Actual communications are performed behind MAGE.

3.4 Reconfiguration-Manager

Reconfiguration-Manager controls component arrangement inside Component-Manager. It includes several reflection algorithms that are dynamically loadable. A client can install reflection algorithm packages at runtime. The reflection algorithm package consists of one reflection algorithm component and several environment monitoring components. Environment monitoring components collect system information and provide this information to a reflection algorithm component. Using this information, the reflection algorithm component changes the installed component list on Component-Manager. A user can install several reflection algorithms at the same time with priority value.

4 Service-Oriented Interface

MAGE is a service-oriented architecture based on dynamically loadable components. A client does not need to know any physical information about components. A client only needs to know names and operations of services. The service information does not contain any physical information such as component location, implementation language, and communication method. Even though physical components are moved to other node, an application can be executed without notice it. MAGE provides common component interfaces to developers. MAGE only interacts with components using this interface.

MAGE provides component interfaces as shown in Table 1. The component interfaces consists of control operations and service operations.

Table 1. MAGE Component Interface

Operation	Description
Component Control Operations	
initProvider	called after loading component
startupProvider	called after service starts
stopProvider	called before service stops
destroyProvider	called before unloading component
Service Operations	
executeMethod	entry point of method invocation
setProperty	entry point of property set
getProperty	entry point of property get

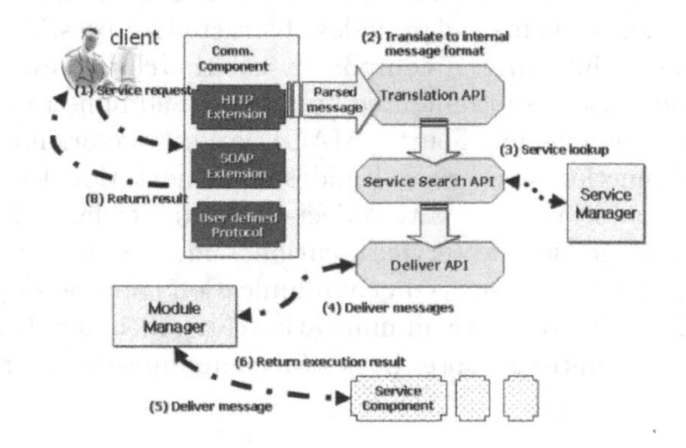

Fig. 4. Process Flow of Service

Component-Manager delivers a message to matching components through MAGE interface. Each interface can implement actual jobs based on the delivered request message. The service results go back to the client through MAGE. This approach provides network neutrality to component developers. MAGE hides its internal processing for the requested message from both clients and components. The client's request message is delivered to the service component through the following steps.

First, a client sends a request message through the network. Then, MAGE accepts the client request by one of the protocol-components. A protocol-component translates the request message into internal format using MAGE translation APIs. Request-Manager searches for a service wanted by the request through Service-Manager. Request-Manager delivers the request message to Component-Manager, which contains the appropriate service component. Component-Manager activates the target component and delivers messages. Component-Manager also collects results from the component and sends them back to Request-Broker. The protocol-component in Request-Broker sends results back to the client.

5 Conclusion

In this paper, we introduced a service-oriented architecture based on component technology. This architecture provides transparency for running platform, implementation languages, and communication protocols. It is also easy to reconfigure application architecture without stopping an application.

MAGE has the following characteristics for the Grid environment. First, MAGE is a component-based architecture. An application composed using a functional unit called a component. Every component works together on MAGE framework while does not know about physical relationships among components. Independency among components provides a convenient way for application developers and system administrators. Second, MAGE provides transparency of languages, platforms, and communication protocols. A component does not show any physical information to clients and provides abstracted services. Third, it is an easily reconfigurable architecture. A component-based architecture can reconfigure itself at any time or can be reconfigured by a system administrator or by the relationship among components. Fourth, MAGE is a service-oriented architecture, which reveals all functions as services. It hides implementation details from clients and shows functions as a service with object-oriented interfaces. MAGE controls the lifecycle of a service such as service creation, find, execution, or terminations.

In the near future, we will focus on communication transparency for client applications. MAGE will provide a communication component for clients and it must be downloadable at runtime before the actual communication starts.

Acknowledgements

This work was supported by the National e-Science Project, funded by the Korean Ministry of Science and Technology (MOST).

References

1. Szyperski, C.: Component Software: Beyond Object-Oriented Programming. Addison-Wesley, Reading (1998)
2. Chen, X., Simmons, M.: Extending RMI to support dynamic reconfiguration of distributed systems. In: ICDC 2002. 22nd International conference on Distributed Computing Systems (2002)
3. Goudarzi, K.M.: Consistency Preserving Dynamic Reconfiguration of Distributed Systems, Ph.D. Thesis, Imperial College (1999)
4. Hillman, J., Warren, I.: Quantative analysis of dynamic reconfiguration algorithms. Design Analysis and Simulation of Distributed Systems (DASD) (2004)

5. Clarke, M., Blair, G.S., Coulson, G., Parlavantzas, N.: An efficient component model for the construction of adaptive middleware. In: IFIP/ACM International Conference on Distributed Systems Platforms (2001)
6. Foster, I., Kesselman, C. (eds.): The Grid: Blueprint for a New Computing Infrastructure. Morgan Kaufmann, San Francisco (1998)
7. Weiser, M.: Some Computer Science Problems in Ubiquitous Computing. Communications of the ACM (July 1993)
8. Globus Toolkit, `http://www.globus.org/`
9. GridLab, `http://www.gridlab.org/`
10. Cai, W., Coulson, G., Grace, P., Blair, G.S., Mathy, L., Yeung, W.K.: The Gridkit Distributed Resource Management Framework. In: Proceedings of the European Grid Conference, Amsterdam, The Netherlands (February 2005)

Netfiles: An Enhanced Stream-Based Communication Mechanism

Philip Chan and David Abramson

Caulfield School of Information Technology
Monash University
900 Dandenong Road, Caulfield East
Victoria 3145, Australia
{pchan, davida}@csse.monash.edu.au

Abstract. *Netfiles* is an alternative API for message passing on distributed memory machines. Based on the communication stream model, *Netfiles* provides enhanced functionality such as broadcasts and gather operations. *Netfiles* overload conventional file I/O primitives enabling parallel programs to be developed and tested on a file system before execution on a parallel machine. *Netfiles* is part of a parallel programming system called FAbrIC. This paper also presents the design and implementation of the FAbrIC architecture and demonstrate the effectiveness of this approach by means of two parallel applications: a parallel shallow water model application and parallel Jacobi method.

Keywords: Pipes, Stream Communication, File I/O, Message Passing, Parallel Programming.

1 Introduction

The stream communication model dates back to the early days of Unix. It encompasses mechanisms such as pipes, named pipes and sockets with TCP/IP. This model is simple, well-understood and has been successfully employed in many parallel programming systems.

This paper presents an enhanced stream communication mechanism called *Netfiles*. The *Netfile* abstraction overloads file I/O primitives with message passing functionality specifically for parallel programming, with minimal deviation from semantics of conventional file I/O. *Netfiles* support point-to-point and collective communication. We have built an implementation of this abstraction and present an evaluation with two parallel applications.

2 *Netfiles*: Enhanced Pipes for IPC

In our proposed mechanism, communication is achieved through the reading and writing of a common *Netfile*, identified by a filename. This filename is a user-specified string that is task/node independent, allowing dynamic binding

J. Labarta, K. Joe, and T. Sato (Eds.): ISHPC 2005 and ALPS 2006, LNCS 4759, pp. 254–261, 2008.

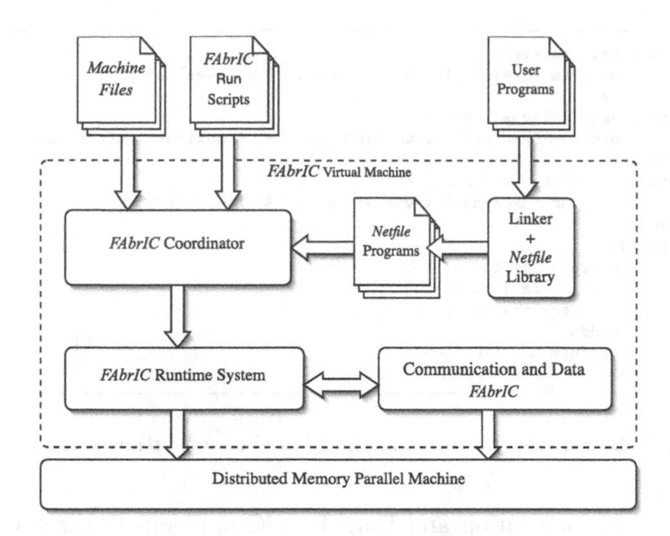

Fig. 1. System architecture of FAbrIC

of process end-points. The sender opens the *Netfile* in `WRITE` mode, while the receiver opens in `READ` mode. When both ends are open, the *Netfile* effectively becomes a uni-directional communication channel. A `write()` behaves like a `send()`, and a `read()` behaves like a `receive()`. Non-blocking *Netfile* semantics is also possible, allowing communication to overlap with computation and potentially improving performance.

No implicit type information is encoded within the stream. Thus, the data written through a *Netfile* do not have message boundaries, enhancing programming flexibility. For example, data sent via a `write()` may actually be read by several consecutive "small" `read()` operations.

Collective operations are also supported in *Netfiles*. A single-writer, multiple-reader *Netfile* behaves as a broadcast, closely resembling multiple reads on a single file. However, when readers share a single logical *Netfile* pointer, this operation behaves as a scatter. Similarly, a gather operation is represented as a single-reader *Netfile* with multiple writers appending data.

Netfiles form part of a proposed parallel programming system called FAbrIC. A parallel program is first written to use files and file I/O for communication. No explicit `send()` or `receive()` are used in the program. Once tested, the program may be relinked with our *Netfile*-library for parallel execution on a distributed memory machine.

3 FAbrIC System Architecture and Implementation

Fig. 1 presents the key components of FAbrIC. User programs are relinked with the *Netfile* library to create FAbrIC-ready applications. Execution is specified via an external run script written in the FAbrIC coordination language. The FAbrIC Coordinator interprets this and starts tasks through the FAbrIC Runtime

```
task sm1 (nslaves)
     program "exe/shallow-master" args "$(nslaves) 0"
endtask
task sm2 (nslaves, maxiters)
     program "exe/shallow-worker" args "-1 $(maxiters) $(nslaves)"
endtask
task sw (slave_id, maxiters)
     program "exe/shallow-worker" args "$(slave_id) $(maxiters)"
endtask
par do
     execute sm1 (16)
     par i from 1 to 16 do
          execute sw (i, 100)
     enddo
     execute sm2 (16, 100)
enddo
```

Fig. 2. Sample run script in the FAbrIC coordination language

System. The Communication and Data FAbrIC implements the FAbrIC Space, a collective abstraction for both stream-like and persistent *Netfiles*. We discuss some implementation details of our prototype.

3.1 Communication FAbrIC: *Netfile* Lookup Service

The *Netfile* lookup service matches the reader and the writer of each *Netfile*. Each unmatched `open()` is recorded in a hash table and marked as pending. When the matching `open()` arrives, this pending request is used to match the pair. Once matched, communication takes place via a TCP/IP connection. For efficiency, connections are reused for different transmissions between the same task pair.

Collective operations are implemented on top of the point-to-point *Netfiles*. For efficiency, broadcasts are performed by establishing tree-structured network of *Netfile* connections between tasks. A gather uses this tree network with the leaf tasks initiating the sending, with all data coalesced at the root task.

3.2 FAbrIC Coordinator and Runtime System

We designed a coordination language with constructs for sequential (`seq`) and parallel (`par`) execution. These two constructs may be arbitrarily nested to created complex execution patterns. Fig. 2 is a script that specifies parallel execution of two masters (scatter and gather) and 16 workers. The `task-endtask` pair specifies details of the task, i.e., the program executable name and how arguments are translated into command-line parameters. The `execute` statement causes a task to be started with a specified set of arguments.

Execution of a FAbrIC application involves running the FAbrIC Coordinator with two input files: (a) the run script; and (b) the machine configuration, a list of host names. This coordinator starts tasks via the FAbrIC Runtime System, implemented as a set of execution daemons, one on each host.

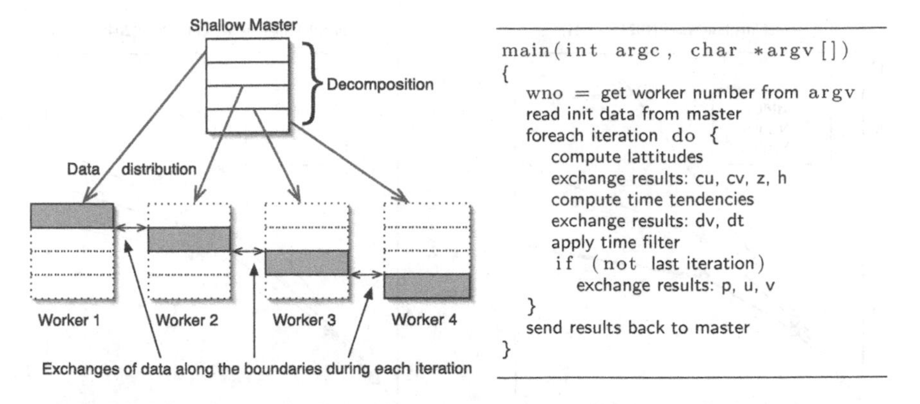

Fig. 3. Data distribution and data exchanges in parallel shallow water application

When a task has to be initiated, the coordinator sends a `task_start` request to the daemon on a selected host. This request contains the logical task ID (TID) used for *Netfile*-lookup for matching tasks. The daemon starts the task and passes the logical TID to the task via an environment variable.

Upon task startup, the daemon replies with the process ID (`pid`). The pair (process ID, host logical ID) is used by the coordinator to uniquely identify a task in the parallel application. When the task completes, its corresponding execution daemon notifies the coordinator by a `task_complete` message. The coordinator terminates when all application tasks have completed and no new ones are to be created according to the script.

4 Case Study 1: Parallel Shallow Water Equations

To evaluate our prototype, we selected a numerical model involving shallow water equations [10,12]. It employs both master-worker and worker-worker communication where data along the boundaries (Fig. 3) are exchanged between adjacent workers during each iteration. At the end of the iterations, workers send their results to the master. We used the MPI code of Abramson, Dix & Whiting [1].

4.1 *Netfiles* Version of Parallel Shallow

With the FAbrIC approach, we first built a parallel shallow application using standard file I/O for interprocess communication. To be descriptive of what the files represented, the variable names of the data structures were used as filenames. For data exchanges involving neighboring elements, we used the row number of the data elements in the filename.

This version was first tested on a local file system using background task execution. Wrapper functions for `open()` and `close()` which provided synchronisation were used. After testing, we relinked the code with our *Netfile* library.

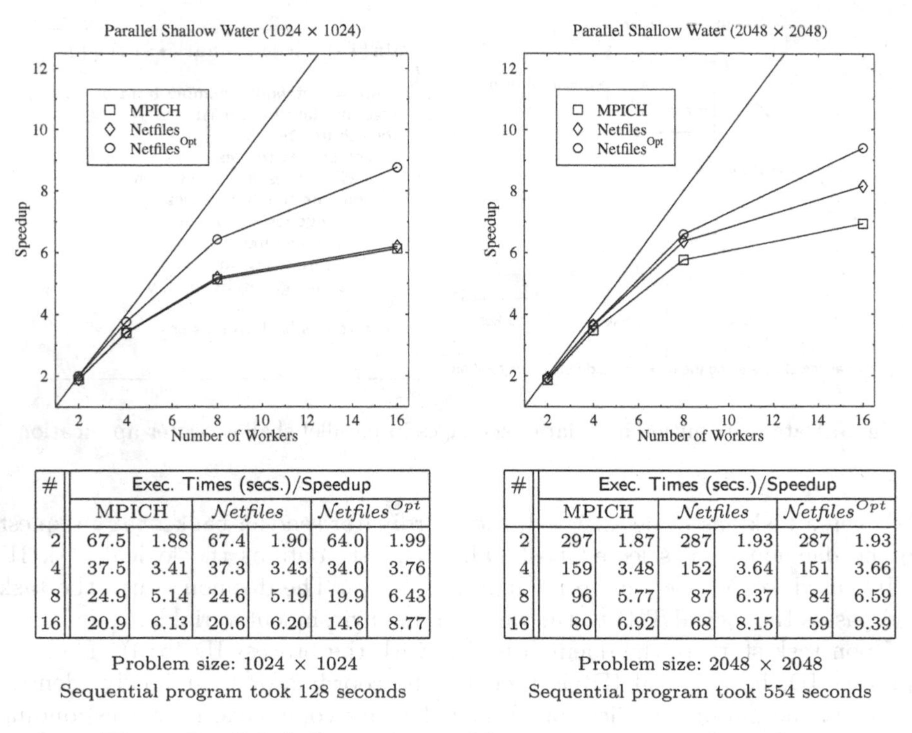

#	Exec. Times (secs.)/Speedup					
	MPICH		*Netfiles*		*Netfiles*^{Opt}	
2	67.5	1.88	67.4	1.90	64.0	1.99
4	37.5	3.41	37.3	3.43	34.0	3.76
8	24.9	5.14	24.6	5.19	19.9	6.43
16	20.9	6.13	20.6	6.20	14.6	8.77

Problem size: 1024 × 1024
Sequential program took 128 seconds

#	Exec. Times (secs.)/Speedup					
	MPICH		*Netfiles*		*Netfiles*^{Opt}	
2	297	1.87	287	1.93	287	1.93
4	159	3.48	152	3.64	151	3.66
8	96	5.77	87	6.37	84	6.59
16	80	6.92	68	8.15	59	9.39

Problem size: 2048 × 2048
Sequential program took 554 seconds

Fig. 4. Parallel shallow water model execution times and speedups

The *Netfile* program is essentially identical to the MPI code in the manner of data and task decomposition and distribution of data between tasks.

4.2 Experimental Results

We considered two grid size configurations: 1024×1024 and 2048×2048. For each run, we executed the model for 100 iterations. And the speedups were based on the execution time of the MPI program with 1 worker. Results from two versions of the *Netfile* applications are presented in Fig. 4. In the first one, *Netfiles* were opened and closed for each message. For small problem sizes, the overhead of requests-replies with the *Netfile*-lookup service dominated the communication cost. In the second version (identified as *Netfiles*^{Opt}), we fine-tuned the program code so that open *Netfiles* for worker-to-worker communications are not closed but are kept open and reused. This minor optimisation resulted in the best performance since it reduced the number of lookups.

These results demonstrate that with our prototype, *Netfiles* applications can perform as well as their MPI equivalents. And in large problem sizes, our *Netfiles* program outperformed its MPI equivalent. We suspect that performance gain is due to our method of reusing the TCP connections between tasks.

Experiments were conducted on a Linux cluster of Intel Xeon processors (each with 3 GHz and 1GB RAM) connected by a Gigabit Ethernet switch. All pro-

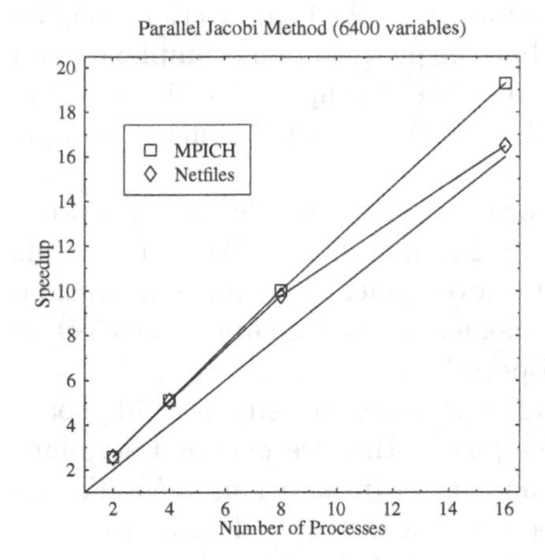

#	Exec. Times (secs.)/Speedup			
	MPICH		*Netfiles*	
2	296.82	2.58	298.28	2.57
4	150.15	5.11	151.06	5.08
8	76.70	10.01	78.58	9.77
16	39.81	19.29	46.57	16.49

Problem size: 6400 × 6400 matrix
Sequential program took 768 seconds

Fig. 5. Parallel Jacobi execution times and speedups

grams were compiled using `gcc` (version 3.3.4). For MPI, we used MPICH (version 1.2.6) from Argonne National Laboratories and Mississippi State University.

5 Case Study 2: Parallel Jacobi Method

We also built a parallel Jacobi iterative method for solving linear equations. This code is characterized by collective communication at the end of each iteration. At the start of the application, the matrix that represents the coefficients of the linear equations is distributed to all other tasks. During each iteration, all tasks perform the Jacobi method and obtain its segment of the solution vector. This solution vector is gathered and broadcast to the other tasks.

Fig. 5 shows the results on a problem size of 6400 variables. The linear equations are stored in a 6400^2 matrix of `doubles` and convergence was at 960 iterations. The superlinear speedup is most likely due to the cache and virtual memory issues with the reference sequential program. Further optimisation in our implementation of collective communication is necessary to match the performance of the MPI version.

We used an IBM POWER5 cluster (4 cpus/node, minimum of 8 GB RAM, Suse Linux) connected via Myrinet. All programs were compiled using IBM `xlc` 64-bit compiler and MPICH (version 1.2.6).

6 Related Work

The idea of using files as a proxy for message passing was inspired from the Nimrod project [2]. Nimrod is a middleware for the execution of parametric modelling applications over a wide-area environment. During an experiment, data files needed by jobs are supplied via file transfers, effectively a form of interprocess communication. Socket programming with TCP connections may be used to couple the application components but the programming complexity can be discouraging. The *Netfile* programming interface is simpler since it abstracts connection details (IP addresses/ports) and uses user-specified names to couple communication endpoints.

The idea of extending standard file I/O for channel communication was probably first proposed by Seevers et al. [11] for Dataparallel C. We were unable to find any subsequent works describing its development and implementation. FAbrIC/*Netfiles*, on the other hand, is an implementation and supports collective communication like broadcast and gather operations.

Similarly, Virtual Shared Files [8] was also proposed to employ a file space model for IPC. The key difference with *Netfiles* is that we overload standard file I/O primitives instead of introducing new ones. In addition, a *Netfile* can behave in two modes: point-to-point stream and a distributed shared object. This shared object semantics for *Netfiles* is currently being investigated.

Message passing libraries (PVM [4] and MPI [9] implementations) are popularly used for parallel programming on a distributed memory system like a network of workstations. FAbrIC/*Netfiles* is essentially message passing within file I/O primitives. This has the benefit that allows applications to be built and tested on a conventional file system, prior to parallel execution.

The channel/pipe abstraction has its early beginnings in the Communicating Sequential Processes (CSP) [5] and Occam [7]. Subsequently, this model has been used in many coordination languages. Recently, this mechanism is also employed in DP [6]. *Netfiles* differs from regular pipe mechanisms by supporting collective operations. Our collective operations employ the file as a metaphor, for example, gather operation is performed using a multiple-writer, single-reader *Netfile*.

Our earlier *Netfiles* paper [3] focused on master-worker parallel programming with no inter-worker communication. And that library and runtime system was implemented on top of PVM. In this paper, we present the FAbrIC system architecture, a coordination language, and its implementation on top of sockets. We also demonstrate the effectiveness of FAbrIC/*Netfiles* for inter-worker communication.

7 Conclusions

We have presented the FAbrIC parallel programming approach using an abstraction called the *Netfile*. While a basic *Netfile* is essentially a pipe that allows IPC across machines on a network, its enhanced form includes multicast capability. FAbrIC provides a coordination language and is implemented using an interpreter and runtime system. Parallel applications may be written in two phases: (a) write and test on a familiar sequential environment using files and file I/O as a means for inter-task communication; and (b) obtain the parallel version by relinking with the *Netfile* library.

FAbrIC is evaluated using a parallel shallow water model and parallel Jacobi method. Results show that with some minor fine-tuning, the FAbrIC parallel shallow water application consistently performed better than the equivalent MPICH program. However, our implementation of collective communications require further optimisation as evident from results in the parallel Jacobi iteration. Overall, this demonstrates viability of this approach and encourages further investigation.

Acknowledgements

We thank the Victorian Partnership for Advanced Computing for the use of their IBM POWER5 cluster.

References

1. Abramson, D., Dix, M., Whiting, P.: A Study of the Shallow Water Equations on Various Parallel Architectures. In: 14th Australian Computer Science Conference, Sydney, Australia, pp. 6:1–6:12 (1991)
2. Abramson, D., Sosic, R., Giddy, J., Hall, B.: Nimrod: A Tool for Performing Parameterised Simulations using Distributed Workstations. In: 4th IEEE Symposium on High Performance Distributed Computing, IEEE Press, Virginia (1995)
3. Chan, P., Abramson, D.: NetFiles: A Novel Approach to Parallel Programming of Master/Worker Applications. In: 5th International Conference and Exhibition on High-Performance Computing in the Asia-Pacific Region, Queensland, Australia (2001)
4. Geist, A., Beguelin, A., Dongarra, J., Jiang, W., Manchek, R., Sunderam, V.: PVM Parallel Virtual Machine: A User's Guide and Tutorial for Network Parallel Computing. MIT Press, Cambridge (1994)
5. Hoare, C.A.R.: Communicating Sequential Processes. Prentice-Hall, Englewood Cliffs (1985)
6. Johnson, B.K., Ram, D.J.: DP: A Paradigm for Anonymous Remote Computation and Communication for Cluster Computing. IEEE Trans. on Parallel and Distributed Systems 12(10), 1052–1065 (2001)
7. Jones, G.A., Goldsmith, M.: Programming in OCCAM 2. Prentice Hall Professional Technical Reference, Englewood Cliffs (1989)
8. Konovalov, A., Samofalov, V., Scharf, S.: Virtual Shared Files: Towards User-Friendly Inter-Process Communications. In: 5th International Conference on Parallel Computing Technologies, St. Petersburg, Russia (1999)
9. Message Passing Interface Forum. MPI: A Message-Passing Interface Standard (1994)
10. Sadourny, R.: The Dynamics of Finite-Difference Models of the Shallow Water Equations. J of Atmospheric Sciences 32(4), 680–689 (1975)
11. Seevers, B.K., Quinn, M.J., Hatcher, P.J.: A Parallel Programming Environment Supporting Multiple Data-Parallel Modules. In: SIGPLAN Workshop on Languages, Compilers, and Run-Time Environments for Distributed Memory Multiprocessors, Boulder, Colorado, pp. 44–47 (1992)
12. Washington, W.M., Parkinson, C.L.: An Introduction to Three-Dimensional Climate Modeling. Oxford University Press, Oxford (1986)

Performance of Coupled Parallel Finite Element Analysis in Grid Computing Environment

Tomoya Niho and Tomoyoshi Horie

Department of Mechanical Information Science and Technology
Kyushu Institute of Technology
680–4 Kawazu, Iizuka, Fukuoka 820–8502, Japan
{niho, hoire}@mse.kyutech.ac.jp

Abstract. Since coupled problem should be solved for multiphenomena, large computational resources are needed for the large scale coupled analysis. In this paper, we propose a coupled parallel finite element analysis method using wide-area distributed computational resources on the Internet. In order for PC clusters located in different places to carry out a coupled parallel finite element analysis, a PC cluster receives the data needed for the coupled analysis from the other through the Internet. To perform the computing efficiently, processes for coupled parallel analysis are allocated based on the estimation of the coupled parallel analysis time taking account of available computer resources and network performance. Parallel finite element analysis of electromagnetic and structural coupled problem was carried out using two PC clusters to discuss the validity of this analysis method and computing environment.

1 Introduction

A target problem of finite element analysis has been extended to coupled problem such as fluid-structure coupled problem and electromagnetic-structural coupled problem. Since large computational resources are required for the coupled analysis which solve multiphenomena, parallel processing techniques are needed for the large scale coupled analysis.

On the other hand, research and development of grid computing [1], which aims to provide seamless and scalable access to wide-area distributed computational resources on the Internet, has become active recently. Since security and computational resource management are required for the grid computing, the Globus Toolkit [2], which offers these services, is developed as grid middleware. MPICH-G2 [3], which is a grid-enabled implementation of MPI, is also developed. Ninf-G [4] and GridSolve [5] are proposed as remote procedure call mechanism for the grid computing. Since available network performance is important to the grid computing, the Network Weather Service[6] which collect the information of network, is developed. The AppLeS system [7] focuses on the development of scheduling agents for parallel metacomputing applications. To determine schedules, agents use the services offered by the Network Weather Service to monitor the varying performance of available resources.

J. Labarta, K. Joe, and T. Sato (Eds.): ISHPC 2005 and ALPS 2006, LNCS 4759, pp. 262–270, 2008.

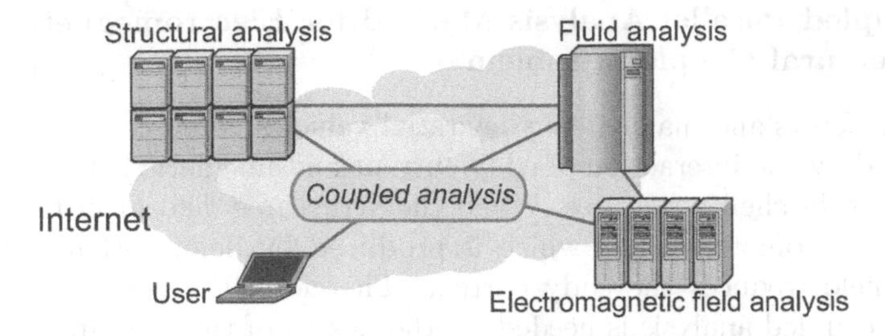

Fig. 1. Coupled parallel analysis method in grid computing environment

In this study, we propose a coupled parallel finite element analysis method in the grid computing environment. In order for PC clusters located in different places to carry out a coupled parallel finite element analysis, a PC cluster receives the data needed for the coupled analysis from the other through the Internet. In order to perform the analysis using wide area distributed computational resources, process allocation, that considered available CPU performance and network speed, is important. In this study, we also propose a process allocation method with estimation of the coupled parallel analysis time taking account of these computer and network information obtained by the Globus Toolkit and the Network Weather Service. The validity of this method is discussed with respect to the total execution time of the coupled analysis for the Internet environment and the LAN environment.

2 Coupled Parallel Finite Element Analysis System in Grid Computing Environment

2.1 Coupled Parallel Analysis in Grid

Since computational granularity is often fine for conventional parallel finite element analysis, high performance network is required to perform the parallel finite element analysis using distributed computational resources on the Internet. However, the coupled analysis by staggered analysis method may be suitable for the grid computing environment because the computational granularity of the coupled analysis is coarse.

In this study, a coupled parallel finite element analysis method in grid computing environment is proposed, which is illustrated in Fig.1. In order for PC clusters located in different places to carry out a coupled parallel finite element analysis, a PC cluster receives the data needed for the coupled analysis from the other through the Internet. By constructing the grid computing environment for the coupled analysis, it may be realized to perform a large scale coupled analysis using large computational resources on the Internet. And various coupled analysis utilizing the existing analysis codes which are developed by various organizations may also be realized.

2.2 Coupled Parallel Analysis Method for Electromagnetic and Structural Coupled Problem

In fusion reactors and magnetically levitated vehicles, large Lorenz force, which is generated by the interaction of eddy current and magnetic field, is applied to conductive thin shell structures. When the structures deform in the magnetic field, the electromotive force, which is produced by deformation velocity and magnetic field, reduces the eddy current. Therefore, the electromagnetic and structural coupled analysis is needed for the design of these components.

In the electromagnetic and structural coupled problem, the matrix equation of the structure [8] is expressed using the normal component $\boldsymbol{T}$ of the current vector potential and the displacement $\boldsymbol{u}$ as

$$\mathbf{M}\ddot{\boldsymbol{u}} + \mathbf{K}\boldsymbol{u} = \boldsymbol{F}^{ex} + \mathbf{C}_s\boldsymbol{T} \tag{1}$$

where matrices $\mathbf{M}$, $\mathbf{K}$, $\mathbf{C}_s$ and $\boldsymbol{F}^{ex}$ are the mass matrix, the stiffness matrix, the coupling sub-matrix by the electromagnetic force and the external mechanical force, respectively. The matrix equation of the eddy current is expressed as

$$\mathbf{U}\dot{\boldsymbol{T}} + \mathbf{R}\boldsymbol{T} = \dot{\boldsymbol{B}}^{ex} + \mathbf{C}_e\dot{\boldsymbol{u}} \tag{2}$$

where matrices $\mathbf{U}$, $\mathbf{R}$, $\mathbf{C}_e$ and $\dot{\boldsymbol{B}}^{ex}$ are the inductance matrix, the resistance matrix, the coupling sub-matrix by the electromotive force and the change of the external magnetic field, respectively.

In the staggered coupled analysis method, both matrix Eqs.(1) and (2) are solved alternately. To solve Eq.(2), $\dot{\boldsymbol{u}}$ obtained by the structural analysis is used to evaluate $\mathbf{C}_e\dot{\boldsymbol{u}}$. The solution $\boldsymbol{T}$ of Eq.(2) is substituted into $\mathbf{C}_s\boldsymbol{T}$ to solve Eq.(1). In this study, Newmark's β method and Crank-Nicolson method are applied to Eqs.(1) and (2) as the time integration, respectively.

Domain Decomposition Method (DDM) [9] are applied to the eddy current analysis and the structural analysis to carry out the parallel processing. Coupled parallel finite element analysis of the electromagnetic and structural coupled problem is performed using PC clusters on the Internet as shown in Fig.2. One of the PC of the PC cluster for the parallel structural analysis receives the current potential $\boldsymbol{T}$ from one of the PC of the PC cluster for the parallel eddy current analysis to make a vector of the electromagnetic force. In order to analyze the eddy current considering the electromotive force, one PC for the parallel structural analysis sends the velocity vector $\dot{\boldsymbol{u}}$ of the structure to the one PC for the parallel eddy current analysis. MPI is used to perform the communication in the coupled parallel analysis.

2.3 Process Allocation Considering CPU and Network Performance

In order to perform the computing efficiently by wide–area distributed computers, process allocation according to available CPU performance and network speed is important. In this study, a process allocation method with estimation of the coupled parallel analysis time taking account of computer and network

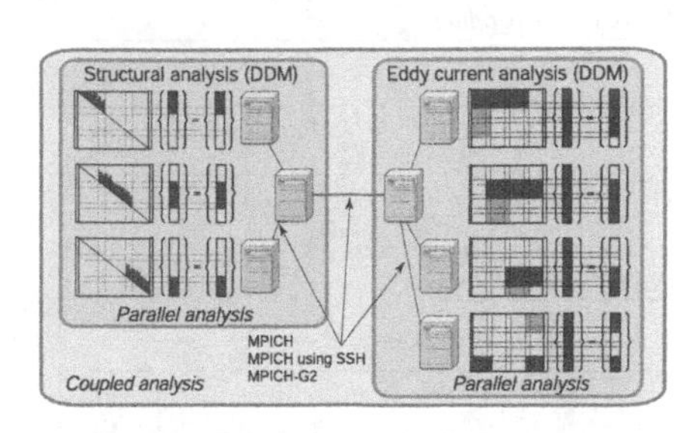

Fig. 2. Coupled parallel finite element analysis method of electromagnetic and structural coupled problem

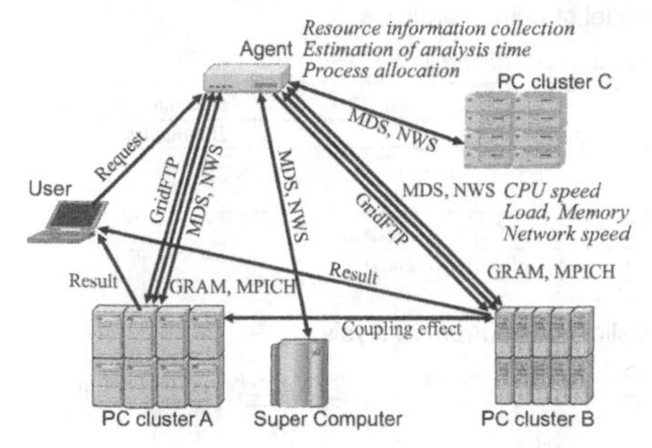

Fig. 3. Coupled parallel analysis system on Grid computing environment

information is proposed. These computational resource information are collected by MDS of the Globus Toolkit and the Network Weather Service.

Coupled parallel analysis system in grid computing environment using the proposed method is shown in Fig.3. In this system, the agent collects the information of computers and network by the MDS and the Network Weather Service. Agent estimates the coupled parallel analysis time under various process allocation cases, and select the optimum process allocation according to the estimated analysis times. Next, the computers that selected by agent are receive the analysis data, and these computer perform the coupled parallel analysis. When the coupled analysis is finished, the result is sent back to the user.

Estimation method of the coupled parallel analysis time is shown in Fig.4. Calculation time of the matrix generation and the matrix vector product of CG method are estimated using CPU speed and load obtained by the MDS. Communication time of the parallel analysis and the coupled analysis are estimated using network bandwidth and latency obtained by the Network Weather Service.

3 Coupled Parallel Finite Element Analysis

3.1 Analysis Models and Conditions

The electromagnetic and structural coupled problem of bending motion [10] is analyzed. A copper rectangular plate rigidly clamped at one end is placed in a transient magnetic field B_z and a steady magnetic field B_x as shown in Fig.5. The interaction between the eddy current, which is produced by the transient magnetic field B_z, and the steady magnetic field B_x causes bending deformation of the plate. While the plate is vibrating, the electromotive force by the deformation velocity and the steady magnetic field B_x influences the eddy current.

The conditions of the analysis as follows: time increment Δt is 0.1msec and number of total time steps is 50.

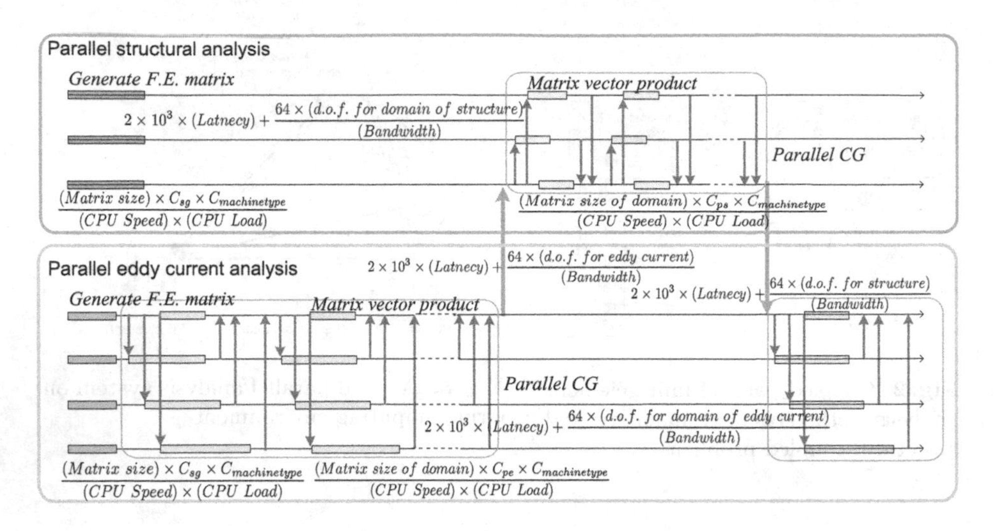

Fig. 4. Estimation of coupled parallel analysis time

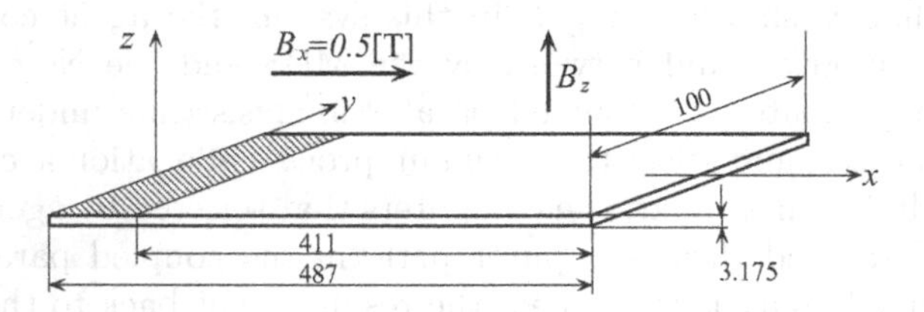

Fig. 5. Schematic diagram of a plate placed in electromagnetic field

3.2 Coupled Parallel Analysis in Grid

Computing Environment. Two PC clusters are used in the coupled analysis. The distance between these PC clusters is about 30km. One PC cluster is Pentium III 850MHz and 866MHz cluster, and the other is Pentium III 1GHz cluster. The communication rate between these PC clusters for each data size is shown in Fig.6. The communication rate in the LAN environment (100Base-TX Ethernet) is also shown in this figure. Although the maximum communication rate of the LAN is about 80Mbps, the maximum rate between PC clusters through the Internet is low, about 8Mbps.

The coupled parallel analyses are carried out in three types of computing environment shown in Fig.7 to examine the effect of the communication rate on the efficiency of the coupled parallel analysis. In Fig.7(a), all processes are carried out by one PC cluster. Although the data of only coupled effect is communicated through the Internet in Fig.7(b), the communications of the parallel eddy current analysis are performed through the Internet in Fig.7(c).

Results and Discussions. The total execution time of the analysis performed in three type computing environment are compared in Fig.8. These analyses

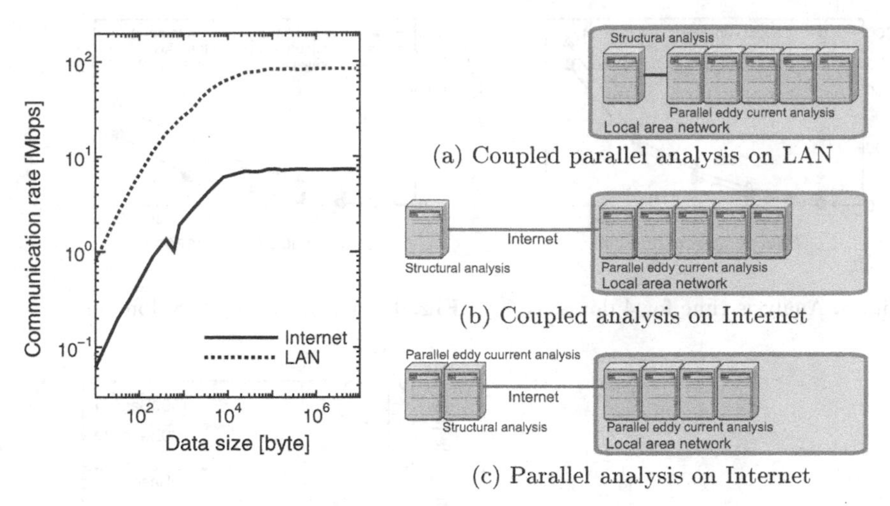

Fig. 6. Communication rate for Internet and LAN

Fig. 7. Composition of computers for coupled parallel analysis

are carried out with the analysis model divided into 4 domains. The difference in total execution time between the coupled analysis in the LAN environment (a) and that in the Internet (b) is only a few percent because the time for communication consumed for the coupled effect is much smaller than that for other processes such as matrix-vector product, communication for the parallel eddy current analysis and so on. Since large data transmissions are frequently carried out in the parallel eddy current analysis, the total execution time may be increase in the coupled analysis if parallel analysis data is transmitted through the Internet (c). However, although the communication rate for the parallel eddy current analysis becomes about 10% of the LAN, the difference in total execution time between the line in Fig.8 for parallel analysis in the Internet and other lines is small. The reason may be that the quantity of communication data is not large for the analysis with 4 domains, and so the coupled analyses are performed increasing the number of domains.

The total execution time of the coupled analysis with 8 domains is shown in Fig.9. By only increasing the number of machine for the eddy current analysis, which requires many operations more than the structural analysis, the total execution time becomes about a half compared to Fig.8 for the case of 4 domains. According to Fig.9, the line for parallel analysis in the Internet differs from other lines. The reason is that the computational granularity becomes small by increasing the number of machines. However, large difference is not observed in the total execution time between two lines such as the coupled analysis in the Internet and the coupled analysis in the LAN environment. Therefore, the coupled parallel analysis by communicating coupled effect through the Internet is suitable for grid computing environment.

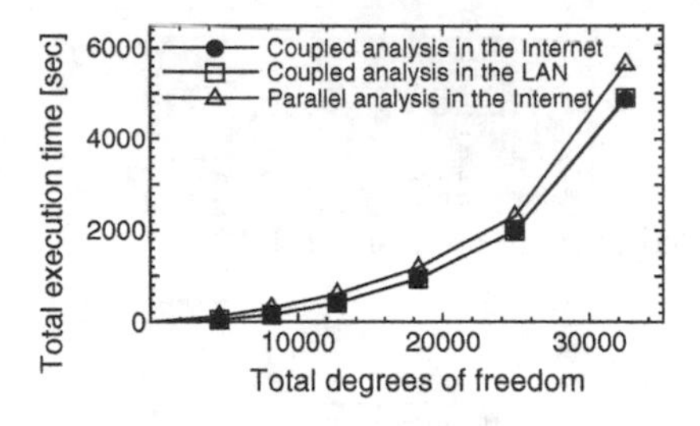

Fig. 8. Analysis time for 4 domains

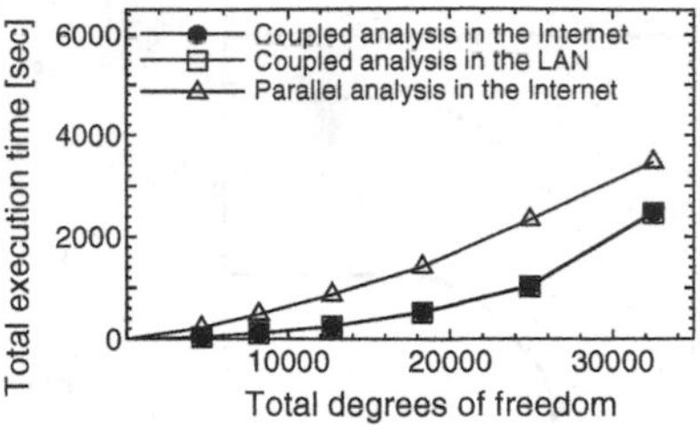

Fig. 9. Analysis time for 8 domains

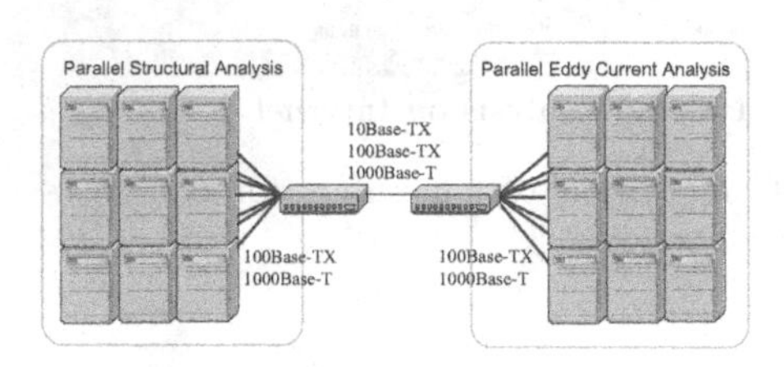

Fig. 10. Analysis environment

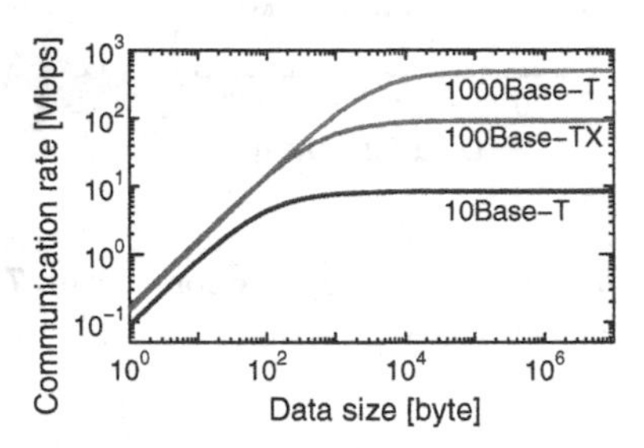

Fig. 11. Network performance obtained by NWS

3.3 Verification of Process Allocation

Computing Environment. To discuss the coupled parallel analysis system with the process allocation, the coupled analyses are performed using Pentium IV 2.8GHz cluster. PCs of the PC cluster are connected by 10Base-T, 100Base-TX or 1000Base-T ethernet as shown in Fig.10. The communication rate obtained by the Network Weather Service is shown in Fig.11. The communication time of the coupled analysis is estimated using these communication rate.

Results and Discussions. The change of estimated and executed analysis time of the coupled analysis with number of total degrees of freedom is shown in Fig.12 using 8 domains for the eddy current analysis and 4 domains for the structural analysis. Since difference between estimated time and executed time is very small, the estimation method of the coupled parallel analysis time can be used for optimum process allocation. The details of estimated and executed analysis time are shown in Fig.13. It is confirm that the communication time for the coupled analysis is small compared to the total analysis time.

The change of analysis time with the number of CPU for the parallel eddy current analysis is shown in Fig.14 using 4 CPU for the parallel structural analysis. The total number of degrees of freedom is 24893. It is also confirm that estimated analysis time and executed analysis time show similar tendency. Therefore, the optimum process allocation becomes possible by the proposed estimation method.

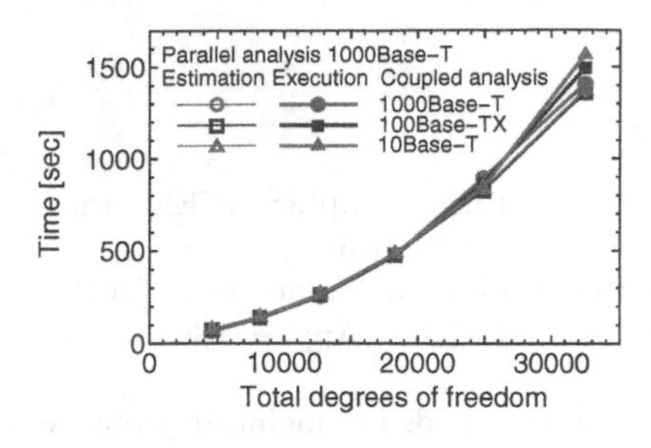

Fig. 12. Estimation and execution analysis time of coupled parallel analysis

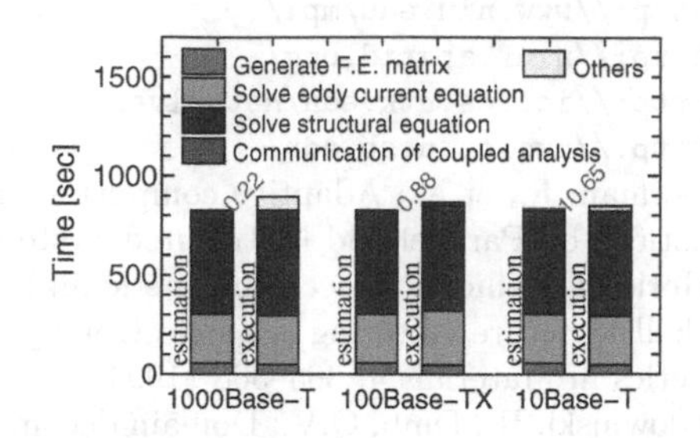

Fig. 13. Details of analysis time (24893 d.o.f.)

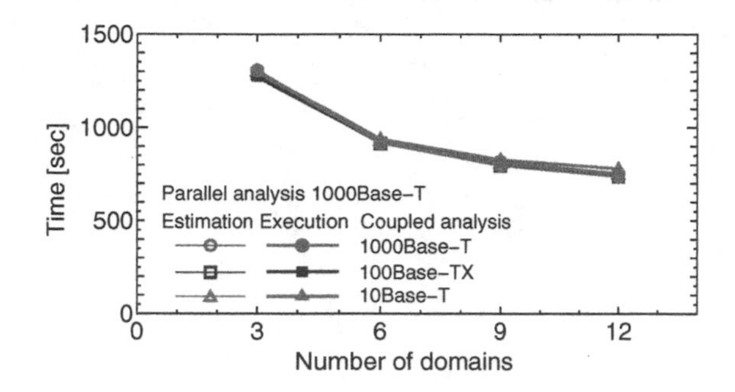

Fig. 14. Estimation and Execution analysis time (24893 d.o.f.)

4 Conclusions

Coupled parallel finite element analysis method especially for grid computing environment is proposed. In this method, PC clusters located in different places perform the parallel finite element analyses, and communicate with the other PC cluster through the Internet to analyze the coupled problem. The electromagnetic and structural coupled analysis was carried out using two PC clusters on the Internet. The difference in the total execution time was only a few percent compared to the coupled analysis in the LAN environment for the coupled problem with a few ten thousand degrees of freedom. The execution time, however, increased for the coupled analysis by performing the parallel analysis in the Internet compared to that by the proposed method. Therefore, the proposed method is suitable for the grid computing environment. It is also confirmed that the analysis time is minimized by using the proposed process allocation method.

References

1. Foster, I., Kesselman, C.: The GRID: Blueprint for a new computing infrastructure. Morgan Kaufmann, San Francisco (1998)
2. `http://www.globus.org/`

3. http://www.niu.edu/mpi/
4. http://ninf.apgrid.org/
5. http://icl.cs.utk.edu/netsolve/
6. http://nws.cs.ucsb.edu/
7. Berman, F., et al.: Adaptive computing on the Grid using AppLeS. IEEE Transactions on Parallel and Distributed Systems 14, 369–382 (2003)
8. Horie, T., Niho, T.: Electromagnetic and mechanical interaction analysis of a thin shell structure vibrating in an electromagnetic field. Int. J. of Applied Electromagnetics in Materials 4, 363–368 (1994)
9. Glowinski, R., Dinh, Q.V.: Domain decomposition methods for nonlinear problems in fluid dynamics. Computer Methods in Applied Mechanics and Engineering 40, 27–109 (1983)
10. Turner, L.R., Hua, T.Q.: Results for the cantilever beam moving in crossed magnetic fields. COMPEL 9, 205–216 (1990)

Photo-Realistic Visualization for the Blast Wave of TNT Explosion by Grid-Based Rendering

Kaori Kato[1], Takayuki Aoki[1], Tei Saburi[2], and Masatake Yoshida[2]

[1] Global Scientific Information and Computing Center, Tokyo Institute of Technology, 2-12-1 O-okayama, Meguro-ku, Tokyo, 152-8550, Japan
[2] Research Center for Explosion Safety, National Institute of Advanced Industrial Science and Technology, 1-1-1 Azuma central 5, Tsukuba-shi, Ibaraki, 305-8565, Japan

Abstract. After the detonation of a solid high explosive, the material has extremely high pressure keeping the solid density and expands rapidly driving strong shock wave. In order to investigate the blast wave propagation driven by the 32-kg TNT explosion of the underground magazine a three-dimensional simulation is performed with a stable and accurate numerical scheme without a special modeling for the expansion process of detonation product gas. The compressible fluid equations are solved by a fractional step procedure which consists of the advection phase and non-advection phase. The former employs the Rational function CIP scheme in order to preserve monotone signals and the latter is solved by IDO (Interpolated Differential Operator) scheme for achieving the accurate calculation. For this simulation results, photo-realistic visualization is achieved with combination of volume rendering with isosurface rendering on grid computer.

1 Introduction

Explosives are very dangerous materials because enormous energy is released at the moment through detonation process. When the detonation process terminated in solid explosives, the pressure reaches tens of thousands times the air and the solid density still remains. The detonation product becomes the gas and expands rapidly driving strong shock wave. They are called blast waves. From the viewpoint of safety study, it is very important to estimate of damages by blast waves. For many years, the numerical schemes and the basic research on blast waves [1] $\sim$ [13] have been developed. However, few schemes can give quantitative estimation for the wide calculation scale. In the early stage of blast wave propagation, the large density jump between the detonation product gas and the air makes the numerical simulation difficult. A negligible numerical error for high-density detonation product gas has serious effect on the air and degrades the shock waves propagation. In order to study such blast waves, a stable and accurate numerical scheme have to be developed.

J. Labarta, K. Joe, and T. Sato (Eds.): ISHPC 2005 and ALPS 2006, LNCS 4759, pp. 271–278, 2008.

2 Governing Equations

The inviscid compressible fluid equations are employed for the blast waves of the detonation product gas expanding rapidly and the strong shock wave propagating into the air. For the detonation product gas, the JWL equation of state $P_{JWL}(\rho, e)$ [15] is applied, and the ideal gas equation of state $P_{air} = (\gamma - 1)\rho e$ is used for the air with the ratio of specific heats for the gas $\gamma = 1.4$. In the following equations, the notation ρ is the density, e is the internal energy and $\mathbf{u}$ is the velocity, where u, v and w are the components of $\mathbf{u}$ in the x-, y- and z-direction, respectively. In order to identify the region of the detonation product gas, we introduce a volume fraction α which is taking a value $0 \leq \alpha \leq 1$. The region occupied by only the detonation product gas is indicated by $\alpha = 1$ and the region for only the air is $\alpha = 0$. The pressure of the mixed region is simply determined by $p = \alpha P_{JWL} + (1 - \alpha) P_{air}$.

$$\frac{\partial \rho}{\partial t} + (\mathbf{u} \cdot \nabla)\rho = -\rho \nabla \cdot \mathbf{u}, \tag{1}$$

$$\frac{\partial \mathbf{u}}{\partial t} + (\mathbf{u} \cdot \nabla)\mathbf{u} = -\frac{1}{\rho}\nabla p, \tag{2}$$

$$\frac{\partial e}{\partial t} + (\mathbf{u} \cdot \nabla)e = -\frac{p}{\rho}\nabla \cdot \mathbf{u}, \tag{3}$$

$$\frac{\partial \alpha}{\partial t} + (\mathbf{u} \cdot \nabla)\alpha = 0. \tag{4}$$

3 Numerical Method

We develop a stable and accurate Eulerian scheme for propagation of strong blast waves. The governing equations (1)-(4) are spilt into the advection phase and the non-advection phase in the same way as the original CIP scheme [14]. The fractional step procedure is introduced for the stability of the scheme. The time advance from the n-th step to the $n+1$-th step consists of the following procedure. The dependent variables of the n-th time step is integrated to the intermediate step by solving the advection equation. Then the variables are advanced to the $n+1$-th step by solving the non-advection equation. In the scheme, not only the values on the grids but also the profiles between the grids satisfy the governing equations because the values propagate together with the spatial gradients of them. Because we need better coupling between the pressure and the velocity for the phenomena with a steep gradient profile in the non-advection phase, we adapt staggered grid.

4 Three-Dimensional Simulation for the Blast Wave Driven by the Explosion in the Magazine

The three-dimensional blast wave simulation is performed on the same condition with the open-air mock experiment of the 32-*kg* TNT explosion of the underground magazine which was done in the exercises field of Japan Ground Self

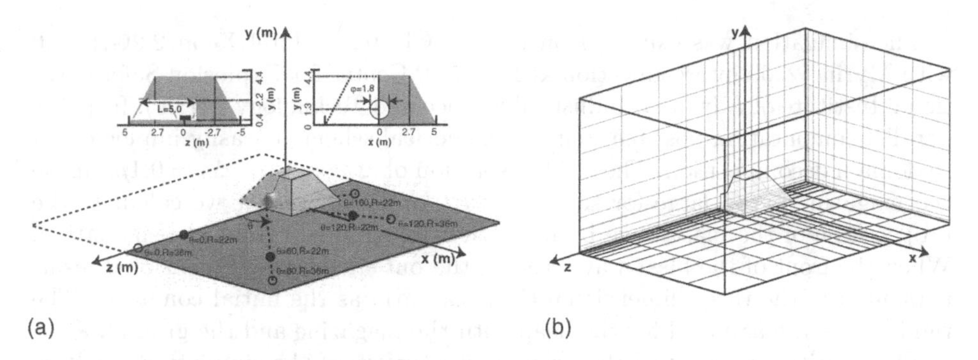

Fig. 1. Configuration of the computational domain for three-dimensional simulation: (a) a schematic of magazine and measurement point; and (b) a schematic of mesh covering the computational domain

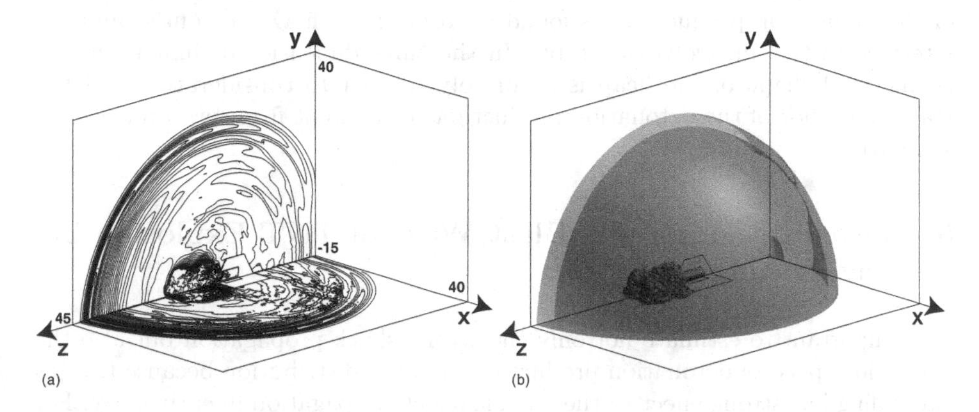

Fig. 2. Blast wave propagation driven by 32-kg TNT explosion at the magazine at t=51.28ms: (a)density contour on z-x and z-y plane with the range of 0.5∼1.35kg/m^3 and the interval of $\Delta\rho$=0.01kg/m^3; and (b)iso-surface for the volume fraction of α=0.1 and the pressure of p=1.05HPa in the range of 63.2∼115.6kPa

Defense Force in 2003 [16]. The heap with an earth was constructed and the magazine was horizontally embedded. The pyramid shape has the base area of $10m \times 10m$ and the height of $4.4m$ and the magazine size is $5.0m$ in length and $1.8m$ in diameter. The schematic of them is shown in Fig.1(a). We set the x-z plane on the ground and the y-axis in the vertical direction and the origin of the coordinate system is located at the magazine outlet. It is assumed that the phenomena are symmetric due to the symmetric geometries of all the objects, so that we calculate only a half of the computational domain in the x-direction. The grid number of $N_x \times N_y \times N_z = 180 \times 180 \times 450$ is assigned to the domain $0m \leq x \leq 40m$, $-20m \leq z \leq 45m$, $0m \leq y \leq 60m$ in the Cartesian grid with non-uniform intervals. Fine meshes of $\Delta x = \Delta y = \Delta z = 0.1m$ are arranged inside the magazine and around the outlet, and distant meshes have larger size to $\Delta x = \Delta y = \Delta z = 0.4m$ as shown in Fig.1(b).

The simulation was executed on the PC-Cluster of Intel Xeon 2.2GHz×40 with Myrinet2000 interconnection at Research Center for Explosion Safety, National Institute of Advanced Industrial Science and Technology (AIST). It makes parallel computation possible that our numerical scheme is easily implemented to domain decomposition. The mesh resolution of $\Delta x = \Delta y = \Delta z = 0.1m$ inside the magazine is too large for the size of 32-kg TNT, so that we calculate the propagation of the blast wave by using two-dimensional cylindrical simulation. When the front of the blast wave reaches the outlet, the two-dimensional result is planted in the three-dimensional Cartesian grid as the initial condition. The rigid bodies are assumed for the heap with the magazine and the ground.

The density contour and the pressure iso-surface of the simulation result at the time of 51.28ms are displayed in Fig.2(a) and (b). Almost spherical shock wave is described also in the three-dimensional simulation and it is qualitatively understood that the shock is stronger in the forward direction. The distribution of the detonation product gas is found in front of the magazine outlet and the interface to the air evolves unstably. In the three-dimensional simulation, the geometrical shape of the heap is accurately taken into consideration and unstable behavior of the detonation product gas is different from two-dimensional simulation.

5 Visualization for the Blast Wave of TNT Explosion by Ray Tracing Method

It is important to estimate not only the strong shock propagation but also the expansion speed of detonation product gas and the distribution because the gas expanding has strong effect to the outside shock propagation in early stage. For evaluation of the gas expanding process, three-dimensional simulation is required due to the importance of fluid instability growing on the gas surface. In order to understand the phenomena from huge data obtained by the simulation using large memory, visualization is one of the must fundamental approaches. The

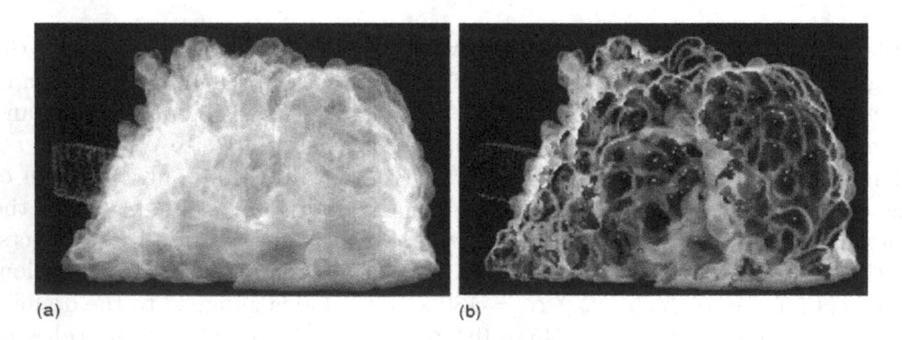

Fig. 3. The sample frame of flame visualized Povray3.5: (a) using only volume rendering; and (b) using volume rendering with iso-surface rendering

Fig. 4. Photo-realistic visualization of the flame expanding in front of the magazine at 5 *msec*

Fig. 5. Photo-realistic visualization of the flame expanding in front of the magazine at 20 *msec*

Fig. 6. Photo-realistic visualization of the flame expanding in front of the magazine at 50 *msec*

Fig. 7. The farming for pov-ray rendering of blast wave on Titech Grid system

detonation product gas expands with complex-shaped surface by fluid instability and intense luminescence based on black body radiation. This luminous body (flame) has not strict material boundary therefore volume rendering is suitable. In simple volume rendering, the flame is expressed transparent because the absorption factor for the incidence light from outside cannot be taken enough, even though it is not actually transparent. It is difficult to observe the expanding process of flame by simple volume rendering. We present an efficient visualization of flame and the key idea is combination of volume rendering with iso-surface rendering by Povray3.5, which employ ray-tracing method.

Povray [17] is a free ray-tracing software to generate high quality frame and the volume rendering function has been added from the version 3.5. The data in df3 format is needed to use this function. The df3 data has the header part of 6 byte and the part of three-dimensional data. To the header part, each number (nx, ny, nz) of three-dimensional data elements in x, y, and z direction must be given. Each element of three-dimensional data is 1 byte variable normalized within the range from 0 to 255 and they must be stored in order of x, y and z. The data pattern of df3 format occupies a unit cube of $< 0,0,0 >< 1,1,1 >$ regardless of the three-dimensional data size. The sample frame of flame visualized by simple volume rendering is shown in Fig.3(a) and the sample frame visualized by volume rendering with iso-surface rendering in Fig.3(b).

From Fig.3, it is found that the frame of expanding flame with complex surface and intense luminescence becomes more realistic by the combination of volume rendering with iso-surface rendering. Next, the frames composite with a background are shown in Fig.4, Fig.5 and Fig.6. In these figures, photo-realistic visualization for the expanding process of flame which is observed by the field experiment is achieved. The consuming CPU resource more than 30 minutes is needed for a frame. For one animation making, hundreds of frames are required and the efficient job management for rendering is necessary. We use the Titech Grid system shown in Fig.7 as the render farm.

6 Conclusions

For the simulation of strong blast waves, a high-accurate numerical scheme was presented. In order to investigate the expanding process of flame by the 32-kg TNT explosion of the underground magazine, a detailed simulation is performed without a special modeling[18] for the expansion process of flame. It is concluded that the scheme proposed in this paper is quit available to solve the compressible fluid equation and applicable to the simulation of strong blast waves. The photo-realistic visualization is achieved with combination of volume rendering with iso-surface rendering on grid computer.

References

1. Boris, J.P., Book, D.L.: Flux-corrected transport. I. SHASTA, a fluid transport algorithm that works. J. Comp. Phys. 11, 38–69 (1973)
2. Roe, P.L.: Approximate Riemann solvers, parameter vectors, and difference schemes. J. Comp. Phys. 43, 357–372 (1981)
3. Osher, S., Chakravarthy, S.: High Resolution Schemes and the Entropy Condition. SIAM J. Num. Anal. 21, 984–995 (1984)

4. Harten, A.: High resolution schemes for hyperbolic conservation laws. J. Comp. Phys. 49, 357–393 (1983)
5. Yang, H.: An artificial compression method for ENO schemes: The slope modification method. J. Comp. Phys. 89, 125–160 (1990)
6. Harten, A., Engquist, B., Osher, S., Chakravarthy, S.R.: Uniformly high order accurate essentially non-oscillatory schemes, III. J. Comp. Phys. 71, 231–303 (1987)
7. Harten, A.: ENO schemes with subcell resolution. J. Comp. Phys. 83, 148–184 (1987)
8. Shu, C.W., Oshert, S.: Efficient implementation of essentially non-oscillatory shock-capturing schemes, II. J. Comp. Phys. 83, 32–78 (1989)
9. Woodward, P., Colella, P.: The numerical simulation of two-dimensional fluid flow with strong shocks. J. Comp. Phys. 54, 115–173 (1984)
10. Colella, P., Woodward, P.R.: The piecewise parabolic method (PPM) for gas-dynamical simulations. J. Comp. Phys. 54, 174–201 (1984)
11. Brode, H.L.: Numerical Solution of spherical Blast waves. Journal of Applied physics 26, 766–775 (1955)
12. Brode, H.L.: Blast wave from a spherical charge. Phisics Of Fluid 2, 217–229 (1959)
13. Kury, J.W., Hornig, H.C., Lee, E.L., McDonnel, J.W., Ornellas, D.L., Finger, M., Starnge, F.M., Wilkins, M.L.: Metal Acceleration by Chemical explosives. In: Fourth Symposium on detonation, pp. 3–13 (1965)
14. Yabe, T., Aoki, T.: A universal solver for hyperbolic equations by cubic-polynomial interpolation I. One-dimensional solver. Computer Physics Communications 66, 219–232 (1991)
15. Dobratz, B.M.: Properties of Chemical Explosives and Explosive Simulants LLNL Explosives Handbook UCRL-52997, Distribution Category UC-45
16. Bakuhatsu-eikyou-hyouka-iinkai-houkoku-sho, All Japan Association for security of explosives (2003)
17. `http://www.povray.org`
18. Rasmussen, N., Nguyen, D.Q., Geiger, W., Fedkiw, R.: Smoke Simulation for Large Scale Phenomena. SIGGRAPH 22, 793 (2003)

Development of an Interactive Visual Data Mining System for Atmospheric Science

Chiemi Watanabe*, Eriko Touma, Kazuko Yamauchi, Katsuyuki Noguchi, Sachiko Hayashida, and Kazuki Joe

Department of Information and Computer Science, Faculty of Science, Nara women's University,
Kitauoyanishi-machi, Nara-city, 630-8506, Nara, Japan
{chiemi,shuro,kazuko,nogu,sachiko,joe}@ics.nara-wu.ac.jp

Abstract. In atmospheric science, 3D visualization techniques have been mainly used to create impressive presentation in recent decades. However, from the viewpoint of utilize for visual data mining, 3D visualization methodology has difficulties in becoming wide spread because most conventional and established way is to make 2D diagrams consisting of two dimensions of a temporal transitional 3D grid. From these observations, we have been developing a quick look tool of atmospheric science data for 3d visual data mining. We expect that scientists can utilize this tool for finding out 2D diagrams from the data by using various 2D or 3D visualization methods, and become accustomed themselves to 3D visualization methods.

Keywords: We would like to encourage you to list your keywords within the abstract section.

1 Introduction

Visualization techniques have been gaining much attention in a broad range of application such as science, medical and finance. In particular, 3D visualization is used in various application fields because users can observe objects of interest from arbitrary viewpoints for obtaining intuitive understanding. Besides being effective presentation tools, visualization techniques are also expected as visual data mining tools that make it possible to explore implicit knowledge from massive data. Among processes in visual data mining, interactive analysis process is considered to be the key because new knowledge is explored by human vision and intuition. More concretely, in an interactive analysis, a user tries to discover new knowledge from 3D visualized simulation and/or real data by changing his/her viewpoints and applying various visualization methods.

In the atmospheric science, 3D visualization techniques have been mainly used to create impressive presentations in recent decades, and many useful visualization tools, such as AVS[1], IDL[2],Vis5D[3] and VisAD[4], are distributed

* Present affiliation: Department of Information Sciences, Ochanomizu University.

J. Labarta, K. Joe, and T. Sato (Eds.): ISHPC 2005 and ALPS 2006, LNCS 4759, pp. 279–286, 2008.

for this field's scientists. In fact, atmospheric science data has 3d special scales and 1d time scale (we name the data as "spatio-temporal 4D grid data") that turn out to be applicable to 3d visualization with animation. However, from the viewpoint of visual data mining, only a few scientists have been used 3d visualization techniques to analyze the phenomenon. We consider the there are several reasons:

(1) It is difficult to completely represent all of spatio-temporal 4D grid data in 2D media, such as paper and 2D monitor. Reseachers usually make 2D graphs and print them out for further analysis. But it is quite difficult to represent all contents of a spatio-temporal 4D grid data in a paper or a 2D monitor because projecting the data incurs lossy compression, and some data will be lost as a consequence. To analyze the grid data, therefore they should try to observe it from various viewpoints and try to visualize it by applying multiple methods by changing various parameters.
(2) The usage of visualization tools are complicated in many cases. Most visualization tools provide many functionalities to meet users various requirements, but such versatility makes the system complex. The consequence is that a user is required to have high knowledge about both visualization techniques and programming, although learning such skills is usually time consuming.
(3) Scientists are not familiar with 3D visualization methods. 3D visualization methods, such as 3D contour and volume visualization, is not so popular in the research area as mentioned above. Most conventional and established way is to make 2D diagrams consisting of two dimensions of a temporal transitional 3D grid, and for this reason, 3D visualization methodology has difficulties in becoming wide spread.

From these observations, we have been developing a quick look tool of atmospheric science data for 3d visual data mining. In fact, we attempt to provide only basic visualization methods in the tool for analyzing 4d grid spatio-temporal data. The user interface is designed to be as simple as possible so that one can operate with only a few instructions. The tool will make a user possible to create a 2D diagrams that he/she would like to obtain by condensing other two dimensions from a temporal transitional 3D grid. More concretely, we attempt to provide a tool that can facilitate the conventional visualization analysis for creating 2D diagrams, consisting of "Overview", "Detail" and "Temporal transition", by providing appropriate methods for each step. As a result, we expect that scientists can utilize this tool for finding out 2D diagrams from the data by using various 2D or 3D visualization methods, and become accustomed themselves to 3D visualization methods. In Section 2, we survey on the visual data mining activities for atmospheric science data. Traditional visual analysis method by condensing dimension of the data is developed in Section 3. In Section 4 we introduce the prototype system and its user interaction flow, and finally Section 5 concludes this paper.

2 Visual Data Mining Activities for Atmospheric Science Data

Atmosphere is the gaseous body which blankets the earth. It consists of four layers, namely, convection sphere, stratosphere, mesosphere and thermosphere. Various astronomical phenomena, such as rain, snow, cloud, typhoon and ascension or dissension of temperature, are caused by various physical phenomena, such as convection, atmospheric constituent, rotation and motion of earth. Atmospheric scientists make analysis of atmospheric science data, captured by radar and satellite, to reveal the mechanism of those phenomena. In recent years NASA[5], ECMWF[6] and NCEP[7] provides atmospheric science data archive center, where user can acquire the data by using web browser. For example, NASA Goddard DAAC[8] archives more than 20 kinds sensor data that cover from 1978 to resent data, and ECMWF provides re-modeled weather monitoring data. Data mining [9] has been widely used to mine important information and/or implicit rules from numerous data. In atmospheric science, explosive data have already been archived in data archiving infrastructure as mentioned above, and even small laboratories or individual scientists are therefore ready for utilizing data mining. Various methods, such as statistical methods and machine learning algorithms, has been proposed as data mining technologies, and we believe that "visual data mining"[10] plays the important role to extract valuable knowledge from underlying data. In fact, visual data mining are methods which promote human visual sense and intuition to discover new knowledge by using visualization techniques. In atmospheric science, visual analysis has been established in several decades and effective visualization and analyzation methods have been confirmed as empirical knowledge. However, in these methods, spacio-temporal 4D grid data have to be projected to 1D or 2D grid data, and scientists then have to pay considerable effort to find appropriate combinations of dimensions and parameters by try-and-select based on their background knowledge. Actually, 3D visualization and animation are attractive to represent 4D spacio-temporal grid data and many scientists are using the 3D visualization systems, their purposes are not for data analysis but just for creation of impressive presentation.

3 Dimensionally Reduction of Atmospheric Science Data

An atmospheric science data can be represented by function whose attributes are 3-dimensional space with time $X = X(x, y, z, t)$, and the data is then projected 1D or 2D surface for further analysis. Consequently, there are 4 patterns of 1D projection such as $X(x; y_0, z_0, t_0), X(y : x_0, z_0, t_0), X(z : x_0, y_0, t_0)andX(t : x_0, y_0, z_0)$, by fixing one of four axis values, Likewise, there are 6 patterns of 2D projection as follows:

(1) Space section diagram: X(x,y; z0, t0), X(x,z: y0,t0), X(y,z: x0, t0)
(2) Time section diagram: X(x,t: y0,z0), X(y,t: x0,z0), X(z,t: x0,y0)

Space section diagram is a surface diagram in that time and one of longitude, latitude or altitude are fixed. The problem is that it is difficult to select fixed values for representing persuasive diagram and the way to select them is effected their purpose of analyze and their background knowledge on the data. Besides, the diagram $X(x, y; z_0, t_0)$ may be used for not only altitude value but also potential temperature and pressure. For example, Fig. 1(a) shows potential vorticity diagram of $X(x, y; z_0, t_0)$ where $z_0 = 850K$ of potential temperature and $t_0 = 4th.Dec.1981$. The potential temperature of a parcel of air at pressure P is the temperature that the parcel would acquire if adiabatically brought to the standard level (1000hPa)[11]. Because air parcels are following on the surface of constant potential temperature, the surface on which z-value is fixed by a potential temperature would be appropriate. When users make a space section diagram, they should select an appropriate map projection. Fig. 1(a) shows equidistant projection for analyzing the stratosphere around the North Pole. Time section diagram is effective for representing motion of atmosphere and it is typically used in the area of atmospheric dynamics. The diagram $X(x, t; y_0, z_0)$, which fixes latitude and altitude, is suitable for representing east-west air motion. Fig. 1(b) shows the time evolution of the ensemble mean anomaly of geopotential height at 500 hPa, averaged over the latitude band 35N-55N during 30 days in Nov. 1945[12]. In this figure, the area marked by transverse lines shows high-pressure area, and the area marked by vertical lines show low-pressure area. You can observe that 5 or 6 piece of high and low pressure areas lies side-by-side and they declines, that turns out to be eastward propagation of atmospheric high and low. From this figure Hovmoeller discovered the feature of traveling anticyclone. Be $X(x, t|y_0, z_0)$ diagram is named as Hovmoeller diagram for this reason.

4 Prototype System

We have been developing a quick look tool for 3d visual data mining of atmospheric science data. This tool is implemented by using JDK1.4.2 powered by Visualization Tool Kit as the visualization API. The tool attempt to provide only basic visualization methods for analyzing 4d grid spatio-temporal data. On the top of calculation engine, we put a simple user interface so that a user can operate with only a few instructions. In the system a scientist makes desired 2D diagrams, like above-mentioned examples, by dimensionally reduction from a temporal transitional 3D grid. We suspect that the following three steps are generally take place for making an desired 2D diagrams from a temporal transitional 3D grid.

Step1: Overview. A user try to observe the overview of temporal transition of 3D grid data to find any clues about the area of interest, such as scalar value and time when some characteristic phenomenon occurred. In this step, we apply volume rendering by applying ray-casting algorithm.

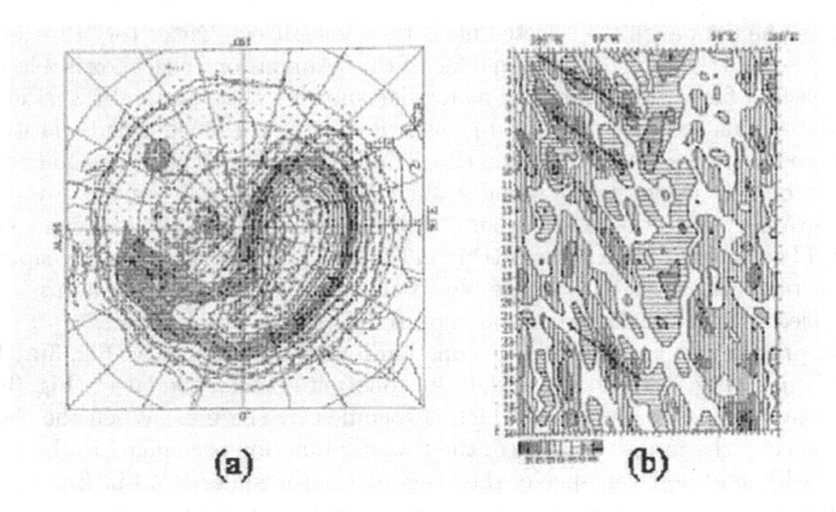

Fig. 1. Dimensionally Reduced Diagrams. (a) Temperature diagram which is expressed $X(x, y; z_0, t_0)$ where $z_0 = 850K$ of potential temperature and $t_0 = 4nd.Dec.1981$. (b) The time evolution of the ensemble mean anomaly of geopotential height at 500 hPa.

Step2: Detail. Based on the clues found in Step1, the user make further analysis on the data when and where the characteristic phenomena appeared. In this step, we apply the following two visualization methods.

1. cross section by three planes paralleled with x, y or z axes respectively.
2. contour by a scalar value

Step3: Analyze the temporal transition. When the user obtains the most appropriate cross section diagram in step2, he/she proceeds this step to examine the temporal transition of the diagram. This tool reconstructs the 3D grid by using the user selection diagram and their transition. For example, suppose one specifies a diagram $X(x, z; y_0, t_0)$ in step2, the 3D grid $X(x, z, t; y_0)$ are then constructed. In this step, this tool uses the same methods as the previous step, and user can generate persuasive time section diagrams.

Fig.2 shows the image capture of this tool. Visualized data in Fig.2 is temperature at 26th September 2002 from European Center for Medium range Weather Forecasting (ECMWF) forecast model. In this month, sudden stratospheric warming phenomena were observed around the North Pole. A sudden stratospheric warming is an event where the polar vortex of westerly winds in the Northern winter hemisphere abruptly (i.e. in a few days time) slows down or even reverses direction, accompanied by a rise of stratospheric temperature by several tens of degrees Celsius. We attempt to explain this systems user interaction with analyzing these phenomena as a running example.

Fig. 2(a) shows the initial state. When a user tries to open files that are going to be analyzed, this tool shows the list of file name in the left of pane, and visualizes the first file by volume rendering. When the user clicks another item on the list panel, the selected file is then visualized. "Next"(or "Previous") button selects the next (or previous) file. If the "Animation" checkbox is selected, the selected file will be animated at regular showing the temporal-transition of the data. This tool has three grid geometries for map the (longitude, latitude) point to other coordinates by using three kinds of map projection methods; they are "cube" by using stereographic cylindrical projection, "cylinder" by using orthographic projection and "sphere" which is like a globe. A user can switch the grid geometry by the buttons ((*1) in Fig. 2.) The "capture" button captures the display and saves it as image file. If the user selects multiple items in file list, selected files are displayed and captured as image files in order.

We provide two kinds of transfer function for color and opacity (Fig. 3(a)(b)). The former (Fig. 3(a)) shows all data by constant opacity, the latter (Fig. 3(b)) shows only values in the range which is specified by the user. When the check-box "specify the range" is checked, the transfer function is changed to the latter (Fig. 3(b)) and user can specify the range using the slider ((*2) in Fig. 2). Although such transfer function can be also customized intricately, we just provide these two transfer functions for the sake of simplicity. In Fig. 3(a) and (b), the user applies cylinder grid structure to observe the sudden warming phenomena around the South Pole. Actually, we may be not able to find any feature of the image due to the fact that it is static snapshot, but we can have further investigation by looking into sight the motion of the atmosphere if it is displayed by animation. Fig. 3(b) visualized the temperature between 180k and 200k. The left image is on 16th. Sep, and the right image is on 25th Sep. We can observe that the atmosphere, whose temperature ranges between 180k and 200k, is in the stratosphere above the Antarctica on 16th Sep, and it is then divided two regions and gets out of the Antarctica on 25th.

Let us have a look at Detail mode. It can be switched by the "Detail" button ((*3) in Fig. 2(a)). Then the tool proceeds the analyze mode of Step2 (Fig. 2(b)). In this step, there are three planes each of which is parallel according to one of X,Y and Z axes. There is also a polygon contoured by a scalar value. There are four slider volumes to specify the value x0 on the plane X(y,z;x0,t0) ((*3) in Fig. 2), y0 on the plane $X(x, z; y_0, t_0)$ ((*4) in Fig. 2), z0 on the plane $X(x, y; z_0, t_0)$ ((*5) in Fig. 2)and a scalar value ((*6) in Fig. 2). When the check-box on the left of each slider is off, the corresponding plane (or the polygon) will be disappeared. One can capture one of planes and polygon by selecting the radio button. Fig. 3(c) shows the plane X(x,y ; z0, t0) where z0=20hPa, t0=16th Sep.(left image) or t0=25th Sep. (right image.) The area with air pressure of 20hPa is about 25km height, and it is in the under part of stratosphere. Since sudden warming phenomena are generally observed in northern hemisphere, the case, which is observed in southern hemisphere, is rare and this phenomenon affected unusual behavior of the Antarctic ozone hole [13].

To proceed to Step 3, a user should select one of the planes and a polygon by radio button, select several files in the file list, and click the button "Timesection". If the user selects the plane $(x, y; z_0, t_0)$ where z=20hPa and the 30 files from 1st. Sep to 30th Sep., the tool reconstructs the 3D grid $X(x, y, t; 20hPa)$

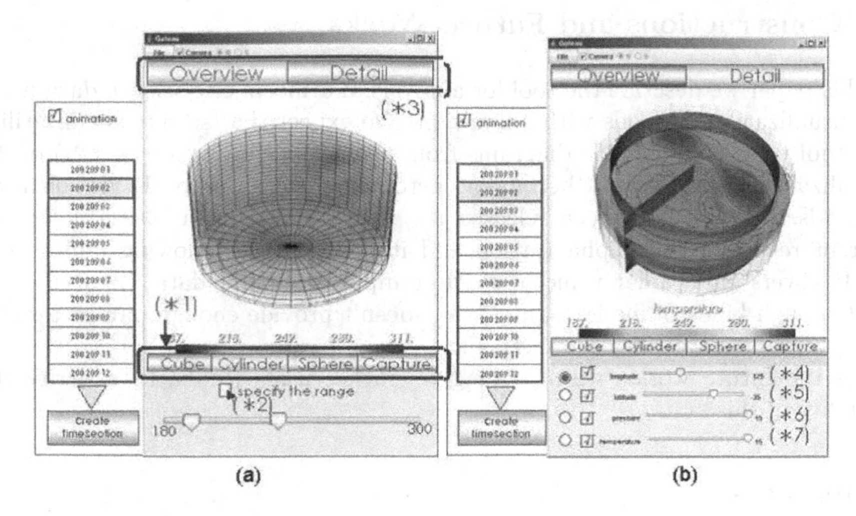

Fig. 2. Captured Image of the prototype tool. (a) The user interface in Step 1, (b) The user interface in Step 2.

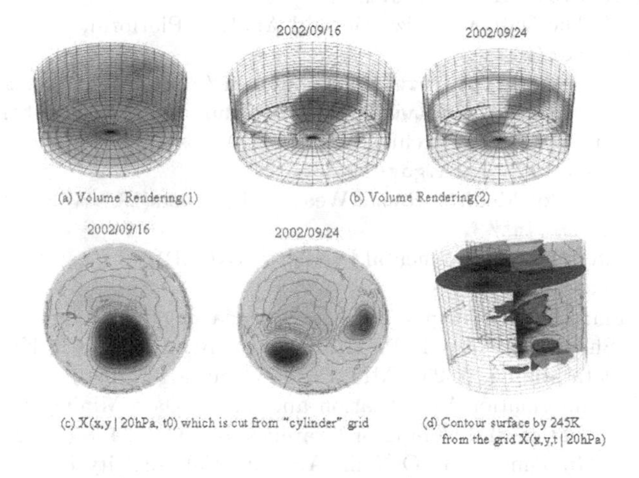

Fig. 3. Visualized ECMWF Temperature Data by Various Methods

by using selected 30 planes. The user interface in this step is almost the same as in the step 2 except that 3rd slider ((*6) in Fig. 2) specifies t0 on the plane $X(x, y; z_0, t_0)$. In this step, one can interactively make two kinds of time-section diagram, which are $X(x, t; y_0, z_0) and X(y, t; x_0, z_0)$. The former is the hovmoeller diagram described in Section 3. When the user moves the time slider ((*6) in Fig. 2), the plane $X(x, y; z_0, t_0)$ is changed and the transition of the plane in 30 days as pseudo-animation. Fig. 3(d) shows contoured polygons by 245K. The contoured polygons can represent the transition regarding a specified value. We can observe three sudden worming phenomena are appearing. The two out of them are small, and the last big one pushes away the low temperature area.

5 Constructions and Future Works

In this paper we describe the tool for analysis of atmospheric science data using 3D visualization methods with animation. We expect that scientists can utilize this tool for finding out 2D diagrams from the data by using various 2D or 3D visualization methods, and become accustomed themselves to 3D visualization methods. Now we have been released the tool [14]. However please note that current release is still alpha version and it has still the following limitations: (1) It covers only scalar value of spatio-temporal 3D grid data. (2) It doesn't load enough kinds of file formats. (3) It doesn't provide enough parameters for configuration of visualization methods.

As the future works, we are going to gather users feedback to verify the usability of this system.

References

1. AVS. Co.: AVS: Advanced Visualization Systems, `http://www.avs.com/index_wf.html`
2. RSI. Co.: IDL The Data Visualization and Analysis Platform, `http://rsinc.com/idl`
3. Hibbard, W.: Vis5D, `http://www.ssec.wisc.edu/~billh/vis5d.html`
4. Hibbard, W.: VisAD, `http://www.ssec.wisc.edu/~billh/visad.html`
5. NASA : Distributed Active Archive Centers (DAACs), `http://nasadaacs.eos.nasa.gov/`
6. European Centre for Medium-Range Weather Forecasts(ECMWF), `http://www.ecmwf.int/`
7. National Centers for Environmental Prediction(NCEP), `http://www.ncep.noaa.gov/`
8. NASA, Goddard DAAC, `http://daac.gsfc.nasa.gov/`
9. Fayyad, U., Shapiro, P., Smyth, P., Uthurusamy, R.: Advances in Knowledge Discovery and Data Mining, p. 251. MIT Press, Cambridge (1996)
10. Daniel, A.K.: Indormation Visualization and Visual Data Mining. IEEE Transactions on Visualization and Computer Graphics(TVCG) 1, 1–8 (2002)
11. Clouth, S.A., Grahame, N.S., O Neill, A.: Potential vorticity in the stratosphere devided using date from satellites. Quart. J. Roy. Meteor. Soc. 11, 335–358 (1985)
12. Hovmoeller, E.: The trough-and-ridge diagram. Tellus 1, 62–66 (1949)
13. Baldwin, M.A., Hirooka, T., O Neill, A., Yoden, S.: Major Stratospheric Warming in the Southern Hemisphere in 2002: Dynamical Aspects of the Ozone Hole Split. SPARC Newsletter No.20 (2003)
14. `http://hpcl.ics.nara-wu.ac.jp/gateau`

A Calculus Effectively Performing Event Formation with Visualization

Susumu Yamasaki and Mariko Sasakura

Department of Computer Science, Okayama University, Japan
{yamasaki,sasakura}@momo.it.okayama-u.ac.jp

Abstract. As a programming technique, we formulate a calculus of illustrating event formation which is effectively performed. An event is visualized as a sequence of abstract charts denoting processes. The calculus contains a set of charts related to basic processes, a set of situations, a semantic function assigning a situation transition to each chart, a logic program with negation-as-failure, and the integrity constraint on the set of situations.

1 Introduction

Even in programming methodologies, techniques containing actions ([14]), or events to cause situations ([11]) are required so that formal methods should be constructed. This comes primarily from programming in distributed environments and procedure construction needs in distributed systems.

Such backgrounds cause us have an aim at an event which is a well-ordered sequence of basic processes, where each basic process consists of some primitive step(s). Each process is supposed to make some situation transition, where the situation transition(s) must satisfy some condition or constraint of (distributed) environments. As event formations, we can design procedures of use in distributed environments by which action, learning, planning and intelligent behaviour are implemented. For the event formation, we formulate a formal calculus of illustrating it as follows.

(1) The calculus supposedly contains the charts each of which contains a process of primitives.
(2) The ordering of charts is required to make an event performance.
(3) The chart causes a situation transition, where predecessor and/or successor feasibility of charts is described by some logical relations of propositions.
(4) An event performance is requested to satisfy some constraint of environments, which is semi-decidable.

In this paper, the event formation is to have a well-ordered sequence of charts so that it is requested for visualization.

The paper is organized as follows. Section 2 describes a formal calculus of illustrating an event formation. Section 3 is concerned with soundness of the procedure for the calculus. An event formation procedure is also given. In Sect.

J. Labarta, K. Joe, and T. Sato (Eds.): ISHPC 2005 and ALPS 2006, LNCS 4759, pp. 287–294, 2008.

4, the event formation procedure is analysed for the visualization technique of events. The event formation is implemented and an event is visualized. Section 5 suggests related topics and future works.

2 A Calculus Forming Events

In this section, we have a calculus for event formation. The calculus is a programming model to effectively perform an event of template charts for processes in order, with the visualization of events.

A calculus of illustrating event formation is a quintuplet

$$\Im = (C, \Sigma, Sem, P, I),$$

where:

(i) C is a set of action charts, and $C^{not} = \{not\ x \mid x \in C\}$.
(ii) Σ is a set of situations.
(iii) Sem: $C \to \Sigma \to \Sigma$ is a semantic injective function.
(iv) $P \subseteq C \times (C \cup C^{not})^*$ is a propositional logic program with negation-as-failure.
(v) $I \subseteq \Sigma \times \Sigma$ is an integrity constraint relation which is recursively enumerable.

We adopt the following notations.

(a) A clause of the program set P is written as

$$x \leftarrow x_1, \ldots, x_l, not\ y_1, \ldots, not\ y_m.$$

(b) A goal set $Goal \subseteq (C \cup C^{not})^*$ is needed, where its member (that is, a goal) is expressed as $\leftarrow y_1, \ldots, y_m, not\ z_1, \ldots, not\ z_n$.
(c) A function $Sem^{-1} : C \to \Sigma \to \Sigma$ is defined for the function Sem such that

$$Sem[\![x]\!]\sigma_1 = \sigma_2 \Leftrightarrow Sem^{-1}[\![x]\!]\sigma_2 = \sigma_1.$$

(d) The relation I is represented in infix notation like:

$$\sigma_1 \Rightarrow_I \sigma_2 \text{ to denote } (\sigma_1, \sigma_2) \in I.$$

Procedure for the program P:

Extending the notion in [10], we define the procedure as the least set closure satisfying the following rules, on the assumption that a calculus $\Im = (C, \Sigma, Sem, P, I)$ is given.

(1) $$\frac{}{move_P(\Box; \sigma; \sigma)}$$

(2) $$\frac{(A \leftarrow G) \in P \quad move_P(\leftarrow G; \sigma_1; Sem^{-1}[\![A]\!]\sigma_2) \quad Sem^{-1}[\![A]\!]\sigma_2 \Rightarrow_I \sigma_2}{move_P(\leftarrow A; \sigma_1; \sigma_2)}$$

(3) $$\frac{failure_P(\leftarrow A)}{move_P(\leftarrow not\ A; \sigma; \sigma)}$$

(4) $$\frac{\text{No clause } A \leftarrow G \text{ with } A \text{ head is in } P}{failure_P(\leftarrow A)}$$

(5) $\dfrac{\text{For all } A \leftarrow G \;\; failure_P(\leftarrow G)}{failure_P(\leftarrow A)}$

(6) $\dfrac{move_P(\leftarrow A; \sigma_1; \sigma_2)}{failure_P(\leftarrow not\ A)}$

(I) $\dfrac{move_P(\leftarrow G_1; \sigma_1; \sigma_2) \;\; move_P(\leftarrow G_2; \sigma_2; \sigma_3)}{move_P(\leftarrow G_1, G_2; \sigma_1; \sigma_3)}$

(II) $\dfrac{failure_P(\leftarrow G)}{failure_P(\leftarrow G_1, G, G_2)}$

3 Soundness of Event Formation

Semantics of an event is defined by the following definition.

Definition 1. The semantic function Sem is extended to be $Sem : C^* \rightarrow \Sigma \rightarrow \Sigma$ by

(1) $Sem[\![\varepsilon]\!]\sigma = \sigma$.
(2) $sem[\![\gamma x]\!]\sigma = Sem[\![x]\!](Sem[\![\gamma]\!]\sigma)$ $(x \in C, \gamma \in C^*)$.

The procedure for the program P in a calculus $\Im$ is consistent in the sense as follows. This is in accordance with model theory in logic programs with negation-as-failure ([5,23,24,26]).

Theorem 1. *Assume that* $move_P(\leftarrow A; \sigma_1; \sigma_2)$ *for some* $\sigma_1, \sigma_2 \in \Sigma$*. Then there is no case that* $failure_P(\leftarrow A)$*.*

Proof. (1) Assume that we have both $move_P(\leftarrow A; \sigma_1; \sigma_2)$ and $fail_P(\leftarrow A)$. By the rules (2) and (5), there exists a clause $A \leftarrow G$ in P such that $move_P(\leftarrow G; \sigma_1; Sem^{-1}[\![A]\!]\sigma_2)$ and $failure_P(\leftarrow G)$. For the relation $failure_P(\leftarrow G)$, there is either some atom B included in the sequence G, or a negation $not\ B$ such that

(i) $failure_P(\leftarrow B)$, or
(ii) $failure_P(\leftarrow not\ B)$, that is, $move_P(\leftarrow B; \sigma_3; \sigma_4)$.

Note that $move_P(\leftarrow G; \sigma_1; Sem^{-1}[\![A]\!]\sigma_2)$. In accordance with case of (i), $move_P$ $(\leftarrow B; \sigma_1'; \sigma_2')$ for some σ_1', σ_2'. In accordance with case of (ii), $move_P(\leftarrow not\ B; \sigma_3'; \sigma_3')$. That is, $failure_P(\leftarrow B)$. Therefore there is some atom B such that we have both $move_P(\leftarrow B; \rho_1; \rho_2)$ for some ρ_1, ρ_2, and $failure_P(\leftarrow B)$.
(2) By the same discussion of (1), if we assume that we have both $move_P(\leftarrow C; \xi_1; \xi_2)$ and $failure_P(\leftarrow C)$ for some atom C, then there is some atom D such that we have both $move_P(\leftarrow D; \eta_1; \eta_2)$ and $failure_P(\leftarrow D)$.
(3) By the reasons (1) and (2), we finally find a sequence of atoms which contains more than one occurrence of the atom A, or an infinite sequence. However, this contradicts the first assumption of the relation $move_P(\leftarrow A; \sigma_1; \sigma_2)$, because the relation cannot be reduced to the form $move_P(\Box; \sigma; \sigma)$ by any way. This contradiction comes from the first assumption, so that the theorem holds.

Assume a calculus $\Im = (C, \Sigma, Sem, P, I)$. We need the procedure to form an event, which is defined as a nondeterministic method.

Event formation with respect to the integrity constraint I:

$$
\begin{array}{l}
Event_I(\leftarrow G; \sigma_1; \sigma_2) \\
\qquad \Leftarrow \textbf{if } \leftarrow G = \Box \\
\qquad\quad \textbf{then} \\
\qquad\qquad \textbf{if } \sigma_1 = \sigma_2 \textbf{ then } \varepsilon \text{ (empty sequence)} \\
\qquad\quad \textbf{else} \\
\qquad\qquad \textbf{if } G = G', A \text{ such that } A \leftarrow G'' \text{ is in } P \\
\qquad\qquad \textbf{then} \\
\qquad\qquad\quad \textbf{if } Sem^{-1}[\![A]\!]\sigma_2 \Rightarrow_I \sigma_2 \\
\qquad\qquad\quad \textbf{then } Event_I(\leftarrow G', G''; \sigma_1; Sem^{-1}[\![A]\!]\sigma_2)A \\
\qquad\qquad \textbf{else } Event_I(\leftarrow G'; \sigma_1; \sigma_2)
\end{array}
$$

By means of the above procedure $Event_I$, we can have an event formation for the relation $move_P(\leftarrow A; \sigma_1; \sigma_2)$.

Theorem 2. *Assume that* $move_P(\leftarrow G; \sigma_1; \sigma_2)$ *for some* $\sigma_1, \sigma_2 \in \Sigma$. *Then* $\exists \beta \in \Sigma^*.[Sem[\![\beta]\!]\sigma_1 = \sigma_2]$.

Proof. Assume that $move_P(\leftarrow G; \sigma_1; \sigma_2)$ for some $\sigma_1, \sigma_2 \in \Sigma$. Take the procedure for events. By structural induction on the goal, we see that if $Event_I(\leftarrow G; \sigma_1; \sigma_2) = \gamma$, then $Sem[\![\gamma]\!]\sigma_1 = \sigma_2$.
(1) If $\leftarrow G = \Box$, then $\sigma_1 = \sigma_2$ so that $Sem[\![\varepsilon]\!]\sigma_1 = \sigma_1$.
(2) If $\leftarrow G = \leftarrow G', A$ for some atom A and $move_P(\leftarrow G', A; \sigma_1; \sigma_2)$, then we must assume that for some G''

$$move_P(\leftarrow G', G''; \sigma_1; Sem^{-1}[\![A]\!]\sigma_2).$$

Assume that $Event_I(\leftarrow G', G''; \sigma_1; Sem^{-1}[\![A]\!]\sigma_2) = \gamma'$, and that

$$Sem[\![\gamma']\!]\sigma_1 = Sem^{-1}[\![A]\!]\sigma_2.$$

By the procedure, $Event_I(\leftarrow G; \sigma_1; \sigma_2) = \gamma' A$. It follows that

$$Sem[\![\gamma' A]\!]\sigma_1 = Sem[\![A]\!](Sem^{-1}[\![A]\!]\sigma_2) = \sigma_2.$$

(3) If $\leftarrow G = \leftarrow G', not\ A$ for some atom A and $move_P(\leftarrow G', not\ A; \sigma_1; \sigma_2)$, then we must assume that $failure_P(\leftarrow A)$ and $move_P(\leftarrow G'; \sigma_1; \sigma_2)$. Assume as induction hypothesis that $Event_I(\leftarrow G'; \sigma_1; \sigma_2) = \gamma'$, and that $Sem[\![\gamma']\!]\sigma_1 = \sigma_2$. In this case, $Event_I(\leftarrow G; \sigma_1; \sigma_2) = \gamma'$. It concludes this case.

4 Event Formation and Visualization

We are now supposed to have the calculus $\Im = (C, \Sigma, Sem, P, I)$. To visualize the event (the sequence of charts) by means of the event formation $Event_I$, we recursively implement two derivations each of which includes each other.

(1) The event formation derivation caused by the relation $move_P$:
  (a) When $Event_I(\leftarrow G; \sigma_1; \sigma_2)$ is replaced by
  $$Event_I(\leftarrow G', G''; \sigma_1; Sem^{-1}[\![A]\!]\sigma_2)A,$$
  the concentric circle for the goal $\leftarrow G', G''$ is described inside the concentric circle for the goal $\leftarrow G$. The chart A is stored into the pushdown stack to denote a (partially completed) event.
  (b) When $Event_I(\leftarrow G; \sigma_1; \sigma_2)$ is, for the equivalent goal $\leftarrow G', not\ A$, replaced by $Event_I(\leftarrow G'; \sigma_1; \sigma_2)$ after the negation-as-failure derivation caused by the relation $failure_P(\leftarrow A)$, the formation derivation is recursively adopted for $Event_I(\leftarrow G')$.
  (c) After $Event_I(\Box; \sigma; \sigma)$ is finally implemented, the pushdown stack denotes the event.

(2) The negation-as-failure derivation caused by the relation $failure_P$:
  (a) The derivation is implementable by means of the procedure rules (4) and (5).
  (b) The derivation is implementable by means of the procedure rule (6) such that the event formation derivation backs the negation-as-failure derivation.

Example 1. Take a calculus
$$\Im = (\{w, x, y, z\}, \{\sigma_1, \sigma_2, \sigma_3\}, Sem, P, I),$$
where:

(i) The program P contains:
$$x \leftarrow not\ w$$
$$y \leftarrow$$
$$z \leftarrow x, y$$

(ii) The function Sem assigns the transition of situations to each chart:
$$Sem[\![w]\!]\sigma_1 = \sigma_3$$
$$Sem[\![x]\!]\sigma_1 = \sigma_2$$
$$Sem[\![y]\!]\sigma_2 = \sigma_2$$
$$Sem[\![z]\!]\sigma_2 = \sigma_3$$

(iii) $\sigma_1 \Rightarrow_I \sigma_2$, $\sigma_2 \Rightarrow_I \sigma_2$, and $\sigma_2 \Rightarrow_I \sigma_3$.

We see that $Event_I(\leftarrow z; \sigma_1; \sigma_2) = xyz$ for:
$$move_P(\Box; \sigma_1; \sigma_1)$$
$$move_P(\leftarrow not\ w; \sigma_1; \sigma_1) \quad \text{(by } failure_P(\leftarrow w))$$
$$move_P(\leftarrow x; \sigma_1; \sigma_2)$$
$$move_P(\Box; \sigma_1; \sigma_2)$$
$$move_P(\leftarrow y; \sigma_2; \sigma_2)$$
$$move_P(\leftarrow x, y; \sigma_1; \sigma_2)$$
$$move_P(\leftarrow z; \sigma_1; \sigma_3)$$

The implementation to form an event xyz is supported as in [20,21] (involving some ideas of [1,8]), such that it is shown in Fig. 1, where the situation transition is omitted.

(1) The left-hand side (concentric) circle denotes the goal $\leftarrow z$, where the already formed event is regarded as the empty. There is no demonstration of its emptiness.
(2) By the right-hand side concentric circles with the innermost circle for the goal $\Box$, we denote the transitions of the goals

$$\leftarrow z; \leftarrow x, y; \leftarrow x; \leftarrow not\ w; \Box,$$

where the negation-as-failure "*not*" is expressed by the symbol "$\sim$". The event formed by the goal transitions is a sequence xyz of charts, which is demonstrated by "XYZ".

To visualize a concentric circle with an event, we take some hierarchical structure caused by the mutual recursion of event formation and negation-as-failure derivations, where such hierarchical structures are implicit and not displayed. See [6,17,21,22] for hierarchical structures. As regards the Web-interface in management task domain, another way of visualization is taken in [9]. The way is feasible for the present event formation.

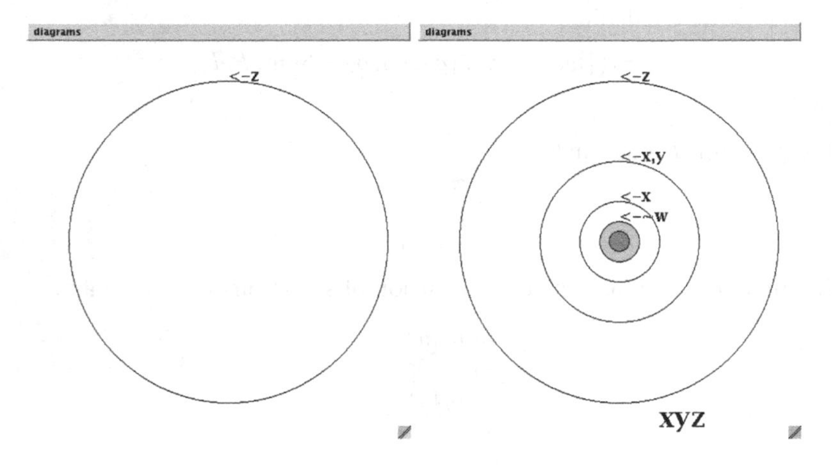

Fig. 1. Event formation evoked by the goal $\leftarrow z$

5 Concluding Remarks

As programming techniques, the calculus is formulated to illustrate and effectively perform event formation for abstraction of action, learning, planning and intelligent behaviour. For the presented calculus, the situation calculus as in [18] may be relevant.

The calculus is itself a way for distributed environments. To be more complex, a concurrent system containing calculi of this paper may be formulated so that the chart or its name can be transferred among communicating calculi. In the context of its extension, the calculus needs to involve distributed programming by means of chart communication. Apart from the established agent and distributed systems as in [3,4,13,15,16], the concurrent system of calculi may contain problems to be solved. A technique for distributed programming as in [19] may be denoted to one of the solutions.

We are now examining whether the present calculus could be a basis framework for specification designs. The logical approaches to the specification are made with reference to the Z language, as in [2,7,12]. In the present calculus, we must design

(1) the process or the procedure assigned to each chart
(2) the logical relation among the charts
(3) the situation transition which a chart causes
(4) the integrity constraint on the situation set

On the basis of the design, we can effectively form an event for the specified goal with the situation transition, as long as the event must exist.

If we prefer to a more expressive system, the logical relation among the charts may be defined in the first-order logic, so that the event formation must need higher-order computing.

As an application, we can consider some managing scheme of event forming, which differs from the producer-consumer adjustment as in [25]. As another application, the event formation is regarded as a process of learning and training so that it may be means for e-learning:

(a) It is needed to learn the chart by assigned basic process of primitive steps
(b) Formation of a sequence of charts is just an event for a requirement under the integrity constraint on the situations.

References

1. Battista, G.D., Eades, P., Tamassi, R., Tollis, I.G.: Graph Drawing. Prentice Hall, Englewood Cliffs (1999)
2. Burke, E., Foxley, E.: Logic and Its Application. Prentice-Hall, Englewood Cliffs (1996)
3. Bruns, G.: Distributed Systems Analysis with CCS. Prentice-Hall, Englewood Cliffs (1996)
4. Ciampolini, A., Lamma, E., Mello, P., Toni, F., Torroni, P.: Cooperation and competition in ALIAS: a logical framework for agents that negotiate. Annals of Mathematics and Artificial Intelligence 37, 65–91 (2003)
5. Dung, P.M.: An argumentation-theoretic foundation for logic programming. J. of Logic Programming 22, 151–177 (1995)
6. Eades, P., Feng, Q.-W.: Multilevel visualization of clustered graphs. In: North, S.C. (ed.) GD 1996. LNCS, vol. 1190, pp. 101–112. Springer, Heidelberg (1997)
7. Galloway, A.J., Stoddart, W.J.: An operational semantics for ZCCS. In: Proc. of IEEE International Conference on Formal Engineering Methods, pp. 272–282 (1997)
8. Harel, D.: On visual formalisms. CACM 31(5), 514–530 (1988)
9. Iwata, K., Sasakura, M., Yamasaki, S.: Visualization for management of electronics product composition. In: Proc. of 9th International Conference on Information Visualization IV 2005, pp. 194–199 (2005)

10. Kunen, K.: Signed data dependencies in logic programs. J. of Logic Programming 7, 231–245 (1989)
11. Kowalski, R.A.: Database updates in the event calculus. J. of Logic Programming 12, 121–146 (1992)
12. Mahony, B., Dong, J.S.: Timed communicating object Z. IEEE Trans. on Software Engineering 26(2), 150–177 (2000)
13. Milner, R.: Communication and Concurrency. Prentice-Hall, Englewood Cliffs (1989)
14. Mosses, P.M.: Action Semantics. Cambridge University (1992)
15. Nielson, F. (ed.): ML with Concurrency, Monograph in Computer Science. Springer, Heidelberg (1996)
16. Pereira, P.M., Nasr, R.: DELTA–PROLOG: A distributed logic programming language. In: Proceedings of the International Conference on Fifth Generation Computer Systems, pp. 283–291 (1984)
17. Raitner, M.: HGV: A library for hierarchies, graphs and views. In: Goodrich, M.T., Kobourov, S.G. (eds.) GD 2002. LNCS, vol. 2528, pp. 236–243. Springer, Heidelberg (2002)
18. Reiter, R.: Knowledge in Action. The MIT Press, Cambridge (2001)
19. Sasakura, M., Yamasaki, S.: An application of Nara View to reasonings for distributed logic programs. In: Proc. of 2003 International Conference on Distributed Processing Techniques and Applications, Las Vegas, vol. 3, pp. 1099–1105 (2003)
20. Sasakura, M., Yamasaki, S.: An explanation reasoning procedure applicable to loop transformation in compiler. In: Proc. of ACM ESEC/FSE International Workshop on Intelligent Technologies for Software Engineering (WITSE 2003), Helsinki, pp. 34–39 (2003)
21. Sasakura, M., Yamasaki, S.: Visualization with hierarchically structured trees for an explanation reasoning system. In: Proc. of 8th International Conference on Information Visualization IV 2004, London, pp. 893–898 (2004)
22. Teoh, S.T., Ma, K.-L.: RINGS: A technique for visualizing large hierarchies. In: Goodrich, M.T., Kobourov, S.G. (eds.) GD 2002. LNCS, vol. 2528, pp. 268–275. Springer, Heidelberg (2002)
23. Van Gelder, A.: The alternating fixpoint of logic programs with negation. J. of Computer and System Sciences 47, 185–221 (1993)
24. Yamasaki, S., Kurose, Y.: A sound and complete proof procedure for a general logic program in no-floundering derivations with respect to the 3-valued stable model semantics. Theoretical Computer Science 266, 489–512 (2001)
25. Yamasaki, S., Iwata, K., Sasakura, M.: Reasoning procedure and implementation for logic programs as managing schemes to extract demand. IPSI Trans. on Advanced Research 1(1), 83–90 (2005)
26. You, J.-H., Yuan, L.Y.: On the equivalence of semantics for normal logic programs. J. of Logic Programming 22, 211–222 (1995)

A Similarity Evaluation Method for Volume Data Sets by Using Critical Point Graph

Tomoki Minami, Koji Sakai, and Koji Koyamada

Kyoto University Graduate School of Engineering, Department of Electrical Engineering, Yoshidanihonmatsutyo, Sakyo-ku, Kyoto city, Kyoto, Japan
{sakai,koyamada}@mbox.kudpc.kyoto-u.ac.jp

Abstract. The ever increasing use of computer simulation has proportionately increased the demands for an efficient method for classification of a large amount of computational results or for searching an arbitrary data set in a given database. In order to classify or to search for a computational simulation result, it is necessary to evaluate the similarity between a given data in respect to the reference data in a database. A similarity estimation method which employs "Critical Point Graph (CPG)" as an index has proven effective, however this method does not support transformation operations such as rotation or scaling. In this paper, we propose a CPG-based similarity estimation method supporting both rotation and scaling transformations for two and three dimensional scalar data sets (volume data sets). We could confirm its effectiveness, and also proved superior to the traditional Contour Tree (CT) based matching technique which uses affine-invariant metrics. Some discussion about the proper use of these matching techniques is also presented to clarify the advantages and disadvantages.

1 Introduction

Advances in computational power of commodity computers have enabled computer simulations to employ high accuracy models, making computer simulation use wide-spread in a variety of fields. Additionally, numerical data generated from these simula-tions has become proportionately larger and more complex. In order to handle this ever increasing size and quantity of numerical simulation results, an effective method-ology for classification in a database or for on demand searching is now required. Such a methodology needs a technique for evaluating the similarity between a requested volume data and a given reference data. Over the past years, a considerable number of studies have been conducted on the database systems which are intended for handling three dimensional objects in Virtual Reality (VR) environment[1]. However, in gen-eral, the size of volume data is larger than that of 3D surface objects. Therefore, when a system evaluates the similarity between two given volume data sets, its computa-tional cost tends to be very high.

To overcome this problem, classification and searching methods which use "Critical Point Graphs (CPGs)" as classification or searching index have received increasing attention. CPGs are generated by connecting critical points

J. Labarta, K. Joe, and T. Sato (Eds.): ISHPC 2005 and ALPS 2006, LNCS 4759, pp. 295–304, 2008.

through the gradient fields[2]. The size of a CPG is considerably small when compared with the original volume data size, and is thus effective as a classification index of volume data. In this paper, we propose a CPG-based method for evaluating the similarity between two volume data sets. Our algorithm uses the "Passed Cell Matching" technique. Passed cell refers to the cells which are passed through the streamlines which connect critical points. We enhanced the original algorithm in order to handle affine transformations. We have confirmed its effectiveness also by comparing with the traditional Counter Tree (CT)[3] based matching technique which uses affine-invariant metrics. Advantages and disadvantages when using these matching techniques are also verified.

In Section 2, we present some related similarity estimation techniques for volume data sets. In Section 3, we will explain our proposed method which enables rotational transformation and scaling of volume data. Section 4 describes the experiment of simi-larity evaluation and a comparison with the CT-based method. Section 5 discusses the experimental results and finally we summarize showing some future directions in Sec-tion 6.

2 Background

In this section, we present some of the similarity evaluation methods for scalar data sets in both two and three dimensions which have been proposed so far. The Contour Tree (CT) based method was firstly introduced by Boyell and Ruston[4]. They sum-marized the evolution of contours on a map. Freeman and Morse also used CT-based technique to find terrain profiles in a contour map[5]. The basic algorithm of CT was detailed by Carr et al. in [3]. CT is a graph that tracks contours of the level set as they split, join, appear, and disappear. The similarity evaluation method is based on com-parison between CTs generated from data sets.

Another method which uses Reeb Graph as an index was proposed by Hilaga et al. [6]. In this technique, the similarity between polyhedral models is quickly, accurately, and automatically calculated by comparing Multi-resolutional Reeb Graphs (MRG). Because this method premises on the Morse function, a perturbation on the data value is needed, and therefore changes the original data slightly. In addition, this method connects the critical points (CPs) by lines, thus the information between CPs is not preserved, and is therefore not reflected in the calculation of the similarity.

In the field of fingerprint recognition, there is a Minutia-based method which is de-tailed in [7,8]. This method estimates the characteristic points (Minutia) inside fin-gerprint images. The similarity between two fingerprint images is estimated by compar-ing the minutias. This technique can only handle two dimensional images, therefore an enhancement is required for handling three dimensional data sets.

3 Proposed Method Using CPG

Our proposed method uses critical point graph (CPG). In this section, we describe the CPG generation, transformation (rotation and scaling), and the CPG-based simi-larity estimation processes.

3.1 Critical Point Graph (CPG)

In this paper, we briefly describe CPG, and more detailed information can be found in [2]. CPG is composed of critical points (CPs) and streamlines which are drawn from points close to CPs in a given field. CPG has been widely used as a method to describe topology of a given field.

Streamlines are tangent to the vectors defined in each given point. In a scalar field, vector can be defined as the gradient of each point. In a tensor field, vector can be defined as the principle direction of the eigenvalue vector of each point. The stream-lines are drawn from points close to CPs. An example of CPG is depicted in Figure 1.

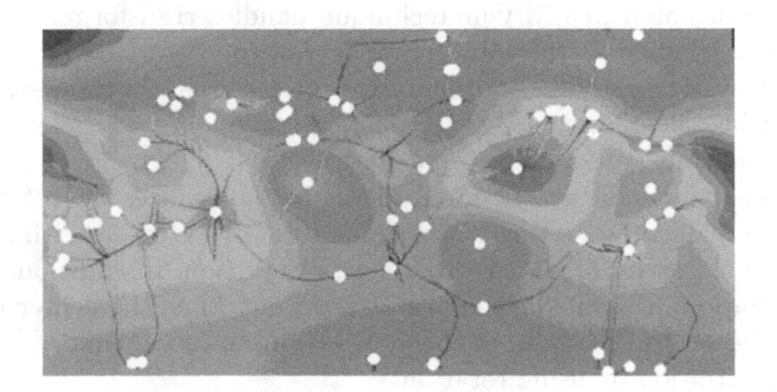

Fig. 1. An example of Critical Point Graph (CPG)

3.2 Proposed Algorithm

Our proposed method enables rotation and scaling transformation of input volume data sets. In order to achieve this, we used normalization for input volume data. The simplified flowchart of the proposed algorithm is described in Figure 2.

Because of the inherent difficulty in illustrating and explaining the behavior of our proposed similarity evaluation method in 3D, we will explain our algorithm applied to a 2D field. The extension to 3D is straightforward.

(1) Read Data
Our method assumes input data as scalar volume data sets defined on rectangular grid.

(2) Calculate Critical Points
This stage calculates the CPs from the input scalar volume data. CPs have coordi-nate information which are used for normalization.

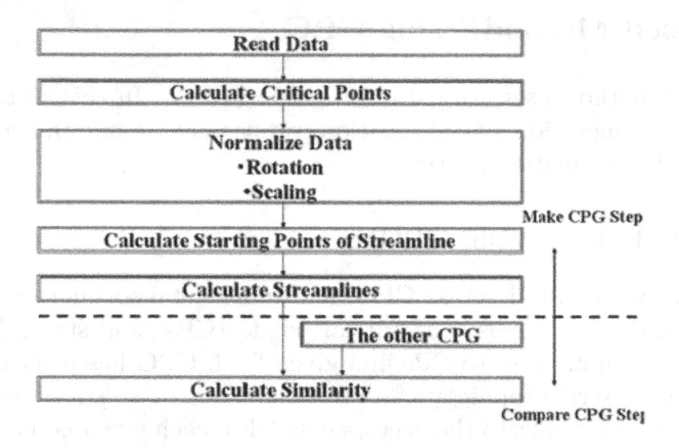

Fig. 2. Flowchart of the Proposed Algorithm

(3) Transformation and Normalization of the input data
This stage transforms and normalizes a given input volume data based on the CPs calculated in (2). Our technique handles transformations such as rotation and scaling. In this version, we are not concerned about symmetrical data such as mirror images. Normalization step consists of three stages which are described as follows:
  (a) Parallel translation of the input data
  In order to handle rotation, our algorithm normalizes the rotation angle of input data by setting a center of rotation. Our algorithm uses the center of gravity of CPs as the center of rotation. In addition, a parallel translation of input data is also carried out, thus the center of gravity may be set to the origin of the world coordinate system.
  (b) Normalization by using rotation
  This stage calculates the rotation angle necessary for normalizing the input volume data. A rotation angle is derived by using the principal component analysis. The vector of the largest principal component is calculated from CP coordinates. After that, it calculates the angle between the vector of the largest principal component and the horizontal axis of world coordinate system. Finally, the angle of rotational transforma-tion is calculated and the input volume data is transformed using this angle. In three dimensions, this stage requires the use of three orthogonal axes for rotation.
  In (a) and (b), the input data is normalized by using rotational transformation.
  (c) Normalization in scale
  This stage normalizes the input data in scale. The farthest distance R from the origin of the world coordinate system is obtained from CP coordinates. After that, the scaling ratio is defined as 1/R. Based on this ratio, the input volume data is normalized.

(4) Calculate starting points of streamlines
It is defined that streamlines are tangent to vector in each point. However, CPs has zero length vectors as definition. As a consequence, a CP can not be the starting point of streamline calculation. Thus, the starting points are defined as points slightly moved from CPs in the direction of the eigenvector of the CPs. In 3D, six different starting points are taken into account for starting the streamline generation.

(5) Calculate streamlines
The streamlines are traced along the vector of the input field from starting point to the point at boundary of data space. Streamlines are drawn until the size of vector becomes near zero or the lines come sufficiently close enough to another CP. From this, we can say that the streamlines represent vaguely the topology of a given input volume data.

(6) Calculate Similarity
In the previous steps, CPG is generated from a given input volume data. In this step, we calculate a similarity between two volume data using CPG as an index for similarity evaluation. CPG will be distributed in the normalized range of each axis in the world coordinate system due to the normalization of input data. The resolution of each axis can be decided by the user. After that, this step calculates streamlines wherever they pass.

Finally, we calculate the similarity by comparing the cells passed by the streamlines in both input data. Figure 3depicts a sample image when comparing two different data sets. Two different data named A and B were used for this comparison. In Figure 3 red cells show the cells which both streamlines pass through. Green and yellow cells are assigned to streamlines which only pass through cells in data A and B, respectively.

We have defined the similarity S as follows.

$$S = ab/(a + b + ab) \tag{1}$$

where a and b are the quantity of cells which are passed by streamlines of only data A and B, respectively. On the other hand, ab means the number of cells which are passed by a streamline in both data A and B.

4 Evaluation

4.1 Comparison Using CPG-Based Method

We evaluated our proposal method by using both two and three dimensional data sets. We used a global daily average air pressure level data set supplied by National Center of Environmental Prediction (NCEP)[9] (Figure 1) as examples of two dimen-sional data. We used a set of fifty images for the evaluating the degree of similarity. These images were selected randomly from a whole set of 365 images of the year 1993. We estimated the similarity between each pair of images mutually for all of fifty data sets. We estimated the similarity between

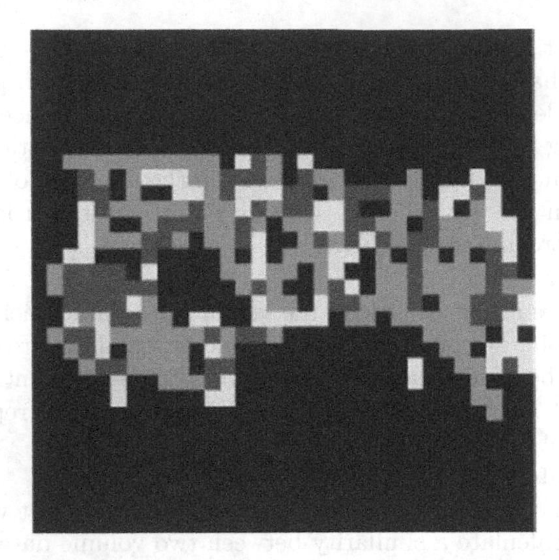

Fig. 3. Example of a Similarity Comparison by using Streamlines

both different and same data sets. The result of these similarity estimations are described in Figure 4 (a). From the obtained results, we can divide them into two distinct groups which indicate different degree of similar-ity. The group in the left (which indicates similarity lower than 0.42) is composed of comparisons executed between different data sets. On the other hand, the group in the right is composed of the results when comparing same data sets. In order to verify the ability for handling rotational and scaling transformations, we prepared an additional set of 36 images generated by adding some rotation and scaling to a particular data. We estimated the similarity by using a combination of all of these 36 sample images. The result of this comparison is described in Figure 4 (b). We can verify from these obtained results that the similarity evaluation is not influenced by the rotation and scaling in two dimensional fields. As an example of three dimensional data, we used the fluid simulation result of wa-ter which is depicted in Figure 5. We obtained a low degree of similarity (under 0.5) as a result of similarity evaluation when using different volume data sets. On the other hand, we obtained higher degree of similarity (near 1.0) when comparing same volume data sets. In order to verify that our proposed method can handle rotational transformation and also scaling, we generated an additional volume data by adding some rotation and scaling (Figure 5). We could also obtain a high degree of similarity (near 1.0) even when comparing these kind of volume data (with added transformation).

4.2 Comparison between CPG and CT-Based Matching Techniques

We conducted a comparison between the proposed CPG-based and the CT-based matching technique which uses affine-invariant metrics. In addition, a detailed study of how these matching techniques can be properly used has also been conducted.

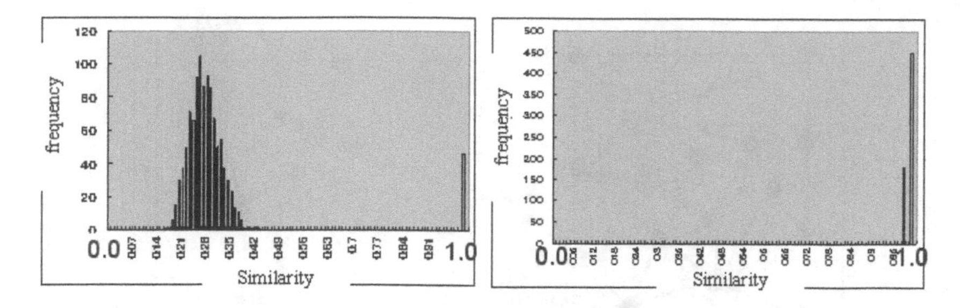

Fig. 4. Histogram of the similarity evaluation (a) On the left is the histogram of similarity between 50 sample data sets. (b) On the right is the histogram of similarity between 36 sample data sets after rotation and scaling.

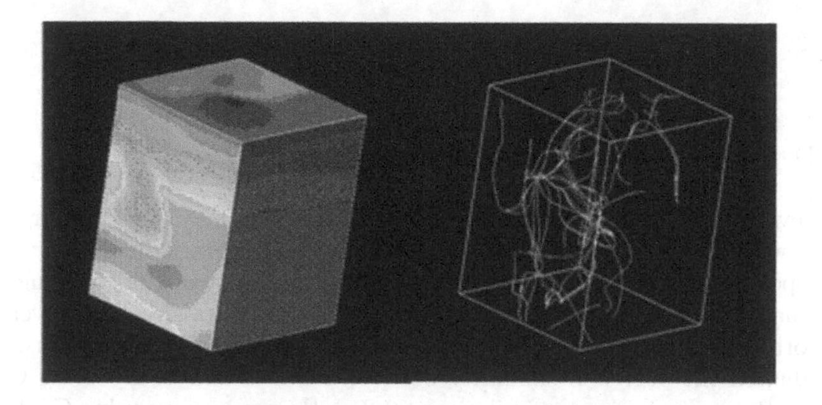

Fig. 5. A sample of utilized 3D volume data

CT-based Similarity Evaluation. Contour tree generation algorithm is detailed in [3]. An example of CT is shown in figure 6. However, the evaluation method by using CT is not documented. Therefore we use the "Elastic Matching" technique for minimizing the cost between saddle points of two Contour Trees. The cost is defined using information of saddle points of CT, scalar value of saddle, connection information, and information of parent of child nodes. As the purpose of this paper is concerned, it is not necessary to discuss the way to evaluate similarity between CTs in detail. An important point to emphasize is that CT-based similarity evaluation technique is affine-invariant metric.

Comparison of CPG and CT-based methods. In order to verify the performance difference between CPG and CT-based similarity evaluation methods, we have evaluated them by using several two dimensional sample data. The utilized data are shown below:

Data A: Global daily average air pressure data of 1/1/1993 (Figure 1).
Data B: Data A rotated 90 degrees.

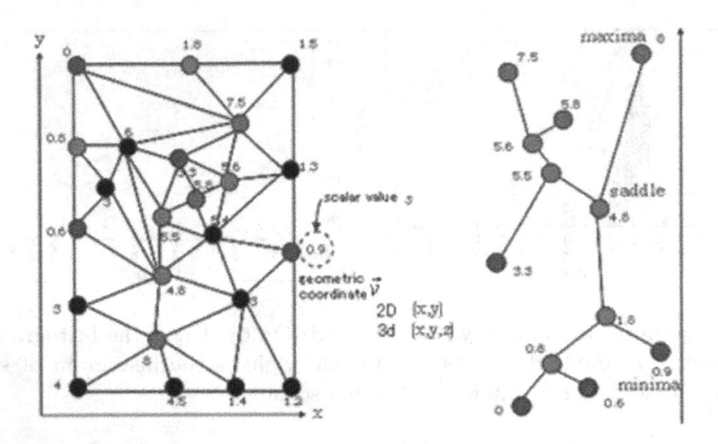

Fig. 6. Example of Contour Tree (in the right)

Data C: Symmetrical data of data A.
Data D: Data A with scalar values reduced to half.

We evaluated the degrees of similarity between the Data A and other data. Firstly, we evaluated the degree of similarity between Data A and B. By using the proposed CPG-based method, we obtained a similarity of 1.0. By using CT-based method, we also obtained a similarity of 1.0. From this result, we verified that both techniques can handle efficiently rotational transformation. Secondly, we evaluated the degree of similarity between Data A and C. By using CPG-based method, we obtained a similarity of 0.316. However, when using CT-based method, we obtained a similarity of 1.0. From this result, we verified that our proposed CPG-based method can classify a symmetric data because CPG use coordinate infor-mation of streamlines. Finally, we evaluated the degree of similarity between Data A and D. By using CPG-based method, we obtained a similarity of 1.0. However, when using CT-based method, we obtained a similarity under 0.01. From this result, we verified that CT-based method can recognize general shifts of scalar values because CT stores the sca-lar values of CPs.

5 Discussion

We can verify from figure 4 (a) that a similarity can be clearly classified into two distinct groups when comparing two of the same data set or two different data sets. This result clearly shows that we can effectively retrieve an objective data from a given database. In addition, this method provides more information facilitating the building of a data retrieval system. The main goal of our proposed method is to handle affine transformation such as translation. From figure 4 (a), we can recognize that our method can efficiently handle such affine translation. The experiment was conducted by calculating the degree of similarity in respect to data generated by rotating and scaling the original. The experi-mental results show that our method can verify a high degree of similarity between these pairs of data. We think calculation errors such as rounding errors are the main cause for not achieving 100% (or 1 in the 0 to 1 scale) of similarity. The proposed method evaluates a similarity by comparing the cells which stream-lines pass through. For this reason, there is also a possibility of some streamlines, espe-cially those passing in the boundaries of the cells, influencing the similarity evaluation. However, even considering all these affecting influences the error margin can be kept small, in the order of 0.05. Therefore, even when some errors are taken into considera-tion, it is reasonable to suppose that this proposed method can efficiently handle affine translation. In three dimensions, the same thing can be said. Our proposed CPG-based method has both advantage and disadvantage. The ad-vantage is that CPG-based method can efficiently classify symmetrical data. For in-stance, in three dimensions, symmetrical mirror isomers of proteins which have both L and D-isomers can divide each other efficiently. However, CT-based methods can not handle such divisions. The disadvantage is that our method can not recognize parallel shifts of scalar values. This is caused by the fact that scalar value is not used in com-parisons between CPGs.

6 Conclusions

We proposed a method to evaluate the similarities between two given volume data sets. Our proposed method permits rotational transformation and scaling of input vol-ume data sets. We evaluated our method by using 50 different data sets and the excel-lent results lead us to the conclusion that our proposed method can sufficiently handle rotational transformation and scaling. In addition, our method has proven highly in-formative for showing the degree of similarity. Moreover, the advantages and disad-vantages of applying our proposed method are made clear when compared to the tradi-tional CT-based method. Future works include the handling of mirror symmetry and a study on more accurate similarity evaluation methods. In addition, further comparison between other correlated evaluation techniques is also being considered.

References

1. Suzuki, M.T.: A Dynamic Programming Approach to Search Similar Portions of 3D Models. The World Scientific Engineering Academy and Society Transaction on SYSTEMS 3(1), 125–132 (2004)
2. Sakai, K., et al.: Classifying Scalar Field by using Critical Point Graph. In: The 10th International Symposium on Flow Visualization, F326 (2002)
3. Carr, H., et al.: Classifying Scalar Field by using Critical Point Graph. In: The 10th International Symposium. Computing contour trees in all dimensions, Computational Geometry, vol. 24, pp. 75–94 (2003)
4. Boyell, R.L., Ruston, H.: Hybrid techniques for real-time radar simulation. IEEE Proceedings Fall Joint Computer Conference 63, 445–458 (1963)

5. Freeman, H., Morse, S.: On searching a contour map for a given terrain elevation profile. J.Franklin Institute 284(1), 1–25 (1967)
6. Hilaga, M., et al.: Topology Matching for Fully Automatic Similarity Estimation. In: ACM SIGGRAPH 2001, Los Angeles, CA, USA, 12-17 August 2001, pp. 203–212 (2001)
7. Maltoni, D., et al.: Handbook of Fingerprint Recognition. Springer, Heidelberg (2003)
8. Kovacs-Vajna, S.M.: A Fingerprint Verification System Based on Triangular Matching and Dynamic Time Warping. IEEE Transactions on Pattern Analysis and Machine Intelligence 22, 1266–1276 (2004)
9. `http://cdc.noaa.gov/cdc/reanalysis/reanalysis.shtml`

Hybrid Parallelization and Flat Parallelization in HPF (High Performance Fortran)

Yasuharu Hayashi and Kenji Suehiro

1st Computers Software Division, NEC Corporation, 1-10 Nisshin-cho, Fuchu-City, Tokyo, Japan

Abstract. We have developed the HPF (High Performance Fortran) compiler HPF/SX V2 as an interface for distributed memory parallel programming. HPF is a de facto standard language for parallel programs. It is possible to write parallel programs just by inserting comment directives into existing serial Fortran programs in HPF. This paper treats two parallelization methods in the HPF/SX V2 on an SMP (Symmetric Multiprocessor) cluster system, each node of which is built by connecting multiple vector PEs (Processor Elements) with a shared memory. The one is hybrid parallelization, which consists of vectorization on a PE, multi-thread parallelization within a node, and distributed memory parallelization across nodes. The other is flat parallelization, which consists of vectorization and distributed memory parallelization. We compare hybrid parallelization with flat parallelization by evaluating several typical codes. The result shows that hybrid parallelization is particularly beneficial, when reduction of memory is expected.

1 Introduction

In recent years, the increasing demand for computational power in scientific, technological, and industrial fields has made mainstream supercomputers multi-node systems. This trend forces users to write distributed memory parallel programs to take full advantage of supercomputers, because a multi-node system is a kind of distributed memory system that is made up of many single-node systems connected via a high-speed network.

There are three troublesome tasks in distributed memory parallel programming: data mapping, computation partitioning, and communication. HPF (High Performance Fortran) is a parallel programming language defined by HPFF (High Performance Fortran Forum) to enable users to write parallel programs easily. In HPF programming, what users mainly have to do in the tasks above is to specify data mapping just by inserting comment directives that designate how to map data onto processors into existing serial Fortran programs. The other two tasks, computation partitioning and communication, are automatically performed by a compiler. This characteristic of HPF enables developers of parallel programming to write and maintain parallel programs easily.

NEC's supercomputer SX series provide the HPF compiler HPF/SX V2. This paper describes two parallelization methods, hybrid parallelization and flat parallelization, in the HPF/SX V2. We show that hybrid parallelization is useful

J. Labarta, K. Joe, and T. Sato (Eds.): ISHPC 2005 and ALPS 2006, LNCS 4759, pp. 305–314, 2008.

for reduction of memory. Reduction of memory is important to solve larger-scale problems, especially in distributed memory parallel programming, because replicated data are usually allocated in equal number to processes.

2 Hybrid Parallelization and Flat Parallelization

2.1 Parallelization in the HPF/SX V2

The HPF/SX V2 conforms to the HPF 2.0 specification [1]. The HPF/SX V2 also supports the major specifications in the HPF Approved Extensions and the HPF/JA 1.0 [2], which include functions for managing computation partitioning and communication. Moreover, unique extensions, such as HALO [3][4], suitable for problems that contain irregular access patterns, such as FEM (Finite Element Method), are available in the HPF/SX V2. Advanced users can improve programs step by step by inserting directives for computation partitioning and communication, in addition to directives for data mapping, or by rewriting a part of a program using MPI (Message Passing Interface), to tune the program ultimately in the HPF/SX V2.

The HPF/SX V2 is a source-to-source translator that inputs HPF source programs and outputs parallelized SPMD (Single Program Multiple Data) Fortran programs. At first, the HPF/SX V2 allocates data among abstract processors, each of which is a logical processor and the unit of parallelization in HPF, according to the data mapping directives specified by users. Then the HPF/SX V2 judges if each loop can be parallelized, and it divides the iterations of the parallelizable loops among abstract processors so that the occurrence of data access to remote abstract processors is minimized. The HPF/SX V2 finally generates parallelized Fortran programs with the invocations of the run-time libraries, which mainly take charge of communication with MPI. Following translation by the HPF/SX V2, the back-end Fortran compiler performs automatic multi-thread parallelization and vectorization, and generates executable programs. Users can also insert directives for multi-thread parallelization and vectorization into HPF programs.

The SX series are offered as single-node and multi-node systems. The single-node system is built by connecting multiple vector PEs (Processor Elements) with a shared memory. The multi-node system is an SMP (Symmetric Multi-processor) cluster composed of multiple single-node systems connected with the IXS (Internode Crossbar Switch).

In the HPF/SX V2, abstract processors, each of which is the unit of parallelization in HPF, have one-to-one correspondence with processes, and MPI libraries are invoked for inter-process communication. Users can choose whether each process corresponds to a node or a PE, with specification at execution-time.

When a process corresponds to a node, parallelization by the HPF/SX V2 is performed on inter-node; and multi-thread parallelization within a node

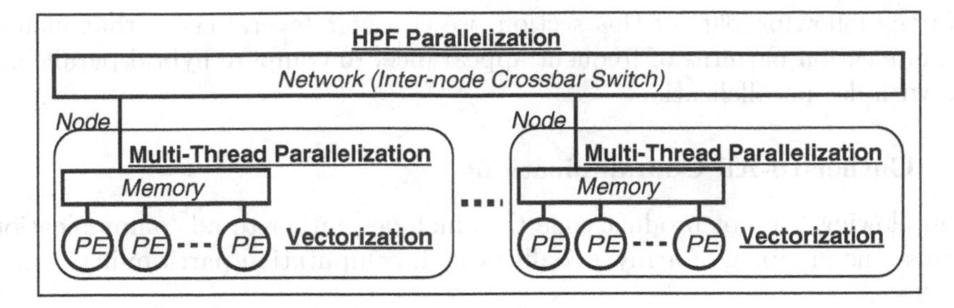

Fig. 1. Hybrid Parallelization with Vectorization

and vectorization on a PE are automatically performed by the back-end Fortran compiler, as shown in Fig. 1. We call this parallelization method hybrid parallelization.

When a process corresponds to a PE, parallelization by the HPF/SX V2 is performed on both inter-node and intra-node, in which the inter-process communication with MPI is generated irrespective of inter-node or intra-node. The back-end Fortran compiler only performs vectorization as shown in Fig. 2. We call this parallelization method flat parallelization.

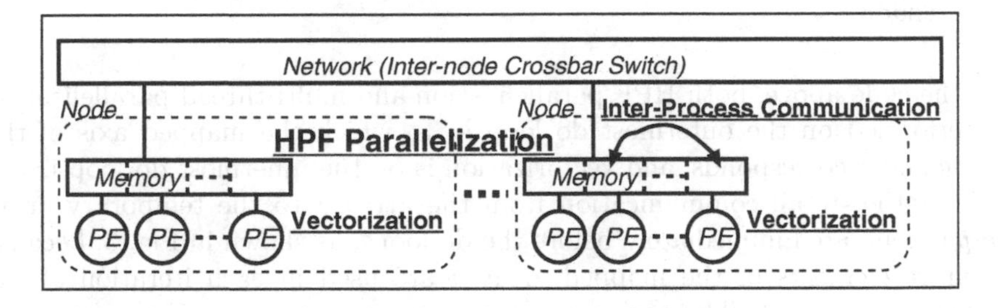

Fig. 2. Flat Parallelization with Vectorization

One significant feature of the HPF/SX V2 is that users can achieve maximum efficiency on an SX multi-node system with the selection between hybrid parallelization and flat parallelization. Flat parallelization without multi-thread parallelization is simpler and easier for users than hybrid parallelization, because users do not have to consider multi-thread parallelization. But the communication cost can become larger when collective communication occurs. The amount of used memory is also larger in flat parallelization. Part of the reason for this is that some memory areas, which include the shadow area in HPF, the communication buffer for MPI, and the I/O buffer used by the system, are allocated in every process. Moreover, this is conspicuous especially when there are arrays to which no mapping directives are specified by users or non-mapped arrays temporally allocated because of data access to remote processes, since the entire size of such arrays is allocated on each process.

In the following part of this section, we consider several codes that include communication patterns of frequent appearance, to compare hybrid parallelization with flat parallelization.

2.2 Gather-to-All Communication

The following is a dot product code that includes gather-to-all communication, because the mapping of array x conflicts with computation partitioning.

```
      double precision a(lm,lm),x(nk,lm),c(lm)
!HPF$ distribute (*,block) :: a,x
!hpf$ align (i) with a(*,i) :: c
!hpf$ shadow (0,0) :: a,x
!hpf$ shadow (0) :: c

      do k=1,lm
        do j=1,nk
          do i=1,lm
            c(k) = c(k) + a(i,k) * x(j,i)
          enddo
        enddo
      enddo
```

In the code above, both HPF parallelization and multi-thread parallelization are performed on the outermost do loop k, to which the mapped axis of the array a and c corresponds, and vectorization is on the innermost do loop i.

The gather-to-all communication from the array x to the temporary array *x_tmp* is generated immediately before the do loops, as shown in Fig. 3, because the entire elements of the mapped array x are used in each iteration of the parallelized do loop k. The non-mapped temporary array *x_tmp* is referred to in the loops instead of the array x as the following code:

```
      tmp_x = x       ! Gather-to-All Communication

      do k=1,lm       ! Parallelization
        do j=1,nk
          do i=1,lm ! Vectorization
            c(k) = c(k) + a(i,k)*tmp_x(j,i)
          enddo
        enddo
      enddo
```

The total amount of memory needed for the temporary array *x_tmp* becomes the product of the size of the entire array x and the number of processes, which is equal to the number of abstract processors.

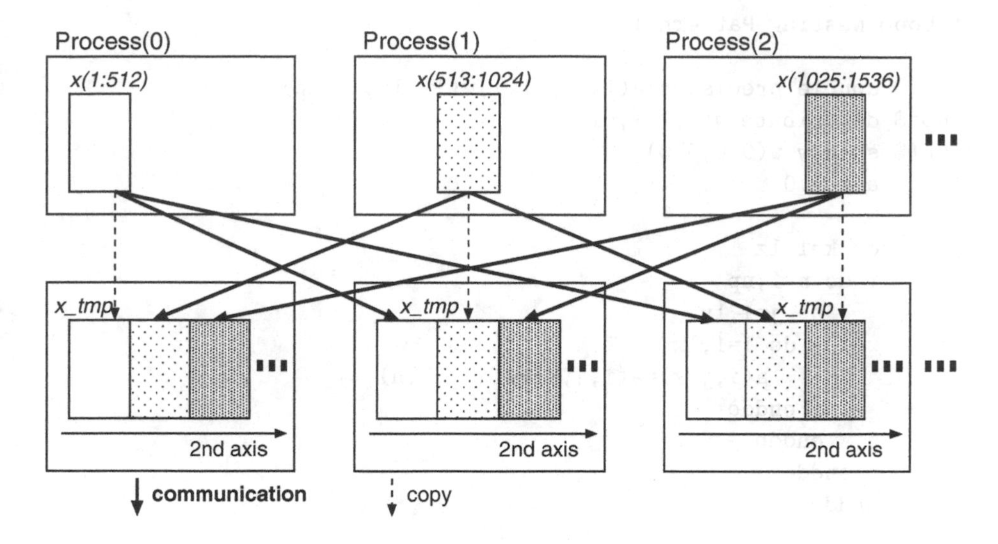

Fig. 3. Gather-to-All Communication

Table 1. The Performance of Dot Product (16-Way Parallel Execution)

Size of x	Method	Num. of Processes	Threads	Time (sec)	Memory (GB)
(16383,16383)	Flat	16	1	74.6	72.45
	Hybrid	2	8	74.6	8.54
(11263,11263)	Flat	16	1	15.6	34.56
	Hybrid	2	8	15.6	4.22
(8191,8191)	Flat	16	1	8.83	18.69
	Hybrid	2	8	8.84	2.40

Table 1 is the result of execution of the code above in different three sizes on the two nodes of SX-8 (16 CPUs).

In this example, the adoption of hybrid parallelization significantly reduces the amount of memory. This is because conversion into multi-thread parallelization of HPF parallelization lessens the total number of copies of the non-mapped temporary array, as a result of reduction of the number of processes. The elapsed time is almost the same, because the communication cost takes only a small part of the execution time.

2.3 Array Reduction

The following is an array reduction code in which the values of *w(i,j,k,:)* accumulate into those of *a(i,j,k)*.

```
! Loop Nesting Pattern 1

      double precision a(lx,ly,lz),w(lx,ly,lz,np)
!hpf$ distribute w(*,*,*,block)
!hpf$ shadow w(0,0,0,0)
      a = 0.0

      do k=1,lz
        do n=1,np
          do j=1,ly
            do i=1,lx
              a(i,j,k)=a(i,j,k)+w(i,j,k,n)
            enddo
          enddo
        enddo
      enddo
```

In the code above, HPF parallelization is performed on the do loop n, which corresponds to the mapped axis of the array *w*: multi-thread parallelization is on the outermost do loop k, and vectorization is on the innermost do loop i.

The entire area of array *a* is allocated on each process, since array *a* is not mapped. The total amount of memory allocated for array *a* by the HPF/SX V2 becomes the product of the size of the entire array *a* and the number of processes.

Collective communication occurs immediately after the do loops, to add up the local values of array *a* on each process.

This example requires careful consideration to the nesting order of the loops in hybrid parallelization. Consider the case in which the do loop n comes to the outermost as follows:

```
! Loop Nesting Pattern 2

      double precision a(lx,ly,lz),w(lx,ly,lz,np)
!hpf$ distribute w(*,*,*,block)
!hpf$ shadow w(0,0,0,0)
      a = 0.0

      do n=1,np
        do k=1,lz
          do j=1,ly
            do i=1,lx
              a(i,j,k)=a(i,j,k)+w(i,j,k,n)
            enddo
          enddo
        enddo
      enddo
```

In the loop nesting pattern 2, both HPF parallelization and multi-thread parallelization are performed on the outermost do loop n, and vectorization is on the innermost do loop i.

Reduction dependence of the loop n makes it impossible to define the values of array *a* concurrently in multi-thread parallelization, because the memory area of array *a* is shared by all threads in a process, whereas each process has separate memory area of array *a* in HPF parallelization. Therefore, temporary area whose size is the same as that of array *a* is allocated in equal number to threads in each process, to store thread-local results of array *a* in multi-thread parallelization.

The values of the temporary areas are accumulated one by one into array *a* with synchronization immediately after the do loops.

Then collective communication occurs, to add up the local values of array *a* on each process, too.

Table 2 is the result of ten iterative execution of the two codes above in different three sizes on two nodes of SX-8 (16 CPUs).

Table 2. The Performance of Array Reduction (16-Way Parallel Execution)

Size of *w*	Method	Loop Nesting Pattern	Num. of Processes	Num. of Threads	Time (sec)	Memory (GB)
(767,767,768,16)	Flat	1	16	1	26.5	111.36
		2	16	1	26.2	111.36
	Hybrid	1	2	8	19.1	49.92
		2	2	8	32.5	117.47
(511,511,512,16)	Flat	1	16	1	7.38	33.54
		2	16	1	7.37	33.54
	Hybrid	1	2	8	5.56	18.72
		2	2	8	9.53	34.85
(255,255,256,16)	Flat	1	16	1	0.902	5.12
		2	16	1	0.906	5.12
	Hybrid	1	2	8	0.689	2.66
		2	2	8	1.195	4.45

In flat parallelization, the nesting order of the loops does not affect the elapsed time or the amount of memory. This is because overhead by HPF parallelization, such as computation for loop division and communication, is automatically hoisted out of the loops by the HPF/SX V2 and no extra memory is needed in both patterns.

On the contrary, the nesting order of the loops is crucial in hybrid parallelization. The loop nesting pattern 1 shows the best performance and the smallest memory use, because the outermost do loop k, which is target of multi-thread parallelization, is perfectly independent. But the loop nesting pattern 2 shows the worst performance, because the outermost do loop n, which is target of multi-thread parallelization, has reduction dependence. Actually, the multi-thread parallel execution with reduction overhead takes almost the same elapsed time as the communication for reduction. The loop nesting pattern 2 also shows the largest memory use in the cases of the largest size and the second largest size, because of the temporary area for reduction operation in multi-thread parallelization. The memory use in flat parallelization is the largest in the case of the smallest size, because buffer areas allocated by the system in every process have greater impact.

These examples show that the adoption of hybrid parallelization reduces both the elapsed time and the amount of memory, so long as target loop of multi-thread parallelization has no dependence. This is because conversion into multi-thread parallelization of HPF parallelization lessens communication cost of array reduction and the total number of copies of array *a*, which is equal to the number of processes. Therefore, it is important to select target loop of multi-thread parallelization, so as not to cause extra overhead in hybrid parallelization.

2.4 Shift Communication

The following is a nearest neighbor computation code that includes shift communication to fetch the array elements *a(:,:,k+1)* on the neighbor abstract processor.

```
!hpf$ template t(lx,ly,lz)
!hpf$ distribute t(*,*,block)
      double precision, dimension(lx,ly,lz)::a,b
!hpf$ align (i,j,k) with t(i,j,k) :: a,b
!hpf$ shadow (0,0,0:1) :: a
!hpf$ shadow (0,0,0) :: b

      do k = 1,lz-1
        do j = 1,ly
          do i = 1,lx
            b(i,j,k)= a(i,j,k+1)-a(i,j,k)
          enddo
        enddo
      enddo
```

In the code above, both HPF parallelization and multi-thread parallelization are performed on the outermost do loop k, and vectorization is on the innermost do loop i.

The shift communication to the shadow area of array *a* is generated immediately before the do loops.

Table 3 shows the result of 100 iterative execution of the code above in different three sizes on two nodes of SX-8 (16 CPUs).

In this example, the performance of flat parallelization is slightly superior to that of hybrid parallelization, because of overhead by multi-thread parallelization. The difference in the performance becomes smaller as the problem size becomes larger, because computation cost becomes relatively larger than overhead by multi-thread parallelization.

Table 3. The Performance of Nearest Neighbor Computation (16-Way Parallel Execution)

Size of a, b	Method	Num. of Processes	Num. of Threads	Time (sec)	Memory (GB)
(1791,1791,1792)	Flat	16	1	22.8	89.09
	Hybrid	2	8	22.9	88.13
(1023,1023,1024)	Flat	16	1	4.29	17.41
	Hybrid	2	8	4.36	16.74
(511,511,512)	Flat	16	1	0.546	3.07
	Hybrid	2	8	0.561	2.43

The communication cost in flat parallelization is larger than that in hybrid parallelization, because the cost of shift communication on three or more processes is larger than that on two processes, though this difference will disappear in execution on more than two nodes. But overhead by multi-thread parallelization has greater impact than the cost of shift communication.

The amount of memory in hybrid parallelization is a little superior to that in flat parallelization, because of the buffer areas allocated in every process.

2.5 Comparison between Hybrid Parallelization and Flat Parallelization

These results show that the adoption of hybrid parallelization can reduce the amount of memory, so long as target loop of multi-thread parallelization does not bring about extra overhead, and the effect is conspicuous especially when large non-mapped arrays appear as user-defined arrays or temporary arrays.

As for improvement of execution performance, the effect depends on the communication pattern. The use of hybrid parallelization is beneficial when communication cost increases as the number of processes increases, such as array reduction, whereas it has little benefit when the communication cost does not depend on the number of processes, such as shift communication, or takes an insignificant part of execution time.

On the other hand, it becomes difficult to get scalable performance in programs whose dominant part is communication, such as array reduction, as the number of processes increases. Therefore, it is difficult to expect significant improvement of performance from hybrid parallelization in highly parallel computation. But the amount of memory depends on the part of a program where the largest size of memory is allocated, even if the part takes just a small part of the total execution time.

Consequently, hybrid parallelization is effective for reduction of memory, when large non-mapped arrays appear as user-defined arrays or temporary arrays, and it enables users to solve larger-scale problems. On the other hand, when reduction of memory with hybrid parallelization is not expected, flat parallelization is easier for users, because hybrid parallelization requires additional consideration to the selection of target loop of multi-thread parallelization, as shown in the example of array reduction.

3 Conclusion and Future Works

We have compared hybrid parallelization with flat parallelization by evaluating several typical codes.

The result shows that hybrid parallelization is particularly beneficial for reduction of memory, when large non-mapped arrays appear as user-defined arrays or temporary arrays. And when reduction of memory by hybrid parallelization is not expected, flat parallelization is usually practical, because it is simpler and easier for users. It also can be said that to derive advantage from hybrid parallelization, it is important to select target loop of multi-thread parallelization that does not have dependence.

In the last few years, the compiling technique of HPF compilers has matured gradually, and the performance of HPF programs bears comparison with that of MPI programs in regular problems[6],[7].

At present, the matter of highest priority is to improve the usability of HPF compilers, because it is still difficult to detect the cause of abortion at execution-time, or that of deficient performance in HPF programs. We are going to provide functions for debugging and tuning, so that users can develop efficient distributed memory parallel programs more readily.

Acknowledgement

We would like to thank all members of the HPF development team in NEC System Technologies, Ltd., for their effort to improve the HPF/SX V2.

References

1. High Performance Fortran Forum: High Performance Fortran Language Specification Version 2.0. (January 31, 1997)
2. Japan Association of High Performance Fortran (JAHPF): HPF/JA Language Specification Version 1.0. (November 11, 1999)
3. Benkner, S.: Optimizing Irregular HPF Applications Using Halos. In: Concurrency: Practice and Experience, vol. 12, pp. 137–155. John Wiley & Sons Ltd, Chichester (2000)
4. Suehiro, K., Hayashi, Y., Hirosawa, H., Seo, Y.: HPF and its Performance on SX-6/SX-7. NEC Research & Development 44(1) (January 2003)
5. Yokokawa, M., Tsuda, Y., Suehiro, K.: An HPF performance of a CFD code on SX-5 SMP nodes. In: HUG 2000, October 19-20, 2000, pp. 124–130 (2000)
6. Sakagami, H., Murai, H., Seo, Y., Yokokawa, M.: 14.9 TFLOPS Three-dimensional Fluid Simulation for Fusion Science with HPF on the Earth Simulator. In: proc. of SC 2002 (November 2002)
7. Okabe, Y.: Performance Evaluation of Large-scale Parallel Simulation Codes and Designing New Language Features on the HPF (High Performance Fortran) Data-Parallel Programming Environment, Annual Report of the Earth Simulator Center(April 2003 - March 2004), pp. 115–118, The Earth Simulator Center (2004)

Mapping Normalization Technique on the HPF Compiler fhpf

Hidetoshi Iwashita[1] and Masaki Aoki[2]

[1] Next Generation Technical Computing Unit, Fujitsu Limited,
140 Miyamoto, Numazu-shi, Shizuoka 410-0396, Japan
[2] Software Unit, Fujitsu Limited,
140 Miyamoto, Numazu-shi, Shizuoka 410-0396, Japan
{iwashita.hideto,m-aoki}@jp.fujitsu.com

Abstract. We propose a technique of mapping normalization which reduces the variety of data and computational mapping representation of HPF into a certain standard form. The base of the reduction is a set of equivalent transformations of an HPF program, using composition of alignment and affine transformation of data and loop indices. The mapping normalization technique was implemented in the HPF compiler fhpf, and made the succeeding processes, such as local access detection and SPMD conversion, much slimmer. The measurement result shows that performance of the MPI code generated by the fhpf compiler is fairly comparable to the one written by a skillful MPI programmer.

Keywords: HPF, compiler, distribution, MPI.

1 Introduction

High Performance Fortran (HPF) has a variety of capabilities of describing data and computational mappings [3]. For example, each different description in Fig. 1 represents the same mapping of a variable A onto three processors with block distribution.

Such rich variation of mapping is the feature of HPF, and the variation is important to improve the portability and reusability of user programs. However, it has made compilers complicated. Compilers can hardly take good care of the entire variation under the limitation of development cost. Identification and relation recognition of the mappings among data, computational loops, and processors are very important for higher-level optimization. For example, if a compiler could not know relationship between the mappings of an array variable and of a loop onto processors, the compiler could not conclude an access of the variable in the loop to be local to the processor, and it would generate a redundant code that decides at runtime whether the access is local or remote.

We are developing the HPF compiler fhpf [5], which is a translator accepting the HPF/JA1.0 [6] program and generating a Fortran program with MPI calls. To obtain higher performance under the limited term and resources of development, we first tried to conquer the huge variation involved in HPF language

J. Labarta, K. Joe, and T. Sato (Eds.): ISHPC 2005 and ALPS 2006, LNCS 4759, pp. 315–328, 2008.

specification. In the result, we propose a technique of *mapping normalization*, which converts various descriptions of alignments, shapes of variables, templates and processor arrays, and indices of loops into a certain standard representation.

In this paper, we describe the mapping normalization technique and the effectiveness. After a brief introduction of the fhpf compiler in Section 2, the mapping normalization technique employed in fhpf is described in Section 3. Optimizations enabled by the normalization is shown in Section 4. Then Section 5 evaluates the fhpf compiler and Section 6 shows related works. Section 7 concludes the paper.

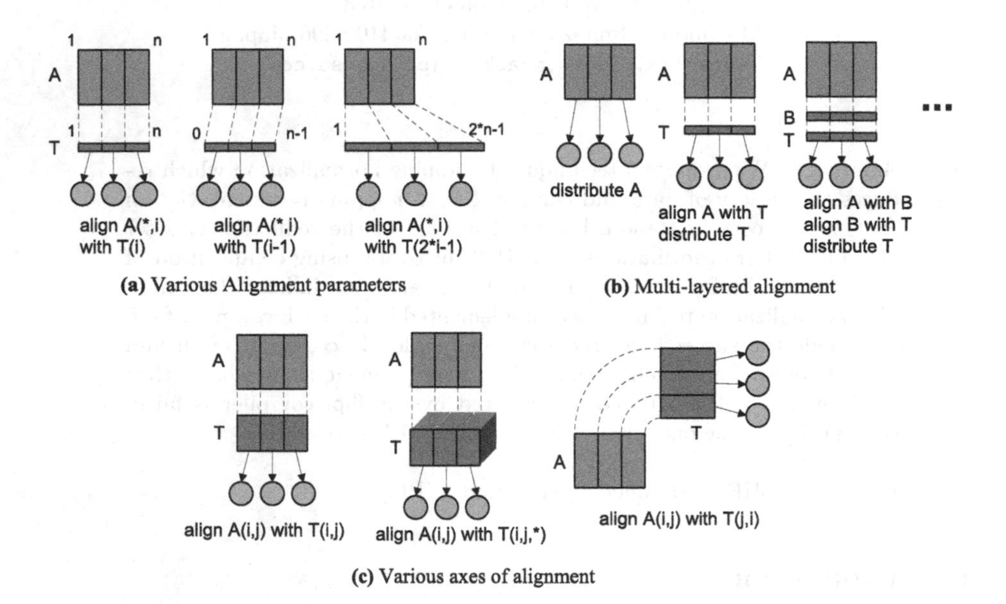

Fig. 1. Various representations for the same mapping of a variable A onto three processors with block distribution. Each description has a different alignment coefficient and offset parameter in (a), a different number of layered alignments in (b), and a different number of dimensions of the alignment target in (c), whose axes mapped from the axes of the variable are also different

2 HPF Translator fhpf

The HPF translator fhpf generates a standard Fortran program code calling MPI library routines and not calling other special libraries. Therefore, the output code can be compiled and can be executed on any platforms, including PC-clusters and blade servers through supercomputers, as long as they can treat standard Fortran 90 and MPI 1.1 [11].

Fig. 2 illustrates how to compile and execute HPF programs with the fhpf. Compilation consists of two processes: The fhpf converts HPF programs into Fortran/MPI programs, and then a Fortran compiler compiles and links the Fortran/MPI programs and Fortran and MPI programs (if any) into an executable code. As such, the fhpf compiler can be regarded as an MPI program generator.

Fig. 3 roughly shows the flow of the fhpf translator. Normalization will be performed in the normalizer before other major conversions including parallelization and optimization. The succeeding SPMD converter converts an intermediate code representing the behavior of the entire system into an SPMD (Single Program/Multiple Data) model representing the behavior of each processor. All modules succeeding the normalizer can assume that the input intermediate code contains just normalized forms.

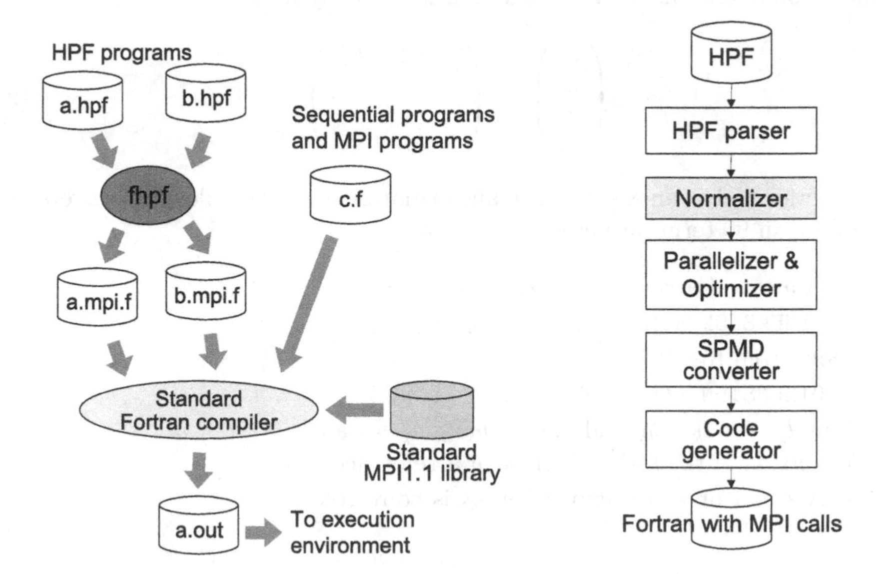

Fig. 2. Compilation using the fhpf compiler

Fig. 3. Processing modules in the fhpf compiler

3 Mapping Normalization

Mapping normalization is an equivalent transformation of an HPF program. This section describes three possible conversion rules as principle, the algorithm we applied in the fhpf, and an example of the normalization.

3.1 Principle

(1) Composition of alignment. Let the alignment of an axis of variable Y with an axis of variable (or template) X be $i \rightarrow ai + b$, and let the alignment of an axis of variable Z with the axis of Y be $i \rightarrow ci + d$. The composed alignment of the axis of Z with the axis of X will be

$$a(ci + d) + b = (ac)i + (ad + b). \quad (1)$$

Using the composition, the alignment of Z with Y can be transformed into the equivalent alignment of Z with X. Similarly, the alignment of loop index described by the ON HOME directive can be transformed. That is, the ON directive

```
ON HOME(Y(c*I+d))
```

can be transformed into the equivalent ON directive

```
ON HOME(X((ac)*I+(ad+b)))
```

using the same composition.

(2) Affine transformation of array shape. Affine transformation without permutation of the shape of an m-dimensional array A

$$\begin{pmatrix} i_1 \\ \vdots \\ i_m \end{pmatrix} \rightarrow \begin{pmatrix} a_1 i_1 + b_i \\ \vdots \\ a_m i_m + b_m \end{pmatrix} \tag{2}$$

can be performed as an equivalent transformation with the following conversion. Some Fortran 90 terms are used here.

- Declaration of array dimensions
  `DIMENSION A(`l_1 `:` u_1`, ···,` l_m `:` u_m`)`
  is converted to
  `DIMENSION A(`l'_1 `:` u'_1`, ···,` l'_m `:` u'_m`)`
  where $l'_k \leq a_k l_k + b_k$ and $a_k u_k + b_k \leq u'_k$ for all $k = 1, \cdots, m$.
- Conversion of the k-th subscript of a reference of A:
  - A scalar or vector expression e_k is converted to

$$a_k e_k + b_k.$$

  - A triplet $s_{k1} : s_{k2} : s_{k3}$ is converted to

$$a_k s_{k1} + b_k : a_k s_{k2} + b_k : a_k s_{k3}.$$

For example, when an array X(1:N) is transformed with parameters $a_1 = 2$ and $b_1 = -2$, the array definition is converted to X(0:2*N-2), accesses X(3), X(1:10) and X(N:1:-2) are converted to X(4), X(0:19:2) and X(2*N-2:0:-4), respectively.

The parameters a_k and b_k are chosen in order to normalize the mapping of the array variable, as shown later.

(3) Affine transformation of loop index. Affine transformation of loop index space on DO loop

```
DO I = e1, e2, e3
...
ENDDO
```

can be performed as an equivalent transformation with the following conversion in the scope of the DO loop.

– Conversion of the three loop parameters

$$e_1 \to ae_1 + b$$
$$e_2 \to ae_2 + b$$
$$e_3 \to ae_3$$

– Conversion of all references of DO variable `I` in the body of the loop

$$\mathtt{I} \to \frac{\mathtt{I} - b}{a}$$

For example, if $a = 3$ and $b = 1$, then

```
DO I = 1,N,2                              DO I = 4,3*N+1,6
J = I                                     J = (I-1)/3
A(3*I) = B(3*J+1)    is converted to      A(I-1) = B(3*J+1)
ENDDO                                     ENDDO
```

The parameters a and b are chosen in order to normalize the mapping of the loop index space. Note that a is always chosen so that the last conversion is divisible.

3.2 Algorithm

Definition:

N The size of the template T in a certain dimension.
P The size of the processor arrangement in the corresponding dimension.
w The block size of block or block-cyclic distribution in the corresponding dimension. If not specified in the case of block distribution, it is defined as $\lceil N/P \rceil$.
W The mapping array of generalized block (gen_block) distribution in the corresponding dimension. The size is P and $\sum_{q=0}^{P-1} W(q) = N$.
M The mapping array of indirect distribution in the corresponding dimension. The size is N. $0 \le M(k) < P$ $(k = 0, \cdots, N-1)$ if M has been normalized.

(1) Normalization of processors. For each processor arrangement, replace the lower and upper bounds P_1 and P_2 by 0 and $(P_2 - P_1)$, respectively, for all dimensions. Then replace all elements of the mapping array of indirect distribution $M(k)$ with $(M(k) - P_1)$.

(2) Normalization of distribution. For each distribution, record the following distribution parameters in the form of the immediate value, or the expression to be evaluated at runtime.

– The block size w in the case of block distribution.
– The mapping array $W(0 : P - 1)$ in the case of gen_block distribution.
– The mapping array $M(0 : N - 1)$ in the case of indirect distribution.

(3) Normalization of alignment and templates

1. For each template, collapse all dimensions that are not distributed.
2. For each template, replace the lower and upper bounds N_1 and N_2 by 0 and $(N_2 - N_1)$, respectively, for all dimensions. Then replace all of the corresponding alignments $i \rightarrow ai + b$ by $i \rightarrow ai + (b - N_1)$.
3. For each variable immediately distributed onto processors, generate a new template with the same shape and distribution as the ones of the variable, and normalize the template and the variable similarly to 2 above.
4. For each variable or loop ultimately but not immediately aligned with the template T, transform the alignment into the immediate alignment with T, using the rule of Section 3.1 (1) recursively.
5. (Optional) Unite templates as much as possible. A template can be replaced by another template if they have the same target processors, the same target axis of the processors, and the same distribution kind and parameters, such as the block size and the mapping array.

(4) Normalization of data objects. For each variable aligned as $i \rightarrow ai + b$ with template T in the corresponding dimension:

1. Replace the lower and upper bounds by the same ones of T, i.e., 0 and $(N_2 - N_1)$.
2. Replace the alignment of the variable with T by the identical alignment $i \rightarrow i$.
3. For each reference of the variable, replace the subscript in the corresponding dimension s by $(as + b)$, using the rule of Section 3.1 (2).

(5) Normalization of loop variables. For each loop aligned as $i \rightarrow ai + b$ with template T:

1. Replace the three loop parameters e_1, e_2, and e_3 by $(ae_1 + b)$, $(ae_2 + b)$, and ae_3, respectively.
2. Replace the alignment of the loop with T by the identical alignment $i \rightarrow i$.
3. Substitute all references of the loop variable `I` by the expression $(\mathtt{I} - b)/a$, using the rule of Section 3.1 (3). Note that the surrounding expression can be arithmetically reduced taking into account that the substituted expression is always divisible.

3.3 Example of Mapping Normalization

An example of mapping normalization using this algorithm is shown in Fig. 4. The input program (a) is short but complicated in mapping. As shown in (b), procedures (1) through (3) normalize the shapes of processors and the generated template, make all alignments immediately with the template, and replace HOME variables in ON directives by the reference of the template. As shown in (c), procedure (4) normalizes the shapes of variables, makes alignments of variables identical, and adjusts the subscripts of the variables. Finally, as shown in

```
!hpf$ processors P(4)
      real A(1:30),C(1:29),D(1:15)
      real F(1:10,1:30)
!hpf$ distribute A(block) onto P
!hpf$ align C(I) with A(I+1)
!hpf$ align D(I) with A(2*I)
!hpf$ template T(30)
!hpf$ distribute T(block) onto P
!hpf$ align F(*,I) with T(I)

!hpf$ independent
      do I=2,29
!hpf$  on home(A(I)) begin
        A(I)=C(I)+F(I,I)
!hpf$  end on
      end do

!hpf$ independent
      do K=1,10
!hpf$  on home(D(K)) begin
        D(K)=K*A(2*K)
!hpf$  end on
      end do
      end
```

(a) Input program

```
!hpf$ processors P(0:3)
      real A(1:30),C(1:29),D(1:15)
      real F(1:10,1:30)
!hpf$ template T1(0:29)
!hpf$ distribute T1(block(8)) onto P
!hpf$ align A(I) with T1(I-1)
!hpf$ align C(I) with T1(I)
!hpf$ align D(I) with T1(2*I-1)
!hpf$ align F(*,I) with T1(I-1)

!hpf$ independent
      do I=2,29
!hpf$  on home(T1(I-1)) begin
        A(I)=C(I)+F(I,I)
!hpf$  end on
      end do

!hpf$ independent
      do K=1,10
!hpf$  on home(T1(2*K-1)) begin
        D(K)=K*A(2*K)
!hpf$  end on
      end do
      end
```

(b) Normalization of processors, distribution, alignment, and template

```
!hpf$ processors P(0:3)
      real A(0:29),C(0:29),D(0:29)
      real F(1:10, 0:29)
!hpf$ template T1(0:29)
!hpf$ distribute T1(block(8)) onto P
!hpf$ align A(I) with T1(I)
!hpf$ align C(I) with T1(I)
!hpf$ align D(I) with T1(I)
!hpf$ align F(*,I) with T1(I)

!hpf$ independent
      do I=2,29
!hpf$  on home(T1(I-1)) begin
        A(I-1)=C(I)+F(I,I-1)
!hpf$  end on
      end do

!hpf$ independent
      do K=1,10
!hpf$  on home(T1(2*K-1)) begin
        D(2*K-1)=K*A(2*K-1)
!hpf$  end on
      end do
      end
```

(c) Normalization of data objects

```
!hpf$ processors P(0:3)
      real A(0:29),C(0:29),D(0:29)
      real F(1:10, 0:29)
!hpf$ template T1(0:29)
!hpf$ distribute T1(block(8)) onto P
!hpf$ align A(I) with T1(I)
!hpf$ align C(I) with T1(I)
!hpf$ align D(I) with T1(I)
!hpf$ align F(*,I) with T1(I)

!hpf$ independent
      do I=1,28,1
!hpf$  on home(T1(I)) begin
        A(I)=C(I+1)+F(I+1,I)
!hpf$  end on
      end do

!hpf$ independent
      do K=1,19,2
!hpf$  on home(T1(K)) begin
        D(K)=((K+1)/2)*A(K)
!hpf$  end on
      end do
      end
```

(d) Normalization of loop variables

Fig. 4. An example of mapping normalization along the algorithm described in Section 3.2

(d), procedure (5) normalizes the loop parameters, makes alignments specified with ON directives identical, substitutes the reference of the loop variables and simplifies the surrounding expressions.

Fig. 5 illustrates the change of data and loop mapping for the same example. It shows that the complicated alignment relationship described in the input program is transformed into an extremely simple representation.

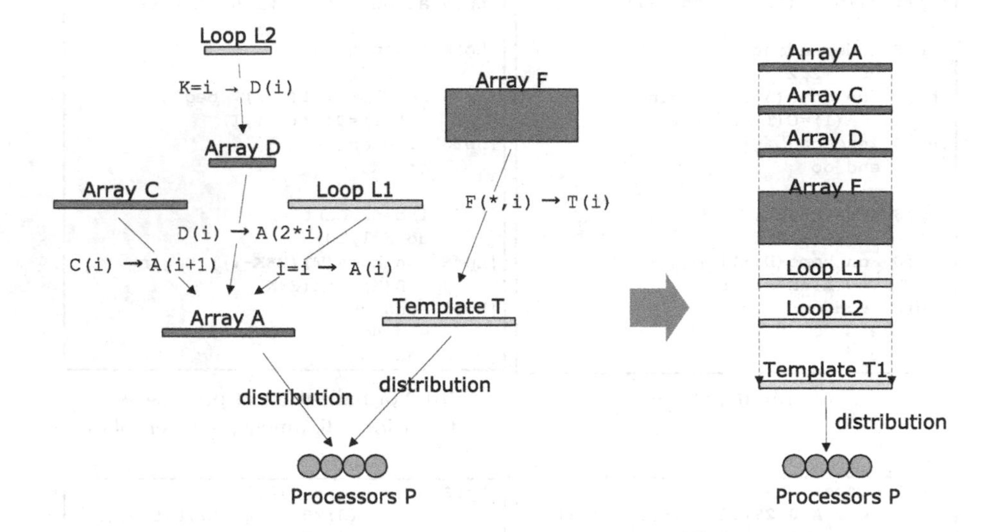

Fig. 5. Illustration of mappings of (a) and (d) of Fig. 4. All mappings of data and loops were transformed into the same identical alignment with a generated template T1.

4 Techniques Enabled by Mapping Normalization

Mapping normalization itself does not optimize the input program. Instead, it encourages the succeeding compiler modules to optimize the program. For instance, the subprogram interface becomes simpler than ever, on the assumption that an argument variable always has the identical alignment and the same shape as the corresponding template. In addition, mapping normalization enables some compiler techniques to be implemented in the fhpf compiler, as shown in this section for example.

4.1 Detection of Local Access

It extremely improves runtime performance, confirming at compile time that the access is *local*, i.e., the data fragment is surely distributed on the processor itself accessing it. The decision requires analysis of alignment matching. If we can assume that mapping of the input code is previously normalized, it is not so difficult:

Let the array A align with the m-dimensional template T and let r the reference of an array element (or subarray) of A appear inside an n-multiple DO-loop L_1, $\cdots$, L_n. We can say r is local if the following condition meets for all $k = 1, \cdots, m$.

- The subscript of r in the dimension corresponding to the k-th dimension of T is the loop variable of L_j itself $(0 \leq j < n)$.
- L_j is aligned with the k-th dimension of T.

In the example of Fig. 4 (d), the reference of `F(I+1,I)` in the first loop can be said on local because its second subscript is corresponding to the first dimension of template T1, the second subscript is the loop variable `I` itself, and the loop is aligned also with the first dimension of `T1`. Similarly, the reference of `A(I)` in the first loop and the references `A(K)` and `D(K)` in the second loop can be said on local. On the other hand, the reference of `C(I+1)` in the first loop cannot to be said on local because the subscript is not the loop variable `I` itself but an expression. From a program like Fig. 4 (a), which is not normalized, it would be much more difficult to make the same decision.

4.2 SPMD Code Generation

An HPF compiler generates SPMD (Single Program/Multiple Data) code as the output describing the behavior of each processor. Fig. 6 shows an example of the SPMD conversion in the fhpf compiler. The parameters for each processor should be determined at runtime as follows:

Processor	*myI1*	*myI2*	*myI3*	*myLB*
P(0)	1	9	1	0
P(1)	0	9	1	10
P(2)	0	8	1	20

As shown in this example, the SPMD converter converts the shape of the distributed data, DO loop parameters, DO loop variables, subscripts of data reference, etc. These conversions are based on the fundamental functions shown below:

- $p = \text{gtop}(k)$
  processor ID p corresponding to global index k
- $i = \text{gtol}(k)$
  local index i corresponding to global index k on the current processor
- $k = \text{ltog}(i)$
  global index k corresponding to local index i on the current processor
- $(i_1, i_2) = \text{gtol2}(k_1, k_2)$
  local duplet $i_1 : i_2$ (or interval $[i_1, i_2]$) corresponding to the subset of global duplet $k_1 : k_2$ (or interval $[k_1, k_2]$) that is mapped onto the current processor
- $(i_1, i_2, i_3) = \text{gtol3}(k_1, k_2, k_3)$
  local triplet $i_1 : i_2 : i_3$ corresponding to the subset of global triplet $k_1 : k_2 : k_3$ that is mapped onto the current processor

```
!hpf$ processors P(0:2)
      real A(0:29)
!hpf$ distribute A(block) onto P

!hpf$ independent
      do I=1,28
!hpf$   on home(A(I))
        A(I)=I
      end do
```

(a) HPF global program

```
real A(0:9)



do I=myI1,myI2,myI3

A(I)=I+myLB
end do
```

(b) SPMD output code

Fig. 6. An example of SPMD conversion in fhpf

On the assumption of the normalized input code, we can express these functions with respect to the distribution types as follows:

	gtol(k)	ltog(i)
block(w)	$k \bmod w$	$i + pw$
gen_block(W)	$k - \sum_{q=0}^{p-1} W(q)$	$i + \sum_{q=0}^{p-1} W(q)$
cyclic	$\lfloor \frac{k}{P} \rfloor$	$iP + p$
cyclic(w)	$w\lfloor \frac{k}{Pw} \rfloor + (k \bmod w)$	$Pw\lfloor \frac{i}{w} \rfloor + pw + (i \bmod w)$

The fundamental functions in the case of indirect distribution are references of tables. Function gtol2 always keeps the form duplet, i.e., contiguous numbers of integer. Function gtol3 can keep the form triplet in the cases of block, gen_block, and cyclic, and it cannot in the cases of block-cyclic and indirect.

Note that these expressions would be much more difficult unless the input code had been normalized previously. Before employing the mapping normalization technique, we used to use runtime library to calculate some of these functions at the cost of runtime overhead because it was too difficult to generate these expressions in the form of Fortran statement. Besides, in the cases of cyclic and block-cyclic, the loop parameters could not be kept in the form of duplet or triplet so the loop used to be expanded into nested loops. Thanks to the normalization, the current fhpf compiler always keeps the single loop single.

5 Evaluation

As shown in Section 4, mapping normalization indirectly contributes to the performance and reduces runtime overhead costs at entrances of parallel loops, array subscripts, entrances of subprograms, etc. They are often summed up in the sequential part of the execution. According to Amdahl's Law, the larger number of processors we use, the more decline of performance they cause.

5.1 Evaluation of the fhpf Compiler

This section evaluates the fhpf compiler using a blade server with the following environment:

CPU	Pentium III, 670MHz
Operating System	Red Hat Linux release 7.2
Memory size	377MB
Compilers	fhpf V1.2.4 for Linux, MPICH1.2.6, and Fujitsu Fortran Compiler (in Linux Parallel Language Package 1.0)

Himeno benchmark is a Poisson equation solver using Jacobi iterative method [4]. We made an HPF version benchmark code from the serial and OpenMP versions, which are being distributed on the Web site. For fair comparison to the MPI version on the Web site, we used the same manner on the HPF version, i.e., the largest elapsed time in all processors, instead of the CPU time on the master processor, is evaluated using the MPI_WTIME function of MPI library. Besides, the reduction operation of the variable `gosa`, calculating the computational error, is performed not outside but inside the loop of `nn` iterations in the subroutine `jacobi`.

Fig. 7 shows the results of MPI and HPF versions of S, M and L models. Fig. 8 shows their performances are almost the same; the fhpf version is 4% slower on average. All data were measured three times and the average was plotted. The average of standard deviation (SD) of the measurements was about 3% both in MPI and fhpf.

5.2 Evaluation of SPMD Code Generation

As shown in Section 4.2, mapping nomarization realized an algorithm of SPMD code generation for various distribution types. This section evaluates the implementation.

Fig. 9 shows the speedup ratios on some distribution types for the LINPACK benchmark program. We used one node of Fujitsu PRIMEPOWER HPC2500 and chose the problem size $N = 4000$. All HPF programs are the same except their distribution types. They has 1-dimensional distribution along J-axis of the $I \times J$ matrix and no special code tuning such as tiling and paneling. The graphs named as block, cyclic, cyclic(1) and cyclic(20) show respectively the measurement results of the HPF programs with block distribution, cyclic distribution, and block-cyclic distributions whose block widths w are 1 and 20.

Fig. 9 shows the following features:

- Cyclic and block-cyclic distributions show better performance than block distribution.
- Cyclic(1), the block-cyclic distribution with $w = 1$, offers almost the same performance as the cyclic distribution despite being made from different fundamental functions shown in Section 4.2.

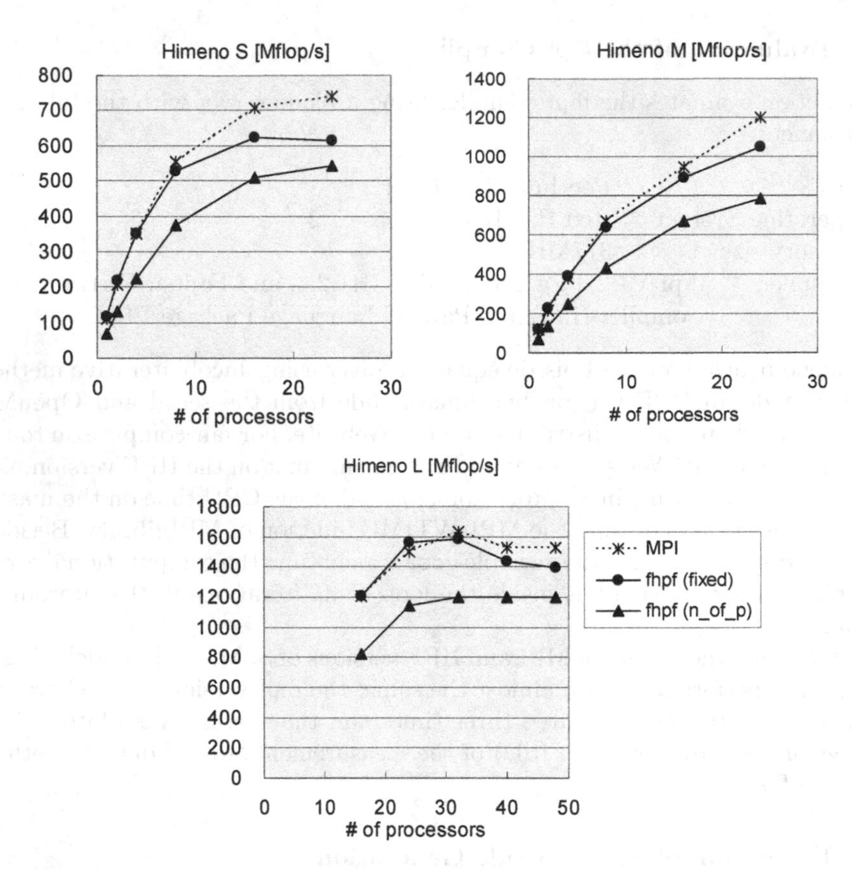

Fig. 7. Result of Himeno benchmark compared with MPI version. "fhpf (fixed)" was evaluated from HPF programs whose numbers of processors are respectively specified at compile time similarly to the MPI version. "fhpf (n_of_p)" was made from single HPF program whose numbers of processors are specified at runtime.

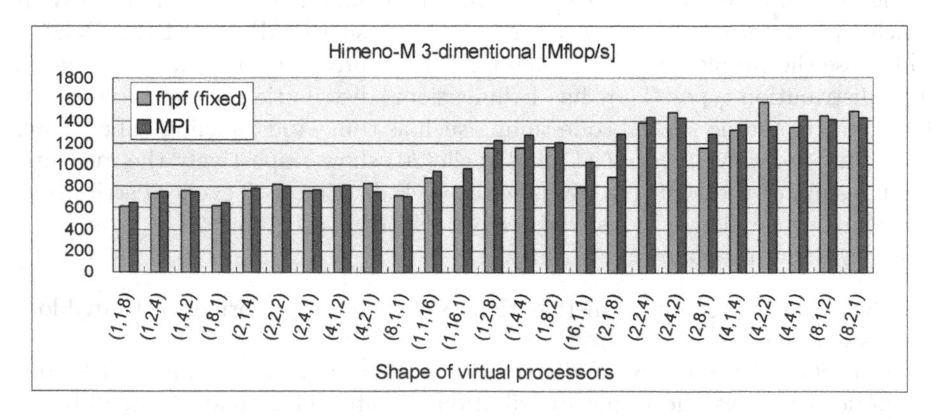

Fig. 8. Detail comparison between two 3-dimensional Himeno MPI codes; one is distributed on the Web page and the other is made from HPF code by the fhpf compiler

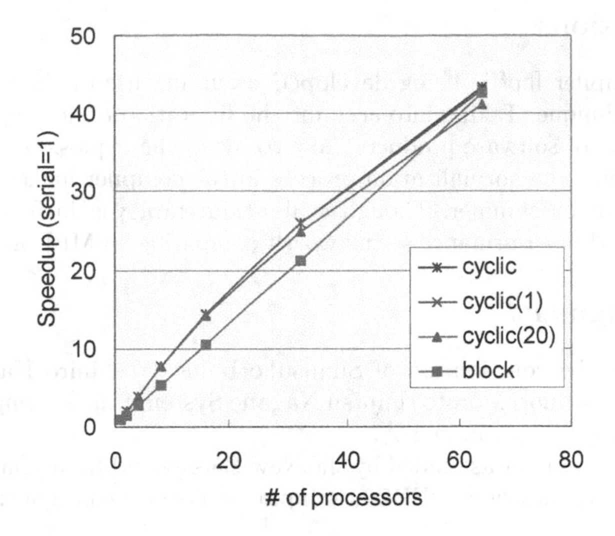

Fig. 9. Speedup ratios on LINPACK comparing block-cyclic distributions with block and cyclic distributions. Thanks to the normalization technique, we have achieved a good implementation of block-cyclic distribution.

The former is because the load balance is not uniform in the LINPACK program. By contrast, on the previous implementation not using mapping normalization, cyclic distribution was much slower than block distribution and block-cyclic distribution was not implemented because of difficulty. As a result, it can be said that mapping normalization technique not only makes various distributions possible but also suppresses the overhead cost of the SPMD conversion.

6 Related Work

There have been many frameworks based on affine loop transformation. Banerjee introduced unimodular transformation [2], and Li and Pingali extended it to non-singular matrix [8]. In Omega Project [12], Kelly, Pugh and Rosser proposed an algorithm on a more general iteration space. Besides, Allen, et al. (Rice Univ.) proposed a loop alignment algorithm which transforms loop carried data dependence into loop independent [1]. All of these aimed for mapping between the iteration space and the data access space, and did not mention mapping of data to the distributed memory.

As an approach taking into account memory hierarchy, Li and Pingali normalized the subscript of the data access to improve data locality [9]. By contrast, we normalize alignment of data and loops, and the subscripts of the data access will become simplified naturally, on the assumption that appropriate data and loop mappings are given.

7 Conclusions

The HPF compiler fhpf is being developed assuming use of the mapping normalization technique. Taking into account the limitation of developers' thought and the quality of software products, how to avoid the explosion of variation is quite important. The normalization process in the compiler makes the succeeding processes much slimmer. Though evaluation employed in fhpf is only just being started, the performance seems worth comparing to MPI programs.

Acknowledgment

We appreciate the contribution of Shunsuke Inoue, Yoshihiro Kasai, Kentaro Koyama, and Masanori Moroto (Fujitsu Nagano System Engeneering) and Masafumi Sekimoto (Fujitsu Limited).

Part of this research was funded by the New Energy and Industrial Technology Development Organization (NEDO), a Japanese governmental agency.

References

1. Allen, R., Callahan, D., Kennedy, K.: Automatic Decomposition of Scientific Programs for Parallel Execution. In: Conference Record of the 14th ACM Symposium on Principles of Programming Languages, pp. 63–76 (January 1987)
2. Banerjee, U.: Unimodular transformations of double loops. In: Proceedings of the Workshop on Advances in Languages and Compilers for Parallel Processing, pp. 192–219 (August 1990)
3. High Performance Fortran Forum: High Performance Fortran Language Specification Version 2.0. (1997), http://dacnet.rice.edu/Depts/CRPC/HPFF/versions/hpf2/hpf-v20/index.html
4. Benchmark, H.: http://accc.riken.jp/HPC/HimenoBMT/index.html
5. Iwashita, H., Hotta, K., Kamiya, S., van Waveren, M.: Towards a Lightweight HPF Compiler. In: Zima, H.P., Joe, K., Sato, M., Seo, Y., Shimasaki, M. (eds.) ISHPC 2002. LNCS, vol. 2327, pp. 526–538. Springer, Heidelberg (2002)
6. Japan Association for High Performance Fortran (JAHPF): HPF/JA Language Specification Version 1.0 (November 1999), http://www.hpfpc.org/jahpf/spec/hpfja-v10-eng.pdf
7. Kelly, W., Pugh, W., Rosser, E.: Code Generation for Multiple Mappings. In: Frontiers 1995. The 5th Symposium on the Frontiers of Massively Parallel Computation, McLean, VA (February 1995)
8. Li, W., Pingali, K.: A Singular Loop Transformation Framework Based on Non-Singular Matrices. Technical Report TR 92-1294, Cornell University, Ithaca, NY (July 1992)
9. Li, W., Pingali, K.: Access normalization: loop restructuring for NUMA computers. ACM Transactions on Computer Systems (TOCS) 11(4), 353–375 (1993)
10. Mellor-Crummey, J.M., Adve, V.S., Broom, B., Chavarria-Miranda, D.G., Fowler, R.J., Jin, G., Kennedy, K., Yi, Q.: Advanced optimization strategies in the Rice dHPF compiler. Concurrency and Computation: Practice and Experience 14(8-9), 741–767 (2002)
11. Message Passing Interface Forum. http://www.mpi-forum.org/
12. The Omega Project. http://www.cs.umd.edu/projects/omega/

Development of Electromagnetic Particle Simulation Code in an Open System

Hiroaki Ohtani[1,2], Seiji Ishiguro[1,2], Ritoku Horiuchi[1,2], Yasuharu Hayashi[3], and Nobutoshi Horiuchi[4]

[1] Theory and Computer Simulation Center, National Institute for Fusion Science, 322-6 Oroshi-cho, Toki 509-5292, Japan
[2] Graduate University of Advanced Studies (SOKENDAI), 322-6 Oroshi-cho, Toki 509-5292, Japan
[3] 1st Computers Software Division, NEC Corporation, 1-10 Nisshin-cho, Fuchu, Tokyo 183-8501, Japan
[4] NEC System Technologies Ltd., 4-24 Osaka, Shiromi 1-Chome, Chuo-ku, Osaka 540-8551, Japan

Abstract. In an electromagnetic particle simulation for magnetic reconnection in an open system, which has a free boundary condition, particles go out and come into the system through the boundary and the number of particles depends on time. Besides, particles are locally attracted due to physical condition. Accordingly, it is hard to realize an adequate load balance with domain decomposition. Furthermore, a vector performance does not become efficient without a large memory size due to a recurrence of array access. In this paper, we parallelise the code with High Performance Fortran. For data layout, all field data are duplicated on each parallel process, but particle data are distributed among them. We invent an algorithm for the open boundary of particles, in which an operation for outgoing and incoming particles is performed in each processor, and the only reduction operation for the number of particles is executed in data transfer. This adequate treatment makes the amount and frequency of data transfer small, and the load balance among processes relevant. Furthermore, a compiler-directive *listvec* in the gather process dramatically decreases the memory size and improves the vector performance. Vector operation ratio becomes about 99.5% and vector length turns 240 and over. It becomes possible to perform the simulation with 800 million particles in $512 \times 128 \times 64$ meshes. We succeed in opening a path for a large-scale simulation.

1 Introduction

Magnetic reconnection plays an important role in high-temperature plasmas; for example, solar corona, high-temperature tokamak discharge, magnetospheric substorm, and reconnection experiments [1,2,3,4,5]. It leads to fast energy release from a magnetic field to plasmas and change of the magnetic field topology. However, the mechanism of fast energy release is not fully understood, and several computer simulations have been performed to study this mechanism[6,7,8,9,10,11].

J. Labarta, K. Joe, and T. Sato (Eds.): ISHPC 2005 and ALPS 2006, LNCS 4759, pp. 329–343, 2008.

To investigate the behaviour of magnetic reconnection, both from the microscale viewpoint for electron and ion, and the macroscale viewpoint for the dynamic change of field, we need to perform a three-dimensional electromagnetic particle simulation on an open system[12,13,14,15,16,17]. In this paper, we develop a simulation code with a distributed parallel algorithm for a distributed memory and multi-processor computer system.

2 Distributed Parallel Computing Method

2.1 High Performance Fortran

For the distributed parallel algorithm, there are some programming paradigms; for example, Message Parallel Interface (MPI) and High Performance Fortran (HPF). Using HPF, programming is easier in comparison with MPI, because it is sufficient to add only instruction directives for parallel calculation and data transfer to the Fortran code. HPF compiler handles the distribution and scheduling of processing, and the control of communication according to the instruction directives. We adopt HPF for the distributed parallel computing method in this paper.

2.2 Estimation of Calculation Cost

Before considering a simulation code in the distributed parallel algorithm, we estimate the calculation cost in a particle simulation. Figure 1 shows a simulation cycle per time step in a particle simulation in an open system [18]. In the first process: "Force from fields," the force on particle position F_j is calculated from the electric and magnetic field defined on the grid $((E,B)_i)$. In the second process: "Particle pusher (Equation of motions)," the integration of equation of motions is performed using the force values F_j. In the third process: "Particle gather," the charge and current densities defined on the grid $((\rho,J)_i)$ are calculated from the particle positions and velocities $((x,v)_j)$. In the last

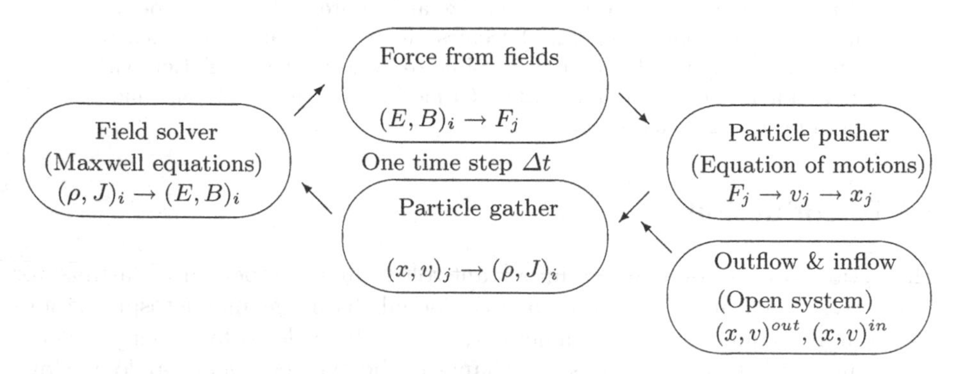

Fig. 1. A simulation cycle per time step in a particle simulation in an open system [18]. i is the grid number and j is the particle number.

and fourth process: "Field solver (Maxwell equations)," the field equations on the grid are integrated using $(\rho, J)_i$. These four processes are the typical cycle, one-time step, in a particle simulation program. Further, in a simulation in an open system, the "Outflow and inflow (Open system)" process is added. In this process, the information of particles which are going out of $((x, v)^{out})$ and coming into $((x, v)^{in})$ the system is obtained under the free boundary condition for particles (See Sec. 3). The calculation costs of three processes ("Force from fields," "Particle pusher" and "Particle gather") are in proportion to the total number of particles, while the calculation cost of "Field solver" is proportional to the number of grids. To investigate the magnetic reconnection, a large-scale simulation is needed, in which the numbers of particles and grids are, for example, one billion and 256^3, respectively. Accordingly, the calculations of the first three operations cost most in the one-time step, and these processes should be distributed in the parallel calculation. The "Outflow and inflow" processes should also be treated adequately in the same way.

2.3 Domain Decomposition and Particle Distribution

It is very important to decide which array is distributed in the distributed parallel algorithm. There are two cases. One case is to decompose the domain [19,20] (Fig. 2(b)). For example, the field variable is defined by three coordinates (x, y, and z), and we distribute it along z-direction. In this case, the processor performs "Field solver" in the mapped domain, and carries out "Particle pusher" for the particles which exist in the domain. However, communication among HPF processes is needed when the particle moves to the neighbouring distributed domain (like the arrow in Fig. 2(b)) and the global calculation is performed (for example, Fourier transformation). Furthermore, as a simulation for magnetic reconnection goes on, the particles are attracted locally due to the physical condition and then the load balance among HPF processes does not become sufficient in the domain decomposition case. From the viewpoints of data transfer and load balance, this case is disadvantageous. Another case is not to decompose the domain, but to distribute only information of particles [21] (Fig. 2(c)). In this case, we do not need the communication frequently, because every HPF process has the same full information of domain and performs "Field solver". As a simulation goes on, the number of particles is changed, because the particles go out and come into the system through the free boundary (See the next section). Because each processor has the same full field information and performs "Outflow and inflow" process, the change of the number of particles is supposed to be uniform among the processors, and the load balance is anticipated to be adequate through the simulation. As mentioned in the previous subsection, the number of particles is much larger than that of grids. It is expected that the simulation code becomes efficient only with distributing the array of particle values. From these reasons, we adopt the latter way in this paper.

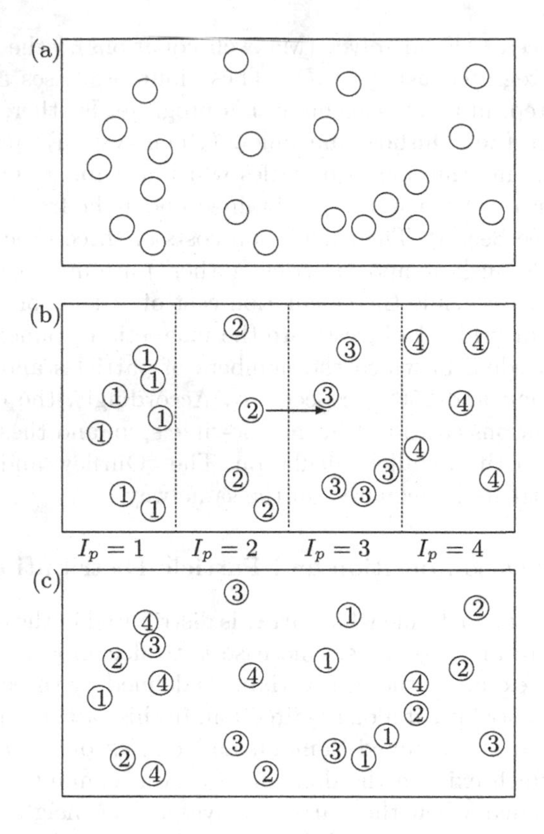

Fig. 2. Schematic illustration of (a) full system, (b) decomposed domain case, and (c) distributed particles case. I_p and the circle number are the index of HPF process.

2.4 Distribution of Particle Array

Figure 3 shows a schematic illustration of a distributed array of particle values, for example, position and velocity, and so on, (a) before and (b) after distribution. I_p and N_p are the index and number of PEs, respectively. The number of PEs equals to the product of the numbers of HPF processes and micro task parallelisation, and we use them together in this program code. The micro task parallelisation is the parallelisation on shared memory in one node. This architecture is available in an NEC SX system. The information of electrons and ions is recorded in the first and latter half of the array, respectively. In each part of electrons and ions, the information of "active" and "reserved" particles is recorded, respectively. "Active" means an array of the particles subject to the calculations of charge and current densities, while "reserved" means an empty array for the particle that will come into the system in the future. Because the number of particles in simulation box increases twice and over as many as the initial one as the simulation goes on, the parts of the array for electrons and ions are initially divided to 40% for 'active' and 60% for 'reserved', respectively.

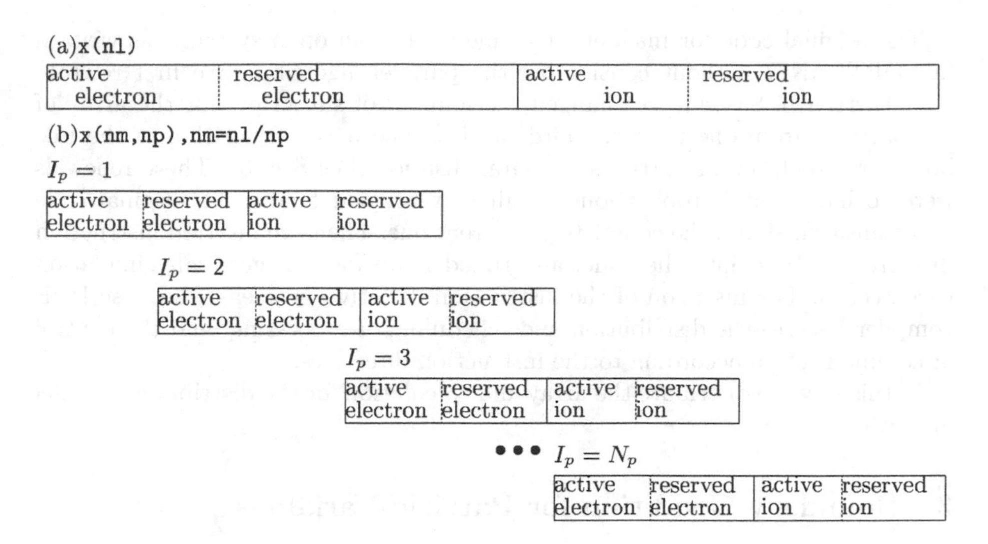

Fig. 3. Schematic illustration of array of particles (position, velocity, mass, and charge) (a) before and (b) after distribution. `nl` is the total number of particles, `nm` is the number of distributed particles, I_p is the index of PEs, N_p and `np` are the number of PEs (the production of HPF processes and micro task parallelisation), and `x(nl)` and `x(nm,np)` are, for example, the particle position

Appearing in Fig. 3, the number of subscripts for the array of the particles is changed from (a) one to (b) two, and the second subscript is applied to the PEs. Using this type of distributed array, we can perform parallel calculation as BLOCK distribution along the second subscript, and attain an adequate load balance among HPF processes, because all of HPF processes have the same-form array. If we simply divide the original array in Fig. 3(a) as BLOCK distribution, some HPF processes may control only "active" particles and the others may control "reserved" particles. In this case, the load balance does not become adequate. Furthermore, because the vector performance gets worse due to the stride access, we do not adopt CYCLIC distribution with using the array type in Fig. 3(a).

A key point in development of a simulation code under a distributed parallel algorithm is that the amount and frequency of data transfer are as small as possible. Fortunately, since the field variables are not distributed and only information of particles is distributed in this code, data transfer of the distributed information of particles is almost not needed as long as load balance is adequate. That is, the number of particles changes due to the outgoing and incoming particles through the free boundary as the simulation goes on, and then the number of particles mapped to the processor will differ among the processors. If the difference of the numbers among the processors becomes large, the load balance is not sufficient and we need to transfer data of particles among the processors (But this situation does not take place almost). The necessary data transfer is the reduction operation when "Open system" and "Particle gather" operations are performed (See Sec. 3 and Sec. 4 in details).

The original code for magnetic reconnection in an open system was written in FORTRAN77 without considering the parallel algorithm. We first rewrote it in Fortran90. Second, we changed the number of subscripts for the array of the particles from one to two. Third, we invented a new algorithm of the free boundary condition of particles for parallelisation (See Sec. 3). These renewals were so hard that it took about a half of year, and finally the original code was transformed into the completely different one. Then, we inserted instruction directives of HPF into the code, and tuned it up for a larger-scale simulation (See Sec. 4). The insertion of the instruction directives was easy, because HPF compiler handles the distribution and scheduling of processing, and the control of communication according to the instruction directives.

In this way, we distribute the array and calculation for the distributed parallel algorithm.

3 Boundary Condition for Particle Variables

The free boundary condition for particles is the key point of the simulation in the open system [14]. Besides, we need to elaborate its algorithm for parallel computing.

3.1 Upstream Boundary

The upstream boundary is supposed to be the ideal magnetohydrodynamic (MHD) region where both ions and electrons are frozen in the magnetic field. Thus, the plasma inflow is supplied with $\boldsymbol{E} \times \boldsymbol{B}$ drift velocity from the upstream boundaries into the simulation domain. The input particle velocity is controlled by a shifted Maxwellian with the initial constant temperature, where the average velocity $\boldsymbol{u}_i^{(d)}$ and density $\rho_i^{(d)}$ are

$$\boldsymbol{u}_i^{(d)} = \frac{\boldsymbol{E}_i \times \boldsymbol{B}_i}{B_i^2}, \tag{1}$$

$$\rho_i^{(d)} = \rho_i^{(d),t=0} \frac{B_i}{B_i^{t=0}}, \tag{2}$$

where i is the grid number. The density $\rho_i^{(d)}$ is calculated at every grid point on the upstream boundary, and the number of input particles $N_i^{(d)}$ is obtained from the product of $\rho_i^{(d)}$ and volume element ΔV. As particles of $N_i^{(d)}$ must obey the specific shifted Maxwellian to preserve locality, data transfer is needed if $N_i^{(d)}$ is distributed. Instead of the distribution of $N_i^{(d)}$, the grid number is distributed cyclically to the parallelised process as follows.

```
real(kind=8),dimension(nx)::n_d,n_d0,b_d,b_d0
real(kind=8),dimension(nx,np)::np_d
!hpf$ distribute (:,block):: np_d
      do i=1,nx
```

```
          n_d(i)=n_d0(i)*b_d(i)/b_d0(i)
        enddo
!hpf$ independent
        do ip=1,np
         do i=1,nx
           np_d(i,ip)=0
         enddo
        enddo
        do i=1,nx
          nn=mod(i,np)+1
!hpf$ on home(np_d(:,nn))
          np_d(i,nn)=n_d(i)
        enddo
```

Where `nx` is the dimension size of space, `np` is the number of parallelised processes, `n_d(nx)` is $N_i^{(d)}$, `n_d0(nx)` is $N_i^{(d),t=0}(=\rho_i^{(d),t=0}\Delta V)$, `b_d(nx)` is B_i, and `b_d0(nx)` is $B_i^{t=0}$. The shifted Maxwellian is also computed on every parallelised process. Using this algorithm, much data transfer does not take place.

3.2 Downstream Boundary

Figure 4 schematically illustrates the free boundary condition at downstream for particles. At the open downstream boundary, particles can not only go out of but also come into the system across it. The information of the outgoing particles (N^{out}) can be obtained directly by observing their motion at the boundary. We must define the number of particles which should go into the system (N^{in}), and the positions and velocities of these incoming particles as boundary condition in the open system.

As the first step, we show how to calculate the number of incoming particles across the open downstream boundary in the distributed parallel algorithm. In every parallelised process (I_p), we first calculate the average particle velocity $\overline{v_x}(I_p)$ and number density $n(I_p)$ in region I, which we choose with width of six grids ($6\Delta x$) on the boundary. Second, we obtain the net number of outgoing particles (N^{net}) passing across the boundary during one-time step Δt in the system under reduction operation, as

$$N^{net} = -\sum_{I_p=1}^{N_p} n(I_p)\overline{v_x}(I_p)\Delta t y_b, \tag{3}$$

where y_b is the width of region I, and N_p is the number of parallelised processes. According to the charge neutrality condition, we assume that the net numbers of electrons and ions are the same. So the numbers of incoming electrons and ions to the system are given as

$$N_e^{in} = N_e^{out} - N^{net}, \quad N_i^{in} = N_i^{out} - N^{net}. \tag{4}$$

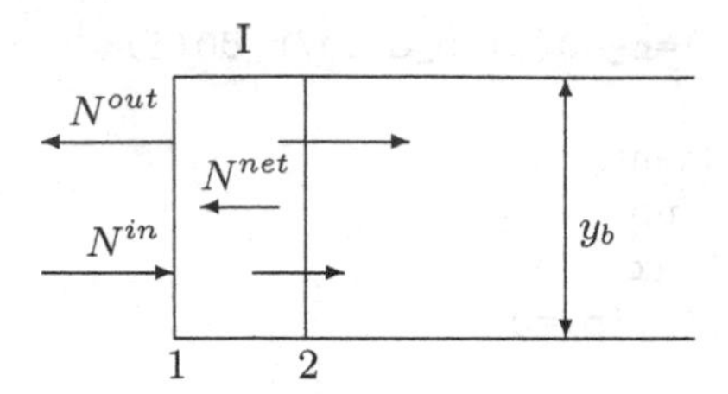

Fig. 4. Illustration of the free boundary condition for particles

Then, we distribute N_e^{in} and N_i^{in} to every parallelised process as

$$n_e^{in}(I_p) = N_e^{in}/N_p, \quad n_i^{in}(I_p) = N_i^{in}/N_p. \tag{5}$$

As the next step, the positions and the velocities of the incoming particles have to be assigned. We assume that the physical state outside is the same as that in region I. In other words, the particle distribution function outside is the same as that in region I. Then, the positions and velocities of the incoming particles can be defined by using the information of particles crossing surface 2 from left to right (The arrows in Fig. 4). In this distributed parallel algorithm, the information of electrons and ions crossing surface 2 is kept in each parallelised process, and we use them for positions $x_j^{I_p}$ and velocities $v_j^{I_p}$ of input electrons and ions, respectively.

Using these algorithms at the upstream and downstream boundary condition, respectively, data transfer is necessary only in the reduction operation, and much data transfer, such as data transfer of particle position and velocity, is not needed.

4 Gather Process

The vector algorithm makes an error in the calculations of the charge and current densities in the particle gather process due to a recurrence of array access, when several particles exist in the same cell. To avoid this error, we introduce a temporary array assigning the particle information to different array indices to avoid any recurrences. This temporary array has three subscripts in the one-dimensional simulation case, such as vector `mv`, grid number `nx`, and parallelised process `np` (`w(mv,nx,np)`). Due to the vector subscript, the vector operation can be performed in this gather process [22]. In the HPF process subscript, because the distributed array `w` is mapped onto the same HPF process on which the distributed arrays of particles (position and velocity) are also mapped, this execution of the loop should be parallelised without data transfer. The concrete code is shown as follows.

```
real(kind=8),dimension(nm,np):: x
real(kind=8),dimension(mv,nx,np):: w
real(kind=8),dimension(nx):: rho
```

```
!hpf$ distribute (:,BLOCK):: x
!hpf$ distribute (:,:,BLOCK):: w
!hpf$ independent
      do ip=1,np
       do m=1,nm,mv
        do k=1,min(mv,nm-m)
         j=m+k-1
         ix=int(x(j,ip)/dx+0.5)
         xx=x(j,ip)/dx-ix
         s1=(0.5 -xx)**2/2
         s2=(0.75-xx**2)
         s3=(0.5 +xx)**2/2
         w(k,ix-1,ip)=w(k,ix-1,ip)+q*s1/dx
         w(k,ix,  ip)=w(k,ix,  ip)+q*s2/dx
         w(k,ix+1,ip)=w(k,ix+1,ip)+q*s3/dx
        enddo
       enddo
      enddo
!hpf$ independent,reduction(+:rho)
      do ip=1,np
        do i=1,nx
          do k=1,mv
            rho(i)=rho(i)+w(k,i,ip)
          enddo
        enddo
      enddo
```

Where `nm` is the number of particles mapped in one PE, `dx` is the grid length Δx, `q` is the particle charge q, `x(j,ip)` is the particle position x_j^{Ip}, and `rho(i)` is the charge density ρ_i. However, in a large-scale three-dimensional simulation case, this temporary array needs a lot of memory size, because the size of the first vector subscript of the array `w` should be 256, for the vector operation with full efficiency, and the number of the grid subscript becomes three. In contrast, when we use the compiler-directive *listvec*, which is available on an NEC SX vector computer, we can decrease the memory size of the temporary array and perform the vector calculation [23,24]. This compiler-directive can vectorise an assignment statement in a DO loop of the immediately following executable statement, when the same array element with the same vector subscript appears on both of the left and right of the assignment statement [25]. As such, we can drop the vector subscript from the temporary array `w`. Moreover, the operation amount of the reduction of temporary array can be reduced along the vector subscript when `rho` is calculated.

```
real(kind=8),dimension(nm,np):: x
real(kind=8),dimension(nx,np):: w
real(kind=8),dimension(nx):: rho
```

```
!hpf$ distribute (:,BLOCK):: x,w
!hpf$ independent
      do ip=1,np
       do m=-1,1
!cdir listvec
        do j=1,nm
          ix=int(x(j,ip)/dx+0.5)
          xx=x(j,ip)/dx-ix
          if(m.eq.-1) s1=(0.5 -xx)**2/2
          if(m.eq. 0) s1=(0.75-xx**2)
          if(m.eq.+1) s1=(0.5 +xx)**2/2
          ix=ix+m
          w(ix,ip)=w(ix,ip)+q*s1/dx
        enddo
       enddo
      enddo
!hpf$ independent,reduction(+:rho)
      do ip=1,np
       do i=1,nx
         rho(i)=rho(i)+w(i,ip)
       enddo
      enddo
```

Note that this vectorisation may cause performance down in the case of the array subscript having some elements with the same value. That is, if particles whose indices are close to one another exist in the same cell, the vector calculation is not performed; rather, the scalar calculation is. As such, we should randomise the particle indices sufficiently, when we make the initial particle distribution. On the application of the compiler-directive *listvec*, a restriction is imposed. Namely, in a DO loop of the immediately following executable statement, an operation to `w` must be performed only one time. As such, a `do m` loop is inserted newly.

To estimate the effect of the compiler-directive *listvec*, we program a code only for the particle gather process and test-run both in the cases of (a) vector subscript and (b) *listvec*. In this simple code, three-dimensional calculation is performed. Calculation parameters are as follows: The number of active particles is 60 million, the grid numbers are $128 \times 64 \times 128$, and the number of calculation times is 100. The computer system we use is the “Plasma Simulator” at the National Institute for Fusion Science [26]. The “Plasma Simulator” is composed of 5 nodes of an NEC SX-7 vector parallel supercomputer. Its total memory size and logical peak performance are 1,280Gbytes and 1,412GFlops, respectively. The number of PE per one node, the number of vector pipe lines, and the data transfer speed between nodes are 32, 4, and 8Gbps, respectively. The number of HPF processes is 1, and the number of tasks for micro parallelisation is 32. The calculation performance is listed in Table 1, where the floating-point arithmetic per second (GFlops) is defined as the ratio of the FLOP count to execution

Table 1. Calculation performance of (a) vector subscript case and (b) *listvec* case

Case	(a)	(b)
Memory (GB)	226.2	7.4
Vector length	254.3	254.5
Vector operation ratio (%)	99.25	99.80
Time (sec)	179.8	137.62
GFlops	24.88	55.74

time. FLOP count, execution time, vector length and vector operation ratio are obtained from run-time information of the NEC SX computer.

The memory size of case (b) is 3.2% of that of case (a). The reason why the ratio of them is not 1/256 as follow: It is sure that a memory size of the temporary array `w` becomes 1/256, but a memory size of the particle values, such as, 6 components of position and velocity, charge and mass, and a memory size of the stack area for the micro task parallelisation do not change. The memory size of case (b) is almost occupied by them. (The memory size of array of the particles and stack area are 3.6Gbytes and 2.2Gbytes, respectively). The vector length is almost the same in the cases (a) and (b). The vector operation ratio of the case (b) is about 0.55 points larger than that of the case (a). The vector operation is vastly improved from the viewpoint of vector operation ratio. The executed time in case (b) is smaller than that in case (a), and the value of GFlops of case (b) also becomes larger compared with that of case (a). However, this increase of GFlops is a little unnatural: GFlops value in case (b) is about 2.24 times as much as in case (a), while the executed time in (b) is only 0.765 times as long as in case (a). The reason is that extra operations are counted in the *listvec* DO loop (FLOP count in case (b) is 1.7 times as much as in case (a)).

As a consequence, it becomes possible to perform a much larger-scale simulation due to using the compile-directive *listvec* because the memory size of temporary arrays dramatically reduces compared with that of typical method.

5 Performance

To investigate the execution performance of the simulation code developed in this paper (the code is tuned up moreover), we test-run under various parallel calculation condition, such as a different number of parallelisation (PE), nodes, and so on (Table 2). Calculation parameters are as follows: The number of particles is 60 million, the number of active particles is 24 million at initial time, the numbers of grids are $128 \times 64 \times 128$ and the number of time steps is 1000. In the Table 2, the load balance of PE is defined as the ratio of the two kinds of concurrent time. One is the time period in which more than one processors are concurrently performed, and the other is for the same number of processors as that of micro tasks in one process, where concurrent time is obtained from run-time information of the NEC SX computer. The load balance of process is

Table 2. Calculation performance

Case	(1)	(2)	(3)	(4)	(5)	(6)	(7)
Compiler	f90	hpf	hpf	hpf	hpf	hpf	hpf
Number of nodes	1	1	1	1	2	3	5
Number of PE	32	32	32	32	32	96	160
Number of HPF processes	-	1	2	4	2	6	10
Number of micro tasks	32	32	16	8	16	16	16
Memory (GB)	44.93	45.18	47.13	53.25	48.26	131.25	222.50
Time/step (sec)	1.808	1.974	1.929	2.022	1.744	1.144	0.996
Vector operation ratio (%)	99.69	99.62	99.60	99.58	99.61	99.28	99.03
Vector length	247.3	247.7	243.7	236.5	243.6	234.14	227.95
Gflops	45.36	42.42	41.97	41.71	46.29	83.09	98.69
Ratio to peak performance (%)	16.1	15.1	14.9	14.8	16.4	9.8	7.0
Load balance of PE (in one process) (%)	85.4	83.8	86.5	87.8	87.6	85.2	82.3
Load balance of process (between processes) (%)	-	-	94.8	95.4	94.2	91.3	89.5

the ratio of the smallest and largest concurrent times, which are the executed times using more than one processors, among processes.

We first make a comparison between f90 and hpf compilers ((1) and (2) in Table 2). The calculation time per step and GFlops in f90 compiler are smaller and larger than that of hpf, respectively. And the load balance of PE in hpf case is smaller than that in f90 case. But the difference between them is not so large. The vector operation ratio and the vector length are almost the same. From the comparison of these values, it is reasonable to suppose that the efficiency of the HPF program bears comparison with that of the f90 program.

Next, we change the number of HPF processes keeping the number of PE in one node (Compare (2), (3), and (4) in Table 2). In the node, a micro task parallelisation is performed. In this test-run, when the number of HPF processes increases, the number of the micro task parallelisation reduces. When the number of HPF processes changes from one to two, the executed time reduces, while when it changes from two to four, the time per step increases. However, the differences among them are very small. From these results, it is found that the efficiency of HPF is almost the same favourable as that the micro task parallelisation, and that 16 is the best suitable for the number of micro tasks in one HPF process in this simulation code. Note that memory size increases when the number of HPF processes increases, because the each process has the full field information.

Third, we consider the influence of data transfer among the HPF processes. To study it, we test-run under the conditions of two HPF processes in one node ((3) in Table 2), and two HPF processes in two nodes ((5) in Table 2). The former case is data transfer in the node, while the latter is data transfer between nodes. The executed time in case (5) is smaller than that in case (3), although the other efficiency values are almost the same. This result is, at a glance, strange, because the speed of data transfer among the nodes in case (5) might be slower than that of within the node in case (3). The reason for this strange result is considered to be that memory conflict takes place in one node in case (3). The only reduction operation is performed when data transfer occurs in this simulation code, as mentioned in Secs. 2, 3 and 4, and it is concluded that data transfer is carried out efficiently.

To sum up the efficiency of this simulation code, when the number of PE increases ((6) and (7) in Table 2), the vector operation efficiency is maintained good, while the load balances of PE and nodes get worse. This is the general tendency. Since all field data are duplicated on each parallel process for data layout, the parallel efficiency becomes worse when the number of PE increases. Consequently, it is reasonable to suppose that this code is efficient.

6 Summary

To investigate the behaviour of magnetic reconnection, both from the microscale viewpoint for electron and ion dynamics, and from the macroscale viewpoint for the dynamic change of field, we developed a three-dimensional full electromagnetic particle simulation in an open boundary system. Moreover, we improve it using High Performance Fortran (HPF) for a distributed memory and multi-processor computer system. Using HPF program language, the distribution and scheduling of processing, and the management of communication are handled by HPF compiler according to the instruction directives. It is considered that this language is suitable for development of simulation code from the view point of programming.

It is of significance in the development of simulation code for a distributed parallel algorithm to decrease the amount and frequency of data transfer. In the simulation code developed in this paper, each process has the same full field information, and the information of particles is distributed. Every process controls only the particles in its own process, which going out of and coming into the system through the boundary. Data transfer takes place when the reduction operation is performed. These algorithms make the load balance proper and reduce the communication between processes.

A temporary array is required in the particle gather process, in which the current and charge densities are calculated from the particle position and velocity for a vector operation without much data transfer. However, this temporary array needs a lot of memory size in a large-scale simulation. In this paper, we succeed in dramatic reduction of the temporary array memory size and improvement of vector operation performance using the compiler-directive *listvec*.

As a consequence, we can make up an efficient simulation code and make it possible to perform a larger-scale simulation. Actually, we have performed the simulation with 800 million particles in $512 \times 128 \times 64$ meshes, and are opening new frontier in this research field.

Acknowledgments

This work was performed with the support and under the auspices of the National Institute for Fusion Science (NIFS) Collaborative Research Program.

References

1. Biskamp, D.: Magnetic Reconnection in Plasamas. Cambridge University Press, Cambridge (2000)
2. Drake, J.F., Lee, Y.C.: Kinetic theory of tearing instabilities. Phys. Fluids 20 (1977)
3. Ono, Y., Yamada, M., Akao, T., Tajima, T., Matsumoto, R.: Ion acceleration and direct ion heating in three-component magnetic reconnection. Phys. Rev. Lett. 76 (1996)
4. Yamada, M., Ji, H., Hsu, S., Carter, T., Kulsrud, R., Trintchouk, F.: Experimental investigation of the neutral sheet profile during magnetic reconnection. Phys. Plasmas 7 (2000)
5. Hsu, S., Fiksel, G., Carter, T.A., Ji, H., Kulsrud, R.M., Yamada, M.: Local measurement of nonclassical ion heating during magnetic reconnection. Phys. Rev. Lett. 84 (2000)
6. Sato, T., Hayashi, T.: Externally driven magnetic reconnection and a powerful magnetic energy converter. Phys. Fluids 22 (1979)
7. Shay, M.A., Drake, J.F., Rogers, B.N., Denton, R.E.: Alfvnic collisionless magnetic reconnection and the hall term. J. Geophys. Res., [Space Phys.] 106 (2001)
8. Kuznetsova, M., Hesse, M., Winske, D.: Collisionless reconnection supported by nongyrotropic pressure effects in hybrid and particle simulations. J. Geophys. Res., [Space Phys.] 106 (2001)
9. Hesse, M., Birn, J., Kuznetsova, M.: Collisionless magnetic reconnection: Electron processes and transport modeling. J. Geophys. Res., [Space Phys.] 106 (2001)
10. Birn, J., Hesse, M.: Geospace environment modeling (gem) magnetic reconnection challenge: Resistive tearing, anisotropic pressure and hall effects. J. Geophys. Res., [Space Phys.] 106 (2001)
11. Ma, Z.W., Bhattacharjee, A.: Hall magnetohydrodynamic reconnection: The geospace environment modeling challenge. J. Geophys. Res., [Space Phys.] 106 (2001)
12. Horiuchi, R., Sato, T.: Particle simulation study of driven magnetic reconnection in a collisionless plasma. Phys. Plasmas 1 (1994)
13. Horiuchi, R., Sato, T.: Particle simulation study of collisionless driven reconnection in a sheared magnetic field. Phys. Plasmas 4 (1997)
14. Pei, W., Horiuchi, R., Sato, T.: Long time scale evolution of collisionless driven reconnection in a two-dimensional open system. Phys. Plasmas 8 (2001)
15. Pei, W., Horiuchi, R., Sato, T.: Ion dynamics in steady collisionless driven reconnection. Phys. Rev. Lett. 87 (2001)
16. Ishizawa, A., Horiuchi, R., Ohtani, H.: Two-scale structure of the current layer controlled by meandering motion during steady-state collisionless driven reconnection. Phys. Plasmas 11 (2004)
17. Horiuchi, R., Sato, T.: Three-dimensional particle simulation of plasma instabilities and collisionless reconnection in a current sheet. Phys. Plasmas 6 (1999)
18. Birdsall, C.K., Langdon, A.B.: Plasma Physics Via Computer Simulation. McGraw-Hill, New York (1985)
19. Fox, G.C., Johnson, M., Lyzenga, G., Otto, S., Salmon, J., Walker, D.: Solving Problems on Concurrent Processors. Prentice-Hall, Englewood Cliffs (1988)
20. Liewer, P., Decyk, V.: A general concurrent algorithm for plasma particle-in-cell codes. J. Comput. Phys. 85 (1989)
21. Martinoa, B.D., Briguglio, S., Vladb, G., Sguazzeroc, P.: Parallel pic plasma simulation through particle decomposition techniques. Parallel Computing

22. Anderson, D.V., Horowitz, E.J., Koniges, A.E., McCoy, M.G.: Parallel computing and multitasking. Comp. Phys. Comm. 43 (1986)
23. Suehiro, K., Murai, H., Seo, Y.: Integer sorting on shared-memory vector parallel computers. In: ICS 1998. Proceedings of the 12th international conference on Supercomputing (1998)
24. Sugiyama, T., Terada, N., Murata, T., Omura, Y., Usui, H., Matsumoto, H.: Vectorized particle simulation using listvec compile-directive on sx super-computer. IPSJ 45 (2004)
25. NEC Corporation: FORTRAN90/SX Programmer's Guide. NEC Corporation (2002)
26. Theory and Computer Simulation Center, National Institute for Fusion Science, Japan(2005), http://www.tcsc.nifs.ac.jp/mission/

Development of Three-Dimensional Neoclassical Transport Simulation Code with High Performance Fortran on a Vector-Parallel Computer

Shinsuke Satake[1], Masao Okamoto[2], Noriyoshi Nakajima[1], and Hisanori Takamaru[3]

[1] National Institute for Fusion Science, Toki, Japan
satake@nifs.ac.jp,
[2] Department of Information Processing Education, Chubu University, Kasugai, Japan
[3] Department of Computer Science, Chubu University, Kasugai, Japan

Abstract. A neoclassical transport simulation code (FORTEC-3D) applicable to three-dimensional configurations has been developed using High Performance Fortran (HPF). Adoption of computing techniques for parallelization and a hybrid simulation model to the δf Monte-Carlo method transport simulation, including non-local transport effects in three-dimensional configurations, makes it possible to simulate the dynamism of global, non-local transport phenomena with a self-consistent radial electric field within a reasonable computation time. In this paper, development of the transport code using HPF is reported. Optimization techniques in order to achieve both high vectorization and parallelization efficiency, adoption of a parallel random number generator, and also benchmark results, are shown.

1 Introduction

In research activities on magnetic confined fusion plasma, one of the basic and important issues is to evaluate the confinement performance of the plasma. The loss mechanisms of plasma can be classified roughly into two categories. One is caused by orbit motion of charged particles and their diffusion by Coulomb collision, and the other is caused by several types of instabilities occurring in plasmas such as MHD instabilities and micro-turbulences. In this paper, we focus on the development of transport simulation code for the former transport process, which is called "neoclassical transport" [1,2] in fusion research activities.

In Fig. 1, two typical configurations of plasma confinement devices are shown. The nested surfaces shown in the figures are called "magnetic flux surfaces", which consist of twisting magnetic field lines around those surfaces. Fig. 1 (a) is a tokamak configuration, which has a symmetry in the toroidal direction, and (b) is a helical configuration of Large Helical Device (LHD)[3] in National Institute for Fusion Science (NIFS), Japan, which is a typical configuration of so-called

J. Labarta, K. Joe, and T. Sato (Eds.): ISHPC 2005 and ALPS 2006, LNCS 4759, pp. 344–357, 2008.

heliotron devices[4]. The coordinate system (ρ, θ, ζ) (magnetic coordinates) we use here is also shown in Fig. 2. Here, $\rho = \sqrt{\psi/\psi_{edge}}$ is a normalized radius, θ is the poloidal angle, and ζ is the toroidal angle. ψ is the toroidal magnetic flux inside a flux surface ρ =const, and ψ_{edge} is the value of ψ at the outermost surface, respectively. The change of magnetic field strength $|B|$ along a field line is illustrated in Fig. 3. In the tokamak case, the modulation of $|B|$ is caused by its toroidicity. In a heliotron configuration, modulation of $|B|$ caused by helically winding coils is superimposed on the toroidal modulation.

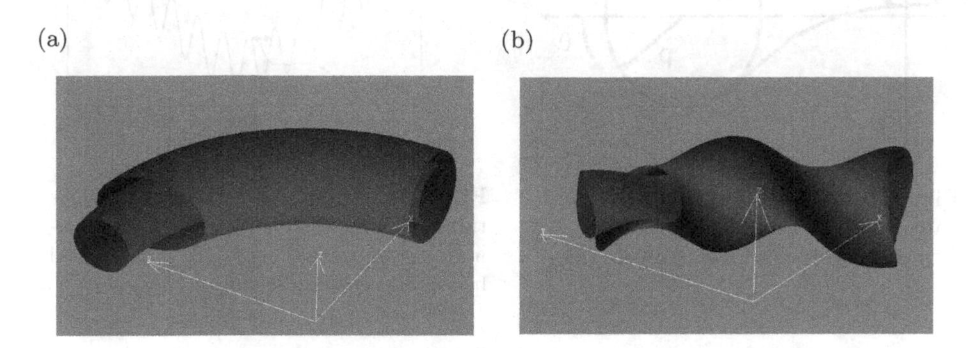

Fig. 1. Cut images of the magnetic flux surfaces (a) tokamak (two-dimensional) configuration, (b) Large Helical Device (LHD) configuration

The guiding center velocity of a charged particle can be decomposed as $\mathbf{v} = v_\parallel \mathbf{b} + \mathbf{v}_E + \mathbf{v}_d$, where $v_\parallel$ is the parallel velocity, $\mathbf{b} = \mathbf{B}/B$, $\mathbf{B}$ is the magnetic field, $\mathbf{v}_E = \mathbf{E} \times \mathbf{B}/B^2$, $\mathbf{E} = -\nabla \Phi$ is the static electric field in plasma, and $\mathbf{v}_d$ is the drift velocity, which is caused by the gradient and curvature of magnetic field. Since Φ is considered to be constant on a flux surface, i. e., $\Phi = \Phi(\rho)$, the $\mathbf{E} \times \mathbf{B}$ direction is along the flux surface. Therefore, only the drift velocity $\mathbf{v}_d$ has a component in the radial direction. Usually the dominant component of particle motion is the parallel motion $v_\parallel \mathbf{b}$. On the other hand, the magnetic moment $\mu = m{v_\perp}^2/2B$ is a constant of the motion, where ${v_\perp}^2 = v^2 - {v_\parallel}^2$, $mv^2/2 = \mathcal{E} - e\Phi$, and $\mathcal{E}$ is the total energy of a charged particle. Since $\mu B \leq mv^2/2$, particles with large μ are trapped in a weak-B region as illustrated in Fig. 3. Compared with passing particles, trapped particles have a large excursion of orbit in radial direction, called "orbit width" Δ_ρ, caused by the drift motion. Neoclassical transport theory treats the enhancement of transport by these trapped particles in torus plasmas.

The radial electric field evolves in time so that the radial ion and electron particle fluxes Γ_i, Γ_e satisfy the ambipolar condition $\Gamma_i(\rho, E_\rho) = \Gamma_e(\rho, E_\rho)$, where $E_\rho = -d\Phi/d\rho$. Compared with tokamak cases, particle orbit in helical configurations becomes complicated as shown in Fig. 3(b), and it is difficult to be treated exactly by an analytic way. Since neoclassical ion flux in helical plasma strongly depends on $\mathbf{v}_E$, the self-consistent radial electric field by treating properly the

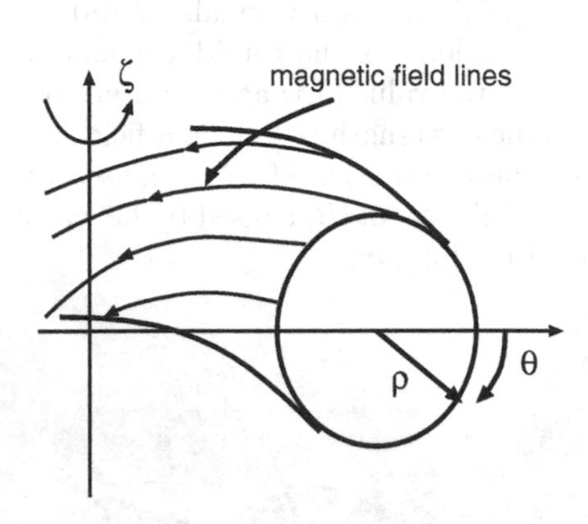

Fig. 2. Illustration of a magnetic coordinates (ρ, θ, ζ). Flux surface ρ =const. is formed by twisting magnetic field lines.

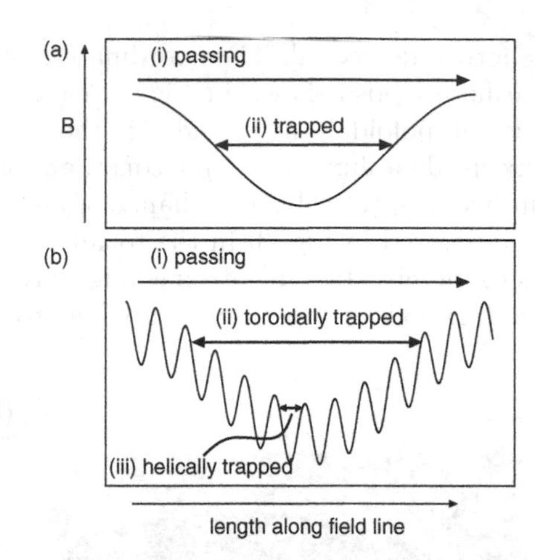

Fig. 3. Illustrations of guiding center motions in magnetic field : (a) in a tokamak configuration and (b) in a helical heliotron-type configuration

particle orbits in a three-dimensional magnetic field is a key to evaluate transport level correctly in helical plasma. Another point which is difficult to treat in analytic way is the non-local effect on transport brought by the finiteness of orbit width (finite-orbit-width (FOW) effect)[5]-[7]. Conventional neoclassical theory has been established under the assumption of the local transport model (small-orbit-width limit) in which Δ_ρ is treated as a higher-order small value. In order to evaluate precisely the neoclassical transport level in a realistic plasma, the FOW effect is needed to be considered.

To simulate the dynamic transport process and formation of an ambipolar electric field, including non-local effects in helical configurations, we have been developing the δf Monte Carlo code FORTEC-3D[8]. The δf method[9,10] directly solves the drift-kinetic equation, which describes the time evolution of plasma distribution function, by using a Monte Carlo technique. The outline of the code is explained in Sec. 2. 1. Since electron motion is much faster than that of ions while the FOW effect on electron transport is negligible, it is inappropriate to treat both ions and electrons by the δf method, in practice. In our simulation model, only the ion transport is solved by the δf method, while the electron transport is obtained from a reduced kinetic equation solver GSRAKE[11,12], which is more compact than FORTEC-3D but does not include the non-local effect. The adoption of this hybrid simulation model enables us to simulate neoclassical transport including the FOW effect of ions and self-consistent evolution of E_ρ within a reasonable computation time.

To carry out a simulation in the full volume of a helical configuration like LHD by the δf method, a large amount of memory and high calculation performance are needed. We have developed FORTEC-3D code by using High Performance Fortran (HPF)[13] in order to achieve both high parallelization and vectorization efficiency as well as to develop the code easily, on the SX-7 supercomputer system (NEC Corporation, Japan) at the Theory and Computer Simulation Center, NIFS. The system has five nodes connected by high-speed network switches, and each node has 32 PE (processor elements) and 256GB memory. The total peak performance is 1412 GFLOPS/160 PE. Parallelization of codes with HPF can be achieved by specifying the data mapping on the distributed memory and instructing the parallel calculation by embedding HPF directives (in the form of `!HPF$ distribute, !HPF$ independent`, etc.), which is easier than programming a parallelized code using Fortran with Message Passing Interface (MPI). We have also adopted a parallelized pseudorandom number generator with a preferable vectorization efficiency because a non-parallelized random number generator will be a bottleneck of Monte Carlo simulation. Since the Monte Carlo method itself is suitable for parallel computing, as it basically treats independent events, we have achieved a high computation performance on the SX-7 system, as fast as 30% of the peak performance at the full five-node, 160 PE. The technique to optimize the code and some benchmark results, both in two-dimensional and three-dimensional simulations, are shown in Sec. 2. 2 and 2. 3. Section 3 contains a discussion and a summary.

2 Development of FORTEC-3D Code with HPF

2.1 Outline of FORTEC-3D Code

In FORTEC-3D, the drift-kinetic equation for the deviation of plasma distribution function from local Maxwellian, $\delta f = f - f_M$, is solved by using the two-weight scheme[9,10]. The flowchart of FORTEC-3D code is shown in Fig. 4. In the two-weight scheme, two weights w and p are introduced which satisfy the relations $wg = \delta f$ and $pg = f_M$, where g is the distribution function of simulation markers. Each marker moves like a charged particle in plasma, and time evolution of marker weights are solved. The effect of collisions is implemented numerically by random kicks of marker velocity. In FORTEC-3D, the time evolution of radial electric field E_ρ is solved according to $\partial E_\rho / \partial t = -e \left[\Gamma_i - \Gamma_e \right] / \epsilon_\perp$, where subscripts i and e describe particle species, $\epsilon_\perp$ is the dielectric constant in the torus plasma. As mentioned in Sec. 1, the hybrid simulation model for evaluating neoclassical particle fluxes is adopted, in which only Γ_i is solved by the δf method while the table of $\Gamma_e(\rho, E_\rho)$, which is prepared by using GSRAKE, is referred to in FORTEC-3D. Since the collisionality of fusion plasma is very low, collision operator is solved once after `nss` times orbit calculation using 4th-order Runge-Kutta method. The field particle operator is introduced to retain the conservation property of the Fokker-Planck collision term. The marker weights w and p are averaged in the phase space $(\rho, \theta, \zeta, v_\parallel, v_\perp)$ to reduce the statistical noise in the simulation. Markers escaped from the simulation region are recycled, and the assignment procedure of new weights for the recycled markers is integrated into the weight-averaging procedure[14]. The procedures with star marks in Fig. 4 contain a part to take some ensemble averages and reflect them on the time evolution of the simulation, which make FORTEC-3D different from

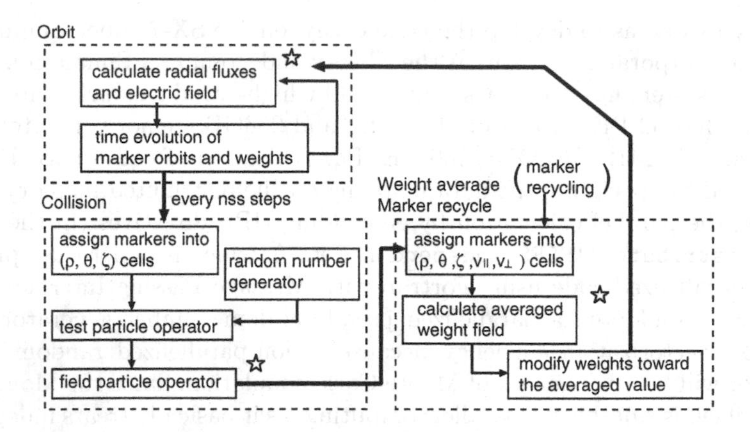

Fig. 4. Flowchart of FORTEC-3D code. Procedures with star mark involve reduction calculation and communication between HPF processes.

a simple Monte Carlo code that treats completely independent phenomena. In parallel computing, as we will explain in the next subsection, communication among parallel processes is needed in these procedures.

2.2 Optimization

Parallelized random number generator. On implementation of test particle collisions in a Monte Carlo way, a long sequence of random numbers $\{X_i\}$ is needed. Though there are many types of pseudorandom number generators used in simulations, we need to use the one which has both a good statistical properties as a random numbers and ability to generate random numbers as fast as possible. Taking account of the above points, we have adopted Mersenne Twister[15] (MT) in FORTEC-3D. We have also adopted the Dynamic Creation scheme[16] to create independent sets of MT, which enables generating independent random number sequences in parallel.

At first, we have tuned the subroutine of MT to achieve high vectorization efficiency. The original source code of MT returns one random number for each calling. We made a subroutine `grnd(rnd,n)` returns `n` sequence of MT random number in the array `rnd(1:n)`. It is known that MT has a long period of pseudorandom number sequence, which is equal to Mersenne prime number $2^p - 1$, where $p = 521$, 4423, 9941, 19937, and so on. It is found that, as the index p becomes larger, the vector length becomes longer. Therefore, we decided to use $p = 19937$ version of MT, which is longest one available for generating 32 bits pseudorandom numbers. The vector length of `grnd` becomes 216 (max=256 on SX-7), and the vector operation ratio is 99.5%.

For parallelization on our SX-7, 160 PE system, we have created 160 data sets which specify the form of recurrences in MT. MT characteristic polynomials for each data set are independent each other, and therefore the sequences of random number are also independent each other. It took about 4 days to create

160 independent MT data sets on an Athlon XP desktop PC. Subroutine **grnd** is parallelized by using HPF. If one needs to generate total **n** pseudorandom numbers in an HPF code, `grnd(rnd,ni)` is called, where `ni` = `n/ncpu`. Then each HPF process refers to different data sets and creates `ncpu` independent sequences of random numbers into `rnd(1:ni,1:ncpu)` at once in parallel, where `rnd` is distributed on `ncpu` HPF processes. The running time of parallel MT to generate total 1.924×10^7 random numbers is shown in Fig. 5, where each PE generates 1.924×10^7/`ncpu` numbers. It can be seen that the overhead time cost for parallelizing MT by HPF is small even if `ncpu` becomes larger. Because of the good parallelization efficiency of the **grnd** routine, the ratio of time consumption of **grnd** on the total simulation time of FORTEC-3D is suppressed as small as 0.2%.

The statistical independence of parallelized MT random numbers was not evaluated well. On purpose to check it, we devised a test scheme, named "checker-board test", which can check the independence of two sequences of random numbers $\{X_i\}$ and $\{Y_i\}$ which are uniformly distributed, as illustrated in Fig. 6. In the test, $N = n \times n$ cells are considered, and the flag at the cell (X_i, Y_i) is turned on at i-th step and the number of cells $m(i)$ which are remain off is counted. If $\{X_i\}$ and $\{Y_i\}$ are independent, the mean and variance of $m(i)$ at i-th step become

$$E[m(i)] = N\left\{1 - \left(\frac{N-1}{N}\right)^i\right\} \simeq N\exp(-t), \tag{1}$$

$$\sigma^2[m(i)] = N\left\{N + (N-1)\left(\frac{N-2}{N}\right)^i - (2N-1)\left(\frac{N-1}{N}\right)^i\right\}, \tag{2}$$

$$d(t) = \{m(t) - E[m(t)]\}/\sigma[m(t)], \tag{3}$$

where t is the normalized time step $t = i/N$. By plotting $d(t)$ as shown in Fig. 7, one can easily check if two sequences $\{X_i\}$ and $\{Y_i\}$ have an inappropriate correlation or not. If $\{X_i\}$ and $\{Y_i\}$ are independent, the amplitude of $d(t)$ will not grow as the time step goes on. We have checked every ${}_{160}C_2$ combinations of independent MT created from Dynamic Creation scheme in this way by using $N = (2^{14})^2$ checker-board and found that there is no combination of random number sequences which has an inappropriate correlation. Moreover, we have also checked some other statistical values such as the average ratio of time steps on which the deviation of $m(i)$ from the mean becomes larger than $\pm\sigma$, $\pm 2\sigma$, $\pm 3\sigma$, and the average steps t_{end} to fill all the N cells. The results are shown in Table 1. One can see that the parallelized MT random numbers follow the expected statistical properties.

Parallelization of FORTEC-3D. On SX-7 5 node system, there are two cases to parallelize a code, as shown in Fig. 8. In Fig. 8(a), an HPF process is executed on a PE with a distributed memory. On the other hand, in the model (b), an HPF process is executed on multiple PEs with a shared memory, and each HPF process is further parallelized among those PEs which shares the memory.

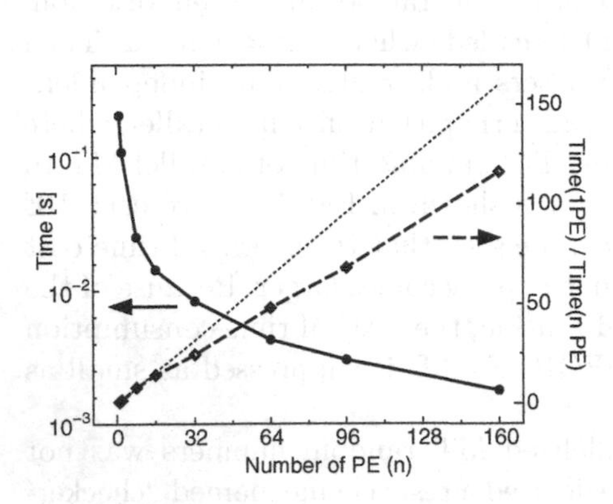

Fig. 5. Time consumption of pseudorandom number generators to generate total 1.924×10^7 random numbers by the parallelized Mersenne twisters of which periods is $2^{19937} - 1$. Solid line is the calculation time and dashed line shows the speedup ratio; Time(1 PE) / Time(n PE), respectively. Dotted line is the ideal maximum of the speedup.

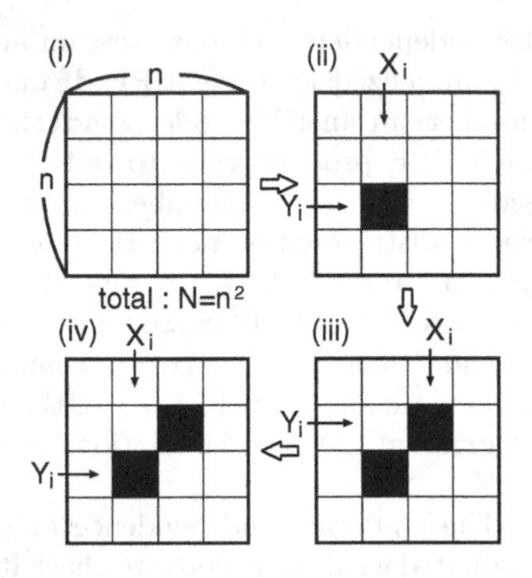

Fig. 6. A diagram of the checker-board test. (i) Initially, all the flags of the cells are off. (ii) Generate two independent random numbers $\{X_i\}$, $\{Y_i\}$ and turn on the flag at (X_i, Y_i). (iii) Continue the procedure and count the number of the flags that are still off. (iv) The flag does not change if (X_i, Y_i) hits the cell of which flag has already been turned on.

Table 1. statistical check for parallel MT

$\|m - E[m]\| \geq$	σ	2σ	3σ	time steps t_{end}
test result of ${}_{160}C_2$ parallelized Mersenne Twisters				
	0.3191	0.04588	0.00288	19.992
expected value if two MT are independent				
	0.3173	0.04550	0.00270	19.408

The shared memory parallelization in an HPF process is automatically done (or using embedded directives like `!CDIR PARALLEL DO`) by SX-7 Fortran compiler. Since non-distributed arrays in HPF program case (b) is shared by several PEs, memory consumption and the number of communication events in the model (b) is expected to be smaller than in the model (a). If we adopt the model (b), however, it is needed to optimize the ratio of HPF processes to shared memory processes in a node (for example 16 : 2, 8 : 4, etc.) to achieve the best running performance of a parallelized code, and coding such a hybrid parallel program is more difficult than in the one-by-one model (a) to tune the parallelization. Moreover, there is a possibility that the bank-conflict will increase in the shared memory parallelization. Therefore, we have adopted the model (a) for FORTEC-3D with HPF. The consumption of memory in the one-by-one model would be

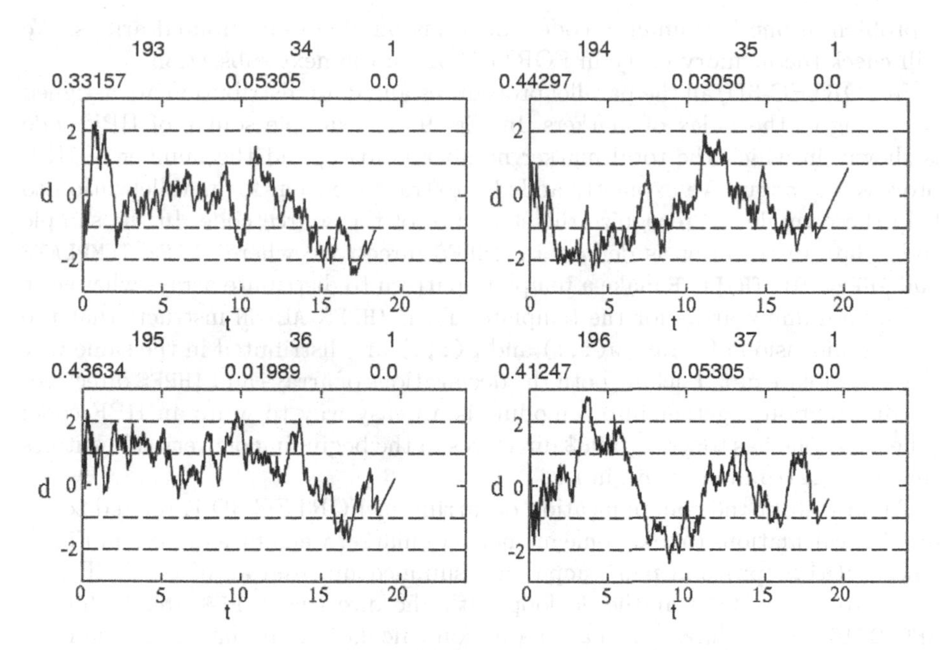

Fig. 7. Example of the results of checker-board test (between parallel MT #1 and #34 to 37). The horizontal axis is the normalized time step $t = i/N$ and the vertical axis is the normalized deviation of the number of remaining cells from the mean $d(t) = (m(t) - E[m(t)])/\sigma(t)$. The triplet on each graph are the proportions of time steps on which $m(t)$ deviates $\pm\sigma$, $\pm 2\sigma$, $\pm 3\sigma$ from the mean, respectively.

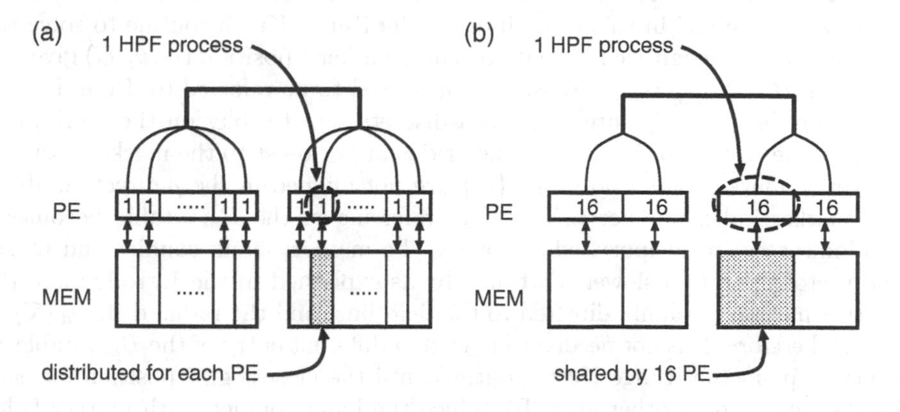

Fig. 8. A diagram of distributing processors and memory in running HPF parallelized code. (a) an HPF process is executed on a PE with a distributed memory.(b) HPF processes with shared memory parallelization on each HPF process. Note that the diagram is truncated to 2 nodes though actual system has 5 nodes.

a problem if one is running a code which has large non-distributed arrays. We will check the memory usage in FORTEC-3D in the next subsection.

In FORTEC-3D, all the parallel procedures and data distribution are assigned according to the index of markers. In Fig. 9, an example source of HPF code is shown, in which the total marker number is `ntot`, and the number of HPF processes is `ncpu`. We explicitly added a extra dimension for parallelization to help the compiler to recognize the structure of the source code. In the sample code, data distribution is defined by `!HPF$` directives, where `!HPF$ TEMPLATE` and `!HPF$ DISTRIBUTE` make a template pattern to distribute arrays where `dst` is only a dummy array for the template. Then `!HPF$ ALIGN` instructs that the second dimension of arrays `w(:,:)` and `p(:,:)` are distributed in the same way as the template `dst`. Packing both the declarations of arrays and `!HPF$` directives for distribution together into a module is an easy way to write an HPF code; otherwise, one has to write `!HPF$` directives in the beginning of every subroutines that uses distributed arrays in it.

Almost all of the communication occurring in FORTEC-3D is related to reduction calculations to take some moments of marker weights, as shown in Fig. 9, where `w(ni,ncpu)` and `p(ni,ncpu)` are summed up from all of the HPF processes to `wpsum(1:2)` at the do loop with the directive `!HPF$ INDEPENDENT, REDUCTION`. Procedures that have reduction calculation are marked in the flowchart in Fig. 4. An optimization of communication is taken in this example by summing up `w` and `p` not into a separate variable, as in the original source, but into the same array `wpsum`. It serves to pack the data to be communicated, and to reduce the overhead time on the communication.

Optimization for vectorization has also been taken. One of the most effective tunings concerns orbit calculation, because about 80% of the total simulation time was consumed in this procedure. In the Runge-Kutta routine to solve the marker motion, magnetic field data on each marker's position $(\rho_i, \theta_i, \zeta_i)$ given by the form $B = \sum_{m,n} B_{m,n}(\rho)\cos(m\theta - n\zeta)$ need to be referred to. Here, Fourier spectrum data $B_{m,n}(\rho)$ are given as a discrete set of tables on the ρ-grid, and each marker refers to the data on the grid that is closest to the marker position. However, marker radial positions $\{\rho_i\}$ are not aligned in the ρ-direction about the marker index number i. Therefore referring of the $B_{m,n}$-table becomes a random access to memory, which causes the memory bank conflict and makes the vector operation slower. Fortunately, as explained in the Introduction, the marker motion is mainly directed to the field line, and the radial drift $\mathbf{v}_d \cdot \nabla\rho$ is slow. Therefore, It is not needed to refer to a different entry of the $B_{m,n}$-table on every steps in the Runge-Kutta routine until the closest grid position for each marker moves to another grid. To reduce the bank conflict, each marker holds the field data $B_{m,n}(\rho)$ on the closest radial grid, and renews it only when the closest grid has changed. This optimization makes orbit calculation time almost twice faster compared with the original version.

Original source

```
module VARIABLES
   real, allocatable(:) :: w,p
   integer ntot
end module

program main
use VARIABLES
        .
        .
read(5)ntot
allocate (w(ntot),p(ntot))
        .
        .
call reduce_wp(wsum,psum)
        .
        .

subroutine reduce_wp(wsum,psum)
use VARIABLES
wsum=0.0
psum=0.0
do i=1,ntot
    wsum=wsum+w(i)
    psum=psum+p(i)
end do
return
        .
        .
        .
```

HPF parallel source

```
 module VARIABLES
    real, allocatable(:,:) :: w,p
    integer ntot,ni
    parameter :: ncpu=32
!HPF$ TEMPLATE dst(ncpu)
!HPF$ DISTRIBUTE dst(block)
!HPF$ ALIGN (*,i) with dst(i) : w,p
 end module

 program main
 use VARIABLES
 real wpsum(2)
        .
 read(5)ntot
 ni=ntot/ncpu
 allocate (w(ni,ncpu),p(ni,ncpu))
        .
        .
 call reduce_wp(wpsum)
        .
        .
 subroutine reduce_wp(wpsum)
 use VARIABLES
 real wpsum(2)
 wpsum(:)=0.0
!HPF$ INDEPENDENT, REDUCTION(+:wpsum)
 do nd=1,ncpu
    do i=1,ni
       wpsum(1)=wpsum(1)+w(i,nd)
       wpsum(2)=wpsum(2)+p(i,nd)
    end do
 end do
 return
        .
        .
```

Fig. 9. Example of HPF source code and its original code to parallelize some reduction calculations. Here, `ntot` is the total marker number and `ncpu` is the total number of HPF processes, respectively. This is a calculation of the sum of marker weights `w` and `p`, which is a typical procedure in FORTEC-3D.

2.3 Benchmark Results

In this section, benchmark results of simulations, both in 2D (tokamak) and 3D (LHD) configurations, are shown. The total marker number is `ntot` $= 1.344\times10^7$ for 2D cases and 3.072×10^7 for 3D cases. The radial electric field is calculated on 60 radial mesh points. $(20, 20, 1)$ meshes in (ρ, θ, ζ) for 2D cases ($(20, 20, 10)$ for

3D cases) and (20, 10) meshes in $(v_\parallel, v_\perp)$ are used. To benchmark FORTEC-3D by changing number of nodes, we have chosen a somewhat small `ntot` here compared with that used in a practical run. In table 2, number of nodes and PEs, marker number per 1 PE (`ni`), total simulation time, total FLOP count, and communication time are shown. These values were measured by using FTRACE and PROGINF run-time options of the Fortran compiler on SX-7. The total performance of floating point operation and total simulation time on each run are shown in Fig. 10, where the GFLOPS value is total FLOP count/total simulation time. One can see that the GFLOPS value is almost linearly growing with the number of PEs, which indicates the good efficiency of parallelization of FORTEC-3D with HPF. The vector length and vector operation ratio of the benchmark runs observed by FTRACE are about 254 and 98.0%, and they hardly change as the number of PEs changes. Therefore, it can be said that HPF has good affinity for a vector computer. The maximum performance of FORTEC-3D reaches 369 GFLOPS on run #9. In fact, GFLOPS value becomes higher if more markers are used, because vector length becomes longer while the amount of data communicated does not change much by changing `ntot`. The fastest run we had ever done was the one in which `ntot` $= 6.4 \times 10^7$ was used in a 3D case, and it reached 417GFLOPS, which is about 30% of the peak performance.

Table 2. Description of benchmark runs

two-dimensional case, `ntot` $= 1.344 \times 10^7$					
	run #1	run #2	run #3	run #4	run #5
# of node	1	2	2	3	5
# of PE (ncpu)	32	32 (16×2)	64	96	160
`ni`$(\times 10^4)$	42	42	21	14	8.4
tot. time (s)	3194	2499	1511	1126	762
tot. FLOP count $(\times 10^{14})$	2.259	2.259	2.260	2.261	2.262
comm. time (s)	502	71	157	206	195
three-dimensional case, `ntot` $= 3.072 \times 10^7$					
		run #6	run #7	run #8	run #9
# of node		2	2	3	5
# of PE (ncpu)		32 (16×2)	64	96	160
`ni`$(\times 10^{14})$		96	48	32	19.2
tot. time (s)		8563	5208	3500	2309
tot. FLOP count $(\times 10^{14})$		8.247	8.255	8.264	8.280
comm. time (s)		101	412	412	365

Next, the ratio of communication time to the total simulation time is shown in Fig. 11. It becomes larger as the number of PEs becomes larger. Although we have optimized the reduction communications that appear in FORTEC-3D by packing the communication data, the increase of the time consumption for communication in simulation runs with many HPF processes is inevitable. It is to be noted that the ratio is very large in run #1, in which only one node is used. It seems that the bank conflict or the imbalance of calculation time among

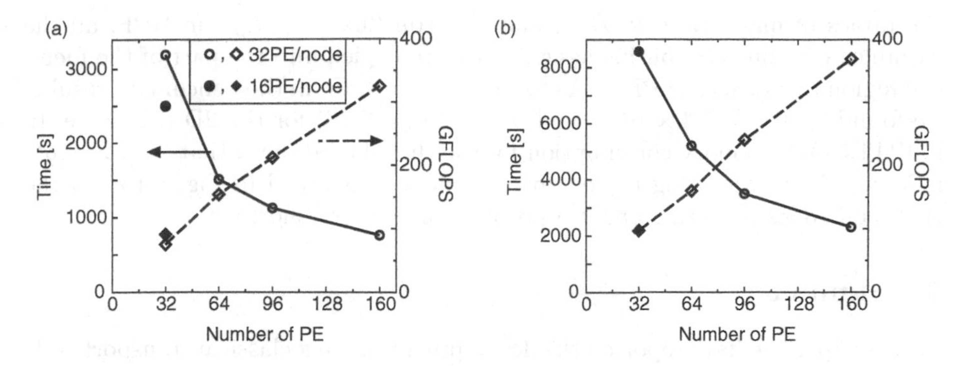

Fig. 10. Total simulation time (solid line) and GFLOPS value (dashed line) of the benchmarks for the two-dimension case (a) and three-dimension case (b). Filled marks are the results of run #2 and #6 in the table 2 in which only 16 of 32 PE in a node were used, while in the other runs full 32 PE in each node were used.

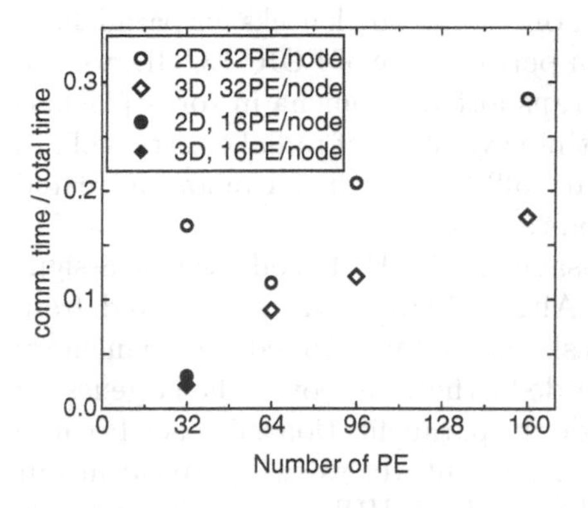

Fig. 11. Ratio of communication time over total simulation time

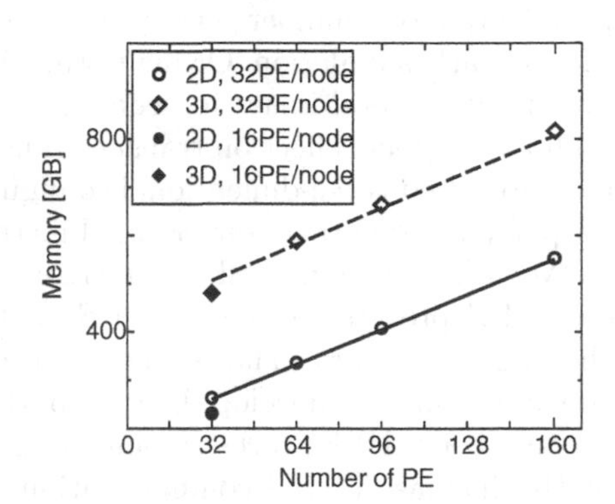

Fig. 12. Memory used in the benchmark calculations. Solid and dashed lines are obtained from the fitting formula eq. (4).

HPF processes at the orbit calculation increases in run #1, which results in the increase of waiting time of the reduction communication at the subroutine that solves the time evolution of radial electric field. We have not found the reason why the bank conflict or the imbalance of calculation time increased so much in run #1, and the problem is still under investigation.

Finally, the memory usage of FORTEC-3D is shown in Fig. 12. One can see that it changes proportional to the number of PEs. Since the total number of markers is fixed in changing the number of PEs, the total amount of memory used in the simulation is expected to be described in the following form

$$\text{total memory (GB)} = \alpha \times \mathtt{ntot} + \beta \times \mathtt{ncpu}, \tag{4}$$

where the coefficients α and β are related to the memory usage by distributed data (*ex.* marker positions and weights) and non-distributed common data (*ex.*

the tables of magnetic field $B_{m,n}$ and electron flux $\Gamma_e(\rho, E_\rho)$; in HPF, all the variables without distributing assignments are copied into the part of the memory region to which each PE refers to), respectively. From the benchmark results, we found that $\alpha = 1.4 \times 10^{-5}$, and $\beta = 2.26$ (= 2.36) for the 2D (3D) case. In FORTEC-3D, memory consumption by the distributed data; that is, $\alpha \times$ `ntot`, is large. Therefore, using the one-by-one model explained in Fig. 8 to run the HPF code does not bring about a problem of memory shortage.

3 Summary

In this paper, we have reported the development of a neoclassical transport code FORTEC-3D on a vector-parallel supercomputer. It is shown that, because of the good parallelization efficiency of the code written in HPF language and the high affinity of HPF language for vector supercomputer, we have succeeded to obtain high performance in running a large scale simulation on the multi-node system of SX-7 supercomputer. We have also adopted and optimized a scheme to generate parallel random number generator to avoid one of the bottlenecks for parallelized Monte Carlo simulation. Owing to the high performance of FORTEC-3D, we can investigate several issues in neoclassical transport phenomena in torus plasmas such as non-local effect on transport and time evolution of radial electric field in a complicated three-dimensional configuration[8,14,17], which cannot be treated properly by the other conventional methods.

We have benchmarked the memory usage in the HPF code which assigns one HPF process to one PE in Sec. 3. Although we have not suffered from shortage of memory so far, some measures should be taken to reduce the memory consumption if we develop the transport code further. Moreover, the largeness of the number of HPF processes would degrade the parallelization efficiency because of the increase of the communication time. In future, we will transform our code by adopting a hybrid parallel model using both HPF and shared memory parallelization as shown in Fig. 8 (b), and the performance of these codes will be compared with the present version of FORTEC-3D.

Acknowledgements

One of the authors (S. S.) would like to thank Prof. C. D. Beidler in Max-Planck Institute for offering us GSRAKE code, and Mr. Yasuharu Hayashi in NEC Corporation for offering helpful information about HPF compiler and SX-7 system. This work is performed under the auspices of the NIFS Collaborative Research Program, No. NIFS05KDAD004 and No. NIFS05KNXN040.

References

1. Helander, P., Sigmar, D.J.: Collisional Transport in Magnetized Plasma. Cambridge Univ. Press, Cambridge (2002)
2. Balescu, R.: Transport Process in Plasmas, vol. 1, 2. Elsevier Science Publishers, Amsterdam (1988)

3. Iiyoshi, A., et al.: Fusion Technol. 17, 169 (1990)
4. Wakatani, M.: Stellarator and Heliotron devices, p. 291. Oxford Univ. Press, Oxford (1998)
5. Shaing, K.C., Houberg, W.A., Strand, P.I.: Phys. Plasmas 9, 1654 (2002)
6. Helander, P.: Phys. Plasmas 7, 2878 (2000)
7. Satake, S., Okamoto, M., Sugama, H.: Phys. Plasmas 9, 3946 (2002)
8. Satake, S., et al.: Conference Proc. 20th IAEA Fusion Energy Conf. Villamoura, Portugal, November 1-6, 2004, TH/P2-18 (International Atomic Energy Agency, Vienna, Austria, 2005) (CD-ROM)(2004)
9. Wang, W.X., et al.: Plasma Phys. Control. Fusion 41, 1091 (1999)
10. Brunner, S., et al.: Phys. Plasmas 6, 4504 (1999)
11. Beidler, C.D., et al.: Plasma Phys. Control. Fusion 37, 463 (1995)
12. Beidler, C.D., Maaßberg, H.: Plasma Phys. Control. Fusion 43, 1131 (2001)
13. High Performance Fortran Forum, High Performance Fortran Language Specification Version 2.0, Springer-Verlag,Tokyo (1997)
14. Satake, S., et al.: J. Plasma Fusion Res. 1, 2 (2006)
15. Matsumoto, M., Nishimura, T.: ACM Transactions on Modeling and Computer Simulation 8, 3 (1998)
16. Matsumoto, M., Nishimura, T.: Dynamic Creation of Pseudorandom Number Generators. In: Monte Carlo and Quasi-Monte Carlo Methods 1998, pp. 56–69. Springer, Heidelberg (2000)
17. Satake, S., et al.: Nuclear Fusion 45, 1362 (2005)

Distributed Parallelization of Exact Charge Conservative Particle Simulation Code by High Performance Fortran

Hiroki Hasegawa*, Seiji Ishiguro**, and Masao Okamoto***

Theory and Computer Simulation Center
National Institute for Fusion Science
322–6 Oroshi-cho, Toki-shi, Gifu, 509–5292

Abstract. A three-dimensional, relativistic, electromagnetic particle simulation code is parallelized in distributed memories by High Performance Fortran (HPF). In this code, the "Exact Charge Conservation Scheme" is used as a method for calculating current densities. In this paper, some techniques to optimize this code for a vector-parallel supercomputer are presented. In particular, methods for parallelization and vectorization are discussed. Examination of the code is also made on multi-node jobs. The results of test runs show high efficiency of the code.

1 Introduction

The Particle-in-Cell (PIC) method is a conventional scheme on plasma computer simulations [1]. In this method, dynamics of full particles (usually, ions and electrons) and time evolutions of self-consistent fields are calculated.

In relativistic, electromagnetic (EM) PIC codes, first, particle positions at the next time step ($\boldsymbol{x}^{n+1}$) are calculated from the old positions $\boldsymbol{x}^n$ and momenta $\boldsymbol{p}^{n+1/2}$. Then, current densities $\boldsymbol{J}^{n+1/2}$ on spatial grids (cells) are computed from $\boldsymbol{x}^n$, $\boldsymbol{x}^{n+1}$, and $\boldsymbol{p}^{n+1/2}$. Further, substituting $\boldsymbol{J}^{n+1/2}$ and $\boldsymbol{B}^{n+1/2}$ in Ampere's law, the self-consistent electric fields at the next time step ($\boldsymbol{E}^{n+1}$) are obtained. Next, using $\boldsymbol{E}^{n+1}$ and $\boldsymbol{B}^{n+1/2}$, the new magnetic fields $\boldsymbol{B}^{n+3/2}$ are given by Faraday's law. Finally, substituting $\boldsymbol{E}^{n+1}$, $\boldsymbol{B}^{n+1/2}$, and $\boldsymbol{B}^{n+3/2}$ into the relativistic equation of motion, particle momenta are advanced. This cycle is continued.

Although there are several methods for calculating time evolutions of the fields, finite-difference Ampere's and Faraday's laws are used in this paper. Further, we apply the "Exact Charge Conservation Scheme [2,3]" for the calculation of current densities. In this scheme, current densities are computed with rigorous satisfaction of the finite-difference continuity equation.

---

* Present address: Earth Simulator Center, Japan Agency for Marine-Earth Science and Technology, 3173–25 Showa-machi, Kanazawa-ku, Yokohama 236–0001, Japan.

** Present affiliation: Department of Simulation Science, National Institute for Fusion Science.

*** Present affiliation: College of Engineering, Chubu University.

J. Labarta, K. Joe, and T. Sato (Eds.): ISHPC 2005 and ALPS 2006, LNCS 4759, pp. 358–364, 2008.

This paper describes a coding method of optimizing the above scheme for a vector-parallel supercomputer. In actual calculation, we use the SX-7 supercomputer (NEC Corporation), which is used at the Theory and Computer Simulation Center, National Institute for Fusion Science (NIFS). This machine has five nodes, and one node consists of 32 processor elements and a shared memory of 256 GB. The total theoretical peak performance of the system is 1.4 TFLOPS.

To adapt multi-nodes, we use not only automatic parallelization in shared memory on each node but also manual parallelization in distributed memories among nodes. For this purpose, we write simulation code with High Performance Fortran (HPF), which is a distributed parallel processing language [4].

In Section 2, we present methods for parallelization and vectorization. In Section 3, the performance of our code is examined. In Section 4, we summarize our work.

2 Optimization

2.1 Parallelization

Our PIC code is formed from four subroutines. The first one calculates particle positions at the next time step and current densities on spatial grids (subroutine CURRENT). The second one computes time evolutions of electric fields through the finite-difference Ampere's law (subroutine AMPERE). The third one executes calculation of the magnetic fields at the next time step by the finite-difference Faraday's law (subroutine FARADAY). The fourth one advances particle momenta via the relativistic equation of motion (subroutine PUSH). In PIC simulation, the required memory size for particle data (positions, momenta) is usually more than 100 times as large as the one for field data. And then, the total calculation for particles needs more expensive costs of computing time than one for fields.

In our code, subroutines CURRENT and PUSH include the computations about particles. Hence, we now especially optimize these subroutines. Although the calculations for fields (subroutines AMPERE and FARADAY) are not parallelized (that is, all of the processors execute the same subroutines simultaneously), the performance of this code is highly improved by only parallelization of subroutines CURRENT and PUSH with HPF and an automatic parallelizing function of FORTRAN90/SX, which is a Fortran90 compiler for the SX supercomputer. We show an overview of the actual source code of subroutine CURRENT in Fig. 1. (Subroutine CURRENT is described in detail in our previous paper [5].) Here, the variables and arrays in the code denote as follows. The parameters `npe` and `npro` are the total numbers of processors and HPF processes, respectively. The parameter `neop` is equal to `neo / npe`, where `neo` denotes the total number of particles. The parameter `nb` satisfies the relation `neo` = `npe` × `nb` × `veclen`, where `veclen` is set at a maximum vector register length of 256. The arrays `x` and `px` are the particle position and momentum, respectively. The array `curx` represents the current densities on grids. The arrays

```
subroutine current(x,px,...,curx,...)
!hpf$ processors pro(npro)
      double precision, dimension(neop,npe) :: x,px
        :
        :
!hpf$ distribute (*,block) onto pro :: x,px
!hpf$ distribute (*,*,*,block) onto pro :: jx0
        :
        :
!hpf$ independent, new(ii,iv,m,lx0,ly,lz,k,j,...)
!cdir parallel do
!cdir& private(ii,iv,m,lx0,ly,lz,k,j,...)
      do ip = 1, npe
      do ii = 1, nb
!cdir nodep
!cdir noinner
!cdir shortloop
        do iv = 1, veclen
          m = iv + (ii-1)*veclen
          lx0(iv)   = int( x(m,ip) + 1.0d0 )
          ly(iv,0)  = int( y(m,ip) + 1.0d0 )
          lz(iv,0)  = int( z(m,ip) + 1.0d0 )
            :
            :
        end do
!cdir novector
!cdir noconcur
        do k = -1, 2
!cdir novector
!cdir noconcur
        do j = -1, 2
!-----------------------------------------------------------------------
!      Loop A
!-----------------------------------------------------------------------
!cdir listvec
          do iv = 1, veclen
            :
            :
            wx0 = ...
            :
            :
            jx0(lx0(iv),ly(iv,j),lz(iv,k),ip) =
     &                        jx0(lx0(iv),ly(iv,j),lz(iv,k),ip) + wx0
            :
            :
          end do
        end do
        end do
      end do
      end do
!
!hpf$ independent, new(i,j,k), reduction(+:curx)
      do ip = 1, npe
      do k = ...
      do j = ...
      do i = ...
        curx(i,j,k) = curx(i,j,k) + jx0(i,j,k,ip) + ...
      end do
      end do
      end do
      end do
```

Fig. 1. Source code of subroutine CURRENT

`lx0`, `ly`, and `lz` denote the nearest indices to particle position. The variable `wx0` and array `jx0` are the work variable and array for the current densities.

As shown in Fig. 1, the extent of last rank of arrays for particle data (positions, momenta) is taken to be the total number of processors, `npe`. Then, inserting DISTRIBUTE directive lines (`!HPF$ distribute`) under declaration statements in the main code and subroutines CURRENT and PUSH, they are distributed with respect to the last rank. Further, writing the INDEPENDENT directive line (`!HPF$ independent`) immediately before the outermost DO statements, these loops are parallelized in each HPF process.

In an HPF process, calculations are automatically parallelized in shared memory by the FORTRAN90/SX compiler. That is, an HPF process includes multi tasks, and a task is allocated to one processor. Thus, the total number of generated tasks is taken to be `ntask` = `npe` / `npro`.

To use the "Exact Charge Conservation Scheme", the double loop that is `k = -1, 2` and `j = -1, 2` before Loop A in Fig. 1 is necessary although it is not needed in a conventional method for calculating current densities. Because of those newly added loops, the values which are calculated before Loop A and used in Loop A (for example, indices for `jx0`) are repeatedly calculated to the same values, and it's wasteful. Thus, we store these values to arrays (e.g. `lx0`, `ly`, and `lz`) before Loop A and just refer them within the loops. As the outermost loop contains complex multiple loops, it is not automatically parallelized by the FORTRAN90/SX. Thus, we inserted the compiler directive to force parallelization (`!cdir parallel do`) [6]. Finally, it is parallelized in both distributed memories and shared memory.

2.2 Vectorization

To make the code more efficient, it is also important to aim at effective vectorization. Therefore, we take the loop count of the innermost loop to be a maximum vector register length. Further, we add some directive lines above some loops, to promote vectorization. (Directive lines shown in Fig. 1 are explained in detail in Ref. [7].)

On the other hand, Loop A in Fig. 1 is generally not vectorized, due to uncertainty of dependency between a particle number and an index of its nearest cell. Thus, in order to vectorize this loop, work arrays, whose first rank is declared equal to the loop count, are usually implemented. This vectorization scheme, however, requires large memories. Hence, in our code, using the LISTVEC compiler directive lines (`!cdir listvec`), we vectorize Loop A saving memories. (This directive option is described in detail in Refs. [7] and [8].)

3 Examination of Code

We then make examinations of the performance of this code. We show, in Table 1, the results of test runs. The parameters of these runs are as follows. The total number of cells in the system is $64 \times 64 \times 64$. Both the numbers of electrons

Table 1. Results of test calculations. Here, the total number of cells, simulation particles, and generated tasks are 64 × 64 × 64, 113,246,208, and 16, respectively. Also, simulation time is 100 steps.

		RUN1	RUN2	RUN3	RUN4	RUN5	RUN6
Total PE (`npe`)		16	32	48	64	96	144
Node		1	1	3	2	3	5
HPF process (`npro`)		1	2	3	4	6	9
Conc. Time (PE$\geq$ 1) (sec)		1149.03	630.25	494.34	324.27	232.55	163.80
Conc. Time (PE= 16) (sec)		1144.76	594.28	466.09	307.61	209.48	149.76
Memory Size (GB)		10.8	12.9	14.8	17.0	21.0	26.9
Vector Length		255.7	255.3	254.4	254.5	253.8	252.7
Vector Ratio (%)		99.94	99.93	99.91	99.92	99.90	99.87
Execution Time	CURRENT	744.33	398.03	311.35	203.90	145.87	108.33
	PUSH	403.05	219.11	173.96	109.06	73.54	50.67
	AMPERE	0.075	0.082	0.097	0.083	0.081	0.120
	FARADAY	0.097	0.105	0.126	0.108	0.105	0.126
Ratio of performance to theoretical peak one (%)		19.7	17.9	15.3	17.5	16.2	15.4

and ions per cell are 216; that is, the total number of simulation particles is 113,246,208. The total number of generated tasks is taken to be `ntask`= 16. The test calculations have been done for 100 time steps. Concurrent times and information of vector operation are obtained from run-time information of the system monitor facility.

The results of calculation time indicate that the distributed parallelization with HPF is very successful. Also, this table shows that the difference between concurrent times (Conc. Time) with one processor or over, and with 16 processors, is quite small. This means that the code achieves the highly parallelized state in shared memory. Since subroutines CURRENT and PUSH occupy most of the calculation time as shown in the table of execution time, the parallelization becomes effective.

Because the last rank extent of the work array `jx0` is `npe`, and the field data are copied to each HPF process, the memory size gradually increases with the total number of processors or HPF processes. Further, the observed vector length is almost equal to a maximum vector register length, and vector ratios are close to 100 percent. The performance is hardly deteriorated as the total number of HPF processes increases.

Fig. 2 shows the dependence of the computing performance (GFLOPS) on the total number of processors. Here, the value of FLOPS is calculated from dividing the total FLOP count by the longest concurrent time in HPF processes, and the total FLOP count is also obtained by program execution analysis information listing function of FORTRAN90/SX [7]. The value of FLOPS linearly increases with the total number of processors.

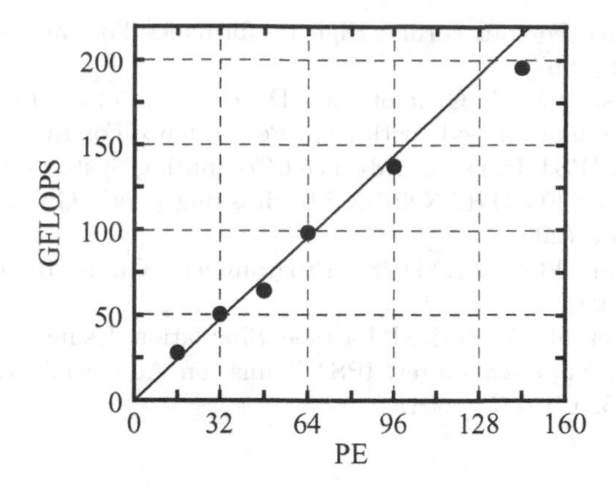

Fig. 2. Dependence of the computing performance (GFLOPS) on the total number of processors (PE). Here, the closed circles are the observed values in RUN1∼6, and the solid line denotes the value of 17% of theoretical peak performance.

4 Summary

We have parallelized a three-dimensional, relativistic, electromagnetic particle simulation code that use the "Exact Charge Conservation Scheme." We have succeeded in optimization of this scheme for vector-parallel supercomputers.

To make the code adequate at multi-node jobs, we have used not only an automatic parallelizing function of FORTRAN90/SX in shared memory but also HPF for distributed parallelization. In vectorization, inserting the LISTVEC directive line, the required memory size can be saved. From the results of test runs, it is found that parallelization and vectorization are quite successful.

Acknowledgments

The authors are grateful to Prof. R. Horiuchi (NIFS), Dr. H. Ohtani (NIFS), and Mr. N. Horiuchi (NEC) for stimulating discussions.

This work is performed with the support and under the auspices of the NIFS Collaborative Research Program (NIFS04KDAT007) and supported in part by a Grant-in-Aid for Scientific Research from the Ministry of Education, Culture, Sports, Science and Technology.

References

1. Birdsall, C.K., Langdon, A.B.: Plasma Physics via Computer Simulation. Adam Hilger (1991)
2. Villasenor, J., Buneman, O.: Rigorous charge conservation for local electromagnetic field solvers. Comput. Phys. Comm. 69, 306–316 (1992)
3. Esirkepov, T.Z.: Exact charge conservation scheme for Particle-in-Cell simulation with an arbitrary form-factor. Comput. Phys. Comm. 135, 144–153 (2001)

4. High Performance Fortran Forum: High Performance Fortran Language Specification Version 2.0 (1997)
5. Hasegawa, H., Ishiguro, S., Okamoto, M.: Development of the Efficient Electromagnetic Particle Simulation Code with High Performance Fortran on a Vector-parallel Supercomputer. IPSJ Trans. on Advanced Computing Systems 46, 144–152 (2005)
6. NEC Corporation: FORTRAN90/SX Multitasking User's Guide, Revision No. 11, NEC Corporation (2002)
7. NEC Corporation: FORTRAN90/SX Programmer's Guide, Revision No. 13, NEC Corporation (2002)
8. Sugiyama, T., et al.: Vectorized Particle Simulation Using LISTVEC Compile-directive on SX Super-computer. IPSJ Trans. on Advanced Computing Systems (in Japanese) 45, 171–175 (2004)

Pipelined Parallelization in HPF Programs on the Earth Simulator*

Hitoshi Murai[1] and Yasuo Okabe[2]

[1] 1st Computers Software Division, NEC Corporation,
10, Nisshin-cho 1-chome, Fuchu, Tokyo 183-8501, Japan
murai@hpc.bs1.fc.nec.co.jp
[2] Academic Center for Computing and Media Studies, Kyoto University,
Yoshida-Honmachi, Sakyo-ku, Kyoto 606-8501, Japan
okabe@i.kyoto-u.ac.jp

Abstract. There is no explicit way for parallelization of DOACROSS loops in the HPF specifications. Although recent advanced HPF compilers such as HPF/ES have been as powerful as MPI in many situations of parallel programming, many of them do not have the capability of pipelining DOACROSS loops. We propose a new extension for pipelined parallelization, the `PIPELINE` clause, and have developed a preprocessor, named *HPFX*, that translates an HPF source program annotated by the `PIPELINE` clause into a normal HPF one, to evaluate the effectiveness of the clause. Evaluation on the Earth Simulator shows that pipelined parallelization in implementations of the NPB LU benchmark with HPFX and HPF/ES outperforms the hyperplane parallelization in the conventional HPF implementations of the benchmark.

1 Introduction

Twelve years have passed since the first High Performance Fortran (HPF) language specification was published [1]. Not a few HPF compilers have been developed and some are provided on *real* supercomputers like the Earth Simulator(ES) [2]. The language itself has been evolving; the specifications of HPF2.0 [3] and its approved extensions were standardized in 1997, and the HPF/JA extensions [4] were published in 1999. HPF/ES, the HPF compiler for the ES, also supports some unique extensions, such as functionality for irregular problems, parallel I/O, etc. [5].

Although recent advanced HPF supercompilers have been as powerful as MPI in many situations of parallel programming [6], there are still some limitations in their parallelization capability. One of such limitations is in parallelization of DOACROSS loops, which is a `DO` loop that has loop-carried dependence in it. It is known that a DOACROSS loop that has only loop-carried dependence of a fixed distance along some axes of arrays in it can be parallelized by *pipelined parallelization* [7]. HPF has a directive `INDEPENDENT` that asserts a loop has no

* This work was a part of Earth Simulator Research Projects in FY2004.

J. Labarta, K. Joe, and T. Sato (Eds.): ISHPC 2005 and ALPS 2006, LNCS 4759, pp. 365–373, 2008.

loop-carried dependence in it and can be parallelized. NEW or REDUCTION clauses can be added on an INDEPENDENT directive to provide additional information on some types of dependence that does not hinder parallelization and to increase a chance of the compiler parallelizing the loop. But there is no explicit way in which a programmer can give information on dependence of general DOACROSS loops to the compiler. Some of the existing HPF compilers automatically detect the dependence in a loop and parallelize it in a pipeline fashion [8,9], but Lewis and Snyder [10] reported that even such compilers failed to parallelize automatically a number of loops, e.g., those which iterate over indices in reverse order.

In this paper we propose the PIPELINE clause, which is a new extension to the HPF specification. The clause asserts that there is a loop-carried dependence of a fixed distance on the target array, and that the compiler can parallelize the loop in a pipeline fashion. In order to evaluate the effectiveness of the extension, we have developed a preprocessor, named *HPFX*, which translates an HPF source program annotated by the PIPELINE clause into a normal HPF one. We have parallelized the NPB LU benchmark with HPFX and HPF/ES, and evaluated its performance on the ES with some other implementations such as by the conventional hyperplane method for comparison.

The remainder of this paper is organized as follows. Section 2 gives syntax and semantics of the PIPELINE clause we propose. Section 3 describes the function of the HPFX preprocessor. The evaluation on the ES is shown in Section 4. Finally, Section 5 provides the concluding remarks.

2 The PIPELINE Clause

A PIPELINE clause on an INDEPENDENT directive is an assertion to an HPF compiler that there is a loop-carried dependence of a fixed distance on the target array in the following loop. The compiler generates communications and synchronizations to parallelize the annotated loop in a pipeline fashion.

2.1 Syntax

The syntax rules of the INDEPENDENT directive (J301 in Section 3.1.1 of the HPF/JA language specification) are modified as follows:

```
X101 independent-directive-x  is INDEPENDENT [ , new-clause ]
                                      [ , reduction-clause-ja-list ] [ , pipeline-clause ]
X102 pipeline-clause          is PIPELINE( pipeline-spec-list )
X103 pipeline-spec            is pipeline-target [ ( pipeline-region-list ) ]
                                      (pipeline-width-list )
X104 pipeline-target          is object-name
X105 pipeline-region          is int-expr
                              or [ int-expr ]:[ int-expr ]
X106 pipeline-width           is int-expr
```

Constraint: The length of a *pipeline-region-list* must be equal to the rank of the *object-name* mentioned as a *pipeline-target*.

Constraint: The length of a *pipeline-width-list* must be equal to the rank of the *object-name* mentioned as a *pipeline-target*.

Constraint: An *object-name* mentioned as a *pipeline-target* must have a shadow width no smaller than the absolute value of an *int-expr* mentioned as a *pipeline-width*, at the low end if the *int-expr* is positive, or at the high end if the *int-expr* is negative, in each of the axes.

2.2 Semantics

The INDEPENDENT directive asserts that the iterations of a DO loop do not mutually interfere (in Section 5.1 of the HPF specification). The PIPELINE clause relaxes the condition of this interference by adding the following exception.

- Exception: If a variable appears in a PIPELINE clause, then operations assigning values to it in separate iterations of the DO loop do *not* interfere.
- Exception: If a variable appears in a PIPELINE clause, then operations assigning values to it in one iteration of the DO loop do *not* interfere with uses of the variable in other iterations.

The compiler can parallelize the INDEPENDENT loop in a pipeline fashion by generating communications of the target array and synchronizations on the basis of the information from a *pipeline-region* and a *pipeline-width*.

The *pipeline-region-list*, if any, represents the array section really accessed in the following loop. Only elements in the section are to be communicated between two neighboring processors at runtime. The *pipeline-width-list* represents the width and the direction of a shift communication to be generated along each axis of the array section. You can specify a multidimensional pipelined parallelization with this clause.

3 The HPFX Preprocessor

The HPFX preprocessor translates an HPF source program annotated by the PIPELINE clause into a normal HPF one; however, the current implementation of HPFX can accept the clause with some restrictions.

Here we describe how HPFX works with a simple example below.

```
      REAL A(100,100)
!HPF$ PROCESSORS P(4,4)
!HPF$ DISTRIBUTE A(BLOCK,BLOCK) ONTO P
!HPF$ SHADOW A(1:0,1:0)

!HPF$ INDEPENDENT, PIPELINE(A(1:50,:)(1,1))
      DO J=2, 100
!HPF$   INDEPENDENT
        DO I=2, 50
!HPF$     ON HOME(A(I,J))
          A(I,J) = A(I,J) + A(I-1,J-1)
        END DO
      END DO
```

The PIPELINE clause in the code means that there is a loop-carried dependence of distance one along the first and the second axes of a two-dimensional array A, and that a row and a column of elements on the distribution boundary over the array section A(1:50,:) is to be sent to and received from the neighbor processor.

HPFX translates the above code into the following one.

```
      CALL RECV2_REAL(A, WDTH, 2, LB, UB)

!HPF$ INDEPENDENT
      DO J=2, 100
!HPF$   INDEPENDENT
        DO I=2, 50
!HPF$     ON HOME(A(I,J)), LOCAL(A)
          A(I,J) = A(I,J) + A(I-1,J-1)
        END DO
      END DO

      CALL SEND2_REAL(A, WDTH, 2, LB, UB)
```

You can see that one call statement is inserted before the loop, and another after the loop. The subroutines RECV2_REAL and SEND2_REAL are generated automatically from a template by HPFX for the type and rank of the target array, and linked with other HPF subroutines later by HPF/ES. They are written as extrinsic procedures of kind HPF_LOCAL, each of which is embedded with calls to MPI subroutines for communication and synchronization. The ON-HOME-LOCAL directive [4] asserts that no communication is required for the loop, so HPF/ES does nothing for parallelizing the loop except distributing its iteration space onto the processors on the basis of the mapping of the HOME array A.

At runtime in the two subroutines, they get information on mapping of the target array from the mapping inquiry subroutines GLOBAL_ALIGNMENT and GLOBAL_DISTRIBUTION, contained by the HPF local routine library; they construct a communication schedule for pipelined parallelization of the width given by the second argument, and perform communication on the schedule.

The runtime sequential behavior of each processor for the example code is as follows:

(0) The processor P(1,1) goes through RECV2_REAL, while the processors in P(1,2:4) or P(2,:) wait here for arriving data from their neighbor. The processors in P(3:4,:) skip all of the following steps and proceed to the next because they have no task in the loop.
(1) P(1,1) performs its local part of the loop to process the array A.
(2) P(1,1) sends A(25,1:25) to P(2,1) and A(1:25,25) to P(1,2) in SEND2_REAL, and proceeds to the next.
(3) These substeps are executed in parallel.
  (3–1) P(2,1) receives A(25,1:25) from P(1,1) and exits from RECV2_REAL.
  (3–2) P(1,2) receives A(1:25,25) from P(1,1) and exits from RECV2_REAL.
(4) These substeps are executed in parallel.
  (4–1) P(2,1) performs its local part of the loop to process the array A.

(4-2) `P(1,2)` performs its local part of the loop to process the array `A`.
(5) ···

Fig. 1 illustrates these steps. Note that the processors that have finished or skipped these steps earlier can execute the next statements asynchronously, in parallel with the other processors still executing these steps.

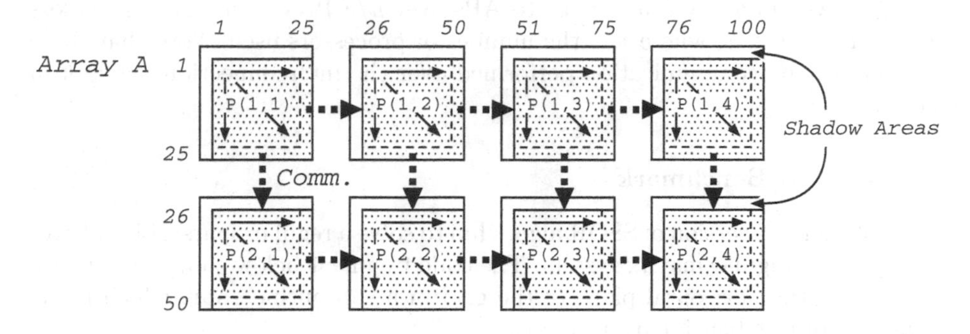

Fig. 1. Pipelined Parallelization on a Two-Dimensional Array

This implementation depends on the fact that HPF/ES does *not* generate global barrier synchronization on the exit of local procedures, such as `RECV2_REAL` and `SEND2_REAL` above, though the HPF specification states that an HPF compiler must do it.

4 Evaluation

We parallelized the LU benchmark (class C) of the NAS Parallel Benchmarks (NPB) [11] with HPFX and HPF/ES in two manners: one is one-dimensional pipelined parallelization, and the other is two-dimensional. We ran the programs on the ES to evaluate the performance of the two manners with those of some other implementations for comparison.

4.1 Evaluation Environment

The ES is a distributed-memory vector parallel supercomputer, composed of 640 processor nodes (PNs) connected by a single-stage crossbar network providing a 12.3GB/s bidirectional bandwidth. Each PN consists of eight vector arithmetic processors (APs) and has a 16GB shared memory. The peak performance of each AP is 8GFLOPS, summing up to the total peak performance of 40TFLOPS.

The software environment used in the evaluation was as follows:

- `HPF/ES: Rev.2.2.2(1020) 2005/03/23`
- `FORTRAN90/ES Version 2.0: Rev.302 ES 17 2004/06/23`
- `MPI/ES: command Version 7.0.4 (15. October 2004)`
- `MPI/ES: daemon Version 7.0.6 (27. October 2004)`
- `MPI/ES: library Version 6.7.3a (22. November 2004)`

HPF/ES translates an HPF program into an intermediate Fortran+MPI program, which is to be compiled and linked with the MPI library by a Fortran compiler to generate an executable. In this evaluation, a set of the compiler options that instructs HPF/ES to do the highest level of optimization was specified.

We ran the programs on the ES through the L-batch queue that is for large-scale batch requests. We adopted flat parallelization: HPF processors were assigned to APs in a PN if $n \leq 8$, or to APs over $n/8$ PNs with eight processors in each PN if $n > 8$, where n is the number of processors used. Note that there is no inter-node communication performed over the interconnection network in the case of $n \leq 8$.

4.2 The LU Benchmark

LU implements a version of SSOR algorithm to solve a regular-sparse, block lower and upper triangular matrix system. The code fragment below shows a sequence of sweeping the horizontal plane of the grid in a subroutine `ssor`, which is the main part of this benchmark program.

```
DO k = 2, nz -1
  call jacld(k)
  call blts(k)
END DO

DO k = nz - 1, 2, -1
  call jacu(k)
  call buts(k)
END DO

call add
call rhs
```

Although the subroutines `jacld`, `jacu`, `add`, and `rhs` are completely data parallel and can be parallelized readily with HPF/ES, both `blts` and `buts` have a limited parallelism because there is a loop-carried dependence of `v(i,j,k)` on `v(i-1,j,k)`, `v(i,j-1,k)` and `v(i,j,k-1)` in `blts`, and on `v(i+1,j,k)`, `v(i,j+1,k)` and `v(i,j,k+1)` in `buts`. Therefore, the `PIPELINE` clauses were inserted into the two subroutines to apply pipelined parallelization to them in both cases of one-dimensional and two-dimensional. Below is a `PIPELINE` clause specified in `buts` in the case of two-dimensional.

```
!HPF$ INDEPENDENT, PIPELINE(v(:,:,:,k)(0,-1,-1,0))
      do j = jend, jst, -1
!HPF$    INDEPENDENT
         do i = iend, ist, -1
!HPF$       ON HOME(v(:,i,j,:)) BEGIN
            ...
            do m = 1, 5
                  tv( m ) = tv( m )
```

```
     > + omega * ( udy( m, 1, i, j ) * v( 1, i, j+1, k )
     >           + udx( m, 1, i, j ) * v( 1, i+1, j, k )
                   ...
             end do
             ...
             v( 1, i, j, k ) = v( 1, i, j, k ) - tv( 1 )
             ...
!HPF$        END ON
          enddo
       end do
```

Without pipelined parallelization, you had to exploit the hyperplane method [12] to parallelize the above loop with HPF. There have been two implementations of LU using the hyperplane method: one is two-dimensional, or a hybrid of hyperplane and pipeline [13]; the other is three-dimensional, adopted in the NPB3.0-HPF [14]. It is known that both of the hyperplane implementations need complicated loop restructuring though neither need any language extensions.

All of our implementations of LU with HPF are based on NPB3.1-SERIAL, and the MPI one is exactly the same as NPB3.1-MPI.

4.3 Evaluation Result

Fig. 2 depicts the evaluation results. The vertical axis of the graph represents the speedup ratio relative to the single-CPU execution of the NPB3.1-SERIAL Fortran implementation.

"HPFX-p1" in the graph is for the result of one-dimensional, and "HPFX-p2" for two-dimensional, both of which are implemented with HPFX. They have little difference in performance in this evaluation.

"HPF-hp2" and "HPF-hp3" illustrate the results of the two- and three-dimensional hyperplane methods, respectively. The former is based on the implementation described in [13] and enhanced by partial `REFLECT` and `EXT_HOME`, which HPF/ES started to support recently; the latter is the same as the NPB3.0-HPF implementation except for a few compiler-dependent modifications. "HPF-hp3" is much faster in the case of a smaller number of processors, because longer vector length on the hyperplane leads to higher vectorization efficiency, but the performance degrades seriously when the number of processors increases. We suppose that it is mainly due to load imbalance.

You can see that both "HPFX-p1" and "HPFX-p2" certainly outperform the two hyperplane implementations in the 64-processors execution; besides pipelined parallelization has the advantage of easy programming over the hyperplane methods, which is revealed by the number of program lines that we added or modified in the subroutines `ssor`, `blts`, and `buts` to parallelize the DOACROSS loops: We added only six HPF directives in each of "HPFX-p1" and "HPFX-p2," while we had to modify or add sixteen Fortran statements, in addition to the six directives, in "HPF-hp2." We found far much more addition and modification of lines in "HPF-hp3." Consequently, these results show that

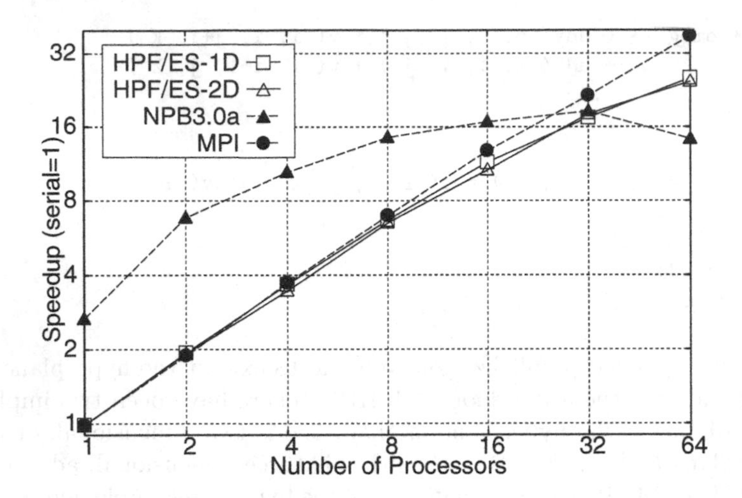

Fig. 2. Evaluation Results

pipelined parallelization gives you a more effective and easier way of parallelizing DOACROSS loops than the conventional hyperplane methods.

On the other hand, neither "HPFX-p1" nor "HPFX-p2" is comparable to "MPI," which is based on the two-dimensional pipeline algorithm that is the same as "HPFX-p2." It is mainly because of the overheads of frequently calling the mapping inquiry subroutines and constructing the communication schedule in the subroutines, such as `RECV2_REAL` and `SEND2_REAL` in Section 3, that HPFX automatically generates before and after a pipelined loop. We have verified this fact from an experimental evaluation, in which we modified `RECV2_REAL` and `SEND2_REAL` by hand so that they should reuse in a run the mapping information and the communication schedule once obtained.

The former overhead would be removed if HPFX could directly refer to the runtime information managed by the compiler, such as an array descriptor, and the latter would if HPFX could exploit the compiler's optimization facility of communication schedule reuse to amortize the cost of schedule construction. However, since HPFX is now not a compiler module but a preprocessor, the overheads are rather difficult to avoid. Such drawbacks of pipelined parallelization should be removed if the compiler supports the functionality of the preprocessor.

5 Concluding Remarks

We proposed the `PIPELINE` clause which is a new extension to HPF for pipelined parallelization, developed a preprocessor that adds the functionality of pipelined parallelization to HPF compilers, and evaluated the effectiveness of such pipelined parallelization in HPF programs on the ES. The evaluation results of the NPB LU benchmark show that pipelined parallelization is superior to the conventional hyperplane methods in both performance and easiness of programming. However, it could not perform as well as the MPI one, because of the preprocessor's limited access to compiler facilities. We expect that this disadvantage is resolved if HPF/ES directly supports the functionality of the preprocessor.

In future work, we plan to evaluate the pipelined parallelization of HPFX on other machines to verify its effectiveness.

References

1. High Performance Fortran Forum: High Performance Fortran Language Specification, Version 1.0 (1993)
2. Habata, S., Umezawa, K., Yokokawa, M., Kitawaki, S.: Hardware system of the Earth Simulator. Parallel Computing 30, 1287–1313 (2004)
3. High Performance Fortran Forum: High Performance Fortran Language Specification, Version 2.0 (1997)
4. Japan Association of High Performance Fortran: HPF/JA Language Specification (1999), http://www.hpfpc.org/jahpf/
5. Yanagawa, T., Suehiro, K.: Software system of the Earth Simulator. Parallel Computing 30, 1315–1327 (2004)
6. Sakagami, H., Murai, H., Seo, Y., Yokokawa, M.: 14.9 TFLOPS Three-dimensional Fluid Simulation for Fusion Science with HPF on the Earth Simulator. In: Proc. of SC 2002, Baltimore, MA (2002)
7. Wolfe, M.: High Performance Compilers for Parallel Computing. Addison-Wesley, Reading (1996)
8. Gupta, M., Midkiff, S., Schonbeg, E., Seshadri, V., Shields, D., Wang, K., Ching, W., Ngo, T.: An HPF compiler for the IBM SP2. In: Proc. of 1995 ACM/IEEE Supercomputing Conference (1995)
9. Nishitani, Y., Negishi, K., Ohta, H., Nunohiro, E.: Techniques for compiling and implementing all NAS parallel benchmarks in HPF. Concurrency and Computation – Practice & Experience 14, 769–787 (2002)
10. Lewis, E.C., Snyder, L.: Pipelining Wavefront Computations: Experiences and Performance. In: Proc of the 5th IEEE International Workshop on High-Level Parallel Programming Models and Supportive Environments (2000)
11. Bailey, D., Barszcz, E., Barton, J., Browning, D., Carter, R., Dagum, L., Fatoohi, R., Fineberg, S., Frederickson, P., Lasinski, T., Schreiber, R., Simon, H., Venkatakrishnan, V., Weeratunga, S.: The NAS Parallel Benchmarks. Technical Report RNR-94-007. NASA Ames Research Center (1994)
12. Lamport, L.: The parallel execution of DO loops. Communications of the ACM 17, 83–93 (1974)
13. Murai, H., Okabe, Y.: Implementation and Evaluation of NAS Parallel Benchmarks with HPF on the Earth Simulator. In: Proc. of SACSIS 2004, Sapporo, Japan (2004)
14. Frumkin, M., Jin, H., Yan, J.: Implementation of NAS Parallel Benchmarks in High Performance Fortran. Technical Report NAS-98-009. NASA Ames Research Center (1998)

Sampling of Protein Conformations with Computers to Predict the Native Structure

Junichi Higo

School of Life Science, Tokyo University of Pharmacy and Life Science,
1432-1 Horinouchi, Hachioji, Tokyo, 192-0392, Japan
Higo@ls.toyaku.ac.jp

Abstract. Native-structure prediction of proteins only from the amino-acid sequential information, without using information from a sequence-structure database of proteins, has not yet been succeeded. Computer simulation is now popular in protein conformational sampling for the prediction. The sampling is, however, hopelessly difficult when a conventional simulation technique (canonical molecular dynamics simulation) is used, because the conformation is frequently trapped in energy minima in the conformational space. This trapping makes the sampling efficiency considerable poor. I explain an efficient conformational sampling algorithm, multicanonical molecular dynamics simulation, recently developed. Results on the sampling of polypeptide chains showed that the conformation easily overcomes the energy barriers between the energy minima with using this method.

1 Introduction

Proteins are polypeptide chains, along which amino acids are connected by peptide bonds. After synthesized in solution, the chains spontaneously fold into their own tertiary structures (i.e., native structure) depending on the amino-acid sequences coded in DNA. The native structure is the most thermodynamically stable conformation (the lowest free-energy conformation) in solution. Thus, the native structure is determined by physico-chemical properties of the amino-acid sequence. Once the tertiary structures made up, the biological function is assigned to the structure. Summarizing, the amino-acid sequence determines the tertiary structure and the function of protein.

According to the quick progress of computer power, a computer simulation of a protein is coming to be popular in the protein-folding research field. The first goal of protein simulation is to physico-chemically predict the native structure only from the amino-acid sequence, without using information from a protein database. If the prediction becomes possible, it is prospective that the folding pathways are understood from the simulation. The second goal (the final goal) is to predict the protein function only from the simulation. After the goals, we can design the protein function starting from the amino-acid sequence, and the simulation technique will be applied to drag design.

J. Labarta, K. Joe, and T. Sato (Eds.): ISHPC 2005 and ALPS 2006, LNCS 4759, pp. 374–382, 2008.

1.1 Difficulty in the Conformational Sampling

In theory, a polypeptide (i.e., protein) chain can take an astronomical number (almost infinity) of possible conformations. This means that the conformational space assessable by the chain is huge. In nature, however, only one conformation is selected as the native structure: the volume assigned to the native structure in the conformational space is small in the vicinity of a point. The gap between the two numbers (one and infinity) leads us to a paradox, known as Levinthal's paradox [1]. If the polypeptide chain searches all of the possible conformations one by one in a human body, the chain cannot reach the native structure before the life of the human is over. Contrary, experiments teach us that many proteins fold into the native structure within a millisecond, starting from a random conformation. This means that the proteins do not search one by one (or randomly) all of the candidates. Many researchers, now, consider that the potential surface of protein has a shape of funnel [2,3], and that the native structure is positioned at the bottom of the funnel (Fig.1). If so, the fluctuating polypeptide conformation in the unfolded state is affected by a bias that carries the conformation toward the bottom of the funnel, and consequently, the protein reaches the native structure in a short time.

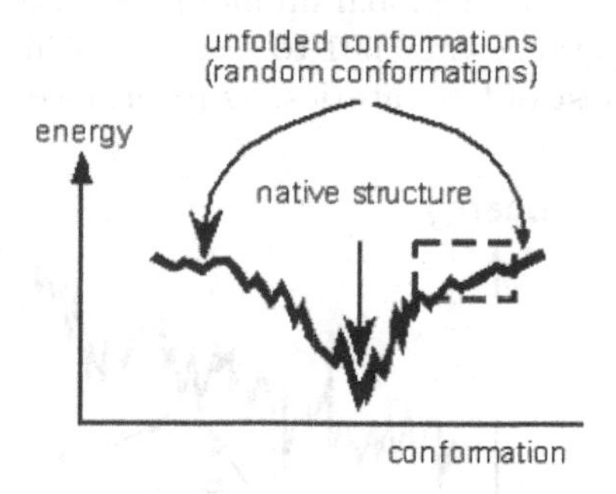

Fig. 1. Schematic drawing of folding funnel as a model of protein potential surface. Polypeptide conformation is originally defined in the multi-dimensional space, although it is represented by x-axis in one-dimensional form. Native structure is at the bottom, and unfolded conformations are on periphery. Rectangular box (*broken line*) is mentioned in caption of Fig.2.

A molecular dynamics (MD) simulation is widely used to numerically solve the Newtonian equation of a system consisting of biomolecules (protein and/or DNA, etc) and solvent molecules (i.e., water molecules and ions). Time step of the simulation is usually set to 1 fs (i.e., 10^{-15} sec), which is a physico-chemical requirement to obtain the simulation trajectory (i.e., series of snapshots) accurate enough. Suppose that the integration of one step of the equation takes a cpu-time of 0.1 sec. This value is likely in many laboratories with using a PC. Then, the entire cpu time for simulating entire protein folding is $0.1 * 0.001/10^{-15} \approx 3,200$ years for a protein, which folds in 1 ms. This estimation indicates that the computer simulation with using the accurate all-atom model is impossible.

There are two approaches to this 3,200-year problem: one is to develop a fast computer, and the other is to develop an efficient sampling algorithm. Besides, the combination of the fast computer and the efficient algorithm is the most effective. The computer speed is growing quickly thanks to efforts of computer companies. The Blue Gene/L or the Earth Simulator is, now, available for simulations of biological systems.

If the potential surface is a perfect funnel, the time scale of protein folding must be shorter by several orders of magnitude than the experimentally observed. Then, the development of the effective sampling method is not important, because the conformation, starting from an unfolded conformation, will straightforwardly reach the native structure. Then, the essentially important task is to develop a fast computer. However, the potential-energy surface is unfortunately highly rugged (Fig.2), even if the global shape of the surface is funnel. It is likely that there are a large number of energy minima in everywhere of the real potential surface of a protein. Then, the conformational sampling encounters the following difficulty: a conformation falling into an energy minimum (point A in Fig.2) spends a long time to escape from the minimum before moving into another minimum (point B in Fig.2). The falling into energy minima is called energy trapping. Due to the ruggedness, a conformation must jump a large number of energy barriers to reach the global minimum. Consequently, the folding time scale of many proteins ranges in 0.001 to 1 sec, for which the folding simulation is not achievable in most of laboratories, as mentioned above.

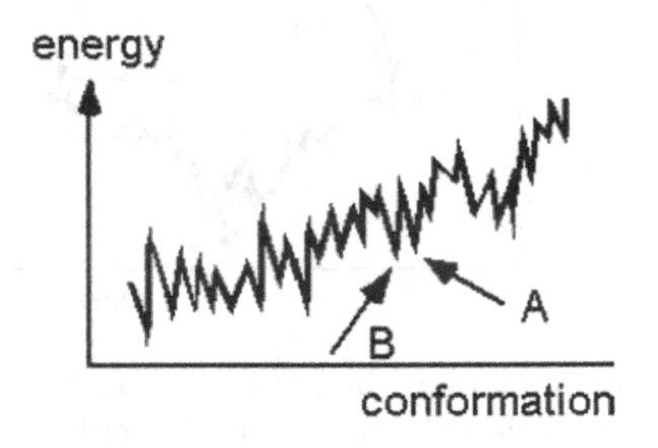

Fig. 2. Schematic drawing of rugged energy surface, which is a close-up of the region in the rectangular box of Fig.1. Two energy minima A and B (*arrows*), which are the nearest neighbor to each other in the conformational space, are mentioned in text.

1.2 Equilibrium Sampling

Suppose that the temperature of the system is set at a room temperature (i.e., a physiologically important temperature), and that the conformation is in the minimum A in Fig.2. The conformation cannot escape from A, if the excitation energy is not enough to climb up the barrier at the room temperature. Heating (i.e., increment of the temperature) can carry the conformation on the top of the barrier. However, the kinetic energy at the high temperature does not give a chance to the conformation to visit energy minima. This means that the conformation never reaches the native structure (or the probability of the conformation folding into the native structure is negligibly small), since the native structure corresponds to the lowest energy (or to ensemble of some low-energy minima). This property of protein is experimentally well known as the heat denaturation. To find the native structure out of a number of energy minima in the conformational space, the sampling should has an ability that the conformation quickly escapes from a minimum and visit another minimum.

One may think that the repetition of heating and cooling may carry the conformation to the native structure. However, in this procedure, the current conformation falls into an energy minimum nearest to the conformation in the conformational space (Fig.3): Starting from a random conformation, which is far from the native structure, the correct folding does not happen, and the conformation repeatedly falls into energy minima belonging to the unfolded state. This process corresponds to a quenching, where the conformation has no time to sample a wide area in the conformational space.

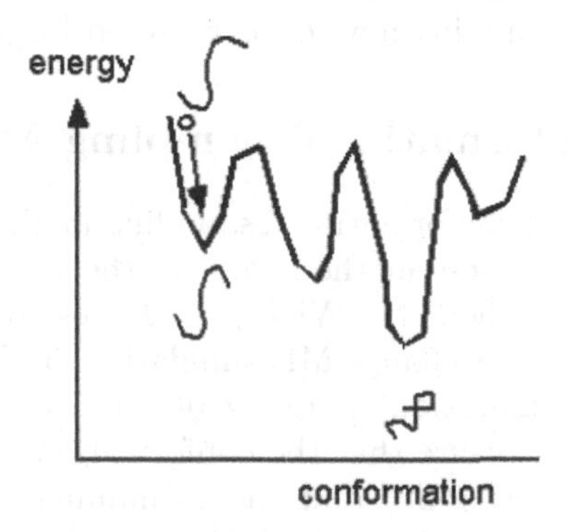

Fig. 3. Schematic drawing of conformational motion (*arrow*) from the current conformation (*open circle*) to an energy minimum nearest to the current conformation. Conformational difference between the current and the nearest minimum-energy conformations is small, which is far from the native structure. The three conformations are also schematically shown.

Imagine a simple conformational space that has two energy minima: an energy minimum M_1 with a wide entrance, where the size of the entrance $= W_1$, and the other minimum M_2 with a narrow entrance, where the size $= W_2$. We suppose, for simplicity of discussion, that the depths of the two energy minima are the same. Tracing the conformational motion for long, the probability, $\rho(M_1)$, of finding the conformation in M_1 should be larger than that, $\rho(M_2)$, in M_2 with a relation of $\rho(M_1)/\rho(M_2) = W_1/W_2$ (Fig.4). The conformational ensemble satisfying this relation is regarded as a physico-chemically acceptable one, called "canonical ensemble".

If a conventional MD simulation is done infinitely long at room temperature, the conformational ensemble generated can be a good approximation of the canonical ensemble, and the ensemble should involve the most thermodynamically stable conformation, the native structure. However, the equilibrium

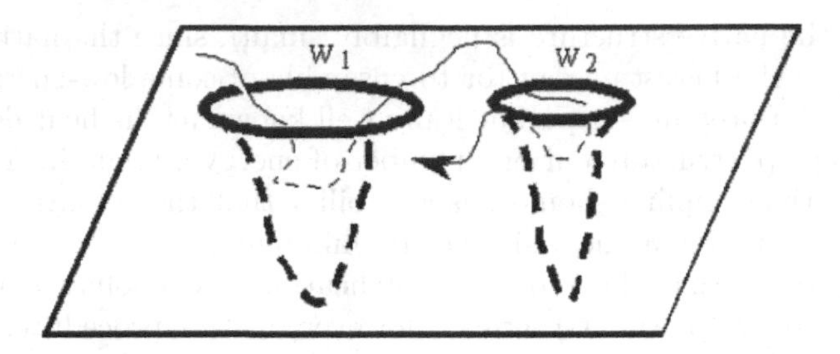

Fig. 4. Schematic drawing of conformational motion (*curved line with arrow*). The conformation first falls into the energy minimum M_1, and second into M_2.

sampling is not equivalent to an effective sampling. Since the conformation is frequently trapped into energy minima in the conventional MD simulation, a long run (i.e., a 3,200-year simulation) is necessarily to reach the native structure. Thus, a new effective sampling method is required for protein folding.

2 Effective Conformational Sampling Methods

Recently, some powerful conformational sampling methods have been developed for the protein folding. These methods are, as the whole, called enhanced conformational sampling methods [4]. Widely used ones are the multicanonical MD simulation [5] and replica exchange MD simulation [6]. Both had been originally developed for the configuration sampling of physical systems, such as a spin system. These methods ensure that the conformation can quickly overcome the energy barriers with visiting different energy minima (i.e., the energy trapping does not occur), and generate the canonical ensemble. I focus on the multicanonical method, below.

2.1 Algorithm

Usually, force f acting on an atom is calculated from a relation $f = -\text{grad } E$, where E is the potential energy of the system, and the derivative is taken with respect to the coordinates of the atom. Usage of this force lead the simulation to the conventional MD one, and the conformation is frequently trapped in energy minima, as explained above. In the multicanonical method, a term is added to E:

$$E_{mc} = E - RT \ln \left[P\left(E, T\right) \right] , \quad (1)$$

where R is the gas constant, T is the simulation temperature, and $P(E, T)$ is the energy distribution of the system, which are calculated from the canonical ensemble. Then, the force $\boldsymbol{f}_{mc}$ acting on the atom is computed by

$$\boldsymbol{f}_{mc} = -\text{grad} \left[E - RT \ln \left[P \right] \right] . \quad (2)$$

The difference between the multicanonical MD and the conventional MD simulations is the difference of the force used. The additional term ensures the high sampling efficiency.

One may notice that Eq.2 involves the energy distribution, P, which is not given *a priori*. In other words, P is a quantity to be obtained from the simulation, not a given quantity. Then, P is usually determined with an iteration of consecutive runs of simulation, through which P converges.

The energy distribution obtained from a simulation with Eq.2 provides a flat distribution [5] in an energy range, where P is estimated accurate enough. The iteration should be continued till the energy range covers a high to low (i.e., room) temperatures to ensure that the conformation overcomes the energy barriers and visit energy minima. The more flat the distribution, the more efficient the sampling.

2.2 Example of the Multicanonical MD Simulation

In this subsection, I report results from the multicanonical MD simulation of a chameleon sequence [7]. This sequence was taken from a MATα2/MCM1/DNA complex [8]. The two proteins, MATα2 and MCM1, forms a dimmer when binding to a DNA, and the chameleon sequence, located at the linker part between the two proteins, took either α-helix or β-strand [8]. It is known that the sequence is unstructured before binding to DNA [9]. Thus, it is likely that the chameleon sequence modulates the binding energy between the DNA and proteins with adopting the structure, because the relative position between MATα2 and MCM1 changes depending on the structure of the linker part. The multicanonical simulation treated the only part of the chameleon sequence, but taking the solvent explicitly. The obtained energy distribution was flat (Fig.5) in a wide temperature range from 290 K to 650 K. The energy profile (i.e., fluctuations of E along time) is shown in Fig.6. The simulation length ($6 * 10^6$ steps) shown in Fig, 6 is a part of the whole simulation ($120 * 10^6$ steps). The figures manifested

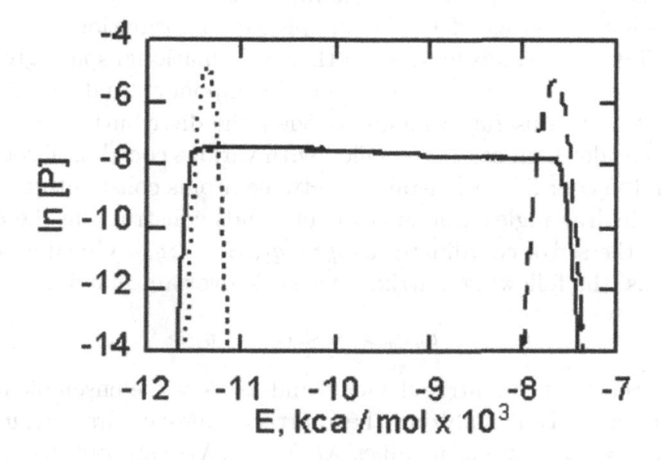

Fig. 5. Flat energy distribution (*solid line*) from a multicanonical MD simulation. Reweighted canonical energy distribution, $P(E)$, at 650 K and 300 K, are shown (*broken* and *dotted lines*, respectively). The y-axis is presented by the logarithm of P.

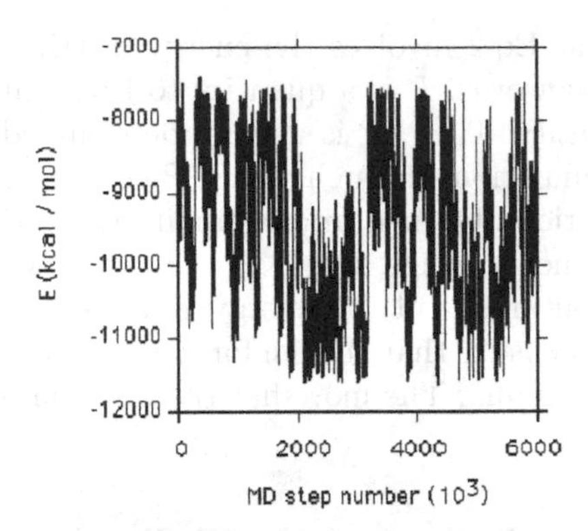

Fig. 6. Energy profile. The shown is part of the whole simulation (see text). The energy fluctuated in the energy range [7,500 to 11,500 kcal/mol], which corresponds to 650 to 290 K.

that the conformation randomly fluctuated in the energy space, with ensuring that the conformation could frequently overcome energy barriers.

3 Biophysical Interpretation of the Results

From the multicanonical MD simulation, the conformational ensemble consisting of a number of polypeptide conformations was generated. Due to the flat energy distribution in the wide energy range, the ensemble was a mixture of conformations at various temperatures. To analyze the conformations, first of all, conformations corresponding to a temperature should be picked from the ensemble, with using a re-weighting procedure [10]. The picked conformations were, then, used for the analysis. For example, the enthalpy of the system at the temperature is the average of E over the picked conformations.

The distribution of conformations in the conformational space gives us an image of the free-energy landscape. A principal component analysis (PCA), briefly explained below, is a useful technique to view the distribution. A conformation of the polypeptide chain can be specified with various coordinate sets: Cartesian (i.e., x, y, and z) coordinates, distances between atoms constituting the polypeptide chain, dihedral angles around covalent bonds constructing the chain. Here, we designate the set of coordinates as $\boldsymbol{q} = [q_1, q_2, \cdots, q_n]$. Given an ensemble of conformations, the following matrix, a variant-covariant matrix, is defined:

$$C_{ij} =< q_i \, q_j > - < q_i >< q_j > \ . \tag{3}$$

where C_{ij} is the (i, j)'th matrix element, and $< \cdots >$ the ensemble average over the conformations. Diagonalyzing the matrix, eigen-vectors $\{\boldsymbol{v}_1, \boldsymbol{v}_2, \cdots, \boldsymbol{v}_{3N}\}$, where $\boldsymbol{v}_i \cdot \boldsymbol{v}_j = \delta_{ij}$, and eigen-values $\lambda_1, \lambda_2, \cdots, \lambda_{3N}$ are obtained. The eigen-vectors construct a multidimensional conformational space, where the sampled conformations from the simulation are distributed. The λ_i represents the standard deviation of the distribution along $\boldsymbol{v}_i$. We arranged the eigen-values in the descending order. In the conformational space, $\boldsymbol{q}$ is expressed by a projection of $\boldsymbol{q}$ on the eigen-vectors: $\boldsymbol{q} = [c_1, c_2, \cdots, c_{3N}]$, where $c_k = \boldsymbol{q} \cdot \boldsymbol{v}_k$. Thus, the eigen-vectors are used as a basis set. The distribution in the conformational space gives the free-energy landscape [7,10].

Figure 7 demonstrates the conformational distribution of the chameleon sequence at 300 K in the conformational space, where the Cartesian coordinates were used to calculate the matrix (Eq.3). In the conformational space constructed by the PCA eigen-vectors $\boldsymbol{v}_1$, $\boldsymbol{v}_2$, and $\boldsymbol{v}_3$, we could find six conformational clusters (A-F). A region with crowded points is a low free-energy region, because the system has a high probability of existence in the region. Clusters A and B were those of α-helical conformations (Fig. 7b), where the helix in cluster B was slightly disordered comparing to that in cluster A. Clusters C-F were those of β-hairpin conformations. A β-hairpin is a conformation in that two strands bind to each other with hydrogen bonds (Fig. 7c). Clusters of C-F correspond to those with different hydrogen-bonding patterns between the strands. The details from the analysis had been reported [7].

The results showed that the chameleon sequence inherently has a structural adaptability to either α or β structure, without the frameworks of the MATα2/ MCM1/DNA complex. Statistical mechanics tells us that a system moves to a state where the free energy becomes lower. The property of the chameleon sequence is, thus, advantageous for the structural modulation in the DNA-protein binding.

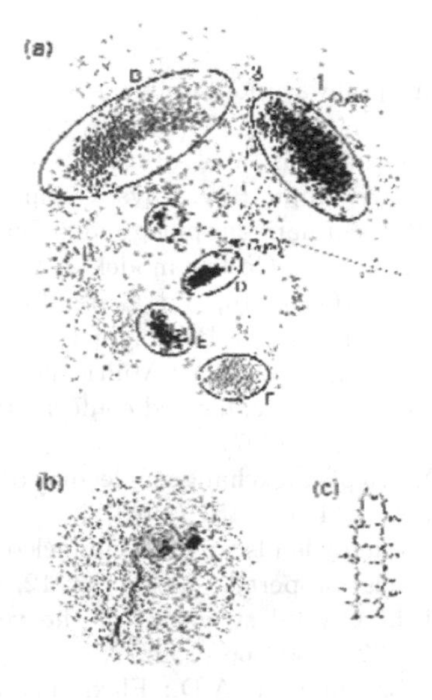

Fig. 7. (a) Distribution of sampled conformations of the chameleon sequence at 300 K in a conformational space constructed by PCA-eigen-vectors $\boldsymbol{v}_1$, $\boldsymbol{v}_2$, and $\boldsymbol{v}_3$ (*solid lines with digits*). Clusters (*capitals A-F*) observed are circled. (b) An a-helical structure picked from the cluster A is displayed, where water molecules surrounding the chameleon sequence in the MD simulation are also shown. (c) A β-hairpin structure picked from the cluster D is displayed, where names of amino acids are also shown.

4 Summary

In this report, I explained the multicanolnical MD simulation method to effectively sample the polypeptide chain. From the sampling, the free-energy landscape is obtained, in which thermodynamically stable conformations at various temperatures are found. Furthermore, the landscape can be regarded as a map to show pathways between the thermodynamically stable clusters [7].

At the moment (June, 2005), we can calculate the free-energy landscape of a polypeptide chain of about 30 amino-acid residues in explicit solvent by one-month computation, for which the result will be reported elsewhere. It is crucially important to increase the chain length. The upper limit of tractable chain length has been gradually extended.

The first goal is the prediction of protein (i.e., longer polypeptide than the peptide) tertiary structure, as mentioned before. This may come true in several years, due to efforts of many scientists in the world. For this purpose, the combination of the fast computers and the effective sampling algorithms is important. To obtain the currently reported results (and the results of the 30-residue polypeptide chain, being reported elsewhere), we used 20 machines of Pentium 4 (2.8 GHz), which is purchasable in many laboratories. A faster computer may open the door to the protein structure prediction.

References

1. Levinthal, C.: Are there pathways for protein folding? J. Chim. Phys. 65, 44–45 (1968)
2. Onuchic, J.N., Luthey-Schulten, Z., Wolynes, P.G.: Theory of protein folding: the energy landscpe perspective. Annu. Rev. Phys. Chem. 48, 545–600 (1997)
3. Koga, N., Takada, S.: Roles of native topology and chain-length scaling in protein folding: a simulation study with a Go-like model. J. Mol. Biol. 313, 171–180 (2001)
4. Mitsutake, A., Sugita, Y., Okamoto, Y.: Generalized-ensemble algorithms for molecular simulations of biopolymers. Biopolymers 60, 96–123 (2001)
5. Nakajima, N., Nakamura, H., Kidera, A.: Multicanonical ensemble generated by molecular dynamics simulation for enhanced conformational sampling of peptides. J. Phys. Chem. B 101, 817–824 (1997)
6. Sugita, Y., Okamoto, Y.: Replica-exchange molecular dynamics method for protein folding. Chem. Phys. Lett. 314, 141–151 (1999)
7. Ikeda, K., Higo, J.: Free-energy landscape of a chameleon sequence in explicit water and its inherent a/b bifacial property. Protein Sci. 12, 2542–2548 (2003)
8. Tan, S., Richmond, T.J.: Crystal structure of the yeast MATα2/MCM1/DNA ternary complex. Nature 391, 660–666 (1998)
9. Sauer, R.T., Smith, D.L., Johnson, A.D.: Flexibility of the yeast a 2 repressor enables it to occupy the ends of its operator, leaving the center free. Genes Dev. 2, 807–816 (1998)
10. Kamiya, N., Higo, J., Nakamura, H.: Conformational transition states of a b-hairpin peptide between the ordered and disordered conformations in explicit water. Protein Sci. 11, 2297–2307 (2002)

Spacecraft Plasma Environment Analysis Via Large Scale 3D Plasma Particle Simulation

Masaki Okada[1], Hideyuki Usui[2], Yoshiharu Omura[2], Hiroko O. Ueda[3], Takeshi Murata[4], and Tooru Sugiyama[5]

[1] Research Institute of Information and Systems, National Institute of Polar Research, 1-9-10 Kaga Itabashi-ku, Tokyo, Japan
mokada@nipr.ac.jp
http://www.nipr.ac.jp/~mokada/

[2] Kyoto University, Research Institute for Sustainable Humanosphere, Gokasho Uji, Kyoto, Japan
{usui,omura}@rish.kyoto-u.ac.jp

[3] Japan Aerospace Exploration Agency, Institute of Space Technology and Aeronautics
2-1-1 Sengen, Tukuba Ibaraki, Japan
ueda.hiroko@jaxa.jp

[4] Ehime University, Center for Information Technology
3 Bunkyo, Matsuyama, Ehime Japan
murata@cite.ehime-u.jp

[5] Japan Agency for Marine-Earth Science and Technology, Earth Simulator Center
3173-25, Showa-machi, Kanazawa-ku, Yokohama, 236-0001, Japan
tsugi@jamstec.go.jp

Abstract. Geospace environment simulator (GES) has started as one of the advanced computing research projects at the Earth Simulator Center in Japan Marine Science and Technology Center since 2002: [1]. By using this computing resource, a large scale simulation which reproduces a realistic physical model can be utilized not only for studying the geospace environment but also for various human activities in space. GES project aims to reproduce fully kinetic environment around a spacecraft by using the 3-dimensional full-particle electromagnetic simulation code which could include spacecraft model inside (NuSPACE). NuSPACE can model interaction between space plasma and a spacecraft by the unstructured-grid 3D plasma particle simulation code embedded in the NuSPACE. We will report current status of the project and our concept of achieving the spacecraft environment in conjunction with the space weather.

1 Introduction

We have developed a 3-dimensional electromagnetic particle simulation code (NuSPACE) as a numerical space chamber. This code solves Maxwell's equations in 3D simulation space. Plasma particles are also traced by solving the equations of motion with the Buneman-Boris method. Thus, this simulation code is suitable for analyzing the plasma environment in the vicinity of a spacecraft especially

J. Labarta, K. Joe, and T. Sato (Eds.): ISHPC 2005 and ALPS 2006, LNCS 4759, pp. 383–392, 2008.

in the region within a Debye length from the surface of the spacecraft as well as the spacecraft charging phenomena. We will show the scheme of this code and also show a couple of results from the test simulation runs taking into account of a realistic shape of a spacecraft.

Plasma particle simulation techniques are widely used in plasma physics among the space plasma physics, the fusion plasma physics and the device plasma engineering. KEMPO proved its wide applicability and precision in space plasma area: [2].

NASA has developed a sophisticated spacecraft charging analyzing software package, NASCAP (NASA Spacecraft Charging Analyzer Program): [3]. This software solves electrostatic potential at the surface of a spacecraft. The code has widely been used in spacecraft design for more than a decade and achieved certain reliability in our community. Though, one deficiency is that this code solves only static potential and not be able to solve time-dependent problem neither nor the electromagnetic problems, such as EM wave propagation analysis.

European research project, SPIS (Spacecraft Plasma Interaction System) is currently developing an open source software of spacecraft-plasma interaction modeling tools. PicUp3D is one of the SPIS software and adopts unstructured-grid to model a spacecraft environment: [4]. This code aims to solve spacecraft charging and not suitable for solving the electromagnetic environment at this moment.

2 Basic Equations and Algorithms

We have developed a 3-dimensional electromagnetic particle simulation code with an unstructured-grid system. This code solves Maxwell's equations which are discretized with tetrahedral elements in 3D simulation space. Plasma particles are also traced by solving the equations of motion with the Buneman-Boris method. The main advantage of this code is the adaptability of modeling more realistic shape of a spacecraft than the orthogonal grid code. Thus, this simulation code is suitable for analyzing the plasma environment in the vicinity of a spacecraft especially in the region within a Debye length from the surface of the spacecraft as well as the spacecraft charging phenomena.

$$\nabla \times \mathrm{E} = -\frac{\partial \mathrm{B}}{\partial t} . \tag{1}$$

$$\nabla \times \mathrm{B} = \mu_o \mathrm{J} + \frac{1}{c^2}\frac{\partial \mathrm{E}}{\partial t} . \tag{2}$$

$$\nabla \cdot \mathrm{E} = \frac{\rho}{\epsilon_o} . \tag{3}$$

$$\nabla \cdot \mathrm{B} = 0 . \tag{4}$$

$$\frac{dv_i}{dt} = \frac{q_s}{m_s}\left(\mathrm{E} + v_i \times \mathrm{B}\right) . \tag{5}$$

$$\frac{dx_i}{dt} = v_i . \tag{6}$$

Equations (1)–(6) are the basic equations solved by the EM particle simulation code. Maxwellfs equations are descritized by unstructured-grid coordinate in the vicinity of a spacecraft. Poissonfs equation (3) can be solved by Finite Element Method (FEM) over the tetrahedral mesh area.

Figure 1 illustrates time step chart used in the NuSPACE. Electric and magnetic fields are solved with the centered differential form, which is called the leap-frog method. The equation of motion is integrated in time with the Buneman-Boris method. Benefit of this method is the conservation of the energy of the particles to be solved. Current is obtained from the particle velocity with the following equation.

$$J = q \sum_i v_i \tag{7}$$

Summation is conducted over all particles located in each grid cell. When calculating the equation (7), there are two types of interpolation. One is the linear interpolation method and the other is the charge conservation method (CCM). The linear interpolation method is usually called Cloud-in-cell method (CIC). The vectorization of the CIC is easier than CCM. As known from its name, charge conservation, as show in equation (8), is fulfilled in grid scale.

$$\frac{d\rho}{dt} + \nabla \cdot J = 0 \, . \tag{8}$$

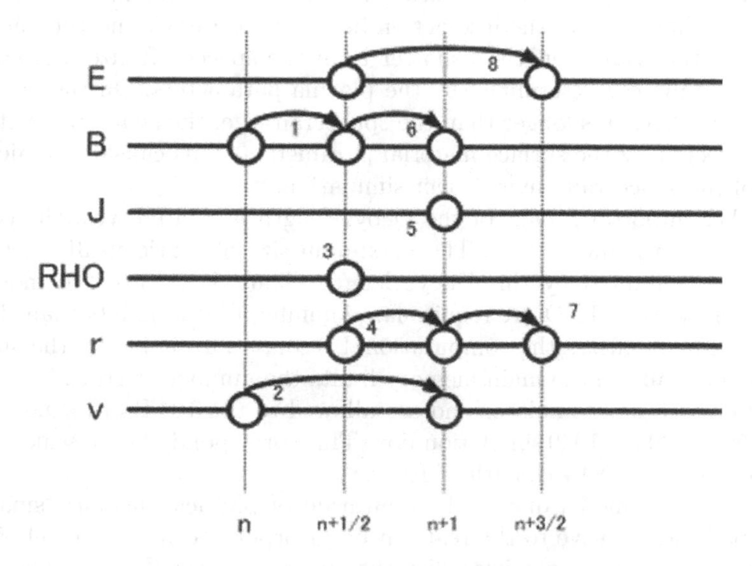

Fig. 1. Time step chart used in the NuSPACE. Numbers depicted in the time chart indicates the sequence of the calculation. E and B fields are calculated with the centered difference form with time. The particle location and velocity are solved with similar way. Current J is solved at the full-time step. The equation of motion is integrated at the half-time step by using the E and B field on the other hand.

3 Models

Our target region of modeling is summarized in table 1. The first is the low earth orbit (LEO), whose altitude ranges from 100 km up to about 6000 km. The Space Shuttle uses the orbit with the altitude of about 100 km. The second one is the geostationary orbit (GEO). This orbit is well-known by the use of Japanese weather satellite, "Himawari". The orbital period of these satellites are synchronized with the Earthfs rotational period. The third one is the polar orbit (PEO). This orbit is useful for observing the Earthfs surface with high spatial resolution. This orbit is known to be used by the LANDSAT, NOAA and DMSP satellites.

Table 1. Characteristic plasma parameters and the spacecraft scale at the orbit of LEO, GEO and PEO. PEO is characterized with not only altitude and the Debye length but also with the aurora precipitation particles.

Orbit	Altitude(km)	Debye Length(m)	Scale(m)
LEO	100-6000	0.001-1	5-100
GEO	36000	10-100	5
PEO(aurora electrons)	600-6000	0.001-0.1	1-10

The most important point in modeling the orbital environment is the Debye length at the orbit. The relation between the Debye length and the scale size of a spacecraft characterizes the interaction between spacecraft and the background plasma. If the Debye length is shorter than the spacecraft size, the spacecraft potential is mainly determined by the plasma parameters. On the other hand, if the Debye length is longer than the spacecraft size, the spacecraft potential is highly affected by the surface material parameters. This causality indicates the needs of the spacecraft environment simulation.

Another important point of the Debye length is related with the computational resource point of view. The maximum size of a grid in 3D EM particle simulation is limited by this Debye length. Thus, if we need to model large simulation space in LEO, we require large number of grid points than the GEO.

Table 2 summarizes the computational resources available by the simulator. As described above, the main memory limits the number of grids. For example, if we use the Earth Simulator and are allowed to use 10TB of memory, we can model $2000\times2000\times1000$ simulation box. This corresponds to the scale of 40 km^3 at GEO with the Debye length of 10m.

In simulation model due to the limitation of physical memory, smaller umber of particles relative to the real can be incorporated in the model. Thus, we substitute real plasma particles with the super-particles. The number of super-particles incorporated in the model is estimated with a minimum number of particles per grid. In 3D EM particle simulations, 64 particles per grid is the minimum number of particles in order to avoid numerical heating. If we perform

Table 2. Relation between the main memory and the modeling size with the major computational resources. Estimated with parameters opened by each sites, the Earth Simulator (ES), the Numerical Simulator III (NS III), Kyoto Denpa-Kagaku Simulator (KDK) and the Polar Science Simulator (PSS).

Computational Resource	Memory (TB)	Grid (Mega Grid)	Particle (Giga-particles)
ES	10	4000	256
NS III	3.6	1000	64
KDK	0.5	100	6.4
PSS	1	200	12.8

lower numerical noise simulation, we need more super-particles in the simulation model. In such cases, we can model less number of grids in space.

In most cases, a single simulation run requires from 1024 to 65536 time steps. Plasma wave analysis and beam interaction analysis require larger number of time steps than the spacecraft charging analysis. For example, if we use the Earth Simulator, one time step requires 1 to 10 seconds depending on the simulation model.

Figure 2 illustrates the physical structure of the spacecraft-plasma simulation. In our simulation code, plasma and spacecraft are solved in time with fully self-consistent manner. The left most loop indicates that the electric and magnetic fields are coupled with Maxwellfs equations. This loop gives the solution of the electromagnetic waves. EM fields and plasma particles are coupled with the

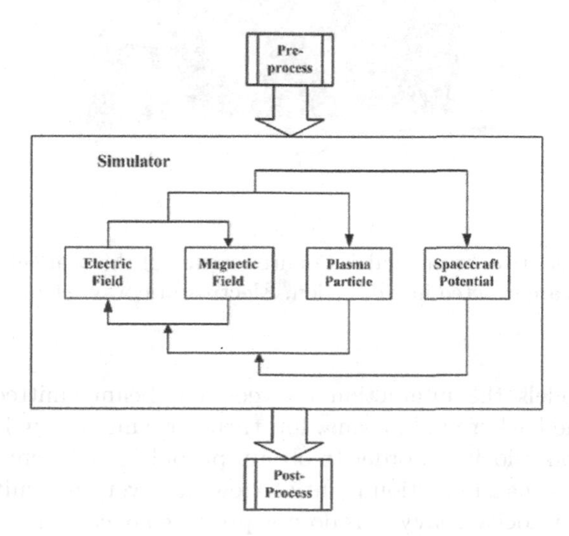

Fig. 2. Fundamental physical property of the spacecraft-plasma interaction process. E and B fields are coupled with Maxwell's equations. Plasma particles move under the equation of motion and creates current. Spacecraft potential is determined by the plasma current and affects the trajectories of plasma particles.

equation of motion and the current created by the motion of plasma particles. This loop gives solutions of plasma waves including electrostatic structure in the plasma. Plasma particles impinging upon the spacecraft surface are absorbed by the spacecraft and changing the potential of the spacecraft by charging. Spacecraft potential modifies the trajectory of the plasma particles and spacecraft potential asymptotically approach to the certain potential.

Figure 3 shows the test simulation results obtained by the Earth Simulator. This model simulates uniform electron beam interacting with the background electrons. This model is not only used as the code check, but also for understanding fundamental plasma phenomena in the Magnetosphere. It is commonly known that the beam electrons cause the two-stream instability and amplifies the electrostatic waves. This phe-nomena is well understood both with analytically and observationally. After decaying the electrostatic waves, the wave is modified to the electrostatic solitary wave. This process is known as non-linear process. Thus, it is difficult to solve analytically but EM particle simulation can simulate the non-linear evolution. Although this figure shows its initial state of non-linear evolution, the first results obtaind by the 3-dimensional electromagnetic simulation.

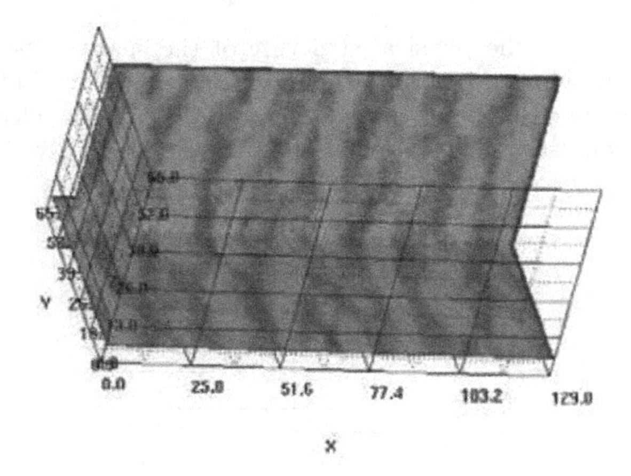

Fig. 3. Uniform electron beam model. Counter stream of the beam electrons and background electrons are injected in x direction. Electrostatic wave structure is observed in y-z plane.

Figure 4 models the interaction between ion beam emitted from the ion thruster and the background plasma. Ion thruster emits heavy ions (Xe) with a high acceleration velocity in order to obtain propulsion efficiently. In spacecraft charging and plasma interaction point of view, it is very difficult to understand why these high velocity heavy ions do not produce spacecraft charging and how large plasma contamination may cause the plasma instabilities. According to the simulations, we can estimate the charging rate depending on the background plasma parameters taking into account of the magnetic field strength and the background plasma density and temperature.

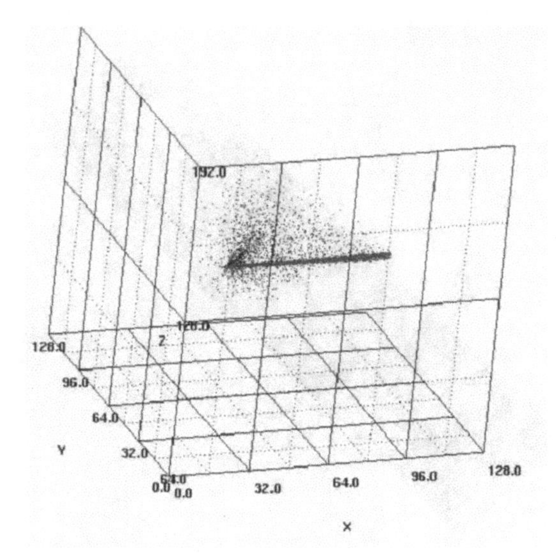

Fig. 4. Ion engine model. Xeon ions (red) are ejected from the ion thruster and the neutralizer elec-trons (blue) rotate around the background magnetic field.

Strictly speaking, the behavior of beam ions and neutralizer electrons are highly affected by the spacecraft potential. Without taking into account the spacecraft potential, it is not appropriate to discuss fully self-consistent feature of spacecraft-plasma interactions. Figure 5 shows the electrostatic environment model around a spacecraft with a set of wire antennas. The background electrons and ions are absorbed by the spacecraft body and wire antennas. By accumulating the charge of absorbed electrons and ions, we obtain spacecraft charging and plasma sheath around the spacecraft.

Figure 6 shows the surface potential profile obtained by a simple spacecraft and solar panel model, like ALOS satellite. From the spacecraft engineering point of view, it is very serious to understand spacecraft surface charging for designing the spacecraft. When an engineer may find a high voltage surface charging during its modeling phase, he must modify its surface material and/or design in order to avoid the charging.

In order to achieving this aim, we do not have to solve electromagnetic waves. Thus, electrostatic simulation is sufficient. By neglecting the electromagnetic waves, we can increase the time step Δt larger than $\Delta x/c$ due to the CFL condition to solve the electromagnetic waves. When we focus on the spacecraft charging, the 3D plasma particle simulation becomes very powerful tool for solving the spacecraft potential in both engineering and scientific use.

At the final stage of the spacecraft design, more precise analysis would be required for spacecraft charging. Fine structure of a spacecraft shape will be able to be incorporated from the CAD. Unstructured-grid electromagnetic particle simulation would be the best for solving the spacecraft plasma environment at this stage. Figure 7 shows the initial results obtained by the unstructured-grid

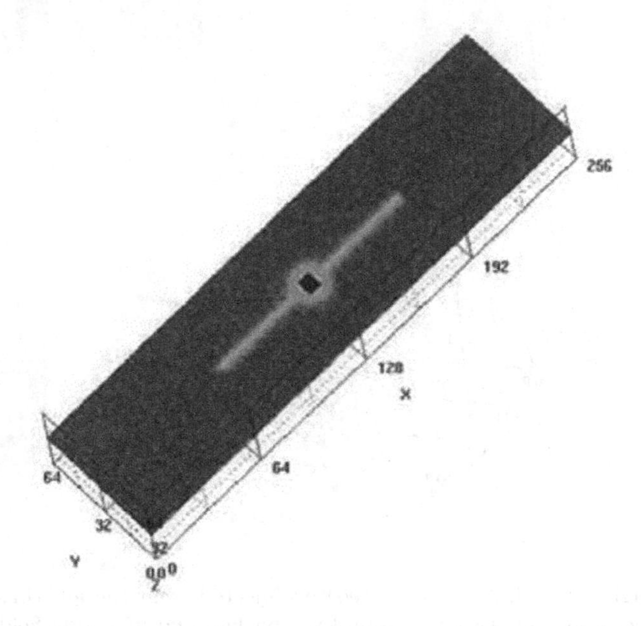

Fig. 5. Spacecraft charging analysis model. Electric field created by the spacecraft body (square region in the middle) and the wire antenna is observed.

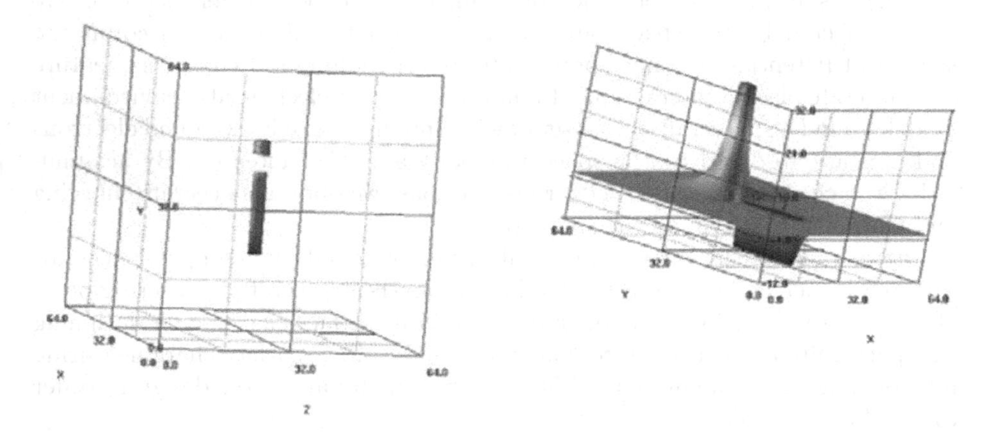

Fig. 6. Spacecraft charging analysis performed with a single spacecraft body and solar panel model. (a) floating potential case and (b) fixed potential case.

EM particle code developed by the authors. The results obtained by the sphere probe model can easily be compared with an analytical model and the chamber experiments. After calibrating the unstructured-grid code by the sphere probe model, more realistic model will be incorporated in the NuSPACE.

Figure 8 shows the realistic model derived from the GEOTAIL satellite shape. This model consists of 11000 nodes and 48000 tetrahedral elements. In order to

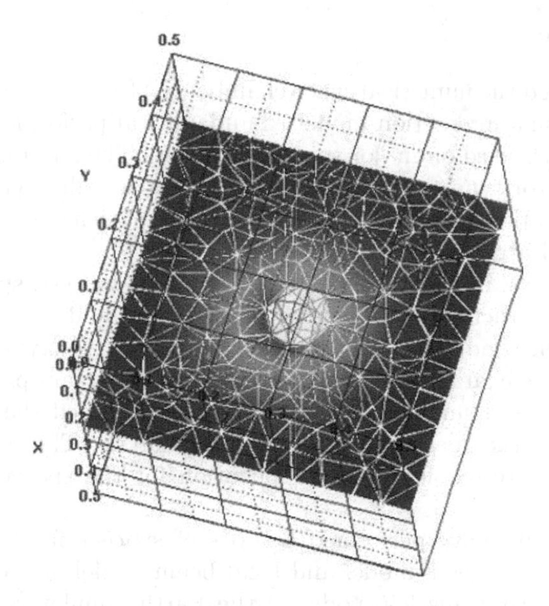

Fig. 7. Sphere probe model tested by the unstructured-grid EM particle code. This results is compared with both theoretical analysis and chamber experiments.

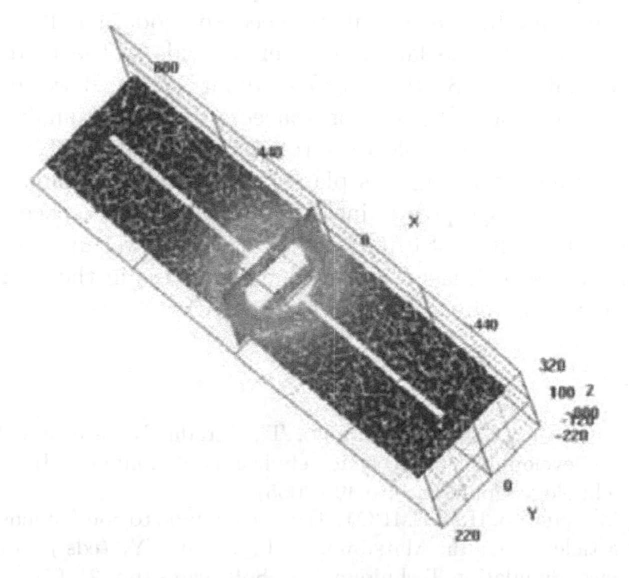

Fig. 8. GEOTAL satellite model. Drum structure in the middle models the GEOTAIL satellite body whose diameter is 2m. The two antenna masts (4m) are attached on the body.

embed this model into the NuSPACE, 3044 cubic elements are allocated for boundary elements. As the computational resource becomes available, the more precise model can be introduced for both engineering and physical analysis.

4 Summary

We have developed the numerical space chamber (NuSPACE) and have tested on 5 spacecraft-plasma interaction models. Fundamental performance of the NuSPACE has been checked on major super-computers. The NuSPACE can be tunable for both vector-type supercomputers and scalar-parallel type supercomputers. The vectorization ratio of 99.7% and parallel efficiency of 100% have been achieved on the Earth Simulator.

5 simulation runs, i.e., uniform beam test, local beam test, spacecraft surface charging test, unstructured-grid sphere probe model and GEOTAIL structure model, have been conducted. The uniform beam test resolves high numerical precision EM wave characteristics by solving the two stream plasma instability. The beam model and surface charging model have proved the applicability of the NuSPACE to the spacecraft environment analysis. The unstructured-grid models have been tested for future application for the spacecraft design and engineering.

In this paper, we have performed 5 types of spacecraft plasma interaction simulations. Uniform beam model and local beam model are conducted on the Earth Simulator. By using 256 nodes of the Earth Simulator, the elapse time was 11 hours to resolve electron beam instability. If we need to resolve ion mode, the more time steps are needed, thus the longer CPU time is necessary.

As for the memory limitations, if we need to model small scale structure of a spacecraft, we need the larger number of grids in the model. The scale of the International Space Station (ISS) is about 200 m. If we model the ISS, 0.1 meter grid would be adequate for spacecraft charging analysis. Then the Earth Simulator may be possible resource for this analysis. In near future, a couple of large scale construction is planed for space development. 1 km scale structure and high voltage power line is necessary for the space development. Thus, we believe that the NuSPACE will become a powerful tool for analyzing the large scale spacecraft plasma environment analysis in the near future with the aid of the peta-flops supercomputers.

References

1. Usui, H., Omura, Y., Okada, M., Ogino, T., Terada, N., Murata, T., Sugiyama, T., Ueda, H.: Development of Geospace environment simulator. In: 9th Spacecraft Charging Technology Conference, p.49 (2005)
2. Omura, Y., Matsumoto, H.: KEMPO1: Technical Guide to one-Dimensional Electromagnetic Particle Code. In: Matsumoto, H., Omura, Y. (eds.) Computer Space Plasma Physics: Simulation Techniques and Soft-wares, pp. 21–65. Terra Scientific, Tokyo (1993)
3. Mandell, M., Katz, I., Cooke, D.: Towards a more robust spacecraft charging algorithm. In: 6th Spacecraft Charging Technology Conference, AFRL-VS-TR-2001578, pp. 251–255 (2000)
4. Forest, J., Eliasson, L., Hilgers, A.: A new spacecraft plasma simulation software, PicUp3D/SPIS. In: 7th Spacecraft Charging Technology Conference, pp. 257–267 (2001)

PetaFLOPS Computing and Computational Nanotechnology on Industrial Issues

Shuhei Ohnishi[1,2] and Satoshi Itoh[3]

[1] NEC Fundamental and Environmental Research Labs.,
Miyukigaoka 34, Tsukuba, 305-8501, Japan
ohnishi@frl.cl.nec.co.jp
[2] Toho University, Fuculty of Science, Miyama 2-2-1
Funabasi, Chiba, 274-8510, Japan
s-ohnishi@sci.toho-u.ac.jp
[3] Toshiba Research and Development Center,
Kawasaki, Kanagawa, 212-8582, Japan
satoshi.itoh@toshiba.co.jp

Abstract. TA prospect of new development by PetaFLOPS computing is discussed for the industrial research and application in the field of materials sciences based on the TeraFLOPS computing in the Earth Simulator. Two examples of simulations for nano-scale materials are presented for metal clusters by the first principles calculation and water droplets by the classical molecular dynamics. A new possibility by PetaFLOPS computing is proposed in terms of a real-time simulator.

1 Introduction

Nanotechnology provides a new and powerful manufacturing strategy for many industrial problems; e.g. high-k and low-ϵ materials for advanced microelectronic, highly efficient catalysts, long-life electro-luminescence materials, MEMS (micro electro-mechanical system) and so on. On devising the strategy, a computer simulation plays an important role at present.

At the beginning of the 21st century, we have entered into possession of the first TeraFLOPS general-purpose supercomputer, called the Earth Simulator (ES), which is not special-purpose computer and can be utilized for many types of numerical calculations. One of the most characteristic properties in the ES is free from the bottleneck of inter-node network communication. We can utilize many CPUs without worrying about a physical location in computers. By using such TeraFLOPS computer, electronic structures of metal clusters with several hundred atoms can be calculated preciously, and classical molecular dynamics simulation for several million atoms can be performed during nanoseconds. These systems, however, are limited compared with experimental systems. If PetaFLOPS supercomputer is developed, bigger and/or longer simulations are realized; i.e. a high-precision electronic structure calculation of metal cluster consisting of 10^{4-6} atoms, a long time molecular dynamics (MD) simulation over milliseconds, and so on. These results would show the fact that the majority

J. Labarta, K. Joe, and T. Sato (Eds.): ISHPC 2005 and ALPS 2006, LNCS 4759, pp. 393–401, 2008.

of developing and designing processes by experiments take the place of that by simulations, and the experiments would be only needed at the final stage of the design and development for confirmation.

In this paper, two simulation results are presented; one is static and electronic properties of metal cluster based on the first principles calculation using the ES [1], and the other is dynamic and dielectric properties of water droplets by using classical MD simulation. The former clues us to develop new catalysts and hydrogen storing alloys, and the latter gives fundamental knowledge in biotechnology. These results are good examples to show the present level of computational nanotechnology. Finally, new usage of the results of computational work in the era of the PetaFLOPS computing will be proposed.

2 Metal Cluster as a Nanotechnology

2.1 Parallel Computation in the ES for the Multi-center Problem in Clusters

It is now well known that the density functional theory (DFT) can describe electronic states in any material system. It provides us a powerful and useful computational scheme, which succeeded to reduce the computational scale to the $O(N^3)$ problem for a system of N atoms. A multi-center problem in the non-periodic system is the most intractable problems in simulating nanostructure materials; atomic basis functions and singularities by the nucleus Coulomb potentials are localized in the atomic site. We can parallelize the non-periodic atomic cluster for each atomic site using MPI in a natural way. The pseudopotential method makes it feasible to get rid of the Coulomb singularity numerically introducing the projection operator to the atomic core states as indicated in the following equations. By solving the single particle Schödinger equation self-consistently of which effective potential field is given by the functional of the electron charge density uniquely, we can obtain the ground state total energy of the cluster based on the DFT.

$$\left\{-\frac{\hbar^2}{2m}\Delta + v_{eff}\left[\rho\left(\vec{r}\right)\right]\right\}\psi_i = \epsilon_i\psi_i \,. \tag{1}$$

$$\begin{aligned} v_{eff}\left[\rho\left(\vec{r}\right)\right] &= \hat{v}_{ext}\left(\vec{r}\right) + v_C\left(\vec{r}\right) + v_{XC}\left[\rho\left(\vec{r}\right)\right] \,, \\ \hat{v}_{ext}\left(\vec{r}\right) &= \textstyle\sum_\alpha \left[\hat{v}_{PS}\left(\mid \vec{r}_\alpha \mid\right) + v_{core}\left(\mid \vec{r}_\alpha \mid\right)\right] \,, \\ \text{where} \quad & \hat{v}_{PS}\left(\mid \vec{r} \mid\right) = \mid l\rangle v_l\left(\mid \vec{r} \mid\right)\langle l \mid \,. \end{aligned} \tag{2}$$

$$v_C\left(\vec{r}\right) = \int \frac{\rho(\vec{r}\,')}{\mid \vec{r} - \vec{r}\,' \mid} d\vec{r}\,'; \; \textit{Coulomb Potential}\,. \tag{3}$$

Our method in dealing with solving eigenvalue problems and the Poisson equation is completely based on the multi-center numerical integration scheme by dividing any physical quantity into each atomic site with a localized fuzzy cell type weight function as follows.

$$F\left(\overrightarrow{r}\right)=\sum_{\alpha=1}^{N}F_{\alpha}\left(\overrightarrow{r}_{\alpha}\right),$$
$$F_{\alpha}\left(\overrightarrow{r}_{\alpha}\right)=w_{\alpha}\left(\overrightarrow{r}_{\alpha}\right)F\left(\overrightarrow{r}\right), \tag{4}$$
$$where\ \overrightarrow{r}_{\alpha}\equiv\overrightarrow{r}-\overrightarrow{R}_{\alpha}.$$
$$\sum_{\alpha=1}^{N}\omega_{\alpha}\left(\overrightarrow{r}_{\alpha}\right)=1\quad\forall\overrightarrow{r}_{\alpha}. \tag{5}$$
$$I=\sum_{\alpha=1}^{N}I_{\alpha}\quad I_{\alpha}=\int F_{\alpha}\left(\overrightarrow{r}_{\alpha}\right)d\overrightarrow{r}_{\alpha}. \tag{6}$$

The numerical integration points are given by one-dimensional radial part and two-dimensional spherical grids; typically 50 times 200 at each atomic site to obtain 6 digit accuracy [2]. The parallelization of computing whole physical quantities is performed by atomic site using MPI. Each physical quantities requires us $O(N^2)$ calculation after determining the Coulomb potential field. It is the most important problem to calculate the accurate Coulomb potential and energy in the DFT because of its non-local property described in Equations 1, 2, and 3. The key point in our treatment is that the Coulomb potential is given by just superimposing each contribution from each atomic site, which is completely parallelized by using MPI.

$$\rho\left(\overrightarrow{r}\right)=\sum_{\alpha=1}^{N}\rho_{\alpha}^{lm}\left(r_{\alpha}\right)y_{lm}\left(\hat{r}_{\alpha}\right);\quad y_{lm}\left(\hat{r}_{\alpha}\right);sphericalharmonic\ function,$$
$$v_{C}\left(\overrightarrow{r}\right)=\sum_{\alpha=1}^{N}v_{C}\left(\overrightarrow{r}_{\alpha}\right),$$
$$v_{C}\left(\overrightarrow{r}_{\alpha}\right)=v_{C}\left(r_{\alpha}\right)y_{lm}\left(\hat{r}_{\alpha}\right), \tag{7}$$
$$v_{C}\left(r_{\alpha}\right)=\sum_{lm}^{l_{max}}\frac{4\pi}{2l+1}\left[\int_{0}^{\gamma_{\alpha}}\frac{r^{l}\rho_{\alpha}^{lm}\left(r\right)}{r_{\alpha}^{l+1}}dr+\int_{r_{\alpha}}^{\infty}\frac{r_{\alpha}^{l}\rho_{\alpha}^{lm}\left(r\right)}{r^{l+1}}\right].$$

Our present code had been completely tuned to be able to have a high vector per-formance by implementing all integration parts into the deepest do-loop; the vector ratio is normally 99.5%. By keeping information of other atoms in each MPI process, we performed the hybrid type parallelization so as to use CPUs as possible as we can. Outer operation for atomic sites is dealt with MPI for inter-node parallelization and middle parts are dealt with the intra-node multi-task treatments. Figs.1 and 2 show the elapse time and parallel performance of LCAO-PS code for nanostructure Pt clusters, respectively. The parallelization ratio is 99.95% and the parallelization efficiency is 64% evaluated from elapse time using 176 CPUs and 1080 CPUs for Pt_{135} after tuning to the ES by the hybrid type parallelization. Fig.2 shows that two kinds of intra-node parallelization by vector and micro-task treatments presents a possibility of O(N) computation in the first principles simulation for the material system.

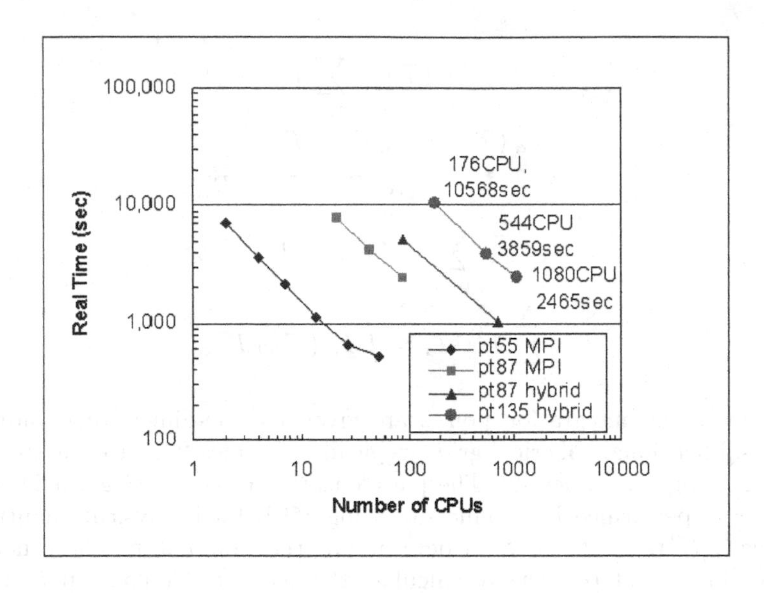

Fig. 1. Elapse time for computation of Pt clusters

2.2 Nano-Metal Clusters

Fig.3 shows electronic structures of Pt_{135}. The characteristic metal like state of the finite system of nano-scale is very important to understand a catalytic high activity. Since electronic energy levels are dense at the highest occupied molecular orbital (HOMO) and the lowest unoccupied one (LUMO), less than 0.01 eV, it is very diffi-cult to obtain good converged total energy especially in noble nano-metal clusters having high catalytic activities while the case of Au_{135} is rather easy because of wide HOMO-LUMO gap, 0.6 eV.

Fig.4 shows another example of the simulation for nano-metal system. Metal-hydrogen (M-H) systems are getting more and more important in a nano-structure materials science associated with energy and environmental problems in terms of the hydrogen storage. It has been recognized that a vacancy plays a key role in hydrogen distribution in M-H alloys; the interaction between a vacancy and hydrogen in metal is strong enough to reduce the formation energy of a vacancy-hydrogen (Vac-H) cluster. Hydrogen atoms trapped by a vacancy are more stable than those in the interstitial sites, which causes the increase of the concentration of Vac-H clusters in the M-H system. Experimental data shows that the superabundant vacancy (SAV) formation is one of the most basic properties of the M-H system [3][4]. Fig.4 shows hydrogen binding energy curves in various atomic elements of the bcc structure. The characteristic feature is that stable hydrogen positions in the vacancy site are almost same in any atomic element. The binding energy differences, eb, given by that of hydrogen of $M_{50}H_6$ (the Vac-H model) and $M_{51}H_6$ (the bulk model) are in good agreement with experiments; 0.42 eV per H (calculated) while 0.46 eV (experiment) in Nb, 0.81 eV

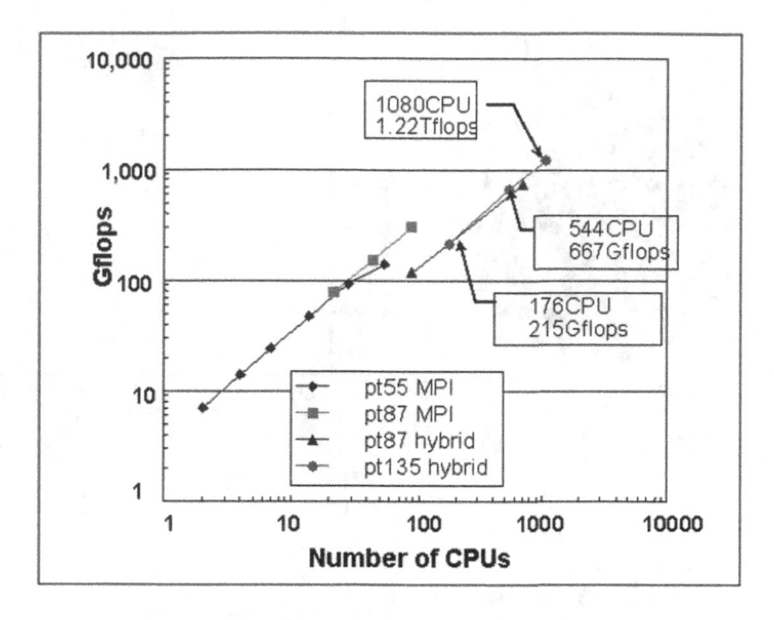

Fig. 2. Parallel performance

(cal.) while 1.03 and 0.80 eV (exp.) in Mo, and 1.095 eV (cal.) while 0.89 and 0.73 eV (exp.).

3 Water at the Nano-Scale

Water is one of the first molecular liquids to be studied intensively by computer simulations because of its importance for human beings. Additionally, water shows unique properties as dielectric materials. By using classical MD simulations, the dielectric properties are investigated, and the relation between the bulk dielectric properties and inter-atomic potentials including charge distribution of water molecules have been discussed in several papers.

Quite recently, the dynamic properties of water at the nano-scale become attractive from the viewpoint of industrial applications such as bio-MEMS (micro electro mechanical system), μTAS (micro total analysis system), and so on. Moreover, the water plays an important role at the nano-scale for ion channels and aquaporin in living cells. In this paragraph, the simulation results of water droplet at the nano-scale are presented, where the relation between the size of the water droplet and its dielectric properties will be discussed.

3.1 Simulation Method

The dielectric properties of liquids are described from total dipole fluctuations at finite temperature. As for water, a dipole moment of a water molecule has defined value and there is no intermolecular charge transfer, so that the total dipole

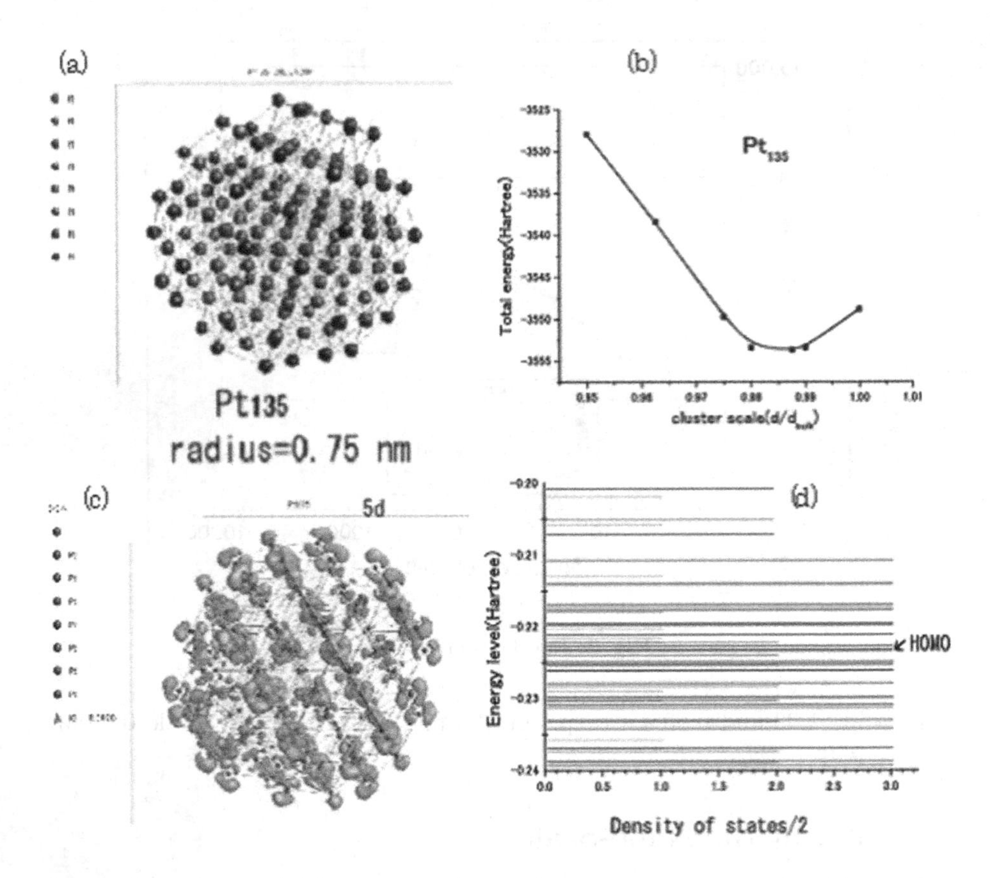

Fig. 3. Pt_{135} cluster. (a) Structure of Pt_{135} : diameter is 1.5 nm. Red and blue sphere indicate bulk and surface like atoms. (b) Total energy change by the cluster size. Small contraction (1.5%) from the bulk crystal. (c) Charge density distribution at the HOMO. It comprises the Pt 5d orbital. (d) Electronic structure near HOMO. Energy difference between HOMO and the LUMO is less than 0.01 eV, which indicates the finite system is almost metallic.

fluctuations are evaluated by using classical molecular dynamics simulation. The static dielectric constant ϵ is expressed as

$$\epsilon = \frac{\epsilon_\infty \left(2\epsilon_{\mathrm{RF}}+1\right)^2 + 6yG_k\epsilon_{\mathrm{RF}}\left(2\epsilon_{\mathrm{RF}}+\epsilon_\infty\right)}{\left(2\epsilon_{\mathrm{RF}}+1\right)^2 + 3yG_k\left(2\epsilon_{\mathrm{RF}}+\epsilon_\infty\right)} \xrightarrow{\epsilon_{\mathrm{RF}}\to\infty} 3yG_k + \epsilon_\infty \,. \tag{8}$$

where G_k and y are Kirkwood G_k-factor and dipole moment strength, respectively [5]. The value ϵ_{RF} is dielectric constant for reaction field, and is set to infinite in this calculation. The Kirkwood G_k-factor is given by the total dipole fluctuation

$$G_k = \frac{\langle M^2\rangle - \langle M\rangle^2}{N\langle\mu^2\rangle} \,. \tag{9}$$

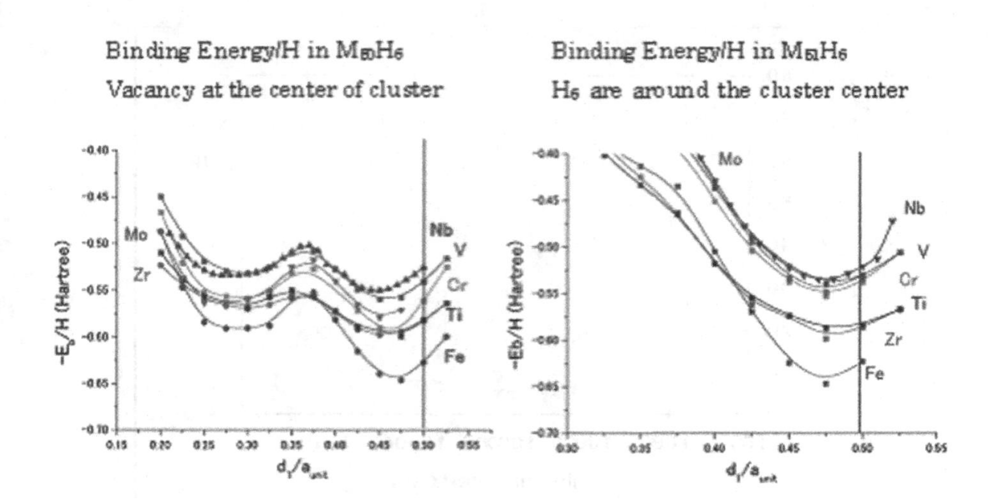

Fig. 4. Hydrogen in bcc metal clusters. d_1/a_{unit} represents normalized distance of hydrogen site; 0.5 indicates the nearest neighbor square surface to hydrogen atom.

where M means total dipole moment of the system, and μ denotes dipole moment of water molecule (= 2.27). The dipole moment strength y is written by

$$y = \frac{4\pi N \langle \mu^2 \rangle}{9 V k_B T} . \qquad (10)$$

where N is total number of molecules, V is total volume, and k_B means Boltzmann constant.

The MD simulations were carried out as isothermal MD (T=300 K), in which the force field of water molecule is SPC/E model. In order to investigate the water confined in a small space, four kinds of water droplets with different size are used, i.e. N=639(d=2 nm), 2685(d=3 nm), 12426(d=5 nm), 99678(d=10 nm), and, additionally bulk system in which N=1024 and the periodic boundary condition are used. Each simulation run was done for 10^6 steps with Δt=2 fsec.

3.2 Numerical Results

By using five simulation runs for the same water droplet size with different initial conditions, the dielectric constants have been estimated. As shown in Fig.5, calculated dielectric constant of bulk water is 65.9, which is underestimated compared with experiments. It is known that the SPC/E water model give an underestimated dielectric constant, and this fact is attributed to the character of the SPC/E force field model. As for small droplets, dielectric constant of water droplet at the nano-scale is dramatically reduced. If this trend is extrapolated to the meso-scale, even at the mm scale, dielectric constant of water in the narrow space is expected to be small value below 40. In recent bio-MEMS, the width of flow channel is order of 100 μm. The change of dielectric properties of water in

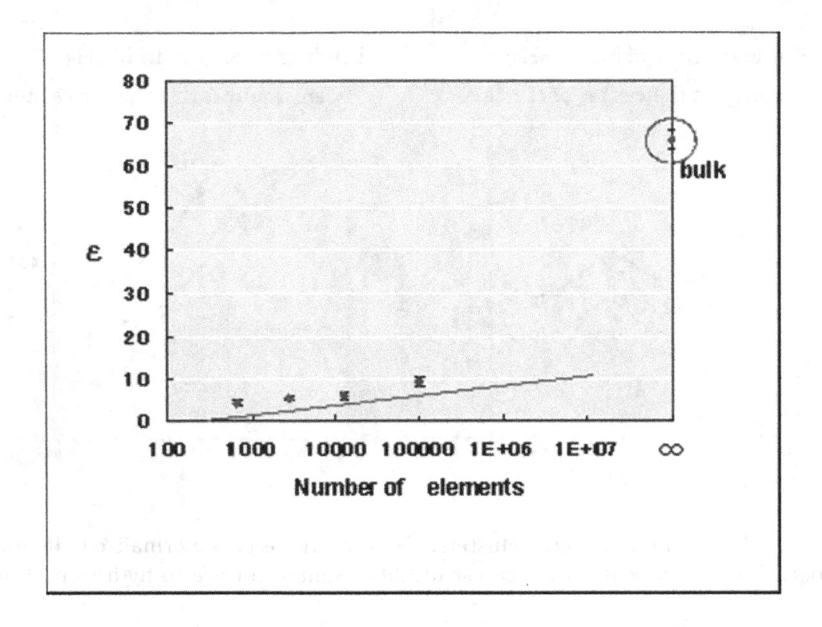

Fig. 5. Calculated dielectric constant of water as a function of droplet size

such narrow channel would produce a considerable problem in bio-MEMS design. Accurate knowledge about material properties at the nano-scale would be useful when developing epoch-making products, e.g. a protein chip which measures all protein existing inside the living cell simultaneously.

4 Concluding Remarks

We have discussed physical properties of nano-materials in sections 2 and 3. The first principles calculations have been done by using the ES system. We could get well converged results for each system within almost one hour, while the standard calculation typically takes a week. It is very efficient not only to obtain very accurate results with a very large number of sampling points for the numerical integration but to consider all possibilities of combination of atomic elements, which will open a new possibility to build up a new kind of nano-materials database. The dynamical simulation needs long time results as possible as we can store whole time series data in both the first principles and the classical calculations; 0.001 fs times 10^3 means the ps order and 1 fs times 10^6 means ns one at the TeraFLOPS stage, respectively. We can, therefore, expect two kinds of breakthrough for the computer aided materials design in the era of the PetaFLOPS computing.

One is the actual computer experiment which is so called *in-situ* theoretical experiment. We can deal with $10^4 \sim 10^6$ atoms estimated from our TeraFLOPS computa-tion in the ES. It will be enough to discuss whole effects of the environment of nano-materials. We will be able to chase the time evolution for quantum and classical sys-tems comparable to *in-situ* experiments.

The other one is the more important issue than the direct real-time simulation. There must be a lot of systems which need a kind of the on demand simulation. The computer simulation can provide us information predicted something described by the basic equations in advance. It is very important to have these predicted samples before anything happens. In the case of nano-materials design for industrial applications, these kinds of knowledgebase created by materials-simulation will become very efficient to have a common platform to share basic atomic elements extracted from the calculated output of materials-simulation. It will be very useful not to waste time to explore new materials at the unknown environment.

References

1. Ohnishi, S.: Annual Report of the Earth Simulator Center, vol. 151 (2004)
2. Watari, N., Ohnishi, S.: Phys.Rev.B. 58, 1665 (1998)
3. Fukai, Y.: J. Alloys Compounds 263, 356–357 (2003)
4. Watari, N., Ohnishi, S., Ishii, Y.: J. Phys.: Condens. Matter 12, 6799–6823 (2000), Watari, N., Ohnishi, S.: J. Phys. : Condens. Matter 14, 769–781 (2002)
5. Valisko, M., Boda, D., Liszi, J., Szalai, I.: Phys.Chem.Chem.Phys. 3, 2995 (2001)

16.14 TFLOPS Eigenvalue Solver on the Earth Simulator: Exact Diagonalization for Ultra Largescale Hamiltonian Matrix

Susumu Yamada[1,3], Toshiyuki Imamura[2,3], and Masahiko Machida[1,3]

[1] Japan Atomic Energy Agency,
6-9-3 Higashi-Ueno, Taito-ku, Tokyo, 110-0015, Japan
{yamada.susumu,machida.masahiko}@jaea.go.jp
[2] The University of Electro-Communications,
1-5-1 Chofugaoka, Chofu-City, Tokyo, 182-8585, Japan
imamura@im.uec.ac.jp
[3] CREST (JST), 4-1-8 Honcho, Kawaguchi-City, Saitama, 330-0013, Japan

Abstract. The Lanczos method has been conventionally utilized as an eigenvalue solver for huge size matrices encountered in strongly correlated fermion systems. However, since one can not obtain the residual during the Lanczos iteration, the iteration count in the Lanczos method is not controllable. Thus, we adopt a new eigenvalue solver based on the conjugate gradient (CG) method in which the residual can be evaluated every iteration step. We confirm that the CG method with an preconditioner shows much more excellent performance than the Lanczos method. We achieve 16.14 TFLOPS on 512 nodes (4096 processors) of the Earth Simulator by the use of the CG method.

1 Introduction

Since the experimental success [1,2,3] of the Bose-Einstein condensation in the trapped atomic Bose gas honored by the Nobel Prize in 2001, the research streamline in atomic physics has been directed toward another difficult condensation, namely, superfluidity of the atomic Fermi gas [4,5]. Thus, we numerically explore a possibility of superfluidity in the atomic Fermi gas [6,7]. Our undertaking model is the fermion-Hubbard model [8,9] with the trapping potential. The Hubbard model describes a many-body fermion system on a discrete lattice, which can be realized by a standing wave created due to interference effect of two laser beams [10].

The Hubbard model is one of the most intensively-studied models by computers because it owns very rich physics although the model expression is quite simple [8,9]. The Hamiltonian of the Hubbard model with a trap potential [6,11] is given as

$$H = -t \sum_{i,j,\sigma} (a_{j\sigma}^{\dagger} a_{i\sigma} + H.C.) + U \sum_{i} n_{i\uparrow} n_{i\downarrow} + \left(\frac{2}{N}\right)^2 V \sum_{i,\sigma} n_{i\sigma} \left(i - \frac{N}{2}\right)^2, \quad (1)$$

J. Labarta, K. Joe, and T. Sato (Eds.): ISHPC 2005 and ALPS 2006, LNCS 4759, pp. 402–413, 2008.

where t, U, and V are the hopping parameter from i-th to j-th sites (normally j is the nearest neighbor site of i), the repulsive energy for on-site double occupation of two fermions, and the parameter characterizing the strength of the trapping potential, respectively, as schematically shown in Fig. 1, and $a_{i,\sigma}$, $a_{i,\sigma}^{\dagger}$ and $n_{i,\sigma}$ are the annihilation, the creation, and the number operator of a fermion with pseudo-spin σ on the i-th site, respectively.

The computational approaches on the Hubbard model are roughly categorized into three types. The first one is the exact diagonalization using the Lanczos method [12], the second one is DMRG and the third one is the quantum Monte Calro [8,9]. The first one directly calculates the ground and a few excited states of the model, and moreover, obtains various physical quantities with considerably high accuracy. However, the numbers of fermions and sites are severely limited because the matrix size of the Hubbard Hamiltonian almost exponentially grows with increasing these numbers. On the other hand, the second and third one have an advantage in terms of these numbers, but confronts dimension limitation due to its theoretical ground and a fatal problem as the negative sign in the probability calculation [8,9], respectively. From these contexts, if computational resources are permitted infinitely, the exact diagonalization is clearly found to be the best way. Thus, one can raise a challenging theme for supercomputing. This is to implement the exact diagonalization code on the present top-class supercomputer, i.e., the Earth Simulator [13], and to examine how large matrices can be solved and how excellent performance can be obtained.

In this paper, we present our progress in both algorithm and technique to solve the eigenvalue problem of the Hubbard Hamiltonian matrix (1) on the Earth Simulator. In terms of algorithm, we propose a new profitable algorithm, i.e., the preconditioned conjugate gradient (PCG) method instead of the conventional Lanczos method and compare between the Lanczos method and the PCG method. On the other hand, in terms of parallelization techniques, we implement two techniques which are the inter-node parallelization using MPI for distributed memory and the intra-node parallelization using the automatic parallelization for shared memory. Combining the inter- and intra-node parallelizations, we suggest a practical memory-saving technique to perform the largest-scale matrix operations on the Earth Simulator.

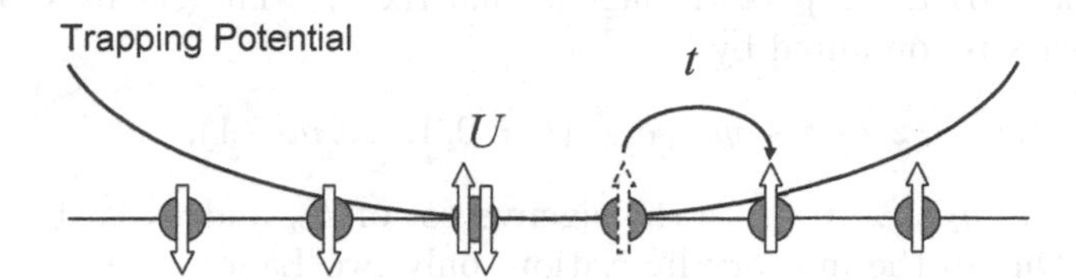

Fig. 1. A schematic figure of the one-dimensional fermion-Hubbard model with a trapping potential. Here, t and U are the hopping parameter and the repulsive energy at the double occupation on a site, respectively. The up-arrow and the down-arrow stand for fermion with up pseudo-spin and down pseudo-spin, respectively.

The contents of this paper are as follows. In Section 2, we introduce three eigenvalue solvers to diagonalize the Hamiltonian matrix of the Hubbard model and compared their convergence properties. Section 3 presents the implementation of two solvers on the Earth Simulator. Finally, we show actual performance in large-scale matrix diagonalizations on the Earth Simulator in Section 4.

2 Numerical Algorithms

Our main target is to calculate the minimum eigenvalue and the corresponding eigenvector of ultra-largescale Hamiltonian matrices on the Earth Simulator. The Hamiltonian matrices are sparse, real, and symmetric. Therefore, several iterative numerical algorithms, i.e., the power method, the Lanczos method, the conjugate gradient method, and so on, are applicable. In this section, we compare three numerical algorithms listed above, in terms of the memory use and the performance on the Earth Simulator.

2.1 Power Method

The power method is one of the most fundamental algorithms in eigenvalue problems. With zero shift determined by the Gerschgorin circle theorem, our problem can be simply solved. The algorithm is described in Fig. 2(a). Obviously, the memory usage to run the algorithm is $2N$ double precision words, where N indicates a dimension of the eignevector. This memory requirement is the minimum among three algorithms. However, as will be shown in Section 2.4, the convergence rate of the power method is quite miserable, and the desired accuracy is not obtained within a CPU time limit. Therefore, we will not touch this algorithm in the following sections.

2.2 Lanczos Method

The Lanczos method is one of the subspace projection methods that creates a Kryrov sequence and expands invariant subspace based on the procedure of the Lanczos principle [14] (see Fig. 2(b)). Eigenvalues of the projected invariant subspace well approximate those of the original matrix, and the subspace can be represented by a compact tridiagonal matrix T_m. The eigenvector z of the original matrix is computed by

$$\boldsymbol{z} \leftarrow \boldsymbol{z} + y_{k+1}\boldsymbol{v}_k \quad (k = 0, 1, \ldots, m-1),$$

where $\boldsymbol{y} = (y_1,\ y_2, \ldots,\ y_m)^T$ is the eigenvector of T_m and $\boldsymbol{v}_k$ is the k-th Lanczos vector. Due to the memory limitation, only two Lanczos vectors are stored and updated by each Lanczos iteration. Therefore, we must totally execute the Lanczos recursion twice to obtain a set of the eigenvalue and the corresponding eigenvector[1].

[1] In the following results, the number of iterations m is defined as that of the first Lanczos recursion.

For the reason given above, the memory requirement of the main recursion is $2N$ words. In addition, an N-word buffer is required for storing an eigenvector. Consequently, the total memory requirement of the Lanczos method is $3N$ words.

2.3 Conjugate Gradient Method

The conjugate gradient (CG) method is originally a popular algorithm, which is frequently used for solving large-scale linear systems. It is also applicable to the eigenvalue calculation (see Fig. 2(c), which is modified from the original algorithm to reduce the operation load related to the calculation S_A). The algorithm requires to store six vectors, i.e., $\boldsymbol{w}_i$, $\boldsymbol{p}_i$, $\boldsymbol{x}_i$, W_i, P_i, and X_i. Thus, the memory usage is totally $6N$ words.

An operator T in Fig. 2(c) indicates the preconditioner, whose choice is crucial for fast convergence. In the present Hubbard-Hamiltonian matrix, we find that the zero-shift point Jacobi preconditioner is the best one from the convergence test. Detailed description of the preconditioning is given in Appendix A.

2.4 Performance Test of Three Algorithms

The performance of the three algorithms, i.e., the power method, the Lanczos method, and the PCG method, are examined. In addition to the basic memory requirement described in Section 2.1-2.3, all algorithms demand two N-word communication buffers and additionally an N-word buffer for diagonal elements of the Hamiltonian matrix to execute parallel calculations. Table 1 summarizes the total memory usage, the number of iterations, the elapsed time, and the FLOPS rate for an eigenvalue calculation of a 1,502,337,600-dimensional Hamiltonian matrix by using 10 nodes (80 processor elements) of the Earth Simulator. The result illustrates that the PCG method is an overwhelmingly powerful algorithm except for the memory requirement.

Table 1. A performance test of three algorithms on 10 nodes of the Earth Simulator

	Power	Lanczos	PCG
Memory Requirement	5N	6N	9N
[Byte]	56.0G	67.2G	100.7G
Iteration Controllability	—	Fixed	Variable
# Iterations	No Convergence	200	91
Residual Error	—	8.3523E-9	1.255E-9
Elapsed Time [sec]	(1800.0)	95.0	28.2
FLOPS	—	269.5G	391.4G
(Peak Ratio)	—	(42.1%)	(61.1%)

```
w_0 := an initial guess.
v_0 := w_0/||w_0||
do i=1,2,... until convergence,
w_i := H v_{i-1}
θ_i := (v_{i-1}, w_i)
v_i := w_i/||w_i||
if (||v_i - v_{i-1}|| < ε||v_0||) exit
enddo
```

(a) Power method

```
x_0 := an initial guess.
β_0 := 1, v_{-1} := 0, v_0 := x_0/||x_0||
do i=0,1,... m - 1
u_i := H v_i - β_k v_{i-1}
α_i := (u_i, v_k)
w_{i+1} := u_i - α_i v_i
β_{i+1} := ||w_i||
v_{i+1} := w_i/β_{i+1}
enddo
```

(b) Lanczos method

```
x_0 := an initial guess, p_0 := 0
x_0 := x_0/||x_0||, X_0 := H x_0, P_0 = 0, μ_{-1} := (x_0, X_0)
w_0 := X_0 - μ_{-1} x_0
do k=0, ... until convergence
W_k := H w_k
S_A := {w_k, x_k, p_k}^T {W_k, X_k, P_k}
S_B := {w_k, x_k, p_k}^T {w_k, x_k, p_k}
Solve the smallest eigenvalue μ
and the corresponding vector v, S_A v = μ S_B v, v = (α, β, γ)^T.
μ_k := (μ + (x_k, X_k))/2
x_{k+1} := α w_k + β x_k + γ p_k, x_{k+1} := x_{k+1}/||x_{k+1}||
p_{k+1} := α w_k + γ p_k, p_{k+1} := p_{k+1}/||p_{k+1}||
X_{k+1} := α W_k + β X_k + γ P_k, X_{k+1} := X_{k+1}/||x_{k+1}||
P_{k+1} := α W_k + γ P_k, P_{k+1} := P_{k+1}/||p_{k+1}||
w_{k+1} := T(X_{k+1} - μ_k x_{k+1}), w_{k+1} := w_{k+1}/||w_{k+1}||
enddo
```

(c) Preconditioned conjugate gradient method

Fig. 2. Algorithms of three eigenvalue solvers

3 Implementation on the Earth Simulator

In this section, we concentrate on a core operation $H\boldsymbol{v}$ common for both the Lanczos and the PCG algorithms and present the parallelization issues including data partitioning, the communication, and the overlap. Furthermore, we give two technical remarks, i.e., a memory saving technique by combining inter- and intra-node parallelizations and an effective usage of vector pipelines.

3.1 Matrix-Vector Multiplication

By using a matrix representation, the Hubbard Hamiltonian H is mathematically given as

$$H = I \otimes A + A \otimes I + D, \tag{2}$$

where I, A, and D are the identity matrix, the sparse matrix due to the hopping between neighboring sites, and the diagonal matrix originated from the presence of the on-site repulsion, respectively.

In the core operation $\boldsymbol{Hv}$, the matrix-vector multiplications are transformed into the matrix-matrix multiplications as

$$\boldsymbol{Hv} \mapsto \begin{cases} D\boldsymbol{v} & \mapsto \bar{D} \odot V \\ (I \otimes A)\boldsymbol{v} \mapsto & AV \\ (A \otimes I)\boldsymbol{v} \mapsto & VA^T \end{cases} \tag{3}$$

where the matrix V is derived from the vector v by the following procedures. First, decompose the vector v into n blocks, and order in the two-dimensional manner as follows,

$$\boldsymbol{v} = (\underbrace{v_{1,1}, v_{2,1}, \ldots, v_{n,1}}_{\text{first block}}, \underbrace{v_{1,2}, v_{2,2}, \ldots, v_{n,2}}_{\text{second block}}, \cdots, \underbrace{v_{1,n}, v_{2,n}, \ldots, v_{n,n}}_{n\text{-th block}})^T.$$

Here, a pair of subscripts of each element v formally indicates a position of row and column of the matrix V. The k-th element of the matrix D, d_k, is also mapped onto the matrix $\bar{D}$ in the same manner, and the operator $\odot$ means an elementwise multiplication.

3.2 Data Distribution, Parallel Calculation, and Communication

The matrix V must be treated as a dense matrices whose dimension is large enough to overfull the memory capacity of a single node. Therefore, the matrix V is columnwisely partitioned. In contrast, since the matrix A is a sparse matrix and its sparsity is high, all non-zero elements of the matrix A can be stored on the memory systems of all nodes. Other large matrices are partitioned columnwisely or rowwisely on each node. The operation $\boldsymbol{Hv}$ including the data communication can be written down as follows:

CAL1: $E^c = \bar{D}^c \odot V^c$,
CAL2: $W_1^c = E^c + AV^c$,
COM1: communication to transpose V^c into V^r,
CAL3: $W_2^r = V^r A^T$,
COM2: communication to transpose W_2^r into W_2^c,
CAL4: $W^c = W_1^c + W_2^c$,

where the superscript c and r mean columnwise or rowwise partitioning, respectively. The above procedure describes twice matrix-transpose operations which are normally realized by all-to-all data communication. In the MPI standards, the all-to-all data communication is realized by a collective communication `MPI_Alltoallv`. However, due to irregular and incontiguous structure of the transferring data, the data-transpose communication should be executed by a point-to-point or a one-side communication function. Moreover, the one-side communication function `MPI_Put` more excellently runs than the point-to-point communication on the Earth simulator. In fact, the use of `MPI_Put` is recommended by the developers in this case [15].

3.3 Communication Overlap

To raise up the performance, it is crucial to hide communication behind computation. In the procedure of the matrix-matrix multiplication in Section 3.2, the calculations **CAL1** and **CAL2** and the communication **COM1** is clearly found to be independently executed. Moreover, although the relation between **CAL3** and **COM2** is not so simple, the overlap is principally possible. Thus, the two communication processes (**COM1** and **COM2**) are expected to be hidden behind the calculations. However, we note that the overlap depends on an implementation level of the MPI library. In fact, the MPI library installed on the Earth Simulator has not provided any functions of the overlap[2]. Thus, we have to use the non-blocking `MPI_Put` as a blocking communication like `MPI_SEND`.

In order to improve the performance under such a situation, we propose a communication strategy to realize the overlap in the use of the blocking `MPI_Put` on the Earth Simulator. The way is as follows: The blocking `MPI_Put` can be assigned to a single PE per node by the intra-node parallelization technique. Then, the assigned processor dedicates only communication (see Fig. 3). Consequently, the calculation load is divided into seven PE's. The intra-node parallelism is described by the automatic parallelization using CDIR compiler directives. This parallelization method, which we call task assignment (TA) method, imitates a non-blocking communication and enable to overlap between the blocking communication and calculation (see Appendix B).

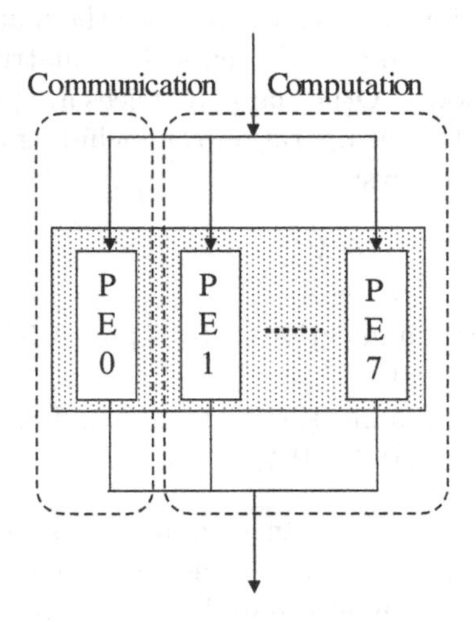

Fig. 3. A schematic figure for the task division. The communication task is assigned to a processor element (ex. PE 0) on each node.

[2] At the present, the situation is different from the time when this paper was accepted, because the MPI library to overlap between the MPI non-blocking communication and computation was released in December, 2005.

Table 2. The dimension of Hamiltonian matrix H, the number of nodes, and memory requirements from Model 1 to 4. Model 4 in the PCG method requires more than 8TB, which is beyond the memory size of 512 nodes of the Earth Simulator.

Model	No. of Sites	No. of Fermions		Dimension of H	No. of Nodes	Memory (TB)	
		↑-spin	↓-spin			Lanczos	PCG
1	24	6	6	18,116,083,216	128	0.8	1.3
2	21	8	8	41,408,180,100	256	1.9	2.9
3	22	8	8	102,252,852,900	512	4.6	6.9
4	22	9	8	159,059,993,400	512	7.1	(10.7)

4 Performance on the Earth Simulator

Let us present the performance of the Lanczos method and the PCG method with TA method for huge Hamiltonian matrices (see Table 2 for the numbers of sites and fermions and the matrix dimension). Table 3 shows the performance of these methods on 128 nodes, 256 nodes and 512 nodes of the Earth Simulator. The performance measurements are made as follows. The total elapsed time and FLOPS rates are measured by using the simple performance analysis routines [16] installed on the Earth Simulator. On the other hand, the elapsed time of the solvers are measured by using `MPI_Wtime` function, and the FLOPS rates of the solvers are evaluated by the elapsed time and flop count estimated according to the following formulae:

Lanczos method: $5 * ndim + 16 * itr * ndim + 2 * itr * (2 * nnz - ndim)$,
PCG method: $35 * ndim + 46 * itr * ndim + (itr + 2) * (2 * nnz - ndim)$,

Table 3. Performances of the Lanczos method and the PCG method with the TA method on the Earth Simulator

a) The number of iterations, residual error, and elapsed time.

Model	Lanczos Method				PCG Method			
	itr.	Residual Error	Elapsed Time(sec)		itr.	Residual Error	Elapsed Time(sec)	
			Total	Solver			Total	Solver
1	200	1.1×10^{-7}	106.905	101.666	105	1.4×10^{-9}	39.325	34.285
2	200	7.7×10^{-7}	154.159	148.453	107	2.3×10^{-9}	55.888	48.669
3	300	3.6×10^{-11}	288.270	279.775	109	2.4×10^{-9}	69.123	60.640
4	300	4.2×10^{-8}	362.635	352.944	—			

b) FLOPS rate.

Model	TFLOPS (Peak Ratio)			
	Lanczos Method		PCG Method	
	Total	Solver	Total	Solver
1	3.062(37.4%)	3.208(39.2%)	4.045(49.4%)	4.607(56.2%)
2	5.245(32.0%)	5.426(33.1%)	6.928(42.3%)	7.893(48.2%)
3	10.613(32.3%)	10.906(33.3%)	14.271(43.6%)	16.140(49.3%)
4	13.363(40.8%)	13.694(41.8%)	–	–

where *ndim*, *itr* and *nnz* are the dimension of the Hamiltonian matrix H, the number of iterations, and the number of the non-zero elements of H, respectively.

As shown in Table 3, the PCG method shows better convergence property and solves the eigenvalue problems about three times faster than the Lanczos method. Moreover, the PCG method overlaps communication tasks with calculations more than the Lanczos method. The reason is that the PCG method includes a routine calculating inner products in which the communication can be efficiently hidden. The best performance of the PCG method is 16.140 TFLOPS on 512 nodes which is 49.3% of the theoretical peak. On the other hand, Table 2 and 3 show that the Lanczos method can solve up to the 159-billion dimensional Hamiltonian matrix on 512 nodes. To our knowledge, this size is the largest in a history of the exact diagonalization method of Hamiltonian matrices.

5 Conclusions

The best performance (16.140TFLOPS) in the present is comparable to those of other ones on the Earth Simulator [17,18,19,20,21]. However, we would like to point out that our application requires a huge amount of communications in contrast to the previous ones. Therefore, we made much effort to hide the communication by paying an attention to the architecture of the Earth Simulator. Furthermore, we suggested a new algorithm (PCG) showing the best performance and succeeded to drastically shorten the total elapsed time. In conclusion, we obtain the best performance by the new algorithm and attain the world record of the large matrix operation. We believe that these results are outstanding achievements in high performance computing.

Acknowledgements

The authors in CCSE JAEA thank G. Yagawa, T. Hirayama, N. Nakajima, C. Arakawa, N. Inoue and T. Kano for their supports and acknowledge K. Itakura and staff members in the Earth Simulator for their supports in the present calculations. One of the authors (M.M.) acknowledges T. Egami and P. Piekarz for illuminating discussion about diagonalization for d-p model and H. Matsumoto and Y. Ohashi for their collaboration on the optical-lattice fermion systems.

References

1. Anderson, M.H., Ensher, J.R., Matthews, M.R., Wieman, C.E., Cornell, E.A.: Observation of Bose-Einstein Condensation in a Dilute Atomic Vapor. Science 269, 198–201 (1995)
2. Bradley, C.C., Sackett, C.A., Tollett, J.J., Hulet, R.G.: Evidence of Bose-Einstein condensation in an atomic gas with attractive interactions. Phys. Rev. Lett. 75, 1687–1690 (1995)
3. Davis, K.B., Mewes, M.-O., Andrews, M.R., van Druten, N.J., Durfee, D.S., Kurn, D.M., Ketterle, W.: Bose-Einstein condensation in a gas of sodium atoms. Phys. Rev. Lett. 75, 3969–3973 (1995)

4. Regal, C.A., Greiner, M., Jin, D.S.: Observation of resonance condensation of fermionic atom pairs. Phys. Rev. Lett. 92, 040403 (2004)
5. Kinast, J., Hemmer, S.L., Gehm, M.E., Turlapov, A., Thomas, J.E.: Evidence for superfluidity in a resonantly interacting Fermi gas. Phys. Rev. Lett. 92, 150402 (2004)
6. Machida, M., Yamada, S., Ohashi, Y., Matsumoto, H.: Novel superfluidity in a trapped gas of Fermi atoms with repulsive interaction loaded on an optical lattice. Phys. Rev. Lett. 93, 200402 (2004)
7. Machida, M., Yamada, S., Ohashi, Y., Matsumoto, H., Machida, et al.: Reply. Phys. Rev. Lett. 95, 218902 (2005)
8. Rasetti, M. (ed.): The Hubbard Model, Recent Results. World Scientific, Singapore (1991)
9. Montorsi, A. (ed.): The Hubbard Model: A Reprint Volume. World Scientific, Singapore (1991)
10. Greiner, M., Mandel, O., Esslinger, T., Hansch, T.W., Bloch, I.: Quantum phase transition from a superfluid to a Mott insulator in a gas of ultracold atoms. Nature 415, 39 (2002)
11. Rigol, M., Muramatsu, A., Batrouni, G.G., Scalettar, R.T.: Local quantum criticality in confined fermions on optical lattices. Phys. Rev. Lett. 91, 130403 (2003)
12. Dagotto, E.: Correlated electrons in high-temperature superconductors. Rev. Mod. Phys. 66, 763 (1994)
13. The Earth Simulator Center Home page, `http://www.es.jamstec.go.jp/index.en.html`
14. Cullum, J.K., Willoughby, R.A.: Lanczos Algorithms for Large Symmetric Eigenvalue Computations. In: Theory, vol. 1, SIAM, Philadelphia (2002)
15. Uehara, H., Tamura, M., Yokokawa, M.: MPI Performance Measurement on the Earth Simulator. NEC Research & Development 44(1), 75–79 (2003)
16. NEC Corporation, FORTRAN90/ES Programmer's Guide, EARTH SIMULATOR User's Manuals, NEC Corporation (2002)
17. Shingu, S., Takahara, H., Fuchigami, H., Yamada, M., Tsuda, Y., Ohfuchi, W., Sasaki, Y., Kobayashi, K., Hagiwara, T., Habata, S., Yokokawa, M., Itoh, H., Otsuka, K.: A 26.58 Tflops Global Atmospheric Simulation with the Spectral Transform Method on the Earth Simulator. In: Proc. of SC 2002 (2002), `http://sc-2002.org/paperpdfs/pap.pap331.pdf`
18. Sakagami, H., Murai, H., Seo, Y., Yokokawa, M.: 14.9 TFLOPS Three-dimensional Fluid Simulation for Fusion Science with HPF on the Earth Simulator. In: Proc. of SC 2002 (2002), `http://sc-2002.org/paperpdfs/pap.pap147.pdf`
19. Yokokawa, M., Itakura, K., Uno, A., Ishihara, T., Kaneda, Y.: 16.4 Tflops Direct Numerical Simulation of Turbulence by Fourier Spectral Method on the Earth Simulator. In: Proc. of SC 2002 (2002), `http://sc-2002.org/paperpdfs/pap.pap273.pdf`
20. Komatitsch, D., Tsuboi, S., Ji, C., Tromp, J.: A 14.6 billion degrees of freedom, 5 teraflops, 2.5 terabyte earthquake simulation on the Earth Simulator. In: Proc. of SC 2003 (2003), `http://www.sc-conference.org/sc2003/paperpdfs/pap124.pdf`
21. Kageyama, A., Kameyama, M., Fujihara, S., Yoshida, M., Hyodo, M., Tsuda, Y.: A 15.2 TFlops Simulation of Geodynamo on the Earth Simulator.In: Proc. of SC 2004 (2004), `http://www.sc-conference.org/sc2004/schedule/pdfs/pap234.pdf`

Appendix A: Preconditioner for CG Method

It is well-known that the preconditioning improves convergence of the CG method. Effective preconditioning needs to identify the mathematical characteristics of the matrix. However, it is generally hard to identify them. In this section, we focus on the following five preconditioners:

1. Point Jacobi,
2. Zero-shift point Jacobi,
3. Block Jacobi,
4. Neumann-polynomial expansion,
5. SSOR-type iteration.

Here, 1, 3, 4, and 5 are very popular preconditioners for the CG method and 2 (zero-shift point Jacobi) is a modified one of 1 (point Jacobi). The zero-shift point Jacobi is a diagonal scaling preconditioner shifted by '$-\mu_k$' to amplify the ground-state eigenvector, i.e., the preconditioning matrix is given by $T = (D - \mu_k I)^{-1}$, where D, I, and μ_k are the diagonal part of the matrix H, the identity matrix, and an approximate of the smallest eigenvalue which appears in the PCG iteration, respectively. In order to solve a huge matrix and achieve higher performance, we select 1 and 2, since they do not require any data communication and any extra storage.

Here, we calculate the same eigenvalue problem as Section 2.4. Table 4 summarizes a performance test of three cases, 0) without preconditioner (NP), 1) point Jacobi (PJ), and 2) zero-shift point Jacobi (ZS-PJ) on the Earth Simulator. This result clearly reveals that the zero-shift point Jacobi is the best preconditioner.

Table 4. Comparison among three preconditioners

	0) NP	1) PJ	2) ZS-PJ
# Iteration	268	133	91
Residual Error	1.445E-09	1.404E-09	1.255E-09
Elapsed Time (sec)	78.904	40.785	28.205
FLOPS	382.55G	383.96G	391.37G

This eigenvalue problem is the same as in Section 2.4.

Appendix B: Overlap between Communication and Calculation

In order to examine the possibility of overlap between communication and calculation by utilizing TA method, we perform the multiplication $H\boldsymbol{v}$ on 10 nodes (80 PE's) of the Earth Simulator. A test matrix H is a 1.5-billion-dimensional matrix derived from the one-dimensional 20-site Hubbard model with 12 fermions ($6 \uparrow$, $6 \downarrow$). We measure the elapsed time of the four calculations **CAL1-4** and the two communications **COM1-2** shown in Section 3.2. We show the timecharts of TA and non-assignment (NA) methods in Fig. 4. As shown in the figure, the calculations and the communications are executed serially in NA method. On the other hand, the calculation and the communication in TA method are executed simultaneously.

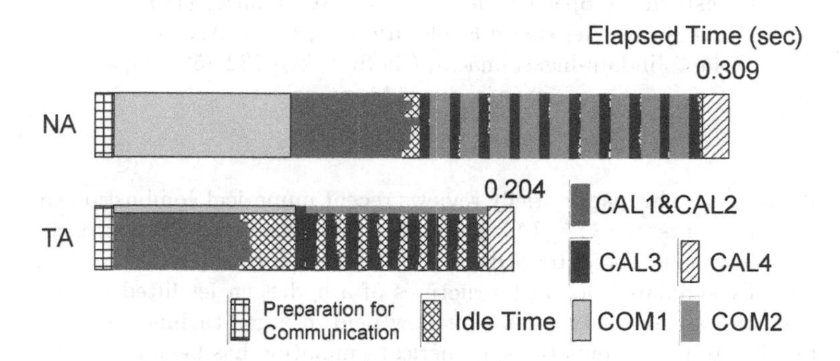

Fig. 4. A comparison of the process timechart on node No.0 for the multiplication Hv between the TA method and the NA method

Numerical Simulation of Combustion Dynamics at ISTA/JAXA

Junji Shinjo, Shingo Matsuyama, Yasuhiro Mizobuchi, and Satoru Ogawa

Institute of Space Technology and Aeronautics (ISTA),
Japan Aerospace Exploration Agency (JAXA)
7-44-1 Jindaiji-higashimachi, Chofu Tokyo 182-8522 Japan
shinjo.junji@jaxa.jp

Abstract. This paper briefly reviews recent numerical combustion simulation results at ISTA/JAXA obtained by DNS and LES approaches, and shows some topics towards future combustion research. We have successfully simulated detailed structures of a hydrogen jet lifted flame by DNS and an unsteady combustor flow field in a gas turbine combustor by LES. In these simulations, numerical simulation has been proved very effective, but its applicability is still limited due to long computational time. It is expected that future progress in computer performance will make this kind of simulation more realistic and useful in combustion research and development.

1 Introduction

Combustion has been utilized by humankind since ancient time to produce heat, light and power, and even today it is one of the most common and familiar energy generation methods in our daily life. In terms of research and development, combustion has been one of the major topics both in scientific and engineering fields because combustion processes involve a lot of complicated physical phenomena, such as chemical reactions, heat transfer, convection and radiation. Even today, there are still many issues to be investigated to understand what is taking place in reacting flows. Furthermore, in practical devices, the flow filed is usually turbulent. This adds one more complexity to the flow field. Turbulence and combustion influence each other and the interaction is non-linear. Numerical simulation has become a powerful tool in understanding combustion processes in several scales. For laboratory-scale flames, the inner flame structure is focused on to understand the balance between chemical reaction, convection, diffusion, turbulence, etc. This kind of analysis is sometimes difficult by experimental measurements. The obtained data will lead to new flame models, novel combustion methods, etc. For practical-scale combustion devices, such as automobile engines, gas turbines, furnaces, etc, resolving fine structures of flame may be difficult, but flame/flow interaction analysis and system evaluation by numerical simulation have become possible and used together with experiments. In the present situation, however, different scales are treated differently. The complexity lies in the very wide range of temporal and spatial scales, and it is

J. Labarta, K. Joe, and T. Sato (Eds.): ISHPC 2005 and ALPS 2006, LNCS 4759, pp. 414–426, 2008.

still impossible to cover the entire range by direct numerical simulation. The computational time is usually large, even for LES, and this limits the number of possible cases for parametric study. It is thus expected that future progress in computer performance will improve the situation. In this paper, we briefly review our recent numerical simulation results and computational environment in combustion research, and discuss future numerical combustion research and demands for computer performance.

2 Numerical Approaches

There are basically three types of numerical approaches to turbulent combustion investigation, direct numerical simulation (DNS), Reynolds-averaged Navier-Stokes simulation (RANS) and large-eddy simulation (LES). In DNS, the smallest scale, the inner heat release layer of flame or turbulent Kolmogorov scale, must be resolved. Even with today's powerful computers, its computational domain must be usually very small. The governing equations are rigorously determined and use no model. DNS is used in combustion research to investigate flame structures, flame/turbulence interaction, flame/wall interaction, etc. RANS is based on ensemble-averaging, so it is very useful in obtaining averaged flow field data. In this paper, however, any simulation result based on this approach will not be shown. LES is a spatially-filtered method, where grid scale structures are directly resolved and sub-grid scale phenomena are modeled. The underlying principle for modeling sub-grid scales is Kolmogorov's similarity law. The flame inner structure is usually smaller than the grid scale, so flame modeling is an important issue in conducting combustion LES. In our simulation of combustion dynamics in a lean premixed combustor, the phenomenon is practical scale and unsteady, so this approach is a natural selection to understand the dynamics. Figure 1 shows schematically the resolved scales in numerical simulation [1].

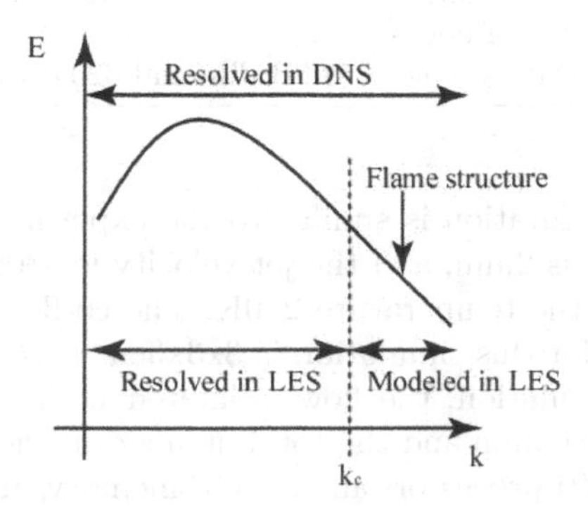

Fig. 1. Turbulence energy spectrum and resolved scales in numerical simulation. kc represents the cut-off wave number. In DNS, all the scales are resolved, while in LES, scales smaller than the cut-off scale are modeled. The flame inner structure usually lies in the modeled region in LES.

In the following section, we briefly present two results from our recent simulation research: one obtained by DNS and the other by LES. The computer resources necessary to conduct these simulations are also shown. Then, the future issues related to numerical combustion simulation will be discussed.

3 Recent Results of Numerical Simulation

3.1 DNS Study

This study is Mizobuchi's work [2,3]. See the references for more detail. In this study, a hydrogen lifted flame is simulated. Hydrogen is injected into air through a nozzle. When the injection speed is low, the flame is attached to the nozzle rim. As the injection speed is increased, the flame detaches from the rim at a certain velocity, and stabilized as a lifted flame with a certain lift-off distance. When the injection velocity is further increased, the flame will be blown off leading to total extinction. The mechanism of flame stabilization may lie in the balance between the flow velocity and combustion processes, but the detailed mechanism has not been known so far. This study is the first to show the complex structures of lifted flame stabilization by numerical simulation. The computation has been carried out on the Central Numerical Simulation System (CeNSS) installed at Aerospace Research Center (ARC), Japan Aerospace Exploration Agency (JAXA) (table 1). The system has 1792 processors and 3TB memory in total. The theoretical peak performance is 9.3TFLOPS.

Table 1. CeNSS performance (Fujitsu Primepower)

CPU architecture	SPARC64V, 1.3GHz
Number of CPUs	1792
Total memory	3TB
Peak performance	9.3TFLOPS
Number of nodes	56
Data storage	57TB disc and 620TB tape

The flow field configuration is similar to the experiment by Cheng [4]. The hydrogen jet diameter is 2mm, and the jet velocity is 680m/s (Mach 0.54). The pressure is 1atm and the temperature 280K. The co-flow velocity is zero. The computation domain for this simulation is 3x3x3cm in real physical volume. At the beginning the calculation, the flow is ignited with additional energy. The finest grid spacing is 0.05mm and the total number of grid points is 200 million. This simulation uses 291 processors and 230GB memory, and computational time is about 1500 hours for 1ms physical time integration. The grid spacing is small enough to resolve the inner heat layer of flame with more than 10 grid points. The computational domain is decomposed into 75 sub-domains and parallelization is based on MPI and OpenMP.

Table 2. Computational conditions for hydrogen lifted flame DNS

Computational domain size	3x3x3cm
Characteristic length	2mm (nozzle diameter)
Number of grid points	200 million
Minimum grid spacing	0.05mm
Number of processors	291
Computational time	1500hrs for 1ms integration

Figure 2 shows the temperature iso-surface of 1000K of the lifted flame. The hydrogen jet is mixed with the surrounding air and the flame stably balances at 5.5D downstream, where D is the diameter of the injection tube. This lift-off height is close to the experimental observation [4]. Using the flame index [2,3], the local flame structure can be classified into two major categories: premixed flame and diffusion flame. When the flame index is positive, the flame is premixed, when it is negative, the flame is diffusive. Figure 3 shows the flame index structure. In the inner flame region, the flame configuration is almost premixed, while the outer flame layer is mostly diffusive. The edge flame at the flame base has a very complicated structure and both premixed and diffusion flames are present. The inner premixed flame is rich (fuel is excessive). Lean premixed flame is only seen around the base flame. In the outer diffusive flame region, the reaction intensity differs locally and it looks that there are several diffusion flame 'islands'.

In the inner premixed region, where the turbulent intensity is high, the heat release layer and the fuel consumption layer are not the same shape as shown in figure 4. This is because the turbulent eddies penetrate into the flame, and the flame structure can be no longer viewed as a flamelet. The flamelet means one-dimensional laminar flame, and this modeling is typically used in LES studies

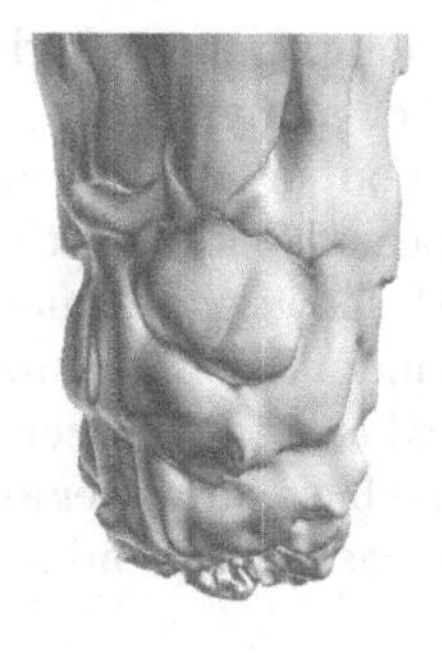

Fig. 2. Temperature iso-surface at 1000K

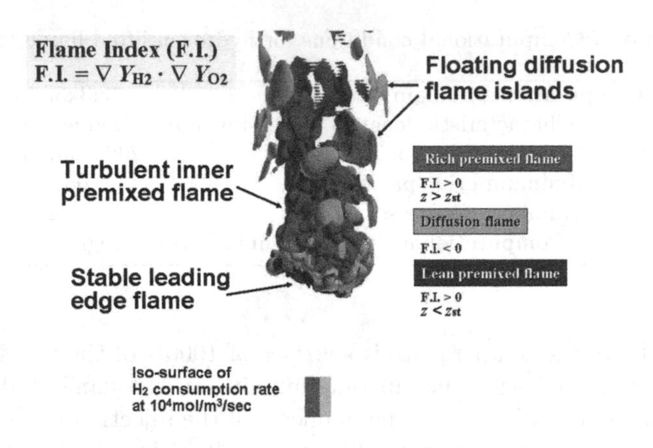

Fig. 3. Flame index distribution

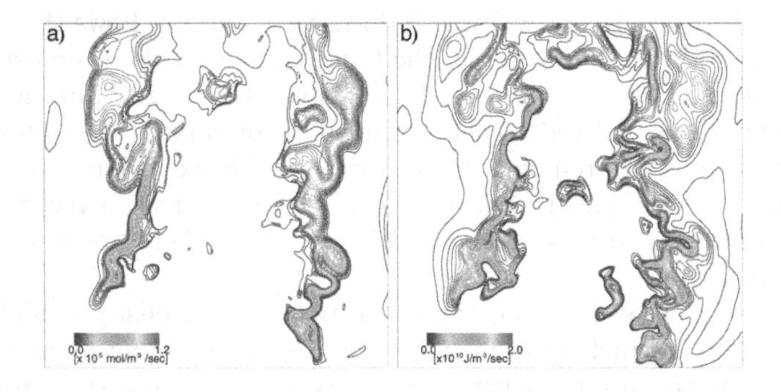

Fig. 4. (a) hydrogen consumption layer and (b) heat release layer in the inner premixed flame region

when the turbulent intensity is not extremely high. Non-flamelet flame modeling is more difficult and needs further analysis using this kind of data.

The formation process of the outer diffusion flame islands has been clarified by time-series analysis. It has been found that there may be two mechanisms of flame islands formation. Figure 5 shows the sequences of flame islands formation. In the downstream region, the inner flame intermittently touches the outer region by turbulent motion. Then, local extinction occurs and the outer part detaches as a diffusion flame island. In the base flame region, a rich gas pocket created by the shear instability attacks the base flame and a part of the base flame becomes a diffusion flame island.

The base flame is very complicated and should be investigated further to elucidate its structure and the stabilization mechanism. The investigation is now under way. As shown above, it has been found that the turbulent lifted flame is not a simple flame, but consists of three flame elements. This has not been found by experiments, but by this numerical simulation.

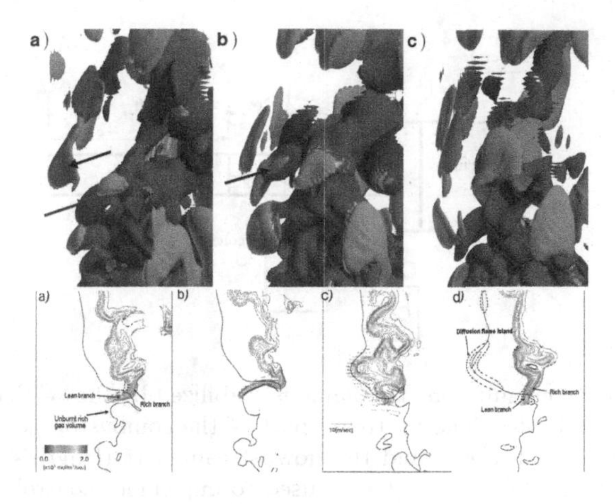

Fig. 5. Sequences of formation of diffusion flame islands (a) at the outer flame region (above) and (b) at the base flame (below)

3.2 LES Study

This study is Shinjo's work [5,6]. This is an engineering study to investigate combustion dynamics in a real-scale combustor. Lean premixed combustion is a promising method to reduce emissions such as NO_x from the exhaust gas, but this combustion process tends to be unstable. Under certain operation conditions, combustion oscillations may occur, sometimes leading to mechanical damage or extinction. Especially for aircraft engines, this is a critical issue and in order to widen the stable operation range, active/passive control strategies have been studied worldwide [7,8]. The objectives of this study are (1) to understand combustion oscillation dynamics in a combustor and (2) to examine active/passive control implementation. This simulation focuses on unsteady combustion oscillation in a practical-scale combustor, so the LES approach is best suited. The combustor considered here is a swirl-stabilized gas turbine combustor used in the experiment of Tachibana, et al. [9]. The combustor length is 63cm and the width is 10cm. The inner diameter of the swirler is 20mm and the outer diameter 50mm. The swirl angle is set at 45 degrees. There are 12 injection holes on the edge of the swirler inner hub to implement secondary fuel injection control. Figure 6 shows the schematic of the combustor.

The fuel used here is methane and it is mixed and heated in the premixing chamber before entering the combustor. The equivalence ratio $(F/A)\,/\,(F/A)_{st}$ (fuel/air mass ratio) is set at 0.5, where the subscript *st* stands for 'stoichiometric condition'. The incoming flow conditions are: temperature of 700K, pressure of 1atm and velocity of 90m/s. Figure 7 shows the computational domain used in this analysis. It includes a swirler inlet section, a downstream region and the combustor. The downstream region is added to implement far-field fixed-pressure condition at the exit boundary. The domain is decomposed into 60 sub-domains for parallel computation. The total number of grid points is around 10 million to

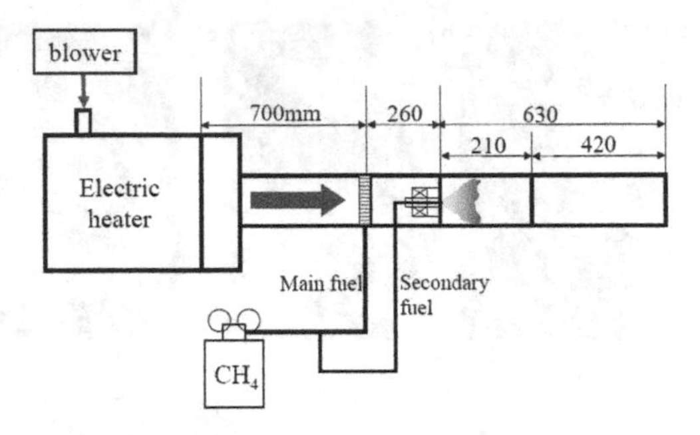

Fig. 6. Combustor configuration. The flame is stabilized by the swirler installed at the entrance of the combustor. The upstream part of the combustor (length 210mm) has optical access for visualization, and the downstream part (length 420mm) is water-cooled. The secondary fuel feed system is used to implement control.

30 million, depending on the flow conditions. The computation has been carried out on the CeNSS supercomputer system at JAXA/ISTA, and the OpenMP and thread parallelization method are used. Total memory used is 150GB for 10 million grid points.

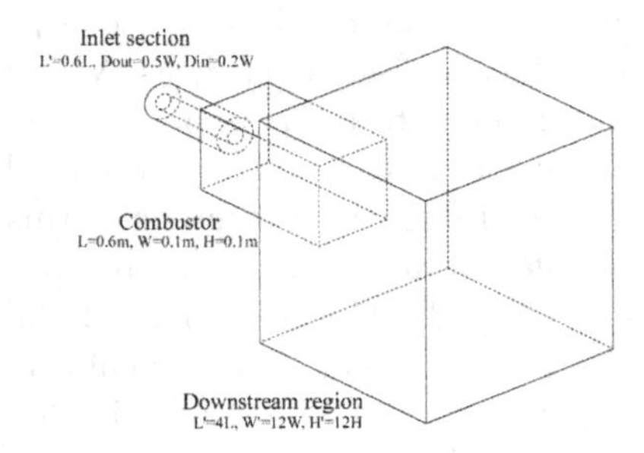

Fig. 7. Computational domain

Under typical gas turbine operation conditions, the flame thickness is thin (0.1-1mm) compared to the scale of the combustor, so it cannot be resolved on an LES mesh. In this study, the flame is modeled as a sheet between unburned and burned gases, and only its geometrical motion is computed to make the calculation less expensive. Chemical reactions are pre-calculated and the result is stored in a look-up table. By doing this, stiffness of chemical calculation is eased. The flame motion is governed by the following G-equation

$$\frac{\partial \rho G}{\partial t} + \frac{\partial \rho G u_j}{\partial x_j} = \rho_u s_L |\nabla G| \tag{1}$$

where $G = 0$ represents the unburned region and $G = 1$ burned region. The inner flame structure is not resolved, and the laminar burning velocity s_L must

Table 3. Computational conditions for combustor LES

Computational domain size	10x10x63cm (combustor)
Number of grid points	10-30 million
Minimum grid spacing	0.3mm
Number of processors	122
Computational time	1400hrs for 100ms integration

be given explicitly. In this study, the CHEMKIN software [10] is used to obtain the burning velocity. After implementing LES filtering, this equation is solved, coupled with the flow field equations of mass, momentum and energy. The LES sub-grid model for flow filed is the dynamic Smagorinsky model and we use a semi-empirical formulation for turbulent burning velocity [1,11] to include the effect of sub-grid turbulent intensity.

$$\frac{\tilde{s}_T}{s_L} = 1 + C\left(\frac{u'_{sgs}}{s_L}\right) \tag{2}$$

$$u'_{sgs} = C_s \Delta \sqrt{\bar{S}_{ij}\bar{S}_{ij}} \tag{3}$$

First, a case without any control is examined to understand the combustion dynamics. Figure 8 shows the instantaneous snapshot of flame shape in the combustor. The flame shape is very complicated due to flame/turbulence interaction. The flame is stabilized behind the swirler due to recirculation zones created by the swirling motion. Figure 8 also shows the time-averaged flow field. The central recirculation zone pushes the burned gas upstream and serves to hold the flame in this position.

The complex flame shape is related to pressure oscillations in the combustor. The temporal pressure trace is measured, as in the experiment, on the combustor wall 10mm downstream from the dump plane. The FFT result shows that the basic frequency of oscillation is about 300Hz, and this frequency corresponds to

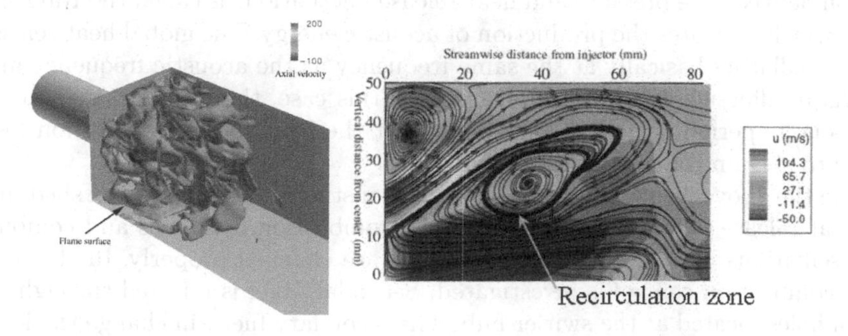

Fig. 8. (a) (left) Instantaneous flame shape. The color on the flame indicates the local axial velocity. (b) (right) time-averaged axial velocity field and streamlines. The upper half from the centerline is shown. The central and corner recirculation zones are formed.

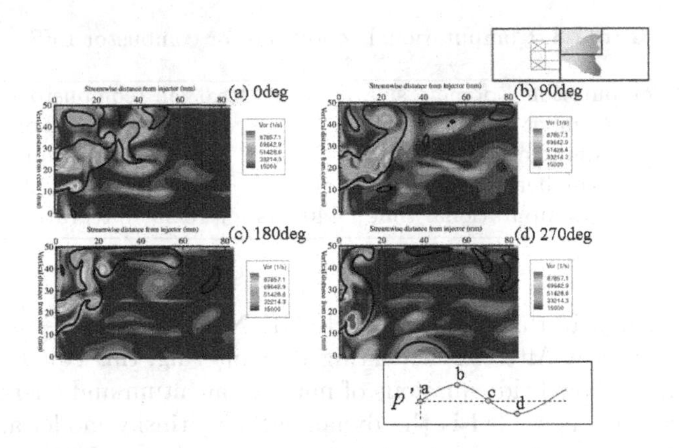

Fig. 9. Vortex/flame interaction in one acoustic cycle. The solid line indicates the flame shape and contours represent vorticity magnitude. The upper half from the centerline is shown.

the quarter-wave mode. The quarter-wave mode is determined by the combustor length L and the speed of sound of the burned gas a as

$$f_{\frac{1}{4}} = \frac{a}{4L} \sim 300Hz \tag{4}$$

In the experiment, we observed a close frequency [9], so this simulation successfully captures the pressure oscillations in the combustor. The pressure oscillations induce velocity fluctuations, thus leading to periodic vortex shedding from the dump plane. Figure 9 shows one cycle of interaction between vortices and flame. During the period, vortices are shed from the combustor dump plane, and they change local velocity field, thus flame shape and position. The interaction is especially strong near the flame base region.

The change in flame shape/location is directly linked to heat release change. Figure 10 shows the temporal traces of pressure and global heat release. The correlation between the pressure and heat release fluctuations is called the Rayleigh index, which indicates the production of acoustic energy. The global heat release rate is oscillating basically at the same frequency as the acoustic frequency and the phase difference is within 90 degrees. In this case, the Rayleigh index over one acoustic period is positive, meaning that the heat release fluctuation feed energy to sustain the combustion oscillations.

From the above result, the links between acoustic oscillations, vortex shedding and heat release fluctuations determine the combustion dynamics and combustion oscillations may be reduced if the links are changed properly. In the next stage, combustion control is investigated. Secondary fuel is injected through injection holes located at the swirler hub. The secondary fuel will change the local and global heat release and expected to reduce the oscillation amplitude. In the experiment by Tachibana et al. [9], both open-loop and closed-loop control methods were tested. The 'open-loop' here means continuous injection and the

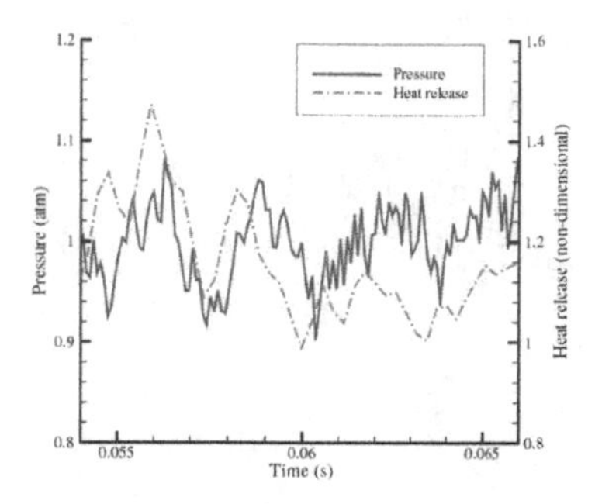

Fig. 10. Temporal traces of pressure and global heat release

'closed-loop' control uses combined H_2/H_∞ feedback injection. The oscillation amplitude was reduced by 17dB by the open-loop control and 28dB by the closed-loop control. In this section, the open-loop control is simulated. The secondary fuel is 100% methane and its amount is 3% compared to the main fuel. In the experiment, the flame shape changes by secondary fuel injection. The flame at the outer rim detaches from the wall. In this analysis, the same situation is simulated. Figure 11 shows a close-up view of the local temperature field when the injection is on. Around the secondary fuel, a diffusive flame is formed and the local temperature becomes higher than the premixed flame temperature. This means that the local heat release is also changed by injection.

The pressure traces with and without secondary injection are plotted in figure 12. The oscillation amplitude becomes smaller when the injection is on, although the long-term trend is not yet examined. The reason for this reduction can be attributed to (1) the flame base is stabilized by diffusive combustion mode and (2) the flame shape extends to downstream region where the coupling with pressure is smaller. The closed-loop control simulation is now under way and will be reported in the near future.

As shown in this section, numerical simulation based on LES with flame models has become a powerful tool to understand what is taking place in practical-

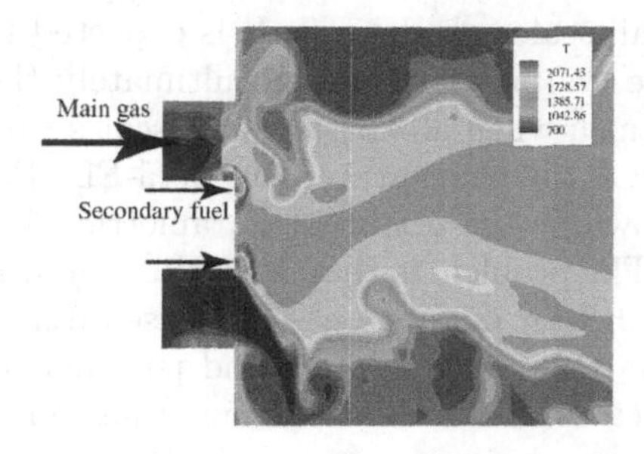

Fig. 11. Temperature field with the injection on

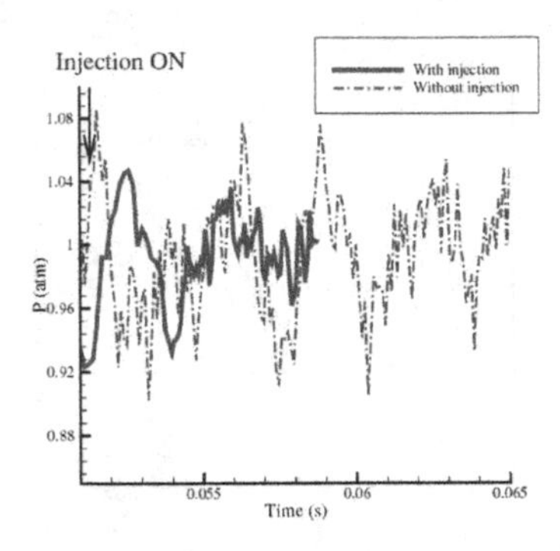

Fig. 12. Pressure traces with and without secondary fuel injection

scale engineering problems. However, flame modeling issues such as local extinction, re-ignition, wall interaction, etc. are not included in the present simulation. The modeling will be future subjects in combustion research.

4 Future Combustion Research and Computer Performance

The effectiveness of numerical simulation in combustion research will continuously grow. To list a few topics expected to be major themes in future numerical combustion research, (1) Simulation with complicated chemical kinetics. In the result shown in 3.1, the fuel used is hydrogen. The chemical kinetics for hydrogen is rather simple. The chemical species involved are 9 species (H_2, O_2, OH, H_2O, H, O, H_2O_2, HO_2 and N_2) and 17 elementary reactions are considered. For methane (CH_4) full kinetics (GRI-Mech3.0 [12]), 53 chemical species and 325 elementary reactions are necessary [12]. For larger hydrocarbons, these numbers dramatically increase. At present, we often use reduced kinetics, where several fast reactions are considered to be in equilibrium and removed, to save computational cost. Future peta-FLOPS machines will make full kinetics DNS of hydrocarbons a realistic option, and physical understanding will be deepened by this. (2) Real-scale DNS. Due to the limit of computational time, the present DNS is limited to small scale calculations. It is expected that peta-FLOPS machines will expand the limitation. Although ultimately the distinction between DNS and LES will vanish when all the physical scales can be resolved in DNS, this will be impossible in the near future. But peta-FLOPS machines will surely make the direction toward that goal. (3) Parametric LES study for designing engineering systems. The problem is that it is still computationally expensive to conduct combustion LES for many cases. One case often takes several-hundred hours, even using today's supercomputers and parametric computation can be only done for limited cases. If this kind of simulation should be used in combustor and combustor system design processes, the turnaround time must be a few days or less. This requires over 100 times faster performance. On peta-FLOPS machines, this will be achieved and combustion LES will become a realistic design tool in engineering applications. The applications include, for example, gas turbine combustors, automobile combustors, rocket engines, power generation systems, etc. (4) Multiphase combustion field. Although it is not discussed in this paper, liquid and solid fuels are commonly used in combustion devices. In relation to chemical kinetics discussed in (1), detailed physical mechanisms of combustion with liquid atomization, phase transition, etc. are expected to be investigated.

As the simulation size enlarges, post-processing of data is also a critical issue. Data analysis, visualization and storage should be considered at the same time with computer performance development.

5 Summary

In combustion research, numerical simulation has become a powerful tool to understand the combustion physics in multi-scales. It provides data that cannot be obtained by experiment, and the significance of numerical combustion research will further increase in the future. However, even using today's supercomputers, the turnaround time is still long, and it is almost impossible to conduct parametric LES or large-scale DNS. And there are still many issues to be investigated, such as chemical kinetics, flame/turbulence interaction, flame/wall interaction, extinction, re-ignition, etc. Future computer development will contribute to solving these problems. Peta-FLOPS supercomputers will dramatically decrease the turnaround time. If the calculation time for one case in LES is reduced to one week or shorter, it can be used in real combustor design and evaluation processes. It will also enable us to conduct larger-scale and more complicated DNS. At the same time, data handling will become more difficult in the future, because the amount of data produced by numerical simulation will further increase. Data storage and processing will be also one of the key issues.

References

1. Poinsot, T., Veynante, D.: Theoretical and numerical combustion. Edwards (2001)
2. Mizobuchi, Y., Tachibana, S., Shinjo, J., Ogawa, S., Takeno, T.: Proc. Combust. Inst. 29, 2009–2015 (2002)
3. Mizobuchi, Y., Shinjo, J., Ogawa, S., Takeno, T.: Proc. Combust. Inst. 30, 611–619 (2005)
4. Cheng, T.S., Wehrmeyer, J.A., Pitz, R.W.: Combust. Flame 91, 323–345 (1992)
5. Shinjo, J., Mizobuchi, Y., Ogawa, S.: CFD J. 13, 348–354 (2004)
6. Shinjo, J., Mizobuchi, Y., Ogawa, S.: Proc. 5th Symp. on Smart Control of Turbulence, Tokyo, pp. 165–173 (2004)
7. Candel, S.: Proc. Combust. Inst. 29, 1–28 (2002)
8. Stone, C., Menon, S.: Proc. Combust. Inst. 29, 155–160 (2002)

9. Tachibana, S., Zimmer, L., Kurosawa, Y., Suzuki, K., Shinjo, J., Mizobuchi, Y., Ogawa, S.: Proc. 6th Symp. on Smart Control of Turbulence, Tokyo. pp. 181–190 (2005)
10. Kee, R.J., Grcar, J.F., Smooke, M.D., Miller, J.A.: SAND85-8240 (1994)
11. Peters, N.: Turbulent Combustion. Cambridge University Press, Cambridge (2000)
12. http://www.me.berkeley.edu/gri_mech/

Realization of a Computer Simulation Environment Based on ITBL and a Large Scale GW Calculation Performed on This Platform

Yoshiyuki Kawazoe[1], Marcel Sluiter[1], Hiroshi Mizuseki[1], Kyoko Ichinoseki[1], Amit Jain[1], Kaoru Ohno[2], Soh Ishii[2], Hitoshi Adachi[3], and Hiroshi Yamaguchi[3]

[1] Institute for Materials Research, Tohoku University, 2-1-1, Katahira, Aoba-ku, 980-8577, Sendai, Japan
{kawazoe, marcel, mizuseki, kyoko}@imr.edu
http://www.kawazoe.imr.tohoku.ac.jp/

[2] Yokohama National University, 79-1, Tokiwadai, Hodogaya-ku, Yokohama,Japan 240-8501
{ohno, ishii-s}@nyu.ac.jp

[3] Hitachi East Japan Solutions, 2-16-10 Honcho, Aoba-ku, Sendai, Japan 980-0014

Abstract. An extraordinarily large GRID environment has been established over Japan by using SuperSINET based on ITBL connecting 4 supercomputer facilities. This new supercomputing environment has been used for a large scale numerical simulations using original *ab initio* code TOMBO and several remarkable results have already been obtained to proof that this newly built computer environment is actually useful to accelerate the speed of designing and developing advanced functional materials expected to be used in nanotechnology.

1 Introduction

The present study aims to create a materials design environment that surpasses the present day standard of local density approximation (LDA) with plane wave and pseudopotentials as a large project with various organizations in Japan. The project has been supported as part of the Grant-in-Aid for Creative Scientific Research Collaboration on Electron Correlations; Toward a New Research Network between Physics and Chemistry. It includes five major research institutions in former national universities and is also part of the SuperSINET Promotion Conference as a branch of nanotechnology [1]. To perform extraordinarily large scale simulations, such a scale of 1000 hours with 100 GB of memory on supercomputer, this project networks multiple supercomputers which are distributed throughout Japan. To achieve this, a VPN (Virtual Private Network) was constructed. It is now called the "Nanotech-VPN" on SuperSINET [2], with the ITBL (IT-Based Laboratory developed by the Japan Atomic Energy Research Institute (JAERI)) environment [3]. For the simulation program, we selected the TOhoku Mixed-Basis Orbitals *ab initio* program (TOMBO) [4] with full

J. Labarta, K. Joe, and T. Sato (Eds.): ISHPC 2005 and ALPS 2006, LNCS 4759, pp. 427–433, 2008.

potential LDA and optional GWA (based on the many body theorem, Green function–Vertex (W) Approximation). We mainly report here the newly obtained numerical results connecting three supercomputer sites, connected via SuperSINET, at ISSP (Institute for Solid State Physics, University of Tokyo), JAERI, and IMR (Institute for Materials Research, Tohoku University). Simulations based on LDA and GWA versions of the TOMBO program were performed, and AVS [5] visualizations were specifically designed for this project and used to display the obtained numerical results.

The ITBL was started in 2001 with the objective of realizing a virtual joint research environment using information technology. We applied the ITBL system to gain a fundamental and comprehensive view of new materials for nanotechnology, via large scale computer simulations, with our specially designed TOMBO program for this environment. The ITBL project is an improvement relative to alternatives, such as, e.g. the Globus collaboration, for several reasons: While the globus GRID toolbox is designed to run on a wide variety of platforms, it is not specifically optimized for the platforms available to us; the present numerical simulations require very large scale individual runs, and therefore a large amount of memory. Thus, toolkits that are optimized for a large number of low performance CPUs with limited memory do not work as well. Also, the ITBL solution is practically available and deployable as a high level top layer for job submission and output visualization whereas toolkits provide functionality through a set of "service primitives". Therefore from the user side, ITBL is far easier to develop necessary environment.

GWA calculations are particularly time consuming and require more nodes and more memory, only performable on TeraFlops machines, and are therefore better suited to the ITBL environment. The ITBL architecture consists of several layers. The computing and data resources are connected by common network that spans the gamut, from the internet up to SuperSINET, and stretches the length of Japan. On this physical network the primitive layer provides scheduling and monitoring facilities, the intermediate layer consists of tools such as the component programming toolkit, and the community toolkit, etc, and finally the top layer is an end user interface that is provided by a Web based technology. The ITBL system is a "meta computing environment", in which we can perform very large scale simulations, e.g. for investigating nanotechnological systems established in the present study.

2 Realization of the SuperSINET Computer Connection Environment

This project started with the activity of the SuperSINET Promotion Conference as a branch of nanotechnology. At that time, there was a proposal from the Ministry of Economy, Trade and Industry, which tried to connect 63 supercomputers distributed throughout Japan (by Dongarra's report) [6]. We started the present project under the Grant-in-Aid for Creative Scientific Research Collaboration on Electron Correlations; Toward a New Research Network between Physics

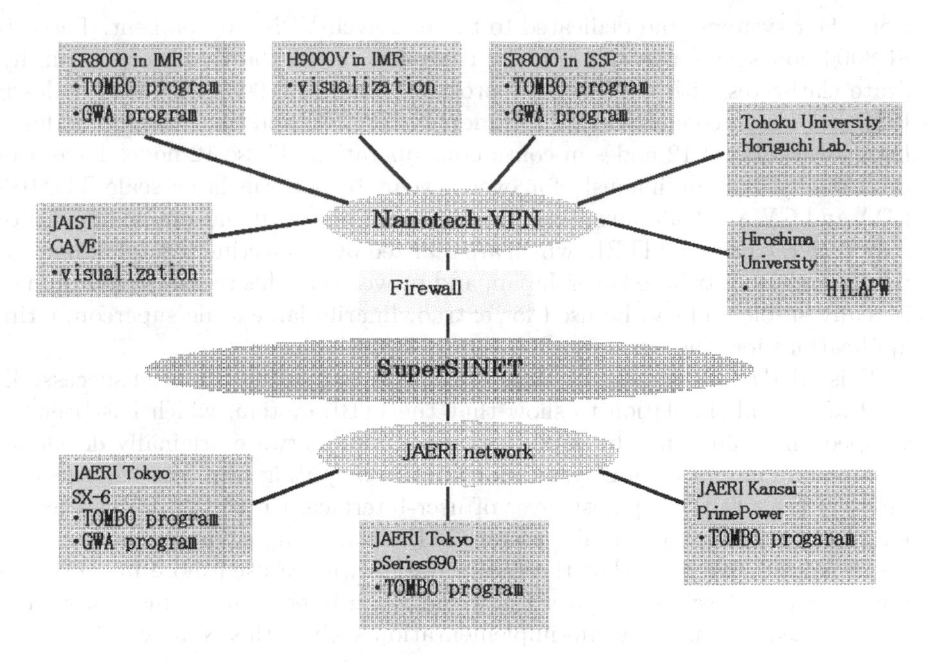

Fig. 1. The present computer network scheme and the connected supercomputers via SuperSINET with ITBL and their purposes in the project

and Chemistry. Five major research institutions, formerly national universities in Japan, participated. YK has been the group leader of the project to realize the first large GRID computing environment on SuperSINET for simulations in nanotechnology.

It was not an easy task, because the project aimed for far more than a simple parallel programming with MPI. It aimed to connect supercomputers throughout Japan, which have site specific security policies. With the help of the SuperSINET Promotion Conference, a new virtual private network (Nanotech-VPN) was installed. Data transfer speed, and SuperSINET efficiency and stability were tested with dedicatedly installed workstations at each site, because of non-restricted usage and security issues. It was immediately known by us that it is very difficult to establish the proposed GRID supercomputer environment, and we have initiated further discussions of a new project to link all GRID computers. This is now established as the NAREGI project [7]. Figure 1 is the structural representation of the computer network with multiple supercomputers distributed throughout Japan for the present project. In the present study, among the available systems in this figure, we mainly used the Nanotech-VPN on SuperSINET, ITBL, and mainly two SR8000 and one SX-6 supercomputers at ISSP, IMR, and JAERI with 4 and 4, and 2 dedicated nodes, respectively.

The first step of the Nanotech-VPN project was to supply a supercomputer-network environment. Eight nodes of the IMR SR8000, and six nodes of the IMS (Institute for Molecular Science, Okazaki) SR8000, were physically separated from their systems, and dedicated to the nanotech-VPN environment. These 14 SR8000 nodes were used for LDA [8] computations of the H_2 absorption in hydrate clathrates with the TOMBO program. Since in 2003, 4 SR8000 nodes in ISSP have been connected to the Nanotech-VPN, in addition to the 8 nodes at IMR for a total of 12 nodes in continuous operation. These 12 nodes have been used stably and continuously for over a year, to perform large scale TOMBO LDA and GWA calculations. We have developed a new virtual supercomputer on SuperSINET based on ITBL, which are realized by connecting several remote supercomputers distributed over Japan, and proved that this research environment is really stable and can be used for extraordinarily large scale supercomputing applications for long time over 1000 hours by wall clock.

This GRID environment for large scale supercomputing has been successfully used for actual simulation to show that the ITBL system, which has been developed and maintained by JAERI, is a superior software originally developed in Japan aiming at a total system for GRID computing from the basic level of physical layer to the topmost layer of user-interface. Up to the present, except for ITBL, this level of actually useful GRID computing environment has never been realized. We hope that the ITBL system will be used more in the future GRID system development over the world as an important original software of Japan based on our successful implementation such as this Nanotech-VPN.

3 Absolute Energy Level Estimation of Fullerene by All-Electron GW Program

Although LDA cannot estimate the excitation energies in the system, the GWA can correctly provide one-particle excitation energies [9,10]. However, GWA calculation demands a lot of CPU time compared with the LDA calculation because CPU time of GWA (LDA) calculation is proportional to N^6 (N^3). Here, N denotes number of electrons. In the present study, we combined the remote supercomputers on the network in the Nanotech-VPN and divided the processes and memories into every supercomputer that are separated in physical space, making it possible to perform large scale computation. We parallelized our GWA code to perform this type of calculations (see Figure 2).

In the present calculation, we used 4 nodes of SR8000 at IMR and ISSP, respec-tively and 2 nodes of SX-6 at JAERI. As a result, the number of floating point calculation per process decreased more than 20 %, usage of memories decreased 65 %. Green's function approach makes it possible to simultaneously evaluate one particle excitation energy spectra such as removing (attaching) one electron from (to) the neutral systems. Although one of the suitable approximations of Green's function approach is, as is well known, the GWA, usual calculations based on the GWA use pseudopotentials, not being able to evaluate the absolute values of excitation energies such as ionization energies and electron affinities. We successfully performed all electron GWA calculations of alkali-metal clusters and clusters of semiconductor after adding our original GWA

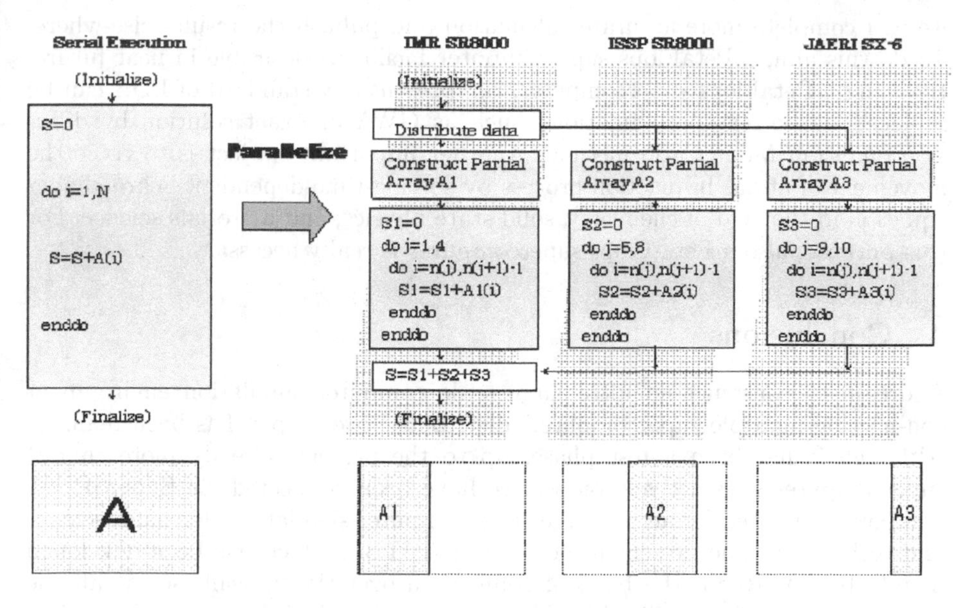

Fig. 2. Distributed execution environment for GW approximation program (Coarse grain parallel number=3, fine grain parallel number=4 for IMR and ISSP, 2 for JAERI)

program to the all electron mixed basis program using plane waves and numerical atomic orbitals within the density functional theory, TOMBO.

In the present report, we tried an all electron GWA calculation of C_{60}. Although the GWA calculation of C_{60} using pseudopotential approach was already performed by Shirley and Louie [11], they use some experimental parameters when evaluating dielec-tric function. That is, their calculation is not a perfect *ab initio* GWA calculation. In the present study, we used an fcc unit cell with a cubic edge of 34 Å. We used cut off energy of 8 Ry when evaluating the LDA eigenvalues and wave functions and 15 Ry evaluating the Fock exchange term. Although at most 1 eV error of the present calcu-lation still remains about the evaluation of correlation-energy, the present results are in good agreement with experiments within the error bar of the present calculation. In the near future,

Table 1. Energies estimated by LDA and GWA compared with experiment

eV	E_{xc}	H_{LDA}	S_x	S_c	E_{calc}	$E_{\exp}$
HOMO	-14.06	-6.61	-14.90	0.0 ± 1.0	-7.4 ± 1.0	-7.6[12]
LUMO	-11.18	-5.26	-7.80	-0.7 ± 1.0	-2.6 ± 1.0	-2.6[13]

In the table, E_{xc}, H_{LDA}, S_x, S_c, E_{calc}, $E_{\exp}$, indicate the expected values of correlation and exchange potential in LDA, eigenvalues in LDA, Fock exchange energy with LDA wavefunctions, correlation energy in GWA, quasiparticle energy in GWA, and experimental values, respectively.

we will complete more accurate calculation and publish the results else-where. To do this aim, a PetaFlops supercomputer facility is desirable in near future. By using a PetaFlops supercomputer, the present day standard of LDA can be replaced by better approximations, such as GWA or exact solution by diffusion quantum MonteCarlo method. Another important subject is to trace the movement of atoms in reaction process by solving time-dependent Schroedinger equation in the field of chemistry, solid state physics, and materials science. For this purpose, also a Peta-Flops supercomputer is really necessary.

4 Conclusions

A dream to construct an ultra large scale computer simulation environment con-necting multiple supercomputers distributed over Japan has been realized. Although it has been a test phase, up to the present, already more than 4 supercomputer sites actively on service have been connected via SuperSINET and have been used for actual large scale computer simulations for nano-science and technology. The present project selected ITBL, which has been developed mainly by JAERI for the basis of connection over the firewalls set at all the supercom-puters fully utilized in the present study, which has reduced the time and human effort to develop such a complex computer networking environment. By applying this newly established environment, atomic structure optimizations for zeolites and hydrate crathlates and absolute energy estimation of quasiparticles in fullerene were conducted by TOMBO program with LDA and GWA. Another important system we have developed is a data visualization environment for remote computers.

Acknowledgments

The present large scale computing environment has been established as a special joint use of the ISSP, JAERI, IMS, and IMR supercomputer systems. Support by JAERI is highly appreciated, since without ITBL this research could never have been realized. We would like to acknowledge the crew of the Center for Computational Materials Science, Institute for Materials Research, Tohoku University for their continuous support of the SR8000 supercomputer and the Nanotech-VPN network systems.

References

1. http://www.sinet.ac.jp/
2. http://www.itbl.riken.go.jp/symp/pdf/kawazoe.pdf
3. http://www.itbl.riken.jp/
4. http://www-lab.imr.edu/~marcel/tombo/tombo.html
5. http://www-vizj.kgt.co.jp/contents/ver1/user/tohoku-u.html
6. http://www.top500.org/

7. http://nanogrid.ims.ac.jp/nanogrid/
8. Sluiter, M., Belosludov, R.V., Jain, A., Belosludov, V.R., Adachi, H., Kawazoe, Y., Higu-chi, K., Otani, T.: High Performance Computing. In: Veidenbaum, A., Joe, K., Amano, H., Aiso, H. (eds.) ISHPC 2003. LNCS, vol. 2858, pp. 330–341. Springer, Heidelberg (2003)
9. Ishii, S., Ohno, K., Kawazoe, Y.: Mat. Trans. 45, 1411–1413 (2004)
10. Noguchi, Y., Ishii, S., Kawazoe, Y., Ohno, K.: Sci. and Tech. Adv. Mat. 5, 663–665 (2004)
11. Shirley, E.L., Louie, S.G.: Phys. Rev. Lett. 71, 133–136 (1993)
12. Muigg, D., Sheier, P., Becker, K., Mark, T.D.: J. Phys. B. 29, 5193 (1996)
13. Wang, X.B., Ding, C.F., Wang, L.S.: J. Chem. Phys. 110, 8217–8220 (1999)

Computations of Global Seismic Wave Propagation in Three Dimensional Earth Model

Seiji Tsuboi[1], Dimitri Komatitsch[2], Chen Ji[3], and Jeroen Tromp[3]

[1] Institute for Research on Earth Evolution, Jamstec
[2] Geophysical Imaging Laboratory, Universite de Pau et des Pays de l'AdourUniversite
[3] Seismological Laboratory, California Institute of Technology

Abstract. We use a Spectral-Element Method implemented on the Earth Simulator in Japan to simulate broadband seismic waves generated by various earthquakes. The spectral-element method is based on a weak formulation of the equations of motion and has both the flexibility of a finite-element method and the accuracy of a pseudospectral method. The method has been developed on a large PC cluster and optimized on the Earth Simulator. We perform numerical simulation of seismic wave propagation for a three-dimensional Earth model, which incorporates 3D variations in compressional wave velocity, shear-wave velocity and density, attenuation, anisotropy, ellipticity, topography and bathymetry, and crustal thickness. The simulations are performed on 4056 processors, which require 507 out of 640 nodes of the Earth Simulator. We use a mesh with 206 million spectral-elements, for a total of 13.8 billion global integration grid points (i.e., almost 37 billion degrees of freedom). We show examples of simulations for several large earthquakes and discuss future applications in seismological studies.

Keywords: seismic wave propagation, 3–D Earth models, spectral-element method.

1 Introduction

Accurate modeling of seismic wave propagation in fully three-dimensional (3–D) Earth models is of considerable interest in seismology in order to determine both the 3–D seismic-wave velocity structure of the Earth and the rupture process during large earthquakes. However, significant deviations of Earthfs internal structure from spherical symmetry, such as the 3–D seismic-wave velocity structure inside the solid mantle and laterally heterogeneous crust at the surface of the Earth, have made applications of analytical approaches to this problem a formidable task. The numerical modeling of seismic-wave propagation in 3–D structures has been significantly advanced in the last few years due to the introduction of the Spectral-Element Method (SEM), which is a high-degree version of the finite-element method that is very accurate for linear hyperbolic problems such as wave propagation. The 3–D SEM was first used in seismology for local

J. Labarta, K. Joe, and T. Sato (Eds.): ISHPC 2005 and ALPS 2006, LNCS 4759, pp. 434–443, 2008.

and regional simulations [1]–[3], and more recently adapted to wave propagation at the scale of the full Earth [4]–[7].

Here we show that our implementation of the SEM on the Earth Simulator in Japan allows us to calculate theoretical seismic waves which are accurate up to 3.5 seconds and longer for fully 3–D Earth models. We include the full complexity of the 3–D Earth in our simulations, i.e., a 3–D seismic wave velocity [8] and density structure, a 3–D crustal model [9], ellipticity as well as topography and bathymetry. Synthetic waveforms at such high resolution (periods of 3.5 seconds and longer) allow us to perform direct comparisons of arrival times of various body-wave phases between observed and synthetic seismograms, which has never been accomplished before. Usual seismological algorithms, such as normal-mode summation techniques that calculate quasi-analytical synthetic seismograms for one-dimensional (1-D) spherically symmetric Earth models [10], are typically accurate down to 8 seconds [11]. In other words, the SEM on the Earth Simulator allows us to simulate global seismic wave propagation in fully 3–D Earth models at periods shorter than current seismological practice for simpler 1-D spherically symmetric models.

The results of our simulation show that the synthetic seismograms calculated for fully 3–D Earth models by using the Earth Simulator and the SEM agree well with the observed seismograms, which illustrates that the current 3–D seismic velocity model captures the general long-wavelength image of Earth's interior with sufficient resolution.

2 Spectral-Element Method

We use the spectral-element method (SEM) developed by Komatitsch and Tromp (2002a, 2002b) [5,6] to simulate global seismic wave propagation throughout a 3–D Earth model, which includes a 3–D seismic velocity and density structure, a 3–D crustal model, ellipticity as well as topography and bathymetry. The SEM first divides the Earth into six chunks. Each of the six chunks is divided into slices. Each slice is allocated to one CPU of the Earth Simulator. Communication between each CPU is done by MPI. Before the system can be marched forward in time, the contributions from all the elements that share a common global grid point need to be summed. Since the global mass matrix is diagonal, time discretization of the second-order ordinary differential equation is achieved based upon a classical explicit second-order finite-difference scheme.

The maximum number of nodes we could use for this simulation is 4056 processors, i.e., 507 nodes out of 640 of the Earth Simulator. Each slice is allocated to one processor of the Earth Simulator and subdivided with a mesh of 48 48 spec-tral-elements at the surface of each slice. Within each surface element we use $5 \times 5 = 25$ Gauss-Lobatto-Legendre (GLL) grid points to interpolate the wave field [12,13], which translates into an average grid spacing of 2.0 km (i.e., 0.018 degrees) at the surface. The total number of spectral elements in this mesh is 206 million, which corresponds to a total of 13.8 billion global grid points, since each spectral element contains $5 \times 5 \times 5 = 125$ grid points, but with points on its

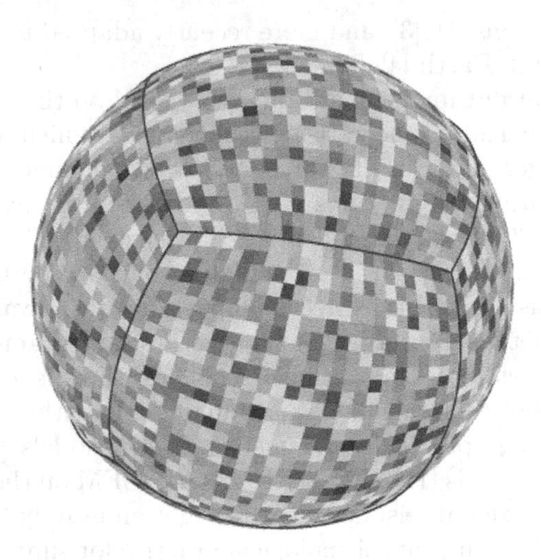

Fig. 1. The SEM uses a mesh of hexahedral finite elements on which the wave field is interpolated by high-degree Lagrange polynomials on Gauss-Lobatto-Legendre (GLL) integration points. This figure shows a global view of the mesh at the surface, illustrating that each of the six sides of the so-called 'cubed sphere' mesh is divided into 26×26 slices, shown here with different colors, for a total of 4056 slices (i.e., one slice per processor)

faces shared by neighboring elements. This in turn corresponds to 36.6 billion degrees of freedom (the total num-ber of degrees of freedom is slightly less than 3 times the number of grid points be-cause we solve for the three components of displacement everywhere in the mesh, except in the liquid outer core of the Earth where we solve for a scalar potential). Using this mesh, we can calculate synthetic seismograms that are accurate down to seismic periods of 3.5 seconds. This simulation uses a total of approximately 7 terabytes of memory. Total performance of the code, measured using the MPI Program Runtime Performance Information was 10 teraflops, which is about one third of the expected peak performance for this number of nodes (507 nodes 64gigaflops = 32 teraflops). Figure 1 shows a global view of the spectral-element mesh at the surface of the Earth. In Figure 2, we compare the vertical component of displacement from synthetic seis-mograms calculated using 507 nodes of the Earth Simulator and observed records for several broadband seismic stations of the F-net array operated by the National Institute of Earth Science and Disaster Prevention in Japan. The earthquake we simulated is a deep earthquake of magnitude 6.3 that occurred in South of Japan on November 12, 2003, at a depth of 382 km.

It is surprising that the global 3–D seismic velocity model used in this simulation still produces fairly good agreement with the observations even at periods of 3.5 sec-onds, because it is supposed that the crustal and mantle structure beneath Japanese Islands are highly heterogeneous and are not captured by the long-wavelength global 3D Earth model. However, Figure 2 also shows that the theoretical seismograms calcu-lated with 507 nodes of the Earth Simulator do not reproduce some of the fine features in the observation and suggests the limitation of this global 3–D seismic velocity model.

For those stations located to the north-east of the epicenter (the azimuth is about 20 degrees), the observed waves show large high-frequency oscillations because the waves travel along the subducting pacific plate, but this feature is not modeled in the theoretical seismograms. This shows that we need to improve our 3–D seismic wave velocity model to calculate theoretical seismic waves that are accurate at 3.5 seconds and longer.

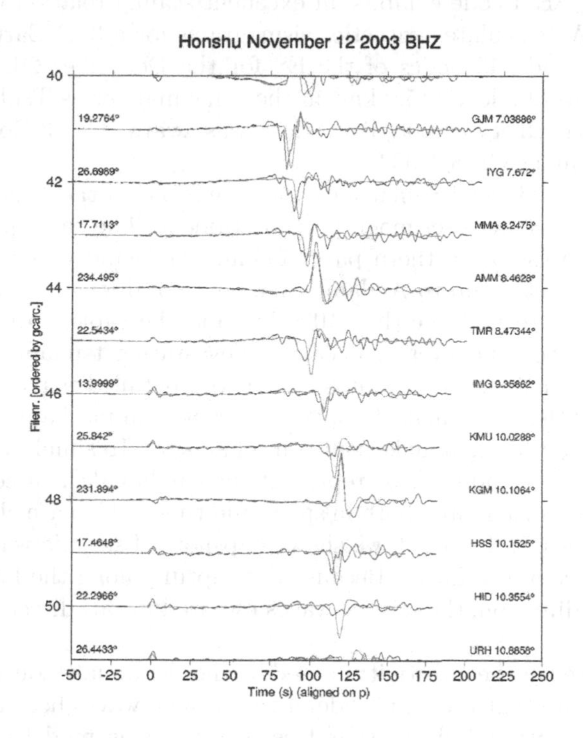

Fig. 2. Broadband data and synthetic displacement seismograms for the 2003 South of Honshu earthquake bandpass-filtered with a two-pass four-pole Butterworth filter between periods of 3.5 and 150 seconds. Vertical component data (black) and synthetic (red) displacement seismograms aligned on the arrival time of the P wave are shown. For each set of seismograms the azimuth is printed above the records to the left, and the station name and epicentral distance are printed to the right.

3 Simulation of the 2004 Sumatra Earthquake

Because we have found that we do not have a 3–D Earth model which has sufficient resolution to simulate seismic wave propagation accurately in regional scale, we de-cide to use 243 nodes (1944 CPUs) of the Earth Simulator for the simulation using the SEM. Using 243 nodes (1944 CPUs), we can subdivide the six chunks into 1944 slices (1944 = 6 × 18 × 18). Each slice is then subdivided into 48 elements in one direction. Because each element has 5 Gauss-Lobatto Legendre integration points, the average grid spacing at the surface of the Earth is about 2.9 km. The number of grid points in total amounts to about 5.5 billion. Using this mesh, it is expected that we can calculate synthetic seismograms accurate up to 5 sec all over the globe. For the 243 nodes case, the total performance we achieved was about 5 teraflops, which also is about one third of the peak performance. The fact that when we double the number of nodes from 243 to 507 the total performance also doubles from 5 teraflops to 10 tera-flops shows that our SEM code exhibits an excellent scaling relation with respect to performance. We calculate synthetic seismograms for a 3–D Earth model using the SEM code and 243 nodes of the ES for the December 26, 2004 Sumatra earthquake (Mw 9.0, depth 15.0 km) in the same manner as Tsuboi et al (2003) [14] and Komatitsch et al (2003) [15], which was awarded 2003 Gordon Bell prize for peak performance in SC2003.

The December 26, 2004 Sumatra earthquake is one of the largest earthquakes ever recorded by modern seismographic instrument. The earthquake started its rupture at the west of northern part of Sumatra Island and propagated in a northwestern direction up to Andaman Islands. Total length of the earthquake fault is estimated to be more than 1000 km and the rupture duration lasts for more than 500 sec. This event has caused devastating tsunami hazard around the Indian Ocean. It is important to know the detailed earthquake fault slip distribution for this earthquake because the excitation mechanism of tsunami is closely related to the earthquake source mechanisms. To simulate synthetic seismograms for this earthquake, we represent the earthquake source by more than 800 point sources distributed both in space and time, which are obtained by seismic wave analysis. In Figure 3, we show snapshots of seismic wave propagation along the surface of the Earth. Because the rupture along the fault propagated in a northwest direction, the seismic waves radiated in this direction are strongly amplified.

This is referred as the directivity caused by the earthquake source mechanisms. Figure 3 illustrate that the amplitude of the seismic waves becomes large in the northwest direction and shows that this directivity is modeled well. Because there are more than 200 seismographic observatories, which are equipped with broadband seismometers all over the globe, we can directly compare the synthetic seismograms calculated with the Earth Simulator and the SEM with the observed seismograms.

Figure 4 shows the results of this comparison for vertical ground motion and demonstrates that the agreement between synthetic and observed seismograms is generally excellent. These results illustrate that the 3–D Earth model and the earthquake rupture model that we have used in this simulation is accurate enough to model seismic wave propagation on a global scale with periods of 5 sec and longer. Because the rupture duration of this event is more than 500 sec, the first arrival P waveform overlapped with the surface reflected wave of P-wave, which is called PP wave. Although this effect obscures the analysis of

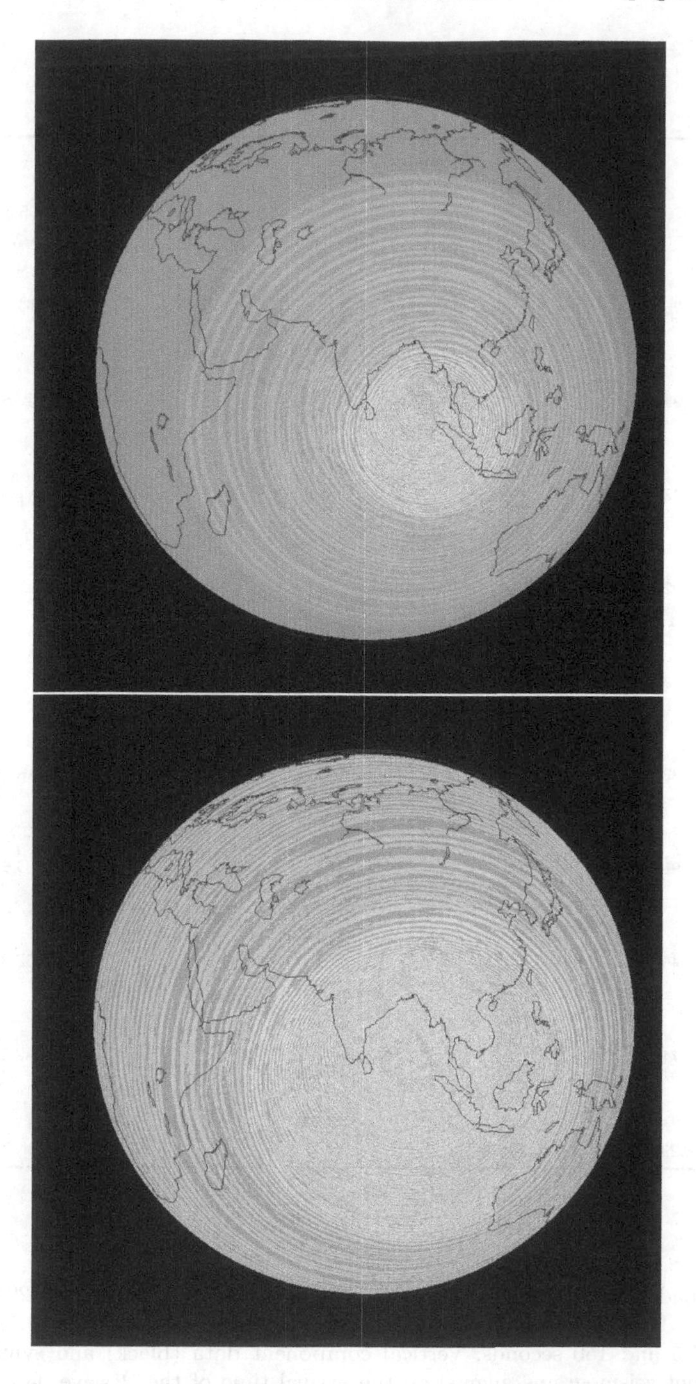

Fig. 3. Snapshots of the propagation of seismic waves excited by the December 26, 2004 Sumatra earthquake. Total displacement at the surface of the Earth is plotted at 10 min after the origin time of the event (top) and at 20 min after the origin time (bottom).

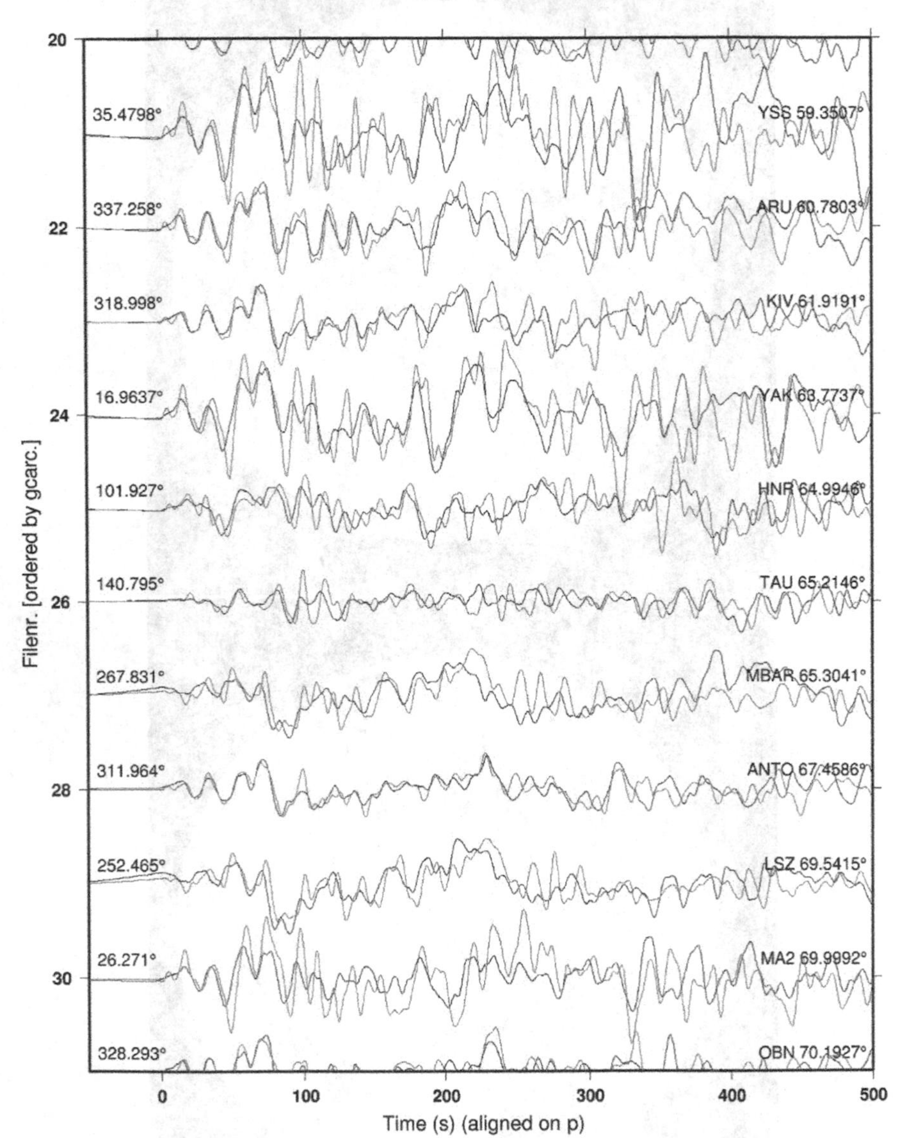

Fig. 4. Broadband data and synthetic displacement seismograms for the 2004 Sumatra earthquake, bandpass-filtered with a two-pass four-pole Butterworth filter between periods of 5 and 150 seconds. Vertical component data (black) and synthetic (red) displacement seismograms aligned on the arrival time of the P wave. For each set of seismograms the azimuth is plotted above the records to the left, and the station name and epicentral distance are plotted to the right.

earthquake source mechanism, it has been shown that the synthetic seismograms computed with Spectral-Element Method on the Earth Simulator can fully take these effects into account and are quite useful to study source mechanisms of this complicated earthquake.

4 Implications for the Earth's Internal Structure

The Earth's internal structure is another target that we can study by using our synthetic seismograms calculated for fully 3–D Earth model. We describe the examples of Tono et al (2005) [16]. They used records of $\sim$ 500 tiltmeters of the Hi-net, in addition to $\sim$ 60 broadband seismometers of the F-net, operated by the National Research Institute for Earth Science and Disaster Prevention (NIED). They analyzed pairs of sScS waves, which means that the S-wave traveled upward from the hypocenter reflected at the surface and reflected again at the core-mantle boundary, and its reverberation from the 410- or 660-km reflectors (sScSSdS where d=410 or 660 km) for the deep shock of the Russia-N.E. China border (PDE; 2002:06:28; 17:19:30.30; 43.75N; 130.67E; 566 km depth; 6.7 Mb). The two horizontal components are rotated to obtain the transverse component.

They have found that these records show clearly the near-vertical reflections from the 410- and 660-km seismic velocity discontinuities inside the Earth as post-cursors of sScS phase. By reading the travel time difference between sScS and sScSSdS, they concluded that this differential travel time anomaly can be attributed to the depth anomaly of the reflection point, because it is little affected by the uncertainties associated with the hypocentral determination, structural complexities near the source and receiver and long-wavelength mantle heterogeneity. The differential travel time anomaly is obtained by measuring the arrival time anomaly of sScS and that of sScSSdS separately and then by taking their difference. The arrival time anomaly of sScS (or sScSSdS) is measured by cross-correlating the observed sScS (or sScSSdS) with the corresponding synthetic waveform computed by SEM on the Earth Simulator. They plot the measured values of the two-way near-vertical travel time anomaly at the corresponding surface bounce points located beneath the Japan Sea. The results show that the 660-km boundary is depressed at a constant level of $\sim$ 15 km along the bottom of the horizontally extending aseismic slab under southwestern Japan. The transition from the normal to the depressed level occurs sharply, where the 660-km boundary intersects the bottom of the obliquely subducting slab. This observation should give important imprecations to geodynamic activities inside the Earth.

5 Conclusions

We have shown that the use of both the Earth Simulator and the SEM has allowed us to reach unprecedented resolution for the simulation of global seismic wave propagation resulting from large earthquakes. We have successfully attempted for the first time an independent validation of an existing 3–D Earth model. Such 3–D calculations on the Earth Simulator reach shorter periods than quasi-analytical 1-D spherically-symmetric solutions that are current practice in seismology. By using the SEM synthetics calculated for a realistic 3–D Earth model, it is possible to determine dif-ferences in the arrival times between theoretical seismograms and observations. As we have discussed in the present paper, these differences in arrival time can be interpreted as depth variations of the discontinuities. This kind of study would not have been possible without the combination of a precise seismic wave modeling technique, such as the SEM, on a powerful computer, such as the Earth Simulator, and a dense seismic observation network. If we extrapolate the numbers we used for our simulation, it is expected that we will get synthetic seismograms that are accurate up to 1 second for fully 3D Earth if we can use 64,896 CPUs of the Earth Simulator.

Acknowledgments. All the simulations were performed at the Earth Simulator Center of JAMSTEC by S.T. Broadband seismograms used in this study were recorded at Global Seismic Network stations and were obtained from the IRIS Data Management Center (www.iris.washington.edu) and F-net and Hi-net stations operated by the National Institute of Earth Science and Disaster Prevention.

References

1. Komatitsch, D.: Spectral and spectral-element methods for the 2D and 3D elastodynamics equations in heterogeneous media. PhD thesis, Institut de Physique du Globe, Paris, France (1997)
2. Faccioli, E., Maggio, F., Paolucci, R., Quarteroni, A.: 2D and 3D elastic wave propagation by a pseudo-spectral domain decomposition method. J. Seismol. 1, 237–251 (1997)
3. Seriani, G.: 3–D large-scale wave propagation modeling by a spectral element method on a Cray T3E multiprocessor. Comput. Methods Appl. Mech. Engrg. 164, 235–247 (1998)
4. Chaljub, E.: Numerical modelling of the propagation of seismic waves in spherical geometry: applications to global seismology. PhD thesis, Universit Paris VII Denis Diderot, Paris, France (2000)
5. Komatitsch, D., Tromp, J.: Spectral-element simulations of global seismic wave propagation-I. Validation.? Geophys. J. Int. 149, 390–412 (2002a)
6. Komatitsch, D., Tromp, J.: Spectral-element simulations of global seismic wave propagation-II. 3–D models, oceans, rotation, and self-gravitation. Geophys. J. Int. 150, 303–318 (2002b)
7. Komatitsch, D., Ritsema, J., Tromp, J.: The spectral-element method, Beowulf computing, and global seismology. Science 298, 1737–1742 (2002)
8. Ritsema, J., Van Heijst, H.J., Woodhouse, J.H.: Complex shear velocity struc-ture imaged beneath Africa and Iceland. Science 286, 1925–1928 (1999)
9. Bassin, C., Laske, G., Masters, G.: The current limits of resolution for surface wave tomography in North America. EOS Trans. AGU. Fall Meet. Suppl., Abstract S12A–03 81 (2000)

10. Dziewonski, A.M., Anderson, D.L.: Preliminary reference Earth model. Phys. Earth Planet. Inter. 25, 297–356 (1981)
11. Dahlen, F.A., Tromp, J.: Theoretical Global Seismology. Princeton Uni-versity Press, Princeton (1998)
12. Komatitsch, D., Vilotte, J.P.: The spectral-element method: an efficient tool to simulate the seismic response of 2D and 3D geological structures. Bull. Seismol. Soc. Am. 88, 368–392 (1998)
13. Komatitsch, D., Tromp, J.: Introduction to the spectral-element method for 3–D seismic wave propagation. Geophys. J. Int. 139, 806–822 (1999)
14. Tsuboi, S., Komatitsch, D., Ji, C., Tromp, J.: Broadband modeling of the 2003 Denali fault earthquake on the Earth Simulator. Phys. Earth Planet. Int. 139, 305–312 (2003)
15. Komatitsch, D., Tsuboi, S., Ji, C., Tromp, J.: A 14.6 billion degrees of freedom, 5 teraflops, 2.5 terabyte earthquake simulation on the Earth Simulator. In: Proceedings of the ACM/IEEE SC2003 confenrence (published on CD-ROM) (2003)
16. Tono, Y., Kunugi, T., Fukao, Y., Tsuboi, S., Kanjo, K., Kasahara, K.: Mapping the 410- and 660-km discontinuities beneath the Japanese Islands. J. Geophys. Res. 110, B03307 (2005), doi:10.1029/2004JB003266

Lattice QCD Simulations as an HPC Challenge

Atsushi Nakamura

RIISE, Hiroshima University, Higashi-Hiroshima739-8521, Japan
nakamura@riise.hiroshima-u.ac.jp

Abstract. We overview the present status of lattice QCD (Quantum Chromodynamics) simulations. Although it is still far from the final goal, the lattice QCD is reaching a level to have a predictive power as a first principle study for the strongly interacting elementary particles, hadron. This is due to many improvements of techniques and rapid development of computational power. We then look into the hot spot of the calculation explicitly. Finally we discuss what kind of achievement can be expected by using Peta-flop computers.

Keywords: QCD (Quantum Chromodynamics), Quark, Strong Interaction, Quantum Field Theory, Sparse Matrix.

1 Introduction

Philosophers in old Greek and China were wondering what the matter is made from: Tree, Fire, Water, .. ? Now we have (probably) the answer for it. The matter is made from quarks and gluons, and their dynamics is governed by QCD. QCD - Quantum Chromo-Dynamics - is a part of the standard theory in modern elementary particle theory. The mesons consist of a quark and an anti-quark, while nucleons are made of three quarks. The force among the quarks are mediated by the gluons. See Fig.1

The synthesis of elementary particles from constituents is controlled by QCD, where the quarks interact through the gluons. QCD has many common characters with quantum electrodynamics, where the electrons interact through the photons. However, there is an essential difference among them: QCD is described by a non-Abelian group SU(3), so-called color group and it is believed that the theory has a remarkable feature, the confinement, i.e., the quarks and the gluons do not appear out of elementary particles.

The only available tool for the quantitative study of the phenomena is the treatment of a discretized Eucledian version of the theory (lattice QCD) by Monte Carlo simulation [1]. Lots of computational resources have been devoted to QCD simulations in the world. In order to get reliable results we need more than several hundred CPU hours with 100 GFLOPS class machines.

The fundamental ingredients of lattice gauge theories are link variables, $U_\mu(n)$, where the four dimensional vector $n = (n_x, n_y, n_z, n_t)$ stands for a site on a lattice and $\mu = x, y, z$ and t represents the direction of the link. In computer simulations the size of a lattice is finite,

J. Labarta, K. Joe, and T. Sato (Eds.): ISHPC 2005 and ALPS 2006, LNCS 4759, pp. 444–451, 2008.

$$i_x = 1, 2, ..., L_x,\ i_y = 1, 2, ..., L_y,\ i_z = 1, 2, ..., L_z,\ i_t = 1, 2, ..., L_t.$$

$U_\mu(n)$ corresponds to gluonic degrees of freedom and is 3×3 complex unitary matrix with unit determinant (SU(3) matrix). A set of link variables is called a configuration and a Monte Carlo step to produce a configuration is called a sweep.

The Feynman path integral for the lattice QCD has the following form,

$$< O >= \int O[U]P[U]\Pi_n \Pi_\mu dU_\mu(n) \tag{1}$$

where

$$P[U] = \exp(-S_G[U]) \det D / \int \exp(S_G[U]) \det D \Pi_n \Pi_\mu dU_\mu(n). \tag{2}$$

Here O is an observable. The sum in (2) is not only on a classical path, but also over those fluctuating around the classical path, since this is a quantum theory.

The action S_G is the sum of so-called plaquette energy, E_{plaq}, which is the products of $U_\mu(n)$ on the minimal closed loops, i.e., squares on the lattice called plaquettes,

$$E_{plaq} = \frac{1}{3} \mathrm{Re\,Tr} U_\mu(n) U_\nu(n + e_\mu) U_\mu^\dagger(n + e_\nu) U_\nu^\dagger(n), \tag{3}$$

where e_μ is a unit vector along the direction μ and $U^\dagger$ is an hermitian conjugate (transpose + complex conjugate) of U. E_{plaq} represents the gluon kinetic energy, and is the product of four SU(3) matrices. This part requires many operations to compute, and was hot spot in early days of lattice QCD. The calculation of the determinant part, which represents the quark-antiquark pair-creation/annihilation in the vacuum, was considered as too heavy and usually ignored. As we will see, later, now lattice QCD simulations are going to include the determinant part, so-called full QCD simulations, because we strongly feel that the full QCD is necessary for the real world simulation and the next-generation computational power endure such hard simulations.

The quark kinetic energy and quark-gluon interaction are included in $\det D$. The rank of D is $3 \times 4 \times L_x \times L_y \times L_z \times L_t$, where 3 is the number of color, and 4 is the spin and particle-antiparticle degrees of freedom. The matrix comes from the discretization of Dirac operator which includes the derivatives with respect to x,y,z and t. As a consequence, it is a sparse matrix. The explicit form of the matrix D is given later.

Determinant is very time consuming numerical object, and therefore usually it is rewritten as

$$\det D = \int d\phi^\dagger d\phi \exp(-\phi^\dagger D^{-1} \phi) \tag{4}$$

The inverse of the matrix, D^{-1} stands for the quark propagator, and for example the nucleon is given by superposition of three D^{-1}. The inversion is usually calculated by solving

$$D\boldsymbol{X} = \boldsymbol{b} \tag{5}$$

Modern lattice QCD consumes most of the computer resource in this calculation.

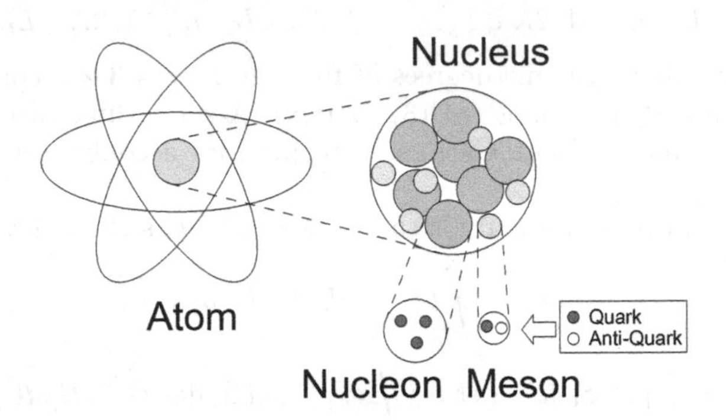

Fig. 1. Hierarchy from atoms to quarks

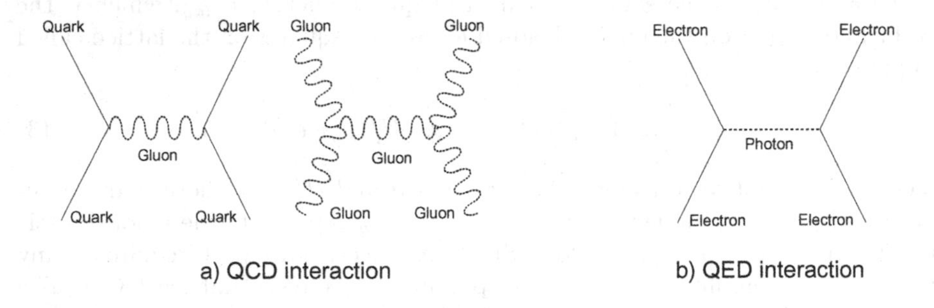

Fig. 2. QCD Interaction and QED Interaction

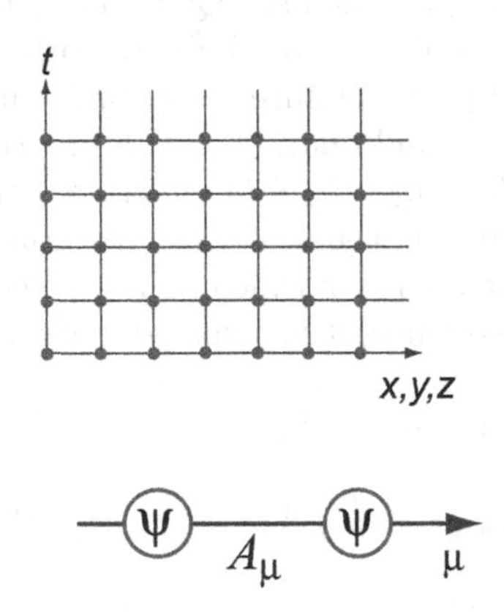

Fig. 3. Quarks and Gluons on the Lattice

2 Lattice QCD – Many Years to Establish Its Reliability

Wilson proposed the lattice formulation of gauge field theories in 1974[1], and since then numerical study of QCD has begun. See Ref.[2] as a good text book. In early days, the quark part was dropped because of its hard computation, and the gluon dynamics were studied. These are without doubt very important studies, especially because QCD includes gluon self energy, which does not exist in the electro-magnetic interaction, and this self energy is expected to contribute the confinement. The result is theoretical and can not be compared with the experiment.

In Fig.4, we show our results for the mass splitting of charmonium mesons in the quench approximation, where $\det D$ is ignored. Using an improved discretization scheme, 'Clover action', we see that the lattice spacing artifact is very small in the regions. Nevertheless, there is a clear deviation from the experiment, which is due to the lack of $\det D$.

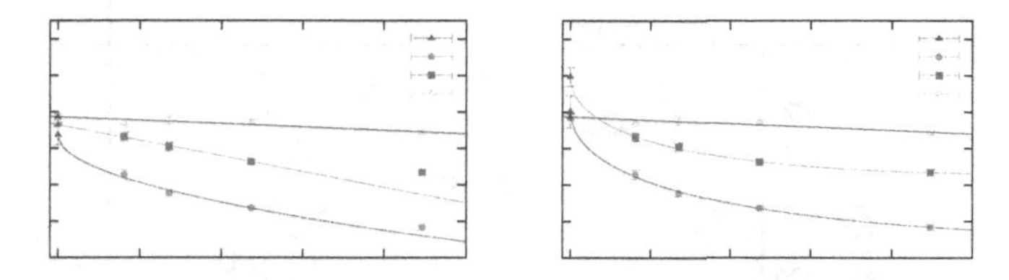

Fig. 4. Charmonium mass simulation results as a function of the lattice spacing. Using clover action, the lattice spacing dependence is small in these regions.

3 Lattice QCD for Exploring New Phenomenon

QCD has been successful to describe many phenomena at very high energy. When the energy is very large, the coupling of the non-Abelian gauge theories such as QCD becomes small because of their asymptotically free character, and the perturbation works well. On the other hand, at low energy the non-perturbative nature dominates and lattice QCD is a tool to investigate elementary particle physics. This is great, but these works are to confirm well-established experiments. Now lattice QCD is well matured, and we hope it is powerful enough to predict new phenomena.

3.1 Finite Temperature and Density

As shown in Fig.5, high energy nuclear experimentalists are performing heavy-ion collision experiments in order to create finite temperature and finite density states. At RHIC (Relativistic Heavy Ion Collider) in USA, they announced the discovery of a "new state of matter"[6,7]. One of the main purposes of these experiments is to recreate the state just after the big-bang.

It is very surprising that this new state of matter behaves like a perfect fluid.[1] Its viscosity as a unit of the entropy is one or two order of magnitude less than water, liquid helium etc.

Very important task for lattice QCD is to simulate the viscosity of gluon matter. In Fig.6, we show our simulation result of the ratio of the viscosity to the entropy. We see, indeed, it is less than one, which is much less than ordinary matter. This calculation was performed on SX-5.

Another important and challenging area is QCD at finite density. There many exciting phenomenon such as quark super conductivity are expected. Lattice QCD simulation is, however, very difficult, because $\det D$ appeared as measure in Eq.(2) becomes complex, and induces the sign problem in Monte Carlo calculation. The detailed analyses and recent overview is given in Ref.[8,9].

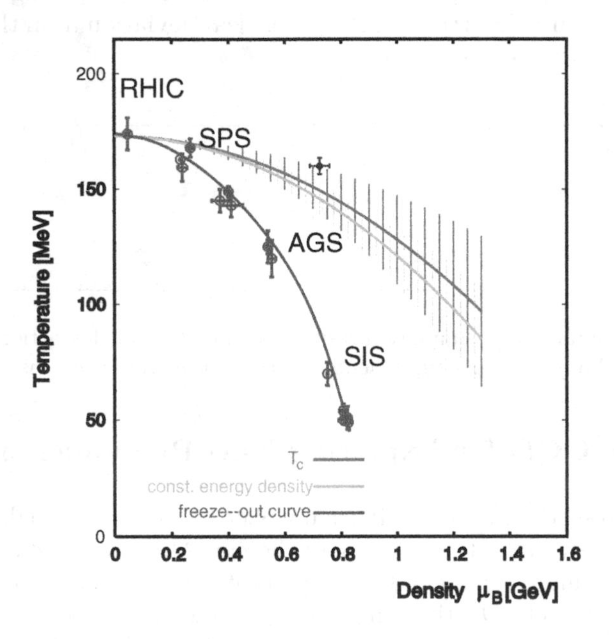

Fig. 5. Temperature-Density Region covered by Present Experiment together with Lattice QCD result. Ref.[5].

3.2 Exotic Hadrons

Until recently, we thought that all hadrons are consist of quark-antiquark or three quarks, and they are well classified in the old quark model. But many experimental activities in such as J-Lab, B-factories provides many new precise data, and we should check what kind of states are allowed in QCD. One of such trials we calculate the scalar meson, σ which once disappeared and reestablished recently. In Fig.7, we show our result of the masses of π, ρ and σ. This calculation

[1] The viscosity of perfect fluid is zero, while that of free gas diverges.

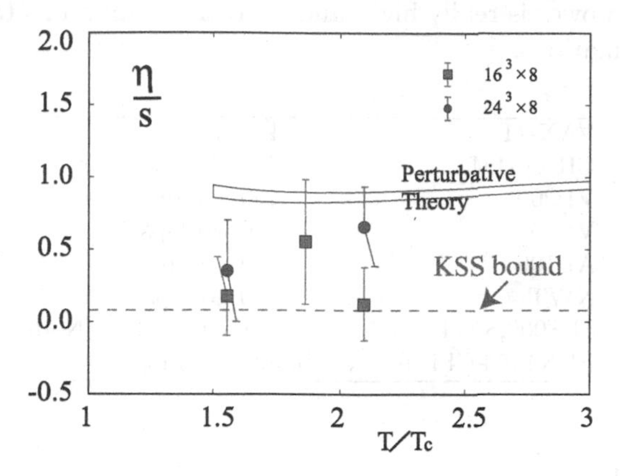

Fig. 6. Transport Coefficient of Gluon Matter

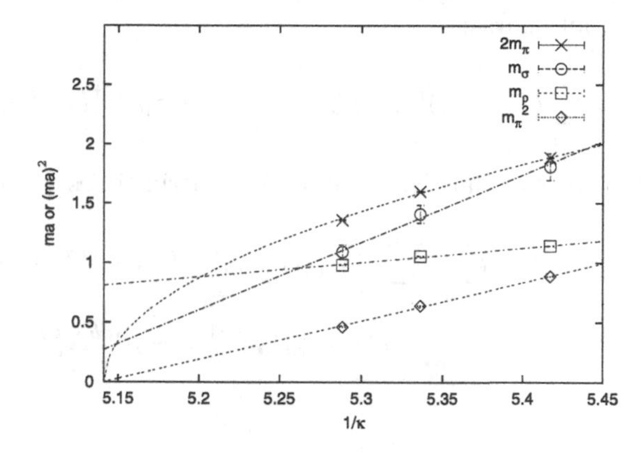

Fig. 7. Masses of the pion, rho and sigma mesons as a function of the hopping parameter (quark mass)

is indeed very heavy because it required the quark propagators from any point to any point, which is difficult by solving Eq.5 and also this calculation should be done in full QCD.

4 QCD as HPC

4.1 History

In the table, we show computers which the author used for lattice QCD simulations. We can see that (1) one can now play lattice QCD simulations even on PC, because its power is comparable with old super-computers; (2) modern super-computer power is really high, and we can be ambitious to challenge real-world QCD simulations.

VAX-11	1 MIPS
CRAY-1M	75 MFlops
VP200	400 MFlops
VP400	800 MFlops
AP1000	5.6G Flops
NWT	280 GFlops
SR8000,SX-4, SX-5	a few G $\sim$ 10G/Node
PEN4x8+GbE in my lab.	3.8x8 GFlops

4.2 Hot Spot

As we stressed in the previous chapters, for the real-world QCD simulations, we need full QCD simulations, where we must solve the inverse matrix, D^{-1}. The standard method is the conjugate gradient type algorithm. This is because our matrix D is large but sparse:

$$D = I - \kappa \sum_{\mu=1}^{4} \left\{ (r - \gamma_\mu) U_\mu(x) \delta_{x', x+\hat{\mu}} + (r + \gamma_\mu) U_\mu^\dagger(x') \delta_{x', x-\hat{\mu}} \right\}, \tag{6}$$

In the conjugate type calculation, we need the matrix times vectors: $\boldsymbol{Y} = D\boldsymbol{X}$

$$Y_\alpha^a(x) = X_\alpha^a(x) - \kappa \sum_{\mu=1}^{4} \sum_{\beta=1}^{4} \sum_{b=1}^{3} \{ (r - \gamma_\mu)_{\alpha\beta} U_\mu(x)^{ab} X_\beta^b(x + \hat{\mu}) \\ + (r + \gamma_\mu)_{\alpha\beta} U_\mu^\dagger(x - \hat{\mu})^{ab} X_\beta^b(x - \hat{\mu}) \} \tag{7}$$

4.3 Future

The operation (7) is very suitable for the vector computers, and many cluster type machines if the network is fast enough. In near future, we can expect the speed-up of CPU, but it is not clear if the network speed is fast enough. Then a new algorithm is desirable for such new machines.

There have been many technical development which make lattice QCD much more high quality, such as improved actions, anisotropic lattice, and chiral fermion formalism. The last one is especially a good news, because it keeps the most important symmetry of QCD, chiral symmetry. But a bud news is that it requires much more computer resources.

The speed of the hot spot (7) in QCD was measured on a prototype of IBM BlueGene/L machine[11]. A good news is that its sustained is over 1 TFLOPS, while not so good news is that it is less than 20 percent of the peak speed. In the real-world simulation, we use a formula which is more complicated than Eq.(6), i.e., there are more operations. Therefore we can expect higher efficiency. In any case, it is now time to prepare the real-world QCD simulations and to design the code to fit the Peta-flops machines.

5 Concluding Remarks

Lattice QCD has reaching a level of quantitative science, which was only a dream twenty years ago. This achievement could not be possible if there were enough development of computational power. Especially the vector type supercomputers match lattice QCD calculations. Lattice QCD has nearest neighbor interaction and therefore is suitable also for parallel computation.

If we can utilize the next generation Peta-FLOPS machines, lattice QCD becomes surely a realistic quantitative tool to study hadron physics. However, there are two obstacles before us:

1. Next generation machines have very high peak performance without question, but it is not yet clear if the communication speed vs. computational power is as good as now.
2. We have now several methods for very high quality lattice QCD calculation, but they require often long range type calculations together with the nearest type interaction. This makes the parallelization inefficient.

In order to overcome these difficulties, it is inevitable to have a good collaboration among lattice QCD physicists and computer scientists.

References

1. Wilson, K.G.: Phys. Rev. D10, 2445–2459 (1974)
2. Montvay, I., Muenster, G.: Quantum Fields on a Lattice, Cambridge monographs on Mathematical Physics. Cambridge University Press, Cambridge (1994)
3. Choe, S., et al.: QCD-TARO collaboration. JHEP 0308, 22 (hep-lat/0307004) (2003)
4. Davies, C.T.H., et al.: Phys.Rev.Lett. 92, 022001 (2004)
5. Braun-Munzinger, P., Redlich, K., Stachel, J.: Particle Production in Heavy Ion Collisions. In: Hwa, R.C., Wang, X.-N. (eds.) Quark Gluon Plasma 3, pp. 491–599. World Scientific Publishing, Singapore (nucl-th/0304013)
6. BRAHMS, PHENIX, PHOBOS, and STAR, Hunting-the-QGP, `http://www.bnl.gov/bnlweb/pubaf/pr/PR_display.asp?prID=05-38`
7. Ludlam, T.: Talk at New Discoveries at RHIC – The Strongly Interactive QGP. BNL (May 14-15, 2004), `http://quark.phy.bnl.gov/~mclerran/qgp/`
8. Muroya, S., Nakamura, A., Nonaka, C., Takaishi, T.: Prog. Theor. Phys. 110, 615–668 (2003)
9. Prog. Theor. Phys. Suppl. 153 (2003), Nakamura, A., et al. (eds.): Proceedings of International Workshop on Finite Density QCD
10. Kunihiro, T., et al.: Scalar collaboration. Phys.Rev. D70, 34504 (2004)
11. Bhanot, G., Chen, D., Gara, A., Sexton, J., Vranas, P.: hep-lat/0409042

Energy-Efficient Embedded System Design at 90nm and Below – A System-Level Perspective –

Tohru Ishihara

System LSI Research Center, Kyushu University,
3-8-33, Momochihama, Sawara-ku, Fukuoka, 814-0001, Japan
ishihara@slrc.kyushu-u.ac.jp
http://www.slrc.kyushu-u.ac.jp/

Abstract. Energy consumption is a fundamental barrier in taking full advantage of today and future semiconductor manufacturing technologies. This paper presents our recent research activities and results on estimating and reducing energy consumption in nanometer technology system LSIs. This includes techniques and tools for (i) estimating instantaneous energy consumption of embedded processors during an application execution, and (ii) reducing leakage energy in instruction cache memories by taking advantage of value-dependence of SRAM leakage due to within-die V_{th} variation.

Keywords: energy estimation, energy characterization, nanometer technology, leakage power, cache memory, compiler optimization, embedded system.

1 Introduction

There is a wide consensus that the energy consumption is a fundamental barrier in taking full advantage of today and future semiconductor manufacturing technologies. As the demands of system integration, high performance, and low power operation have pushed chip vendors down to 90nm and below, NRE (non-recurring engineering) costs and design complexity have increased significantly. A remedy for the NRE explosion is to reduce the number of developments and sell tens of millions of chips under a fixed hardware design. In such a situation, embedded software plays much more important role than today. Our research focus is mainly on software-oriented approaches to estimating and reducing the energy consumption of embedded real-time systems. In this paper, we present our recent research activities and results in the following two categories: estimating software energy consumption and reducing leakage energy of the memory subsystem. Firstly, we present a technique to estimate instantaneous energy consumption of embedded processors during an application execution. We train a per-processor energy-model which receives statistics from the processor instruction-set simulator (ISS) and gives the instantaneous energy consumption. Secondly, we show our

J. Labarta, K. Joe, and T. Sato (Eds.): ISHPC 2005 and ALPS 2006, LNCS 4759, pp. 452–465, 2008.

technique for reducing leakage energy in instruction cache memories by taking advantage of value-dependence of SRAM leakage due to within-die V_{th} variation. We propose a technique to reduce the leakage power of ultra-leaky SRAM cells by storing less-leaky logic values to the corresponding SRAM cells in the cache.

The rest of this paper is organized as follows. The instantaneous energy estimation technique is presented in Section 2. Section 3 presents a techniques for reducing the leakage power of instruction cache memory by exploiting the value-dependence of SRAM leakage due to within-die V_{th} variation. Section 3 summarizes our approaches and concludes the paper.

2 Software Power Estimation

This section shows an overview of our energy characterization tool which helps designers in developing a fast and accurate energy model for a target processor-based system. We use a linear model for energy estimation and find the coefficients of the model using multiple linear regression analysis. For more detailed information of our tool see [1].

2.1 Energy Characterization

The energy consumption of a processor can be estimated using the following linear formula,

$$E_{estimate} = \sum_{i=0}^{N} c_i \cdot P_i \tag{1}$$

where P_i's, c_i's and N are the parameters of the model, the corresponding coefficients and the number of parameters, respectively. The first step for the modeling is to find P_i's required for estimating the energy consumption of the target processor system. The P_i's should be parameters whose values can be easily obtained using a fast simulator like an instruction-set simulator (ISS). For example, P_i's can be the number of load and store instructions executed, the number of cache misses, etc. Once the required set of parameters is obtained, the next step is to find a training bench for the energy characterization. More detailed explanation for our method to generate the training bench is given in [1]. The final step is to find the coefficients, c_i's corresponding to the P_i's. This is done by using multiple linear regression analysis. The energy consumption $E_{estimate}$ is then calculated using Equation (1). Figure 1 shows an overview of our energy characterization flow. To obtain the reference energy values, we simulate the processor system at gate-level using the training bench. The training bench is divided into sub-sequences as shown in Fig.2. We refer to this sub-sequence as the instruction frame. The width is the same for all instruction frames. Since we perform gate-level simulation and calculate the energy consumption values for all instruction frames, this step is time-consuming. However, it needs to be done only once for the characterization. In addition to this, the number of cycles simulated for the training bench is much smaller than that for target application programs. More specifically, our training bench is simulated only about 500,000 cycles while the

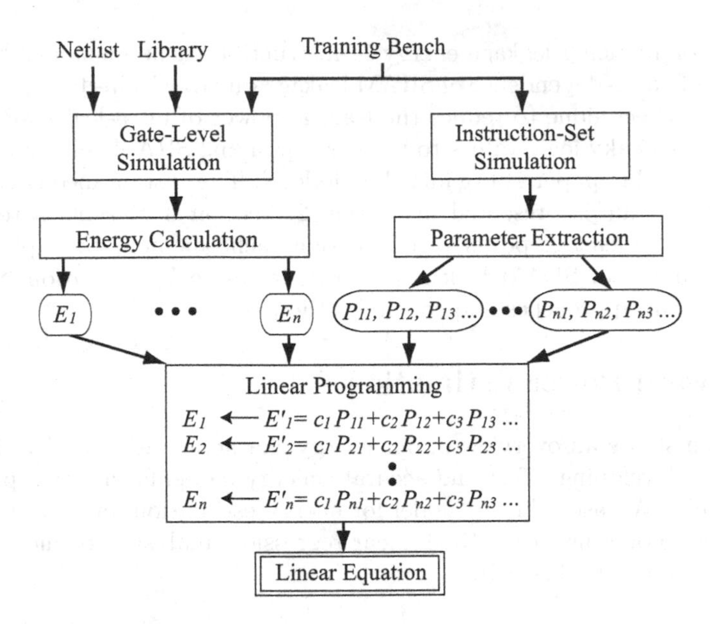

Fig. 1. Overview of Energy Characterization

full simulation of the target application programs needs billions of cycles. Thus, total time required for the energy estimation using our approach is much smaller than that of gate-level or RT-level simulation-based approaches.

We, next, obtain an instruction trace for each application program using an instruction-set simulator. The traces are divided into small segments corresponding to instruction frames. P_i's should be parameters that can be easily extracted from instruction traces. For a set of P_i's, we find coefficients which minimize $\sum |E_{estimate}(i) - E_{gate-level}(i)|$, where $E_{gate-level}(i)$ and $E_{estimate}(i)$ are the energy consumption values obtained by gate-level simulation and Equation (1) for the ith instruction frame, respectively. Once the energy model is developed, the energy consumption of software running on the processor system can be estimated using a cycle-*inaccurate* instruction-set simulator (ISS) with the speed of 300,000 instructions per second.

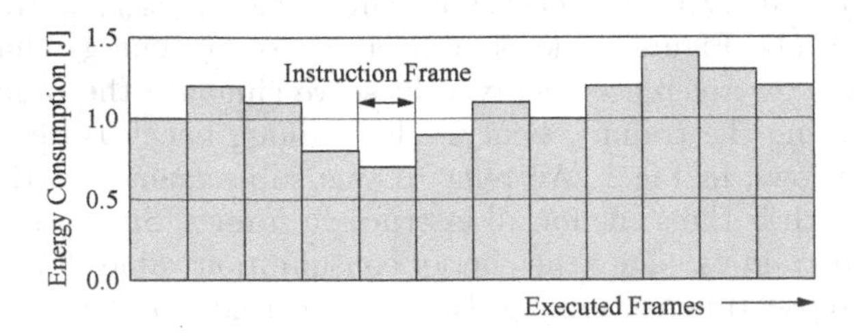

Fig. 2. An Example of Instruction Frame

2.2 Experimental Results

We target a system which consists of a CPU core, on-chip cache memories, and SDRAM as an off-chip main memory as shown in Fig.3. For the off-chip main memory, we assumed a Micron's SDRAM. We used an M32R-II processor and an SH3-DSP processor as CPU cores as follows.

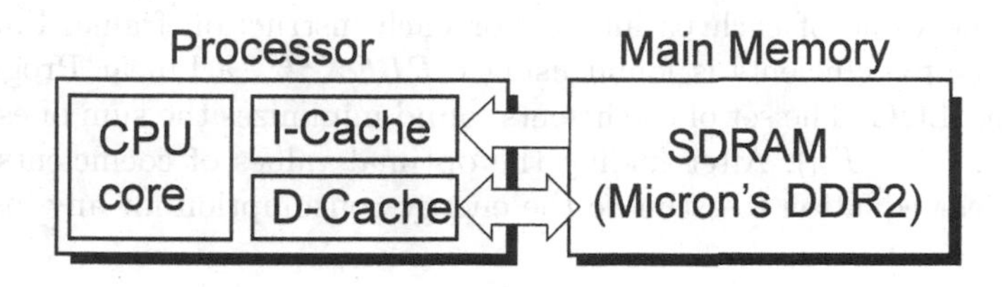

Fig. 3. A Target System Model

- M32R-II processor A 32-bit RISC microprocessor with 5-stage pipeline developed by Renesas Technology Corporation. It has 8KB 2-way set associative separate caches for instruction and data, a 32KB SRAM, and a 16-entry TLB on the chip.
- SH3-DSP processor A 32-bit RISC microprocessor developed by Renesas Technology Corporation. It has a digital signal processor core, a 32KB 4-way set associative unified cache, a 128KB SRAM, and a 18KB SRAM on the chip.

We use 0.18 m CMOS standard cell library and SRAM module library for synthesizing the above two processors. Five benchmark programs shown in Table 1 are used in our experiment. Each benchmark program was simulated 1,000,000 instructions for evaluating our approach. Each instruction frame was 5,000 instructions long and there were total of 200 instruction frames.

Table 1. Description of Benchmark Programs

Benchmark Program	Program Description
JPEG	JPEG encoder version 6b
MPEG2	MPEG2 encoder version 1.2
compress	File compression program
FFT	Fast Fourier Transform
DCT	Discrete Cosine Transform

First, we generate the Switching Activity Interchange Format (SAIF) file through gate-level simulation using $NC\text{-}Verilog^{TM}$ from Cadence design systems. The SAIF file has the information about the values of the signals that change during simulation. Then, the energy consumption, E_i, is calculated for the ith instruction frame using $DesignPower^{TM}$, a gate-level power calculation tool from SYNOPSYS. The average energy consumption per access for the instruction cache and the data cache are calculated using library data sheets. We used the Micron System Power Calculator [2] for calculating the energy consumption of SDRAM. Similarly, we generate an instruction trace using GNU C debuggers for M32R-II and SH3-DSP processors. Note that the many of ISSs used in the GNU C debugger is cycle-*inaccurate*. We divide the instruction trace into small sub-traces each of which corresponding to an instruction frame and calculate the value of each parameter for each instruction frame. Finally, the optimal set of coefficients is found using $CPLEX^{TM}$, a Linear Programming solver form ILOG. The set of coefficients found minimizes the sum of estimation errors (i.e., $|E_i - E_i'|$). After finding the optimal values of coefficients, we can use the linear equation to estimate the energy consumption for any instruction trace.

Table 2. CPU-Time for Characterization (Minutes)

Target Processor	M32R-II	SH3-DSP
Gate-Level Simulation	127	328
Power Calculation	32	41
Instruction-Set Simulation	< 1	< 1
LP Solver	< 1	< 1
Total CPU Time	160	370

Table 2 shows the characterization results. The characterizations for M32R-II and SH3-DSP took 160 minutes and 370 minutes, respectively. Although this step is time-consuming, it needs to be done only once for a target processor system. We start with a set of predetermined parameters which include 82 parameters and select some of them for a given microprocessor. We generate the training bench so that the standard deviations of every predetermined parameter values are large enough and every correlation factors between any two parameters are small enough. The generated training benches are simulated 475,000 instructions and 140,000 instructions for M32R-II and SH3-DSP processors, respectively. If the value of the parameter multiplied by its corresponding coefficient is very small compared to the other values, the parameter will not be used due to its weak impact on the energy estimation. In addition to this, several parameters are merged into a single parameter if corresponding coefficient values are very close to each other. As a result, we chose 30 and 19 parameters for M32R-II and SH3-DSP processors, respectively. The parameters include the following:

- The number of the following classes of instructions executed: 1) multiply, 2) divide, 3) multiply-add, 4) the other arithmetic operations, 5) logic, 6) shift, 7) register transfer, 8) load, and 9) store operations.
- The number of taken and untaken branches executed.
- The number of data and instruction cache misses.
- The number of times the instruction and data caches simultaneously miss.

– The number of times the read-after-write hazard occurs.
– The numbers of other events which cause a pipeline stall occur.

Table 3. Energy Estimation Results for M32R-II Processor

	Average Error	Maximum Error	Standard Deviation of Error Percentage
JPEG	2.70%	10.32%	2.76
JPEG_Opt	0.69%	16.46%	6.17
MPEG2	1.54%	3.97%	0.94
MPEG2_Opt	1.78%	5.15%	0.96
compress	5.00%	6.41%	1.19
compress_Opt	4.35%	7.18%	0.93
FFT	1.55%	6.87%	0.92
FFT_Opt	1.45%	5.59%	0.89
DCT	1.42%	8.58%	0.72
DCT_Opt	1.47%	8.07%	0.69
Total	2.74%	16.46%	2.82

Table 4. Energy Estimation Results for SH3-DSP Processor

	Average Error	Maximum Error	Standard Deviation of Error Percentage
JPEG	3.17%	11.89%	3.11
JPEG_Opt	6.33%	10.02%	2.79
MPEG2	1.32%	3.41%	0.98
MPEG2_Opt	1.31%	5.63%	0.97
compress	5.73%	10.84%	1.37
compress_Opt	1.73%	15.15%	1.27
FFT	1.27%	3.25%	0.76
FFT_Opt	1.15%	4.75%	0.88
DCT	1.12%	2.20%	0.46
DCT_Opt	1.51%	3.04%	0.52
Total	2.47%	15.15%	2.45

Average, maximum, and standard deviation of energy estimation errors for M32R-II and SH3-DSP processors are shown in Table 3 and 4, respectively. A suffix of each benchmark program "_Opt" represents that the program is compiled with a "-O3" option. The energy estimation error of our approach is on an average 2.7% and worst case 16.5% for M32R-II processor. For SH3-DSP processor, the error is on an average 2.5% and worst case 15.2%. The accuracy of energy estimation is overall very good. The notable point is that the standard deviation of error percentage is very small. This shows that our estimation results have a similar trend to the gate-level results even though absolute errors are not very small in some cases.

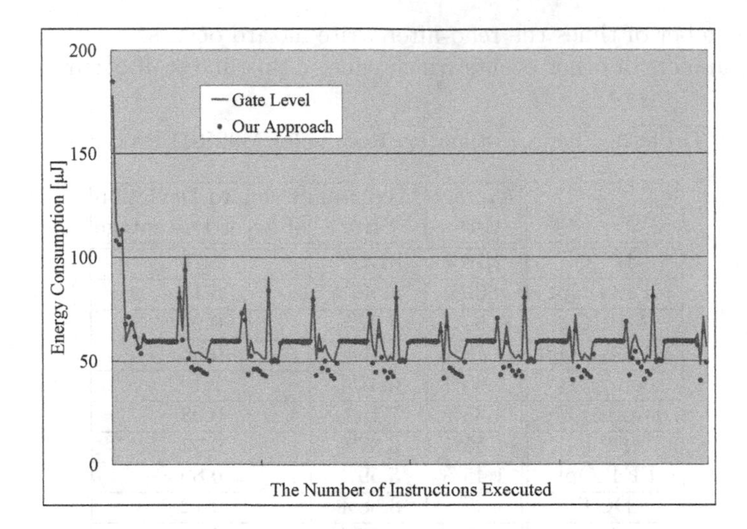

Fig. 4. Results for JPEG encoder run on M32R-II

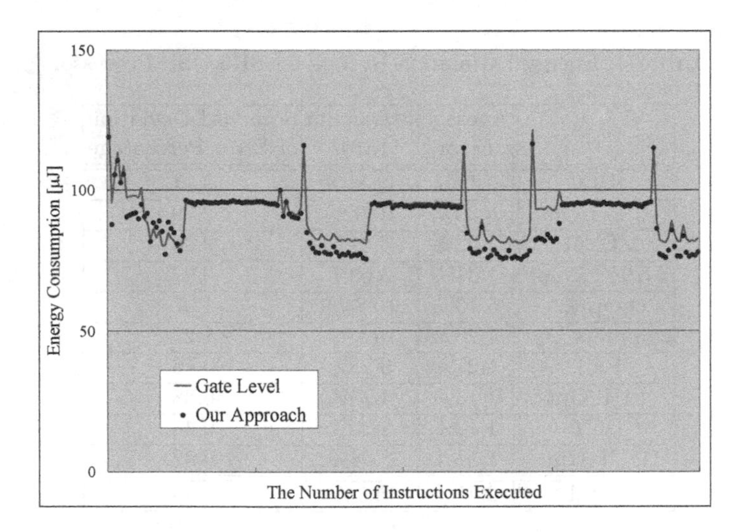

Fig. 5. Results for JPEG encoder run on SH3-DSP

Figures 4 and 5 show the detailed results for JPEG encoder which runs on M32R-II and SH3-DSP processors, respectively. Horizontal and vertical axes represent instruction frame number and energy consumption per instruction frame, respectively. The energy consumption includes the energy for a CPU core, on-chip caches, and off-chip SDRAM. As one can see, the estimation errors for every instruction frames are very small.

2.3 Summary

An energy characterization framework for processor-based embedded system is proposed. Experimental results obtained with two commercial microprocessors with their on-chip instruction and data caches, and an off-chip SDRAM demonstrated that the error of our technique is on an average 3% and worst case 16% compared to the gate-level estimation results. Our energy estimation method works very well even with a cycle-*inaccurate* simulator like a GNU debugger which is a de facto standard of software debugger. Once the model has been obtained, the energy consumption can be calculated with the speed of 300,000 instructions per second.

Today's SoC chips are usually implemented with off-the-shelf processor IPs. Even for those SoC chips, our method can accurately model the energy consumption since our tool does not need to know a detailed internal architecture nor the RTL description of the target processor. SoC vendors can easily generate an accurate energy model using our tool and provide it to their customers. This helps compilers or programmers to customize software codes to meet customers' needs for low power. Our future work will be devoted to extending the current framework to consider multi-core processor systems.

3 Cache Leakage Reduction

Random Dopant Fluctuation (RDF) [3] within the same die results in changes in the V_{th} of transistors. Transistors of cache SRAM cells are more affected since they have minimal physical channel area. The mismatch among V_{th} of transistors of a single SRAM cell results in different leakage currents depending on the value stored in the cell. Thus leakage of the cache memory can be reduced if the values with less leakage can be more often stored in each SRAM cell. We propose techniques to reduce instruction-cache leakage using this phenomenon.

3.1 Value Dependence of SRAM Leakage

When the SRAM cell is storing a 1 (Fig.6) only three transistors contribute to the total leakage (M1, M2, M5); when storing a 0, the other three transistors leak [4]. Since subthreshold leakage exponentially depends on V_{th}, total leakage can be significantly different in the two states. Figure 7 shows probability distribution of SRAM leakage differences between two states. The standard deviation divided by the mean value of SRAM access latency is assumed to be a 5%. The variation of SRAM access latency is closely corresponding to the V_{th} variation since the latency of an SRAM read access is almost linear to the V_{th} of an SRAM cell. Note that the values in Fig.7 are obtained from SPICE simulation for our original SRAM modules designed with a commercial 90nm process technology. In our case, the total leakage power of a 4KB SRAM module is $206\mu W$ at 75 degree Celsius if more-leaky values are stored to every SRAM cells. Contrary, if less-leaky values are stored to the every cells, the leakage power dissipated is only $53\mu W$. If the logic values stored in the SRAM is random, the typical case leakage

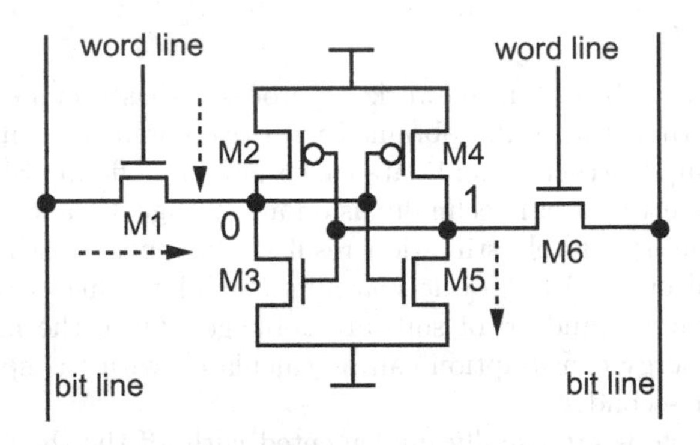

Fig. 6. Different transistors leak based on cell value

power is about 130 μW which is an intermediate value between $206\mu W$ and $53\mu W$. Therefore, we can save about 50% of the leakage power by storing less-leaky values to the SRAM cells. Another important observation is that only 10% of cells consume more than 50% of the total leakage power if there is 5% of delay variation (i.e., there is a 5% V_{th} variation). In case of a 10% delay variation in the SRAM cells, more than 65% of the total leakage power is dissipated by 10% of the SRAM cells. This means that we can save a large amount of leakage power by storing less-leaky values to the small number of ultra-leaky SRAM cells.

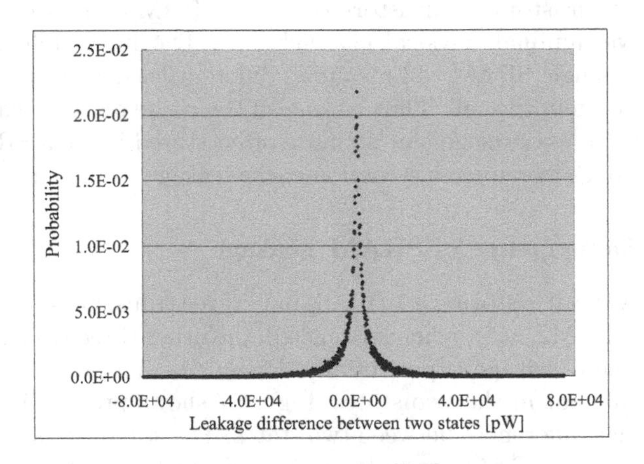

Fig. 7. SRAM leakage variation

3.2 Cache Leakage Power Reduction

Our approach exploits an unused combination of existing flag bits (i.e., valid bit=0 and lock bit=1) to indicate ultra-leaky SRAM cells in a specific cache-line [5]. Suppose we have a 4-way set associative cache with lock and valid bits as shown in Fig.8. If the lock bit of the way1 in the 5th cache-set is "1", the corresponding cache-line will not be used for replacement even in case of a cache miss. If the valid bit of way1 in the 5th cache-set is "0", accessing to the corresponding block will always cause a cache miss. Thus, correct cache operation is guaranteed even in case that invalid values (i.e., less-leaky values) are locked in several cache lines. The basic idea of our approach is to reduce the leakage power of SRAM cells by locking less-leaky values in several cache lines which contain the ultra-leaky SRAM cells.

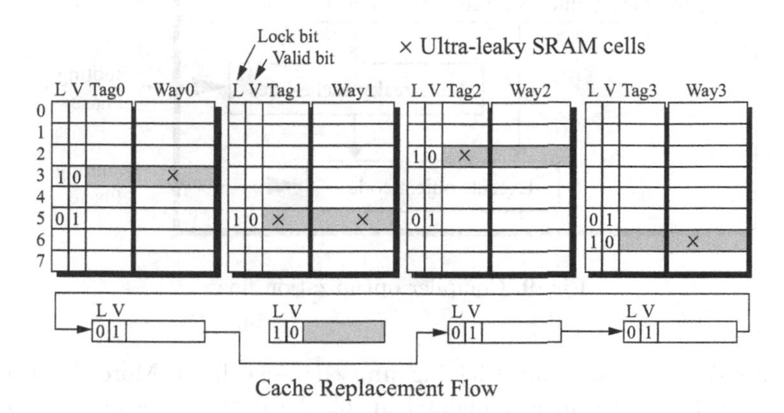

Fig. 8. Cache locking function

Table 5 shows the results of leakage savings by our approach compared to the typical case where the values stored in the SRAM are random. The "power-saving cache-line" in Table 5 represents a cache line which stores less-leaky values in every bits in the cache line. The results show that we can save more than 20% of leakage power if the delay variation is more than 8% and if we store less-leaky values to the 25% of cache-lines.

Table 5. Leakage power reduction

Standard deviation divided by mean latency	4%	8%	12%	16%
Percentage of power-saving cache-lines = 6.25%	4.20%	5.89%	6.48%	6.54%
Percentage of power-saving cache-lines = 12.50%	7.98%	11.05%	12.33%	12.53%
Percentage of power-saving cache-lines = 18.75%	10.79%	16.81%	17.78%	18.34%
Percentage of power-saving cache-lines = 25.00%	14.87%	21.38%	23.37%	23.91%

3.3 Cache Performance Improvement

As shown in Table 2, the more cache lines are power-saving lines, the more leakage power can be saved. However, this leads to an increase of cache misses due to a decrease of an effective cache capacity. In this sub-section, we present a leakage-aware code placement technique which reduces the performance degradation of a partially unused cache memory. Our approach is to modify the placement of basic blocks or functions in the address space so that the number of cache misses

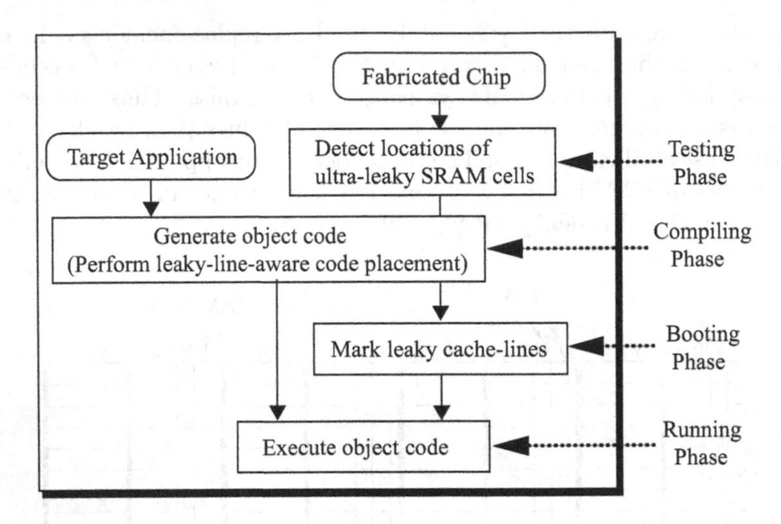

Fig. 9. Compiler optimization flow

is minimized for a given cache having unused cache lines. More detail of our code placement algorithm is explained in [6]. An overview of our approach is depicted in Fig.9. First, we detect the locations of ultra-leaky SRAM cells in a cache memory. Next, our code placement technique generates the object code such that the number of cache misses can be lower than a given number even if several cache lines which contain the ultra-leaky cells are unused. Therefore, we perform recompilation only if the original object code does not satisfy the required performance for a specific chip. If the original object code or an object code previously generated for another chip can satisfy the required performance, we use it. Every time the chip is turned on, it executes an initialization step during which based on the information collected during test, ultra-leaky cache-lines are marked using lock and valid bits. Then the chip executes the compiled code.

3.4 Experiments and Results

We used three benchmark programs; *compress* version 4.0, *JPEG encoder* version 6b, and *MPEG2 encoder* version 1.2. All programs are compiled with "-O3" option. We used GNU C compiler and debugger for ARMv4T architecture to generate address traces. We used the following four types of cache memories:

Cache-1. A 32Kb direct mapped cache with 128 cache-sets whose cache-line size is 32 bytes.
Cache-2. A 32Kb 2-way set-associative cache with 64 cache-sets whose cache-line size is 32 bytes.
Cache-3. A 32Kb 4-way set-associative cache with 32 cache-sets whose cache line size is 32 bytes.
Cache-4. A 16Kb 2-way set-associative cache with 32 cache-sets whose cache line size is 32 bytes.

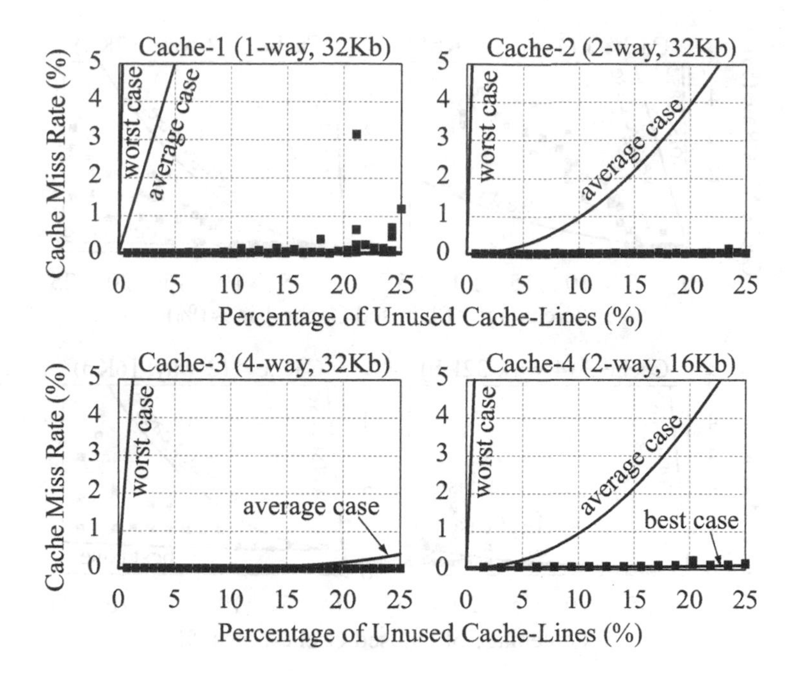

Fig. 10. Results for *compress*

In this experiment, we randomly chose ultra-leaky cache-lines (i.e., unused cache lines), and after that, we applied our leaky-line-aware code placement. We tried 480 different patterns of locations of ultra-leaky cache-lines for each benchmark program. Figures 10, 11, and 12 show the results for *compress*, *JPEG encoder*, and *MPEG2 encoder*, respectively. Black dots represent results of our leaky-line-aware code placement. In case of *compress*, there is no noticeable performance degradation even if 20% of total cache lines are unused. The results for the 4-way set associative cache show that 25% of cache lines can be deactivated for all benchmark programs without noticeable performance degradation. In this case, we can save about 20% of leakage power consumption. This implies that it is better idea to deactivate several cache lines in a highly associative cache and to save leakage power of these deactivated cache lines by storing less-leaky values if there is a large process variation.

3.5 Summary

In this section, a leaky-line-aware code placement technique is presented. Experiments demonstrate that our code placement technique offsets the impact of the reduced cache capacity on performance in most cases for 4-way set-associative caches even if 25% of cache lines are unused. In this case, about 20% of cache leakage power can be saved by locking less-leaky values in the unused lines.

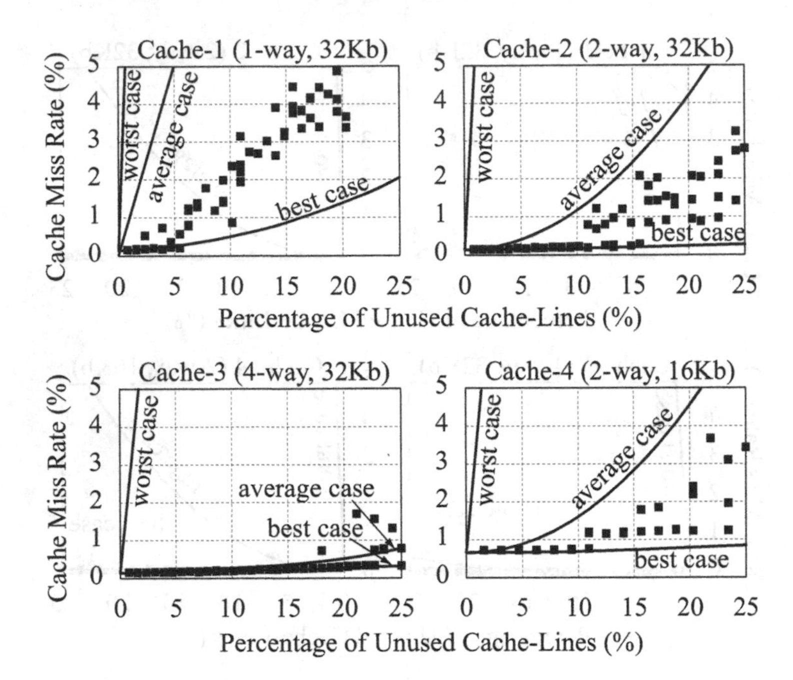

Fig. 11. Results for *JPEG encoder*

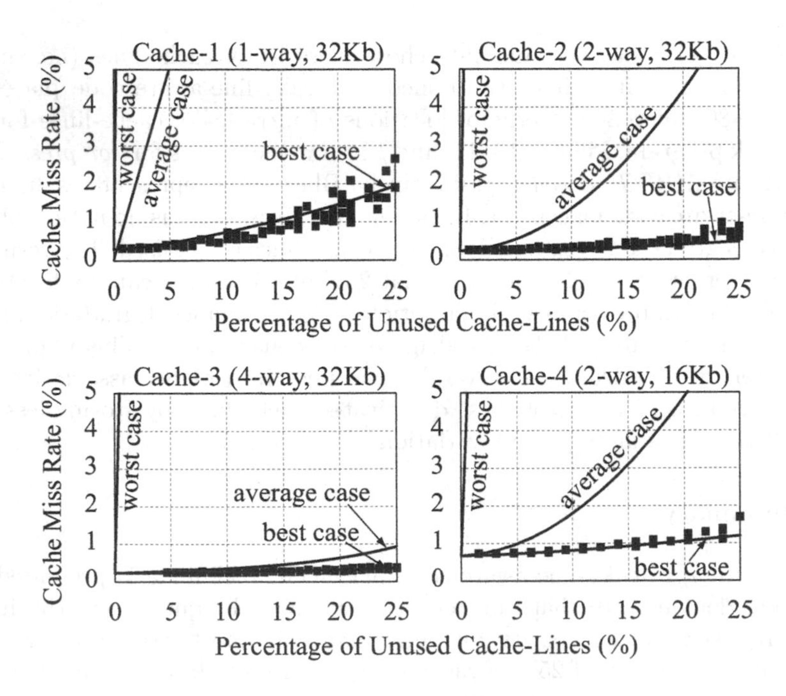

Fig. 12. Results for *MPEG2 encoder*

4 Conclusion

As the transistor size scales down to 90nm and below, non-recurring engineering (NRE) costs associated with mask set costs increase significantly. A remedy for the NRE explosion is to reduce the number of developments and sell tens of millions of chips under a fixed hardware design. In such a situation, embedded software plays much more important role than today. Our main focus is on software-oriented approaches to estimating and reducing the energy consumption of embedded real-time systems. We believe our research activities presented here will be a major trend in nanometer SoC design.

Acknowledgments. This work is supported by VDEC, the university of Tokyo with the collaboration of Renesas Technology, STARC, Panasonic, NEC Electronics, Toshiba, ROHM, Toppan Printing, Cadence Design Systems, Synopsys and Mentor Graphics. This work is also supported by CREST program of JST.

References

1. Lee, D., Ishihara, T., Muroyama, M., Yasuura, H., Fallah, F.: An Energy Characterization Framework for Software-Based Embedded Systems. In: ESTIMedia 2006. Proc. of IEEE workshop on Embedded Systems for Real-Time Multimedia, pp. 59–64 (October 2006)
2. The Micron System Power Calculator, http://www.micron.com/support/designsupport/tools/powercalc/powercalc.aspx
3. Taur, Y., Ning, T.H.: Fundamentals of Modern VLSI Devices. Cambridge University Press, Cambridge (1998)
4. Goudarzi, M., Ishihara, T., Yasuura, H.: A Software Technique to Improve Yield of Processor Chips in Presence of Ultra-Leaky SRAM Cells Caused by Process Variation. In: ASPDAC 2007. Proc. of Asia and South Pacific Design Automation Conference, pp. 878–883 (January 2007)
5. Ishihara, T., Fallah, F.: A Non-Uniform Cache Architecture for Low Power System Design. In: ISLPED 2005. Proc. of International Symposium on Low Power Electronics and Design, pp. 363–368 (August 2005)
6. Ishihara, T., Fallah, F.: A Cache-Defect-Aware Code Placement Algorithm for Improving the Performance of Processors. In: ICCAD 2005. Proc. of International Conference on Computer Aided Design, pp. 995–1001 (November 2005)

Empirical Study for Optimization of Power-Performance with On-Chip Memory

Chikafumi Takahashi[1], Mitsuhisa Sato[1], Daisuke Takahashi[1], Taisuke Boku[1], Hiroshi Nakamura[2], Masaaki Kondo[2], and Motonobu Fujita[2]

[1] Center for Computational Sciences, University of Tsukuba
{takahasi, msato, daisuke, taisuke}@ccs.tsukuba.ac.jp
[2] Research Center for Advanced Science and Technology, The University of Tokyo
{nakamura, kondo, mfujita}@hal.rcast.u-tokyo.ac.jp

Abstract. Power-performance (performance per uniform power consumption) recently has become a more important factor in modern high-performance microprocessors. In processor design, it is a well-known that off-chip memory access has a large impact on both performance and power consumption. On-chip memory is one solution for this problem, so that many processors such as the Renesas SH-4 and some ARM architecture type processors adopt on-chip memory, which resides on the same layer as the cache memory. In this study, the effectiveness of the on-chip memory in an SH-4 processor was quantitatively examined by directly measuring the real power of the processor. For these experiments, we proposed a method that made use of the on-chip memory for power reduction. The experimental results show that the optimization of data transfer using on-chip memory reduces EDP(energy delay product) by up to 15.2%. As an extension of on-chip memory, we have proposed an on-chip RAM architecture called SCIMA (software controllable integrated memory architecture) which enables DMA (direct memory access) transfer to the on-chip memory. According to the empirical data from the SH-4 processor, it was found that the additional DMA transfer using SCIMA reduces EDP by up to 26.3%.

1 Introduction

In recent years, the clock frequency of modern microprocessors has been increased, due to advances in device technology and architecture. However, a high clock frequency causes extreme heat generation so that greater improvement of performance is prevented. Therefore, it is important to pursue improvement of performance related to power consumption (power performance), not only computational power.

In processor design, it is a well-known problem that the off-chip memory access has a large impact on both performance and power consumption. Modern processors have a cache used to transfer data between the register and the main memory. Cache utilization may reduce power consumption due to the reduction of off-chip main memory access. However, the cache holds data that is automatically selected by the hardware according to an access pattern. Furthermore,

J. Labarta, K. Joe, and T. Sato (Eds.): ISHPC 2005 and ALPS 2006, LNCS 4759, pp. 466–479, 2008.

the cache holds data in a fixed unit size. These limitations cause useful data to drop from the cache and non-useful data to be acquired from the main memory. Transfer of non-useful data requires extra execution time, resulting in unnecessary energy consumption. Use of an on-chip memory is one solution to this problem. On-chip memory resides on the same layer as the cache memory, and is accessed as a part of the main memory. Many processors, such as the Renesas SH-4 and some ARM architecture processors adopt on-chip memory.

In this study, we quantitatively examine the effectiveness of the on-chip memory by directly measuring the real power of the processor. We then propose a method to use on-chip memory for the power reduction. Utilizing on-chip memory can improve power performance, because of software controlled memory access. As an extension of the on-chip memory, we have proposed an on-chip RAM architecture called SCIMA (software controllable integrated memory architecture) which enables DMA (direct memory access) transfer to the on-chip memory. Power performance improvement for this SCIMA architecture is then predicted based on the experimental data.

The contributions of this paper are as follows:

- We quantitatively examine the effectiveness of the on-chip memory in a real processor, SH-4, by directly measuring the real power of the processor.
- We propose a method to make use of on-chip memory for power reduction.
- We demonstrate that SCIMA can provide greater power-performance improvement by addition of DMA transfer from off-chip to on-chip memory.

Note that many quantitative evaluations of power performance have been performed by simulation, but few have been reported using real power measurements within processor architecture research.

The next section provides an overview of on-chip memory architecture as the background of our research, in addition to a method for use of on-chip memory. Section 3 details the experimental settings used for the evaluation of power performance of the SH-4 processor. The experimental results are given in section 4, and in section 5, improvement of power performance is discussed with consideration for the SH-4 processor with DMA as an assumption of SCIMA. Concluding remarks and future work are described in section 6.

2 Optimization of Power Performance with On-Chip Memory

2.1 On-Chip Memory Architecture

The on-chip memory is a type of memory located on the same silicon chip as the processor core. Many on-chip memory architectures have been proposed. We proposed an on-chip memory architecture called SCIMA[1,2]. SCIMA is an architecture which has SRAM on the same memory hierarchy layer as the cache. Kogge et al. [3] proposed a SIMD architecture called PIM (Processor-In-Memory), that has a processing unit on the memory module. Draper et al. [4] proposed to connect many PIM, in the so-called Data-IntensiVe Architecture (DIVA). Patterson

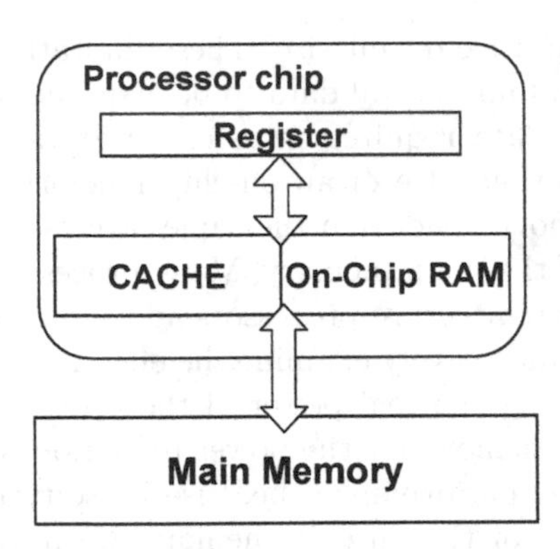

Fig. 1. On-chip memory model

et al.[5] proposed Intelligent RAM (IRAM), that has on-chip DRAM. In addition, some production processors have on-chip memory. For example, the L3 cache of a PowerPC processor in BlueGene/L can be used as on-chip memory[6]. The Renesas SH-4 processor has an on-chip memory mode, which allows half of the data cache to be reconfigured as on-chip memory[7]. There are also several chips that store intermediate data on the scratch pad memory[8,9].

In this paper, we propose to optimize power performance with on-chip memory that resides on the same layer as the cache. The on-chip memory model is shown in Figure 1. Efficient management of data transfers between the on-chip memory and main memory decreases power consumption. On-chip memory has the following advantages.

(1-1) By explicitly specifying the transferred data necessary for computation, unexpected cache misses and non-useful data transfers caused by fixed cache line size data transfer can be avoided.
(1-2) Prefetching the data transfer enables overlap with computation by the DMA function, so that data transfer time can be hidden as in cache prefetch.

On the other hand, there are disadvantages of the on-chip memory, as follows:

(2-1) The requirement for software support to insert codes for explicit data transfer.
(2-2) The data transfer codes may increase the number of instructions.

To make use of on-chip memory, a programmer or compiler must insert data transfer instruction(s) explicitly.

2.2 Metrics of Power Performance

The power consumed in a processor can be divided into two parts. One is static power consumption, which is derived from leakage power consumption and is consumed constantly. The other is dynamic power consumption, which arises from switching activity on the bus and memory, and is consumed when the processor is working.

In this paper, we define the power consumption in the idle state as static power in order to simplify the discussion. This is different from the above definition. For example, although the power consumption in the clock line should be classified as dynamic power consumption, according to our definition, it is classified as static power consumption. Therefore, static power consumption is defined as stationary consumed power, and the dynamic power consumption is that after subtraction of static power from the working power.

The total power consumption is defined by:

$$E_{all} = P_{static} * t + E_{dynamic}$$

where P_{static} is the static power, $E_{dynamic}$ is the dynamic energy, and t is working time.

Reduction of data transfer by optimizing memory access results in a decrease of the switching activities of the bus and memory modules. Therefore, optimization of memory access results in the reduction of dynamic power consumption $E_{dynamic}$. In addition, the improvement of performance by (1-1) and (1-2); that is, shortening the working time t, can reduce the static power consumption P_{static}. Therefore, it is expected that improved use of on-chip memory will enable reduction of power consumption.

The model used in this study is similar to the SCIMA model. The power performance of SCIMA has already been evaluated by simulation, and the effectiveness is reported in [8,10]. In this study, we examine the effectiveness of on-chip memory usage with a real processor.

2.3 Optimization Methods for On-Chip Memory

In this subsection, methods to optimize programs for on-chip memory on the same layer as the cache are presented. The basic strategy of the optimization is similar to that for the cache, which employs the concept of temporal locality. The steps for optimization of on-chip memory are as follows:

1. Find data which can be re-used, and transfer the data to the on-chip memory from the main memory.
2. Execute computation using data in the on-chip memory.
3. Restore dirty data from the on-chip memory to the main memory.

If the data set size is larger than the on-chip memory size, the data should be divided into several chunks of smaller data.

Another possible optimization method is to hide the data transfer time. If the on-chip memory supports Direct Memory Access (DMA) and it can be overlapped with the computation, data pre-fetching can avoid stalling in the pipeline by waiting for data transfer completion. In this situation, performing software pipelining for data transfer and computation is effective.

The optimizations are effective when the access pattern for the main memory is predictable. It is difficult to optimize unpredictable memory access, because it is not possible to specify data transfer instruction(s) at the programming or compilation phase. Therefore, on-chip memory is used only for memory access with a predictable access pattern, and unpredictable memory access is performed by the traditional cache.

In this work, the optimizations are coded manually by hand. Fujita presented the automatic optimization by a compiler for SCIMA, which can be applied to our model. [11]

3 Experimental Setting

3.1 Target Processor

We have examined the power performance of the Renesas SH7751R SH-4 32-bit super scalar RISC processor, which is a commodity product. The SH7751R can use the 32 KB 2-way set associative data cache as 16 KB on-chip memory and 16 KB direct map cache. There are two modes in this processor: "OCR mode" (on-chip RAM mode) is a mode where on-chip memory is available, and "CACHE mode" is a mode where all the memory is used as a cache. The detailed specifications for the SH4 on-chip memory are provided in Table 1. In OCR mode, half of the data cache is mapped on a specific address area. This area can be accessed by the move instruction and so on. A Hitachi Ultra LSI system Solution Engine was used as the mother board for the evaluation. The installed OS is Linux. Details of the processor are given in Table 2.

3.2 Power Measurement Environment

A power measurement environment was built using a Hall device, which measures the electric current on a conducting wire by observing changes in the magnetic field. Sensors were set on the ATX power supply output line, input and output of a 3-terminal regulator. Two types of power data can be obtained as follows:

CORE. This is the power supplied to the SH-4 processor core. It is measured at the output of the 3-terminal regulator(1.5 V), which connects to the processor core. This power is consumed at the registers, cache, on-chip memory, and other processor core, except the chip I/O.

Table 1. SH4 (SH7751R) On-chip Memory

Mode	D-Cache	On-chip memory
CACHE mode	32 KB (2- way)	0 KB
OCR mode	16 KB (1-way)	16 KB

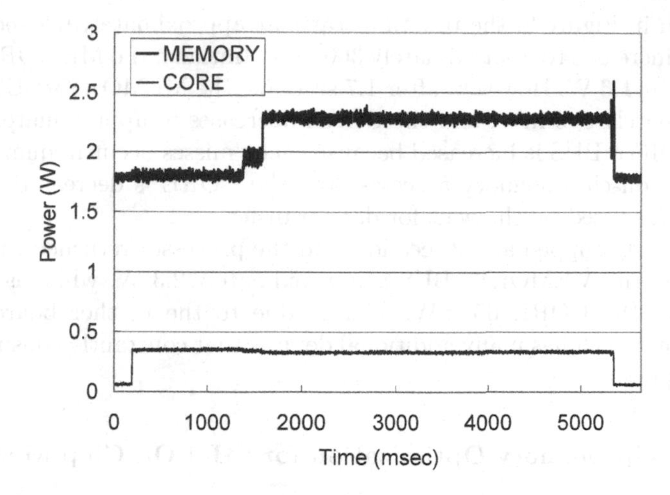

Fig. 2. Power Consumption from the Data Scan Program

MEMORY&BUS. This is the power for memory and I/O. It is calculated by subtracting the 3-terminal regulator input from the ATX 3.3 V output. This 3.3 V power is supplied to the memory modules, the processor I/O (bus), and other on board devices. Changes in power consumption for these memory related devices are observed from this power data.

The motherboard consumes ATX 5.0 V power. However, almost all of the power is consumed in unrelated devices. Therefore, only the CORE and MEMORY&BUS are discussed, not the ATX 5.0 V consumption.

3.3 Basic Power Characteristics

As a preliminary experiment, the power consumption was measured by executing a simple data scan program. The results are shown in Figure 2. The program sequentially scans the data in the memory area , and repeats the scan by gradually increasing the accessed area . Using this experiment, we can observe the behavior of power consumption related to the cache, because by gradually increasing the access area a cache miss is eventually caused.

Table 2. Solution Engine detail

Name	SuperH Solution Engine
Type	MS7751RSE01
CPU	SH7751R (SH-4) 240 MHz
Memory	64 MB SDRAM 60 MHz
I/O	CompactFlash, IDE, etc
OS	Linux 2.4.18
Compiler	gcc 3.2.3

As shown in Figure 2, the program starts at approximately 0.2 seconds, and the CORE increases to approximately 360 mW. Initially, the MEMORY&BUS is maintained at 1.6 W. However, after 1.7 seconds, the MEMORY&BUS increases to approximately 500 mW, and the CORE decreases to approximately 15 mW. The MEMORY&BUS is increased because cache misses occur frequently, resulting in many off-chip memory accesses, and the CORE is decreased due to the pipeline stall caused by the wait for data transfer.

The program stopped at 5.3 seconds, and the processor returned to idle state. At this time, the MEMORY&BUS is approximately 2.3 W, which is very large compared to the CORE 65 mW. This is due to the mother board used for development, which has many additional devices that constantly consume a large amount of power.

3.4 On-Chip Memory Optimization for SH-4 On-Chip Memory

For the SH-4, which has no DMA support from main memory to on-chip memory, data transfer between the on-chip memory and main memory is performed by transfer instructions. In addition to this, the instructions support only transfer between registers and the main memory or registers and the on-chip memory, so that transfer of data between the on-chip memory and the main memory should occur via a register. Furthermore, the size of the data transfer from the main memory to the on-chip memory is fixed to 32 bytes on the bus, because transferred data must go through the cache. This may cause transfer of unnecessary data from the main memory to the cache in the case of data transfers smaller than 32 bytes.

In the experiment, we have optimized three programs: matrix multiplication, NPB CG (sequential version) and FFT. As shown in Table 1, the on-chip memory size is only 16 KB, which is half of the cache in CACHE mode. Consequently, in some cases, not only on-chip memory but also cache is used even in OCR mode.

Matrix Multiplication. We have optimized the matrix multiplication (MM) program, expressed by $C = A \times B$, where A, B and C are square matrices of the same size, with on-chip memory.

In the traditional cache architecture, cache blocking (or tiling) [12] is usually used to make the working set small enough to fit the cache size and to improve the data locality. However, unfortunate line conflicts may occur and degrade performance. In matrix multiplication, data access to non-contiguous addresses (row-ordered access to matrix B) may cause line conflicts. Therefore, these line conflicts should be avoided in order to obtain higher performance. Although this problem is partially solved by the set associativity of the cache, the data access pattern must be modified with software to completely avoid these conflicts. However, such optimization requires complicated programming to fit the properties of the cache size, line size, associativity, etc.

In OCR mode, the data of matrices A, B and C are blocked (as in the tiling transformation for cache). Next, the portion of matrices A and B are transferred to on-chip memory via registers. Therefore A and B do not cause cache conflicts.

A portion of C is transferred to the cache to utilize the cache area. However, C has hardly any cache line conflict because the cache is only used for matrix C.

A matrix size of 500×500 was used for these experiments. The block size used in the blocking optimization was selected with consideration for the capacity of the on-chip memory or cache and the results obtained by the trial run. The following block sizes were carefully selected: 36×36 for CACHE mode and 32×32 for OCR mode.

NPB Kernel CG. The second workload is the CG (Conjugate Gradient) kernel of the NPB (NAS Parallel Benchmarks) [13]. This is a kernel loop to compute the eigenvalues of a large-scale symmetric random sparse matrix using the CG method. The kernel consists of double-nested loops, which include the multiplication of a sparse matrix and a vector. This part dominates the total execution time.

To perform multiplication of a sparse matrix and dense vector efficiently, the actual computation is expressed as $q = \sum_i (A(i) \times p(colidx(i)))$, where q is an element of the resultant vector, A is a packed array of non-zero elements of the target matrix, *colidx* is an array for match-making between elements of the matrix and vector, and p is the target vector. As for the data access characteristics , A and *colidx* are sequentially accessed, whereas p is randomly accessed.

Because matrix A and *colidx* are sequentially accessed, they have spatial locality. However, they do not have data reusability. On the other hand, vector p has significant reusability. Data access to A and *colidx* strongly degrades the cache-hit ratio for data access of vector p, since it is accessed randomly under capacity pressure on the cache. To improve the cache-hit ratio for p, one solution is to use cache blocking for p. By the application of blocking, vector p is divided into several portions, and each portion is calculated with partial data of matrix A that is associated to that portion. The elements of A and *colidx* can be pre-sorted before multiplication to make a set of data that restricts the data accesses of p to each block. This is possible because the contents of A and *colidx* are never modified through calculation in the CG method and this modification is only required once in the entire calculation.

In OCR mode, the same blocking optimization can be applied. The block of p is transferred to on-chip memory. In CACHE mode, the cache miss caused by the random access of vector p may degrade performance. However, in OCR mode, such an unexpected data swap-out never occurs and the reusability of vector p can be utilized. Each of the blocks of A and *colidx* are transferred using only the cache. There is the possibility that A and *colidx* are in conflict with each other. However, the conflict may be insignificant, because the access patterns of these matrices are sequential. Furthermore, the effect of cache miss is very small, because these matrices have no reusability.

A problem size of class-W was used. With a problem size of class-W, the size of the original (not-packed) matrix is 7000×7000 and the sparsity (ratio of non-zero elements) is approximately 1%. The block size of vector p is set at 1750 elements (elements of vector p are divided into four portions). The optimal size was determined by trial run.

Fast Fourier Transform. The third workload is the Fast Fourier Transform (FFT). The FFT is an algorithm used to compute the discrete Fourier transform quickly, and is used for various purposes. In this study, a three-dimensional double precision complex FFT program was used.

The 3-D FFT is computed by data pair of each dimensional axis. In this computation, stride memory access is caused at two directions of the dimensional axis . The stride access affects only 16 Bytes, which is smaller than the cache line size. This fine grain data access is inefficient, because half of the transferred data is non-useful data, and therefore unusable for the computation. To improve this problem, blocking optimization is applied for the sequential address direction of both modes. The optimization packs data to a coarse grain data size, which is larger than the cache line size[14]. Cache line conflicts caused by stride memory access can be improved by padding with non-useful data. However, this padding is not a perfect solution for improvement. In OCR mode, using on-chip memory clearly improves the cache line conflicts. However, OCM mode has a disadvantage because the on-chip memory size is half of the cache in CACHE mode, so that the blocking size is reduced.

A data size of 64×64×64 was used, and repeated twice. The block size used in the blocking optimization was selected by preliminary experiment. The following data sizes are selected: 16 for CACHE mode and 8 for OCR mode.

4 Power-Performance on SH-4

In this paper, the Energy Delay Product (EDP) was employed as an index of power performance. EDP is a widely used index in the field of low power research. EDP is defined by $EDP = P \times t \times t$, where P is average power, and t is time.

The experimental results are shown as Table 3. The EDP ratio, which is normalized by the EDP of CACHE mode, is shown as Figure 3. Figure 3 shows that EDPs are decreased when using on-chip memory for any application. The EDP is decreased to approximately 12.4% for MM, 7.4% for CG, and 15.2% for FFT.

Table 3 shows that the dynamic power consumption $E_{dynamic}$ is reduced in OCR mode for all applications. A reduction of MEMORY&BUS was observed, which results from the reduction of transferred data. It is expected that this will effect an improvement of cache line conflicts. The reduction of the CORE power is considered to be a result of decreasing cache renewal caused by cache misses. A reduction of the EDP results from these effects and results in the shortening of execution time.

The percentage of power reduction for E_{static} is larger than that for $E_{dynamic}$. This is because of the larger static power of the MEMORY&BUS, which is referred to in the previous section. If a different production motherboard (not the evaluation board) was used, the contribution to dynamic power reduction would be larger, because the power consumption of external devices would be smaller than the current environment.

Table 3. Power-Performance of SH-4

		Matrix Multiplication		NPB CG		3D FFT	
		CACHE	OCR	CACHE	OCR	CACHE	OCR
EDP ($W * sec^2$)	CORE	35.5	31.4	761.0	699.8	2.892	2.561
	MEMORY&BUS	207.0	181.0	4615	4258	17.10	14.38
	Total	242.5	212.5	5376	4958	19.99	16.95
$E_{dynamic}$ ($W * sec$)	CORE	2.848	2.675	14.16	13.48	0.826	0.801
	MEMORY&BUS	1.876	1.470	22.54	20.89	0.766	0.669
	Total	4.723	4.145	36.70	34.36	1.592	1.471
E_{static} ($W * sec$)	CORE	0.621	0.585	2.781	2.681	0.177	0.164
	MEMORY&BUS	18.32	17.27	80.20	77.44	5.163	4.751
	Total	18.94	17.86	82.98	80.13	5.340	4.915
Execution time (*sec*)		10.25	9.66	44.93	43.30	2.88	2.66

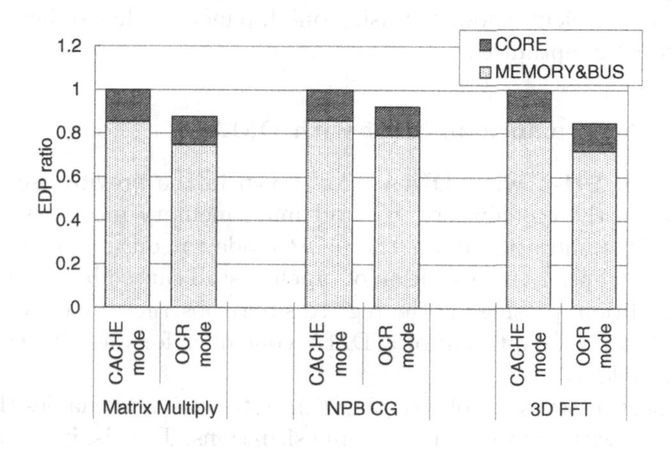

Fig. 3. EDP Ratio of SH-4

5 Discussion

5.1 Improvement of Power Performance by SH-4 On-Chip Memory

The experimental results show that utilizing on-chip memory improves power performance for the SH-4 processor. Considering the number of execute instructions, OCR mode may have some disadvantage for power consumption, because the OCR mode requires extra instructions to transfer data between the on-chip memory and main memory, and this causes an increase in the dynamic power consumption of the CORE. However, Table 3 shows a power consumption that is smaller than that of the CACHE mode . This indicates that the OCR mode performs better than CACHE mode, because off-chip memory access is reduced and the program runs faster due to less off-chip memory accesses.

The CORE dynamic power consumption shown in Figure 3 indicates that OCR mode has a lower value compared with CACHE mode. The on-chip memory reduces non-useful data transfer caused by cache misses. However, since data transfer between on-chip memory and main memory must go through the cache, the CORE power on OCR mode should include power consumed at the cache. This indicates the gain from removing non-useful data transfer exceeds the loss caused by passing the cache. As a result, the OCR mode has better power-performance than the CACHE mode.

It should be noted that the block size for the OCR mode became smaller than the CACHE mode, because the size of the on-chip memory is smaller than the full size of the cache. For example, for MM, while the maximum block size is 36×36 in in CACHE mode, it is only 32×32 in OCR mode. That is approximately 79% of the block size compared to the CACHE mode. For FFT, the OCR mode has only 50% block size. This is unavoidable because sufficient data size must be guaranteed for unpredictable memory access. However, the results of our experiments show that using on-chip memory has sufficient benefits when appropriately optimized.

5.2 Power Performance on SH-4 with DMA

Assumption of SH-4 with DMA. As shown in the previous section, data transfer between the on-chip memory and main memory must pass the cache and register. This limitation forces the OCR mode to consume more power by non-useful data transfer and execution of pipeline stall caused by non-useful data transfer. In addition, passing via the registers may obstruct the computation. It is expected that the adaptation of a DMA controller for on-chip memory will avoid these problems.

DMA support removes the obstructing computations, and enables the overlap of computations with data transfer in some situations. That is, it can be used in software pipelining. This is not only advantageous, but it also allows selection of data granularity and flexible transfer of data. Coarse grain data transfer can reduce the data transfer overhead, and fine grain data transfer reduces non-useful data transfer. These advantages might reduce power consumption through the reduction of data transfer and cause shorter execution time. The power performance using FFT was investigated under the assumption that DMA support is available for the SH-4 processor.

Assumption and Method. We assumed DMA support for the SH4 (SH7751R) processor, and DMA supports direct data transfer between the on-chip memory and main memory with block stride data transfer. Preliminary examination confirmed the data transfer time was sufficiently shorter than the computation time for the FFT program. Therefore the program was optimized by software pipelining. For pipelining, the on-chip memory is divided into two portions, which are

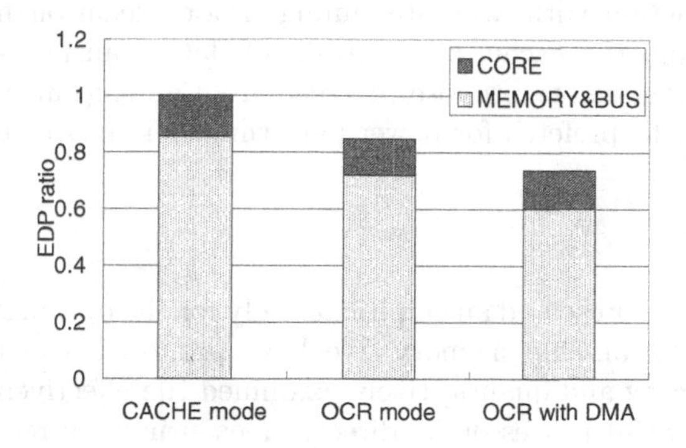

Fig. 4. EDP Ratio of SH-4 with DMA

alternately used for computation or transfer. This optimization makes the block size small, and it is only "2". The rate of pipelined data transfer is shown as:

$$\frac{\begin{array}{c}5 * NX * \log_2 NX * (NY - NBLK * 2) * NZ + \\ 5 * NY * \log_2 NY * (NX - NBLK * 2) * NZ + \\ 5 * NZ * \log_2 NZ * (NX - NBLK * 2) * NY\end{array}}{5 * NX * NY * NZ * \log_2(NX * NY * NZ)}$$

where the element size is $NX \times NY \times NZ$, and the block size is $NBLK$.

This equation shows that 93.75% of data transfer can be pipelined for this evaluation. Power performance is calculated by this ratio and the evaluation results of the actual SH-4 processor. We assume that the static power consumption can be reduced by shortening the execution time, which is enabled by software pipelining. The time required for data transfer on the SH-4 processor with DMA is calculated by the time difference between the pipelined program and the program without data transfer.

Results. The calculated EDP ratio is shown as Figure 3. The figure shows that OCR mode with adopted DMA decreases EDP by 26.3% compared with the CACHE mode, and by 13.1% EDP compared with OCR mode without DMA. As this estimation does not regard the influence of memory traffic reduction caused by variable grain data transfer, the power consumption for a real DMA is expected to be smaller than this result.

The result of the evaluation is worse than that for the SCIMA results. The SCIMA evaluation shows a 70% decrement of EDP for NPB CG [10], which is a larger effect than the result in this study. This is because SCIMA has a larger number of cache / on-chip memory way than SH-4 processor and it can switch every way to on-chip memory or cache. It reduces the disadvantage of OCR mode in SH-4 processor.

The technique of cache prefetching can achieve a similar efficiency to software pipelining for the on-chip memory. Cache prefetching can also reduce EDP in CACHE mode. However, to prefetch data precisely, it is necessary to execute the prefetch instruction with accurate timing. Poor execution may eliminate prefetched data from the cache, and a grain of data transfer is fixed to the cache line size. Therefore, we propose that use of an on-chip memory has more advantages than cache prefetch for power performance improvement.

6 Conclusion

In this paper, we have presented an empirical study for the optimization of power performance with an on-chip memory. We have proposed a method to make use of on-chip memory and quantitatively examined the effectiveness of the on-chip memory in a SH-4 processor by directly measuring the real power usage of the processor, The experimental results show that utilizing on-chip memory enables optimum data transfer, and reduces EDP by up to approximately 15.2%. According to our empirical evaluation of the SH-4 processor, we evaluate the effectiveness of DMA transfer between the on-chip memory and main memory. It was found that DMA transfer can reduce EDP by up to 26.3%.

Future work will include the investigation of other recent processors. Currently, we are planning to make a detailed examination of the SH-4A processor, which is a new processor that has DMA supported on-chip memory.

Acknowledgment

This research is supported by "Research & Development Project for Enabling Technologies for Future Supercomputing" project in "Research and Development For Next-Generation Information Technology" program, which is funded by the Ministry of Education, Culture, Sports, Science and Technology.

References

1. Kondo, M., et al.: SCIMA: Software controlled integrated memory architecture for high performance computing. In: Proc. ICCD 2000, pp. 105–111 (2000)
2. Kondo, M., et al.: Software-controlled on-chip memory for high-performance and low-power computing. ACM SIGARCH Computer Architecture News 30, 7–8 (2002)
3. Sunaga, T., et al.: A processor in memory chip for massively parallel embedded applications. IEEE J. of Solid State Circuits, 1556–1559 (October 1996)
4. Draper, J.T., et al.: The architecture of the DIVA processing-in-memory chip. In: Proc. ICS 2002, pp. 14–25 (2002)
5. Patterson, D., et al.: A Case for Intelligent RAM: IRAM. IEEE Micro 17(2), 34–44 (1997)
6. Almasi, G., et al.: Unlocking the performance of the bluegene/l supercomputer. In: Proc. SC 2004, vol. 57 (2004)
7. Renesas: SuperH RISC engine SH-4 Programming Manual, 5th edn. (2001)
8. Diefendorff, K.: Sony's emotionally charged chip. Microprocessor Report 13(5) (April 1999)
9. Turley, J.: Strongarm speed to triple. Microprocessor Report 13(6) (May 1999)

10. Kondo, M., et al.: Reducing Memory System Energy by Software-Controlled On-Chip Memory. IEICE Trans. on Electronics (E86-C)4, 550–588 (2002)
11. Fujita, M., et al.: Data Movement Optimization for Software-Controlled On-Chip Memory. In: INTERACT-8. Workshop on Interaction between Compiler and Architecture (2004)
12. Lam, M., et al.: The cache performance and optimizations of blocked algorithms. In: Proc. ASPLOS-IV, pp. 63–74 (1991)
13. Bailey, D.: et al.: The NAS parallel benchmarks 2.0. In: NASA Ames Research Center Report. NAS-05-020 (1995)
14. Takahashi, D.: Efficient implementation of parallel three-dimensional FFT on clusters of PCs. Computer Physics Communications 152, 144–150 (2003)

Performance Evaluation of Compiler Controlled Power Saving Scheme

Jun Shirako[1], Munehiro Yoshida[1], Naoto Oshiyama[1], Yasutaka Wada[1], Hirofumi Nakano[1], Hiroaki Shikano[1,2], Keiji Kimura[1], and Hironori Kasahara[1]

[1] Dept. of Computer Science, Waseda University, Tokyo, 169-8555, Japan,
{shirako, kimura, kasahara}@oscar.elec.waseda.ac.jp
[2] Central Research Laboratory, Hitachi, Ltd., Kokubunji-shi, Tokyo 185-8601, Japan

Abstract. Multicore processors, or chip multiprocessors, which allow us to realize low power consumption, high effective performance, good cost performance and short hardware/software development period, are attracting much attention. In order to achieve full potential of multicore processors, cooperation with a parallelizing compiler is very important. The latest compiler extracts multilevel parallelism, such as coarse grain task parallelism, loop parallelism and near fine grain parallelism, to keep parallel execution efficiency high. It also controls voltage and clock frequency of processors carefully to reduce energy consumption during execution of an application program. This paper evaluates performance of compiler controlled power saving scheme which has been implemented in OSCAR multigrain parallelizing compiler. The developed power saving scheme realizes voltage/frequency control and power shutdown of each processor core during coarse grain task parallel processing. In performance evaluation, when static power is assumed as one-tenth of dynamic power, OSCAR compiler with the power saving scheme achieved 61.2 percent energy reduction for SPEC CFP95 applu without performance degradation on 4 processors and 87.4 percent energy reduction for mpeg2encode, 88.1 percent energy reduction for SPEC CFP95 tomcatv and 84.6 percent energy reduction for applu with real-time deadline constraint on 4 processors.

1 Introduction

Multicore processors are attracting much attention, since they allow us to realize low power consumption, high effective performance, good cost performance and short hardware/software development period, with compiler supports. For example, Fujitsu FR-V[1], ARM MPCore[2], IBM, SONY and Toshiba Cell[3], Intel Xeon dual-core[4] and IBM Power5+[5] have been developed for consumer electronics, PCs, servers and so on. In order to achieve efficient parallel processing on multicore processors, cache and local memory optimization to cope with memory wall problem and minimization of data transfer among processors using DMAC (Direct Memory Access Controller) are necessary, in addition to extraction of parallelism from an application program. There have been a lot of researches to extract parallelism for multicore processors in the areas of loop parallelizing compilers [6,7,8]. However, loop parallelization techniques are almost matured and new generation of parallelization techniques like multi-grain parallelization are required to attain further speedup. There are a few compilers trying to

J. Labarta, K. Joe, and T. Sato (Eds.): ISHPC 2005 and ALPS 2006, LNCS 4759, pp. 480–493, 2008.

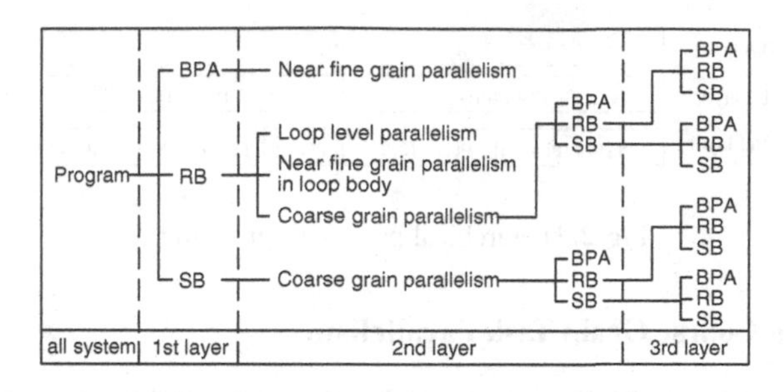

Fig. 1. Hierarchical Macro Task Definition

exploit multiple levels of parallelism, for example, NANOS compiler[9] extracts multi-level parallelism including coarse grain task parallelism by using extended OpenMP API. Also, OSCAR multigrain parallelizing compiler [10,11,12] extracts coarse grain task parallelism among loops, subroutines and basic blocks and near fine grain parallelism among statements inside a basic block, in addition to loop parallelism. Furthermore, OSCAR compiler automatically determines the suitable number of processors for each part of a program in consideration for processing overhead and applies global cache memory optimization over different loops.

Improving processing performance has been one of the most important problems for a long time. Recently, the need of power reduction of computing systems has increased rapidly. For power saving, various methods by hardware and OS have been proposed. Adaptive Processing[13] estimates workloads of computing resources using counters for cache misses and instruction queues and powers off unnecessary resources. On-line Methods for Voltage and Frequency Control [14] settles on the fitting voltage and frequency for each domain of processors using instruction issue queue occupancies as feedback signals.

This paper describes compiler controlled power saving scheme for a multicore processor, which realizes voltage/frequency (V/F) control and power shutdown under constraints of the minimum time execution or real-time execution with deadline.

2 Multigrain Parallel Processing

The developed power saving scheme in OSCAR compiler[15,16] is mainly used with coarse grain task parallelization in multigrain parallel processing. This section describes an overview of coarse grain task parallel processing.

2.1 Generation of Macro-tasks [10,11,12]

In multigrain parallelization, a program is decomposed into three kinds of coarse grain tasks, or macro-tasks (MTs), such as block of pseudo assignment statements (BPA), repetition block (RB) and subroutine block (SB)[12]. Macro-tasks can be hierarchically defined inside each un-parallelizable repetition block, or sequential loop, and a subroutine block as shown in Figure 1. Repeating macro-task generation hierarchically, a source program is decomposed into nested macro-tasks as in Figure 1.

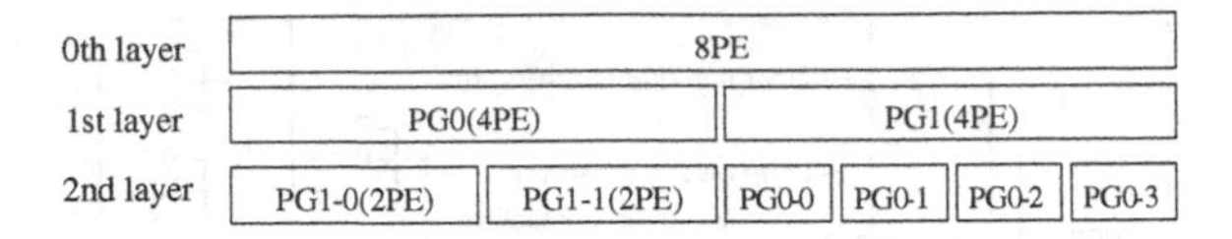

Fig. 2. Hierarchical processor grouping

2.2 Extracting Coarse Grain Task Parallelism

After generation of macro-tasks, data dependency and control flow among macro-tasks are analyzed in each nested layer and hierarchical macro flow graphs (MFG) representing control flow and data dependencies among macro-tasks are generated [10,11,12]. In order to extract coarse grain task parallelism among macro-tasks, Earliest Executable Condition analysis [10,11,12], which analyzes control dependencies and data dependencies among macro-tasks simultaneously, is applied to each Macro flow graph. Earliest Executable Conditions mean conditions on which macro-task may begin its execution earliest. By this analysis, a macro-task graph (MTG)[10,11,12] is generated for each macro flow graph. Macro-task graph represents coarse grain parallelism among macro-tasks.

2.3 Hierarchical Processor Grouping

OSCAR compiler groups processors hierarchically to execute hierarchical macro-task graphs efficiently. This grouping of processor elements (PEs) into Processor Groups (PGs) is performed logically, and macro-tasks are assigned to processor groups in each layer. Figure 2 shows an example of a hierarchical processor grouping with 8 processors. For execution of a macro-task graph in the 1st nest level, or 1st layer, 8 processors are grouped into 2 processor groups each of which has 4 processor elements. This is represented as (2PGs, 4PEs). The macro-task graph in the 1st nest level is processed by the 2PGs. 4 processors are available in each macro-task graph in the 2nd nest level, hence (4PGs, 1PE) grouping is chosen for the left PG and (2PGs, 2PEs) is chosen for the right PG.

2.4 Automatic Determination Scheme of Parallelizing Layer

In order to improve performance of multigrain parallel processing, it is necessary to determine the optimal number of PGs and PEs for coarse grain and loop parallelism of each nested macro-task graphs. OSCAR compiler with Automatic Parallelized Layer Determination Scheme [17,18] estimates parallelism of each macro-task graph and determine suitable (PGs, PEs) grouping. It also determines the suitable number of processors executing each macro-task, considering trade-off between parallelization and scheduling and data transfer overhead. Therefore, OSCAR compiler doesn't assign tasks to excessive processors to reduce parallel processing overhead.

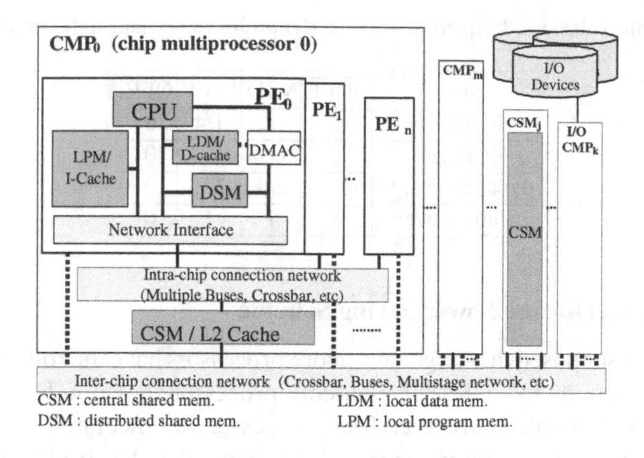

Fig. 3. OSCAR multicore processor

2.5 Macro-Task Scheduling

In coarse grain task parallel processing, a macro-task in a macro-task graph is assigned to a processor group. At this time, static scheduling or dynamic scheduling is chosen for each macro-task graph. If a macro-task graph has only data dependencies and is deterministic, static scheduling is selected. In this case, OSCAR compiler schedules macro-tasks to processor groups. Static scheduling is effective since it can minimize data transfer and synchronization overhead without runtime scheduling overhead. If a macro-task graph is un-deterministic by conditional branches among coarse grain tasks, dynamic scheduling is selected to handle runtime uncertainties. Dynamic scheduling routines are generated by OSCAR compiler and inserted into a parallelized program code to minimize scheduling overhead.

This paper evaluates the power reduction static scheduling scheme of OSCAR compiler.

3 Compiler Control Power Saving Scheme

Multigrain parallel processing can take full advantage of multi level parallelism in a program. However, there isn't always enough parallelism for all available resources through a whole program. In such a case, shutting off power supply to idle processors, which execute no task, can reduce static and dynamic power consumption. Furthermore, execution at lower voltage and frequency may reduce the total energy consumption in real time processing with deadline constraints. Compiler controlled power saving scheme realizes the following two modes of power saving. The first is the fastest execution mode that doesn't apply power reduction to the critical path of a program to guarantee the fastest processing speed. The second is real-time processing mode with deadline constraint that minimizes the total energy consumption within the given deadline.

Table 1. Rate of frequency, voltage, dynamic energy and static power

state	FULL	MID	LOW	OFF
frequency	1	1/2	1/4	0
voltage	1	0.87	0.71	0
dynamic energy	1	3/4	1/2	0
static power	1	1	1	0

3.1 Target Model for the Power Saving Scheme

In this paper, it is supposed that target multicore processors have the following functions with hardware supports like OSCAR multicore processor shown in Figure 3. The OSCAR (Optimally Scheduled Advanced Multiprocessor) architecture has been proposed to support optimization of multigrain parallelizing compiler [19,10,11], especially static and dynamic task scheduling [20,19,21]. In the OSCAR architecture, a simple processor core (PE) has local and/or distributed shared memory which can be accessed by DTCs (Data Transfer Controllers), or DMACs, of remote processor cores. Processor cores are connected by interconnection network like multiple busses or cross bar switches to control shared memory (CSM) [20,19,21,22]. In addition to the traditional OSCAR architecture, the target multicores of this paper have the following power control functions.

- Frequency for each processor can be changed in several levels separately.
- Voltage can be changed with frequency.
- Each processor can be powered on and off individually.

Here, memories, DMACs and networks are not the target of the power saving scheme described in this paper. The developed power saving scheme assumes frequency changes discretely, and the optimal voltage is fixed for each frequency. Table 1 shows an example of the combinations of voltage, dynamic energy and static power at each frequency, which supposes FULL is 400MHz, MID is 200MHz and LOW is 100MHz at 90nm technology. In this table, dynamic energy rate means the rate of energy consumption at each frequency to energy at FULL. Power supply is shut off completely at OFF, hence static power at OFF becomes 0. According to architectures and technology, these parameters and the number of V/F states can be changed. This scheme also considers state transition overhead that is given for each state.

3.2 Target MTG for the Power Control Scheme

OSCAR compiler selects dynamic scheduling or static scheduling for each MTG, as to whether there is runtime uncertainty like conditional branches in the MTG. The developed scheme can be only applied to static scheduled MTGs. However, separating static parts without branches from dynamic scheduled MTG, this scheme is applied for static scheduling parts of MTGs. In static scheduling at compile-time, OSCAR compiler estimates execution cost and consumed energy of each MT at each frequency using instruction cost and energy tables which are previously prepared for each target multicore processor.

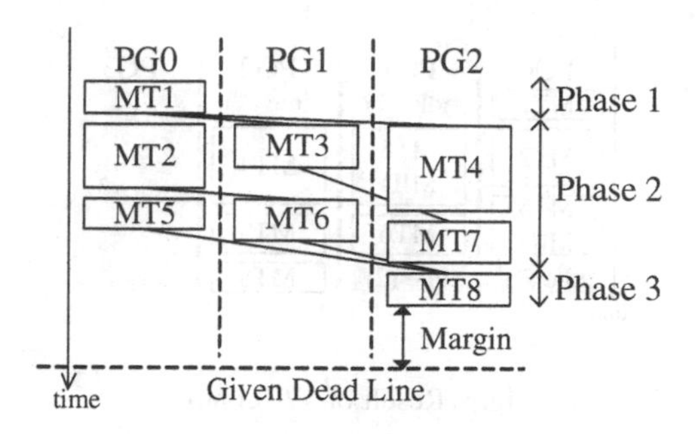

Fig. 4. Static scheduled MTG

3.3 Deadline Constraints for Target MTG

The developed scheme determines suitable voltage and frequency for each MT on a MTG based on the result of static task scheduling. Here, OSCAR compiler uses some different static task scheduling algorithms, such as CP/DT/MISF, DT/CP, ETF/CP and so on, in order to minimize processing time including data transfer overhead. The best schedule is chosen among different schedules generated by the different heuristic scheduling algorithms. Figure 4 shows MTs 1, 2 and 5 are assigned to PG0, MTs 3 and 6 are assigned to PG1 and MTs 4, 7 and 8 are assigned to PG2. Edges among tasks are data dependence.

The following is defined for MT_i in order to estimate execution time of a target MTG to which the developed scheme is applied.

T_i : execution time of MT_i after V/F control

T_{start_i} : start time of MT_i

T_{finish_i} : finish time of MT_i

At the beginning of the developed scheme, T_i is unknown. The start time of the target MTG is set to 0. If MT_i is the first macro-task executed by a PG and has no data dependent predecessor. T_{start_i} and T_{finish_i} are represented as shown below.

$T_{start_i} = 0$

$T_{finish_i} = T_{start_i} + T_i = T_i$

For instance, MT_1 is the entry node of MTG, so it is the first and has no data dependent predecessor. Hence, $T_{start_1} = 0$, $T_{finish_1} = T_1$. In other cases, a previous macro-task which is assigned to the same PG as MT_i is represented as MT_j. Data dependent predecessors of MT_i are defined as $\{MT_k, MT_l, ...\}$. MT_i starts when MT_j, MT_k, MT_l, ... finish.

$T_{start_i} = max(T_{finish_j}, T_{finish_k}, T_{finish_l}, ...)$

$T_{finish_i} = T_{start_i} + T_i$

According to these rules,the finish time of MT_8 which is the exit node is represented as

$T_{finish_8} = T_1 + T_8 + max(T_2 + T_5, T_6 + max(T_2, T_3), T_7 + max(T_3, T_4))$

The finish time of exit node is generally represented by

$T_{finish_{exit}} = T_m + T_n + ... + max_1(...) + max_2(...) + ...$

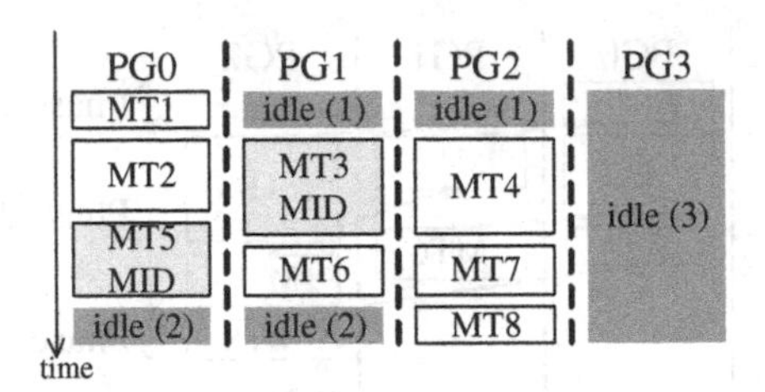

Fig. 5. Result of V/F control

The start time of the entry node is 0, therefore $T_{finish_{exit}}$ expresses the execution time of the target MTG, defined as T_{MTG}. The given deadline for the target MTG is defined as $T_{MTG\_deadline}$. Now therefore, the next condition should be satisfied.

$T_{MTG} \leq T_{MTG\_deadline}$

The developed scheme determines suitable clock frequency for MT_i to satisfy this condition.

3.4 Voltage / Frequency Control

This paragraph describes how to determine suitable voltage and frequency to execute each MT using next conditions. The execution time of MT_i is T_i, the execution time of target MTG is T_{MTG}, the real-time deadline of the target MTG is $T_{MTG\_deadline}$, hence

$T_{MTG} = T_m + T_n + ... + max_1 + max_2 + ...$ - - - (a)

$T_{MTG} \leq T_{MTG\_deadline}$ - - - (b)

For sake of simplicity, MTs corresponding to each term of expression (a), such as T_m, T_n, ..., max_1, max_2, ..., are called Phase. Each term represents a different part of T_{MTG}. Therefore, different Phases are not executed in parallel on any account as shown in Figure 4. The following parameters for $Phase_i$ at frequency F_n are defined.

$T_{sched_i}(F_n)$: scheduling length at F_n

$Energy_i(F_n)$: energy consumption at F_n

$T_{sched_i}(F_n)$ represents the execution time when all MTs in $Phase_i$ are processed at F_n. $T_{sched_i}(FULL)$ is the minimum value of the correspondent term of expression (a). $Energy_i(F_n)$ expresses the total energy consumption as $Phase_i$ is excuted at F_n.

Here, it is considered to change frequency from F_n to F_m. The scheduling length is increased from $T_{sched_i}(F_n)$ to $T_{sched_i}(F_m)$. The energy is decreased from $Energy_i(F_n)$ to $Energy_i(F_m)$. Using these values, $Gain_i(F_m)$ is defined as

$$Gain_i(F_m) = -\frac{Energy_i(F_m) - Energy_i(F_n)}{T_{sched_i}(F_m) - T_{sched_i}(F_n)}$$

$Gain_i(F_m)$ represents reduction rate of energy on scheduling length when F_n is changed into F_m. Therefore, if the increases of scheduling length are same, the more energy reduction can be expected by applying V/F control to $Phase_i$ with larger $Gain_i(F_m)$.

Next, to estimate the margin of the target MTG, the minimum value of T_{MTG} is calculated as the summation of $T_{sched_i}(FULL)$. Using this minimum value and $T_{MTG\_deadline}$, T_{MTG_margin} is defined as

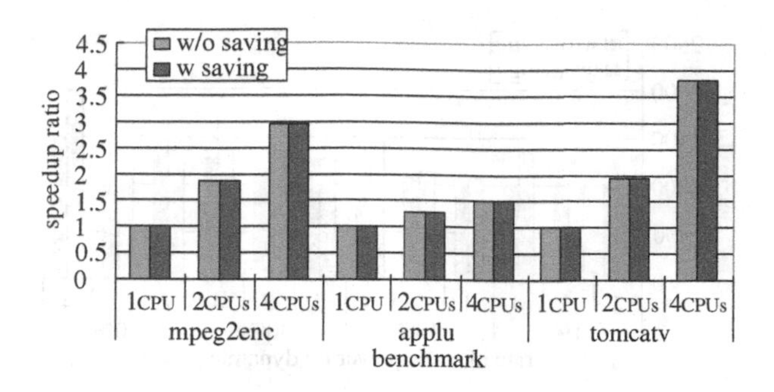

Fig. 6. Speedup in the fastest mode

$T_{MTG\_margin} = T_{MTG\_deadline} - \sum T_{sched_i}(FULL)$

As the target MTG must finish in minimum execution time, or $T_{MTG\_margin} = 0$, each Phase has to be executed at FULL. When $T_{MTG\_margin} > 0$, the developed scheme turns down voltage and frequency of each Phase, according to $Gain_i(F_m)$. If a Phase has a single MT, the frequency of the MT is the same as the Phase. If a Phase includes some MTs and corresponds to the max term, the developed scheme also defines sub-Phases for each argument of the max term and determines voltage frequency of these sub-Phases. The algorithm to determine frequency for each sub-Phase and MT is described in [15,16].

3.5 Power Supply Control

Next, power supply control to reduce unnecessary energy consumption including static power consumption by idle processors is applied. Idle time occurs, when a PG (processor group) is waiting for other PGs to execute their MTs (1), finished all scheduled MTs (2) or has no MTs (3). Gray parts of Figure 5 are the idle in each case. Power supply to a PG is turned off, if its idle time is longer than the frequency transition overhead and its energy becomes lower by power shutdown considering the overhead.

3.6 Applying Power Saving Scheme to Inner MTG

If a MT_i includes a MTG_i inside, it may be more effective to control each MT_{i_j} in MTG_i than to process the whole MT_i at the same clock frequency. Therefore, the deadline for MTG_i is defined as $T_{MTG_i\_deadline}$, which is given by T_i. The power saving control described in paragraph 3.4 and 3.5 is applied to MTG_i. Comparing the case to execute whole MT_i at the same frequency with the case to apply power saving control to MTG_i, the more effective one is selected.

4 Performance Evaluation

This section describes performance of OSCAR multigrain parallelizing compiler with compiler controlled power saving scheme. Evaluation was performed by using static

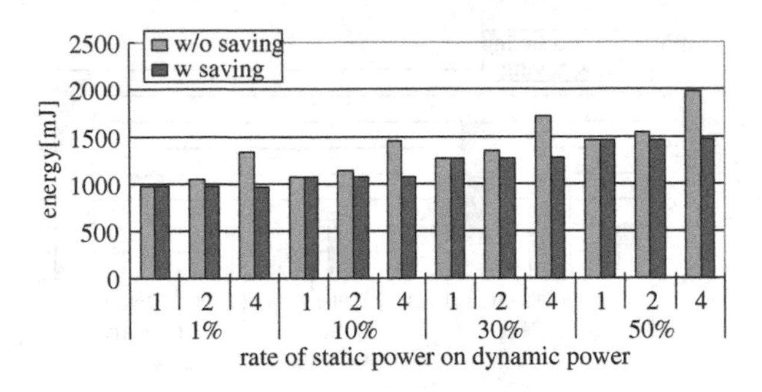

Fig. 7. Energy of mpeg2encode (fastest)

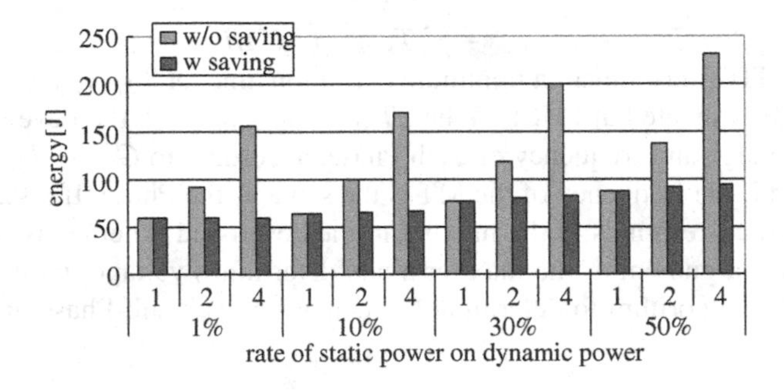

Fig. 8. Energy of applu (fastest)

scheduler in OSCAR compiler. Parameters used for this evaluation, such as frequencies, voltages, dynamic energies, and static powers, are shown in Table 1 were used. In this paper, only energy for processors was evaluated. The state transition overhead of V/F control is 0.1[ms] and overhead of power shutdown is 0.2[ms]. Dynamic power at FULL frequency is 220[ms]. It was measured by using Wattch[23]. Cooperative Voltage Scaling[24] was referenced to determine parameters like transition overhead, attribute of voltage/frequency and dynamic power at MID and LOW frequency. Static power is set to 2.2[mW] (1% of dynamic power), 22[mW] (10% of dynamic power), 66[mW] (30% of dynamic power) or 110[mW] (50% of dynamic power), supposing various type of multicore processors from low power oriented multicores to high performance multiprocessors. In this evaluation, MediaBench mpeg2encode which was rewritten in Fortran[25], SPEC95 CFP applu and tomcatv were used. For applu, inline expansion and loop aligned decomposition for the data localization scheme[26] were applied. Also, the main loop in applu was divided into a static part without conditional branch and dynamic parts with branches, in order to apply the power saving scheme.

4.1 Performance in the Fastest Execution Mode

Figure 6 shows the speedup ratio of each program for 1, 2 and 4 processors in the fastest execution mode, when static power is equal to 1% of dynamic power. Left bars repre-

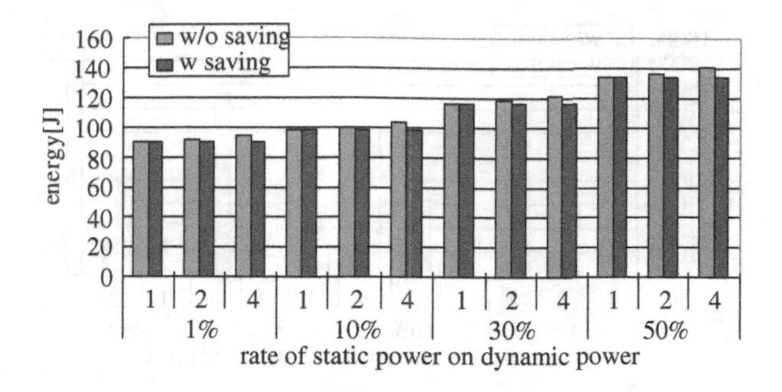

Fig. 9. Energy of tomcatv (fastest)

Table 2. Energy reduction for 4 CPUs(fastest)

program	static	w/o saving	w saving	reduction
mpeg2	1 %	1336[mJ]	973[mJ]	27.2 %
	10 %	1455[mJ]	1071[mJ]	26.4 %
	30 %	1720[mJ]	1278[mJ]	25.7 %
	50 %	1985[mJ]	1476[mJ]	25.6 %
applu	1 %	156[J]	58.5[J]	62.4 %
	10 %	170[J]	65.9[J]	61.2 %
	30 %	201[J]	81.9[J]	59.2 %
	50 %	231[J]	95.1[J]	58.9 %
tomcatv	1 %	94.8[J]	90.4[J]	4.66 %
	10 %	103[J]	98.4[J]	4.65 %
	30 %	122[J]	116[J]	4.65 %
	50 %	141[J]	134[J]	4.64 %

sent the results of OSCAR compiler without the power saving scheme and right bars show the results of OSCAR compiler using compiler controlled power saving scheme. As shown in Figure 6, there is no performance degradation by using the power saving scheme in the fastest execution mode. When static power was changed to 22[mW] (10% of dynamic power), 66[mW] (30% of dynamic power) or 110[mW] (50% of dynamic power) assuming high performance processors, there were also no performance losses. Figure 7, 8 and 9 show the total energy consumption of mpeg2encode, applu and tomcatv for 1, 2, and 4 processors, changing the rate of static power on dynamic power to 1%, 10%, 30% or 50%. In mpeg2encode and applu, the power saving scheme using 2 or 4 processors reduces energy consumption at any rate of static power. These applications have sequential parts which can't be parallelized, hence there is a certain amount of processor idle time. The power saving scheme applied V/F control and power shutdown, using this idle time. The developed scheme reduced consumed energy by 5.5 % (from 1138[mJ] down to 1075[mJ]) for 2 processors, 26.4 % (from 1455[mJ] down to 1071[mJ]) for 4 processors in mpeg2encode and 35.4 % (from 101[J] down to 65.2[J]) for 2 processors, 61.2 % (from 170[J] down to 65.9[J]) for 4 processors in applu, as-

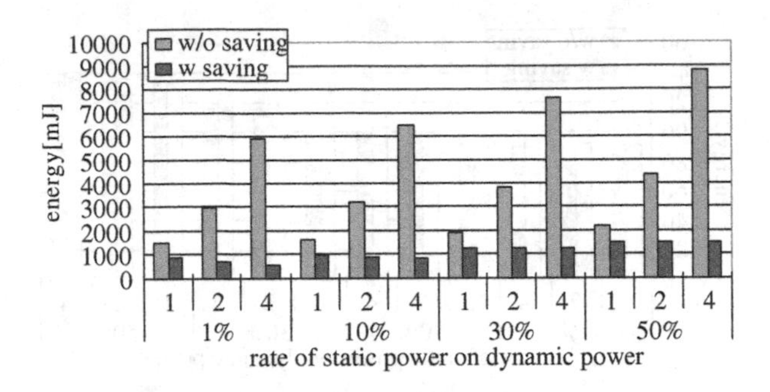

Fig. 10. Energy of mpeg2encode (deadline)

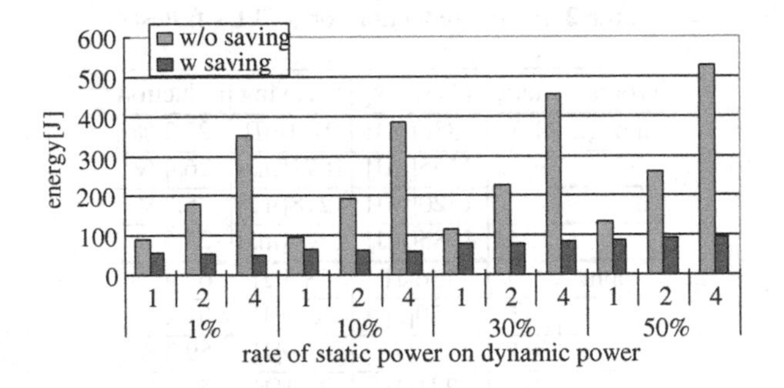

Fig. 11. Energy of applu (deadline)

suming the rate of static power on dynamic power is 10 %. The energy reduction rate for 4 processors changing static power is shown in Table 2.

On the other hand, tomcatv has large parallelism to run all processors almost every time during the program execution. Therefore, all processors must execute at full speed to attain the minimum execution time. The parallel execution time of tomcatv with 4 processors is about one quarter of the sequential execution time. Therefore, though the power consumption is quadrupled by using 4 processors, the total energy consumption is almost equal to the energy of the sequential execution.

4.2 Performance in Real-Time Execution Mode with Deadline Constraints

Next, evaluation results of real-time execution mode with deadline constraint are described. Figure 10, 11 and 12 show the total energy consumed until their real-time deadline. Here, deadline was set to 150% of sequential execution time. Left bars represent the results of OSCAR compiler without any power saving scheme. In this case, all processors run at FULL frequency until the deadline. Right bars show the results of OSCAR compiler using the developed power saving scheme in real-time deadline mode. These figures show the power saving scheme drastically reduced energy consumption, because the developed scheme applied V/F control and power shutdown as

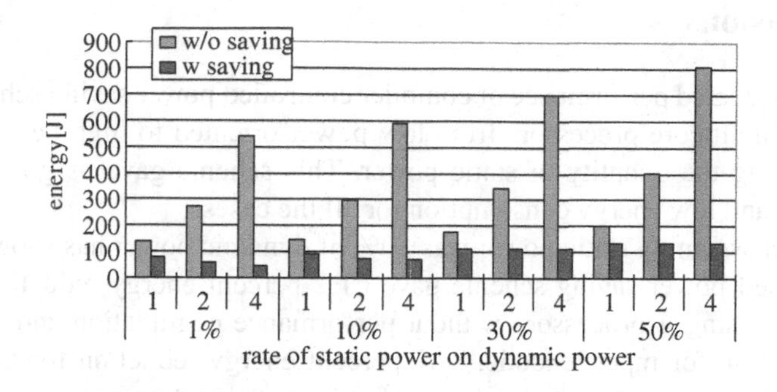

Fig. 12. Energy of tomcatv (deadline)

Table 3. Energy reduct. for 4 CPUs(deadline)

program	static	w/o saving	w saving	reduction
mpeg2	1 %	5929[mJ]	592[mJ]	90.0 %
	10 %	6458[mJ]	815[mJ]	87.4 %
	30 %	7632[mJ]	1262[mJ]	83.5 %
	50 %	8806[mJ]	1476[mJ]	83.2 %
applu	1 %	354[J]	49.5[J]	86.0 %
	10 %	385[J]	59.5[J]	84.6 %
	30 %	455[J]	81.7[J]	82.1 %
	50 %	525[J]	95.1[J]	81.9 %
tomcatv	1 %	542[J]	48.3[J]	91.1 %
	10 %	591[J]	70.2[J]	88.1 %
	30 %	698[J]	114[J]	83.7 %
	50 %	805[J]	134[J]	83.3 %

far as execution time didn't exceed the deadline. Furthermore, the energy consumption of mpeg2encode or tomcatv executed in parallel is lower than the energy of sequential execution, when static power is set to 2.2[mW], 22[mW] or 66[mW]. The developed power saving scheme in real-time processing mode reduced energy by 73.3 % (from 3229[mJ] down to 861[mJ]) for 2 processors, 87.4 % (from 6458[mJ] down to 815[mJ]) for 4 processors in mpeg2encode, 68.9 % (from 193[J] down to 59.8[J]) for 2 processors, 84.6 % (from 385[J] down to 59.5[J]) for 4 processors in applu and 73.8 % (from 295[J] down to 77.3[J]) for 2 processors, 88.1 % (from 591[J] down to 70.2[J]) for 4 processors in tomcatv, assuming the rate of static power on dynamic power is 10 %. Table 3 shows the energy reduction for 4 processors changing static power.

Execution time with the developed power saving scheme was less than the deadline in all the cases where static power was changed. This means the developed scheme could satisfy the given deadline constraints.

5 Conclusions

This paper evaluated performance of compiler controlled power saving scheme for various type of multicore processors from low power oriented to high performance oriented, changing the quantity of static power. This scheme gave us good processing performance and low energy consumption for all the cases.

Evaluation assuming static power was 10% of dynamic power has shown that compiler controlled power saving scheme gave 61.2 percent energy reduction for SPEC CFP95 applu using 4 processors without performance degradation and 87.4 percent energy reduction for mpeg2encode, 88.1 percent energy reduction for SPEC CFP95 tomcatv and 84.6 percent energy reduction for applu using 4 processors with real-time deadline constraint.

The power saving scheme described in this paper only controls processors in static scheduling. Development of the power saving scheme for dynamic scheduling and power saving methods for resources other than processors are future works.

Acknowledgments

A part of this research has been supported by NEDO "Advanced Heterogeneous Multiprocessor", STARC "Automatic Parallelizing Compiler Cooperative Single Chip Multiprocessor" and NEDO "Multi core processors for real time consumer electronics".

References

1. Suga, A., Matsunami, K.: Introducing the FR 500 embedded microprocessor. IEEE MICRO 20, 21–27 (2000)
2. Cornish, J.: Balanced energy optimization. In: International Symposium on Low Power Electronics and Design (2004)
3. Pham, D., et al.: The design and implementation of a first-generation CELL processor. In: Proceeding of the IEEE International Solid-State Circuits Conference. (2005)
4. Intel Multi core. http://www.intel.com/multi-core/
5. Kalla, R., Sinharoy, B., Tendler, J.: IBM Power5 chip: a dual-core multithreaded processor. IEEE Micro 24(2), 40–47 (2004)
6. Wolfe, M.: High Performance Compilers for Parallel Computing. Addison-Wesley Publishing Company, Reading (1996)
7. Eigenmann, R., Hoeflinger, J., Padua, D.: On the automatic parallelization of the perfect benchmarks. IEEE Trans. on parallel and distributed systems 9(1) (January 1998)
8. Hall, M.W., Anderson, J.M., Amarasinghe, S.P., Murphy, B.R., Liao, S., Bugnion, E., Lam, M.S.: Maximizing multiprocessor performance with the SUIF compiler. IEEE Computer (1996)
9. Gonzalez, M., Martorell, X., Oliver, J., Ayguade, E., Labarta, J.: Code generation and runtime support for multi-level parallelism exploitation. In: Proc. of the 8th International Workshop on Compilers for Parallel Computing (January 2000)
10. Honda, H., Iwata, M., Kasahara, H.: Coarse grain parallelism detection scheme of a fortran program. Trans. of IEICE J73-D-1(12), 951–960 (1990)
11. Kasahara, H., et al.: A multi-grain parallelizing compilation scheme on OSCAR. In: Proc. 4th Workshop on Language and Compilers for Parallel Computing (1991)

12. Kasahara, H.: Advanced automatic parallelizing compiler technology. IPSJ MAGANIE (April 2003)
13. Albonesi, D.H., et al.: Dynamically tuning processor resources with adaptive processing. IEEE Computer (December 2003)
14. Wu, Q., Juang, P., Martonosi, M., Clark, D.W.: Formal online methods for voltage/frequency control in multiple clock domain microprocessors. In: Eleventh International Conference on Architectural Support for Programming Languages and Operating Systems (October 2004)
15. Shirako, J., Oshiyama, N., Wada, Y., Shikano, H., Kimura, K., Kasahara, H.: Compiler control power saving scheme for multi core processors. In: Ayguadé, E., Baumgartner, G., Ramanujam, J., Sadayappan, P. (eds.) LCPC 2005. LNCS, vol. 4339, Springer, Heidelberg (2006)
16. Shirako, J., Oshiyama, N., Wada, Y., Shikano, H., Kimura, K., Kasahara, H.: Parallelizing compilation scheme for reduction of power consumption of chip multiprocessors. In: CPC. Proc. of 12th International Workshop on Compilers for Parallel Computers (January 2006)
17. Obata, M., Shirako, J., Kaminaga, H., Ishizaka, K., Kasahara, H.: Hierarchical parallelism control for multigrain parallel processing. In: Proc. of 15th International Workshop on Languages and Compilers for Parallel Computing (August 2002)
18. Shirako, J., Nagasawa, K., Ishizaka, K., Obata, M., Kasahara, H.: Selective inline expansion for improvement of multi grain parallelism. In: PDCN 2004 (February 2004)
19. Kasahara, H., Honda, H., Iwata, M., Hirota, M.: A compilation scheme for macro-dataflow computation on hierarchical multiprocessor system. In: Proc. Int Conf. on Parallel Processing (1990)
20. Kasahara, H., Narita, S., Hashimoto, S.: Architecture of OSCAR. Trans of IEICE J71-D(8) (August 1988)
21. Kasahara, H., Honda, H., Narita, S.: Parallel processing of near fine grain tasks using static scheduling on OSCAR. In: Proceedings of Supercomputing 1990 (November 1990)
22. Kimura, K., Ogata, W., Okamoto, M., Kasahara, H.: Near fine grain parallel processing on single chip multiprocessors. Trans. of IPSJ 40(5) (May 1999)
23. Brooks, D., Tiwari, V., Martonosi, M.: Wattch: A framework for architectural-level power analysis and optimizations. In: Proc. of the 27th ISCA (June 2000)
24. Kawaguchi, H., Shin, Y., Sakurai, T.: uITRON-LP: Power-conscious real-time os based on cooperative voltage scaling for multimedia applications. IEEE Transactions on multimedia (February 2005)
25. Kodaka, T., Nakano, H., Kimura, K., Kasahara, H.: Parallel processing using data localization for MPEG2 encoding on OSCAR chip multiprocessor. In: Proc. of International Workshop on Innovative Architecture for Future Generation High-Performance Processors and Systems (January 2004)
26. Ishizaka, K., Miyamoto, T., Shirako, M.o.J., kimura, K., Kasahara, H.: Performance of OSCAR multigrain parallelizing compiler on SMP servers. In: Proc. of 17th International Workshop on Languages and Compilers for Parallel Computing (September 2004)

Program Phase Detection Based Dynamic Control Mechanisms for Pipeline Stage Unification Adoption

Jun Yao[1], Hajime Shimada[1], Yasuhiko Nakashima[2], Shin-ichiro Mori[3], and Shinji Tomita[1]

[1] Graduate School of Informatics, Kyoto University, Kyoto 606-8501, Japan
{yaojun,shimada,tomita}@lab3.kuis.kyoto-u.ac.jp
http://www.lab3.kuis.kyoto-u.ac.jp/
[2] Graduate School of Information Science, NAIST, Nara 630-0192, Japan
nakashima@is.naist.jp
[3] Graduate School of Engineering, Fukui University, Fukui 910-8507, Japan
moris@fuis.fuis.fukui-u.ac.jp

Abstract. To reduce the power consumption in mobile processors, a method called Pipeline Stage Unification (PSU) is previously proposed to work as an alternative for Dynamic Voltage Scaling (DVS). Based on PSU, we proposed two mechanisms which dynamically predict a suitable unification degree according to the knowledge of the program behaviors. Our results show that the mechanisms can achieve an average Energy Delay Product (EDP) decrease of 15.1% and 19.2%, respectively, for SPECint2000 benchmarks, compared to the processor without PSU.

Keywords: power consumption, pipeline stage unification, program phase detection.

1 Introduction

Recently, considering power consumption has shown its importance in the modern processor design, especially for portable and mobile platforms such as cellular phones and laptop computers. To reduce the total energy, a method called Dynamic Voltage Scaling (DVS) is currently employed. Basically, DVS decreases the supply voltage while the processor is experiencing low work load. This saves energy consumption for program execution.

However, Shimada et al. [1,2] and Koppanalil et al. [3] have presented a different method, which is called Pipeline Stage Unification (PSU), to reduce the processor power consumption via inactivating and bypassing the pipeline registers and using shallow pipelines during the program execution. PSU can save power in the following ways:

1. Energy can be saved because of the clock gating of some pipeline registers;

J. Labarta, K. Joe, and T. Sato (Eds.): ISHPC 2005 and ALPS 2006, LNCS 4759, pp. 494–507, 2008.

2. After pipeline stage unification, a pipeline will become a shallow one with fewer stages. Usually, a shallow pipeline will have a better IPC due to decreased branch misprediction penalties and functional unit latencies compared to the deep pipeline, as illustrated in [4] and [5].

Such designs make PSU still applicable when the efficiency of DVS is restricted by the process technology advancement, as described in paper [1].

Our research described is focused on controlling PSU hardware to achieve good power savings. Currently there is only one research related to PSU control [6] and it described execution with predefined throughput. The study did not consider the effect of different program behaviors during the execution. In this paper, we propose online mechanisms to adjust the pipeline to a suitable unification degree according to the program behavior change, so as to achieve better Energy Delay Product (EDP). By using the two different mechanisms described in this paper, we obtained an average decrease of 15.1% and 19.2% in EDP, respectively, compared to the EDP of processors under normal configuration. And compared these two mechanisms with unification degree 2, which usually has a good EDP efficiency, we achieved a decrease of 1.41% and 4.82%.

The rest of the paper is organized as follows: Section 2 describes the background techniques of this paper. Section 3 introduces the dynamic prediction mechanisms for a PSU enabled system. Simulation methodology and metrics to evaluate the efficiency of different unification degrees can be found in section 4. In Section 5 we show the experiment results, together with some analyses. Section 6 concludes the paper.

2 Related Works

This section describes the background techniques related to our research. Section 2.1 describes briefly Pipeline Stage Unification and Section 2.2 introduces the working set signature method.

2.1 Pipeline Stage Unification

In paper [1,2], Shimada et al. proposed an energy consumption reduction method called Pipeline Stage Unification (PSU) to reduce the power consumption in mobile processors as an alternative for Dynamic Voltage Scaling (DVS). PSU is a pipeline reconfiguration method. Different from DVS, PSU unifies multiple pipeline stages by bypassing pipeline registers when the processor runs with low clock frequency, instead of scaling down the supply voltage.

Our work introduced in this paper is based on Shimada's previous architecture [1], in which a pipeline of 20 stages was adopted. We assume three unification degrees in the latter part of this paper.

1. U1: The normal mode without bypassing any pipeline registers.
2. U2: Merge every pair of two adjacent pipeline stages by inactivating the pipeline register between them. The new pipeline now contains 10 stages.

3. U4: Based on U2, merge the adjacent stages one step further. It becomes a 5-stage's pipeline.

2.2 Working Set Signature

Dhodapkar [7] and Sherwood [8] have shown that programs can be divided into phases in which a program would have similar behaviors including the cache miss, IPC and power consumption. It is described as program phase, which may contain a set of instruction intervals, regardless of temporal adjacency. This theory gives us an opportunity to study the pipeline reconfiguration at a high level, i.e., from the view of the program behavior.

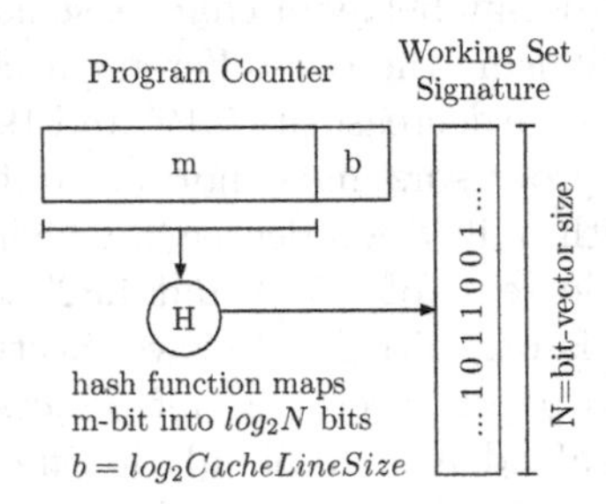

Fig. 1. Mechanism for collecting a working Set Signature [7]

In order to detect the phase changes during the program execution, Dhodapkar designed a working set signature to work as the compacted representation for a program interval. The method to form a working set signature is shown in Fig. 1. b in Fig. 1 is the number of bits which are used to index a instruction in the cache block. If an instruction cache block contains 4 instructions, b is set to 2. During the program execution, Dhodapkar selected m bits from the program counter and used these bits to address 1 bit in the *N-bit* signature via a hash function. The signature is cleared at the beginning of an instruction interval. After the interval begins, a bit in the signature is set if the corresponding instruction cache block is touched.

Dhodapkar used a 1024-bit signature in his paper. The hash function he described is based on the C library *srand* and *rand*. He chose 100k instructions as the instruction interval length.

After collecting working set signatures, a method to calculate the distance between the two signatures S_1 and S_2 is given in [7] to classify intervals into groups. The distance δ is calculated as follows:

$$\delta = \frac{num\_of\_\text{``1''}bit\_in(S_1 \bigoplus S_2)}{num\_of\_\text{``1''}bit\_in(S_1 + S_2)} \tag{1}$$

Where $num\_of\_\text{``1''}\_bit\_in()$ represents the function that counts the number of "1" bits in the bit vector. If the distance δ is larger than a predefined threshold, the two instruction intervals are of different program phases. Dhodapkar used 0.5 as threshold in his paper.

3 Dynamic PSU Control Mechanisms

Based on the background in Section 2, it can be assumed that, since the energy consumption keeps nearly flat in a stable program phase, we can use the same pipeline stage unification degree in that phase and tune a new pipeline stage unification degree at the phase switching point. In the following sections, we design our algorithms as a framework of an interval based loop. The word *interval* here refers to a large bundle of instructions. At each iteration, a calculation of the energy and performance over the current interval is made and passed to the algorithm core. The calculation is used to compare with the results of other configurations to predict a suitable unification degree for the next interval based on the history result of comparison.

Tow different control methods on the original PSU system are applied in this paper. The first control method is basic phase detection method which needs only phase switching detection hardware. The second is history table based method in which we add table-structured hardware to store additional history information for unification degree prediction.

3.1 Basic Phase Detection Method

The algorithm is altered slightly from Balasubramonian et al. [9] and Dhodapkar, et al. [7] to work with the PSU system, as shown in Fig. 2. Fig. 3 outlines the execution of this algorithm. There are three states in this algorithm:

1. *stable*: The adjacent intervals are of same phases;
2. *unstable*: A phase switching in program occurs and current interval is of different phase with last interval;
3. *tuning*: The period when the adjacent intervals become similar again and different unification degrees are being explored.

Fig. 3 is a sample of execution. First, we suppose that the program starts from a *stable* state. After each program interval, we compare the signature of current interval with the signature of the previous interval. If the distance is larger than the threshold, the state is changed to *unstable*. For simplicity, U1 is used as the unification degree for the *unstable* phase. The next intervals are *unstable* until the distance becomes smaller than the threshold again. Then, state is changed to *tuning*, which will try different unification degrees in the following three intervals and collect the corresponding power/performance data. After *tuning*, if the interval is still under the same program phase, we can choose a best unification degree for this phase and set the state to *stable*. This algorithm is based on the assumption that program will show same behavior, including energy and performance, in the same program phase.

```
After each interval I_k:
δ = signature distance of I_k and I_{k-1};
if (state == stable)
   if (δ > threshold)
      state = unstable;
      unification_degree = U1;
else if (state == unstable)
   if (δ ≤ threshold)
      state = tuning;
      unification_degree = U1;
else if (state == tuning)
   if (δ > threshold)
      state = unstable;
      unification_degree = U1;
   else if (unification_degree == U4)
      state = stable;
      unification_degree=best from tuning;
   else
      unification_degree
         = next tuning unification degree;
```

Fig. 2. Algorithm of basic phase detection method

Because we only compare the signatures of each consecutive interval pair, this method is of low cost. The corresponding control hardware will also show the advantage of simplicity.

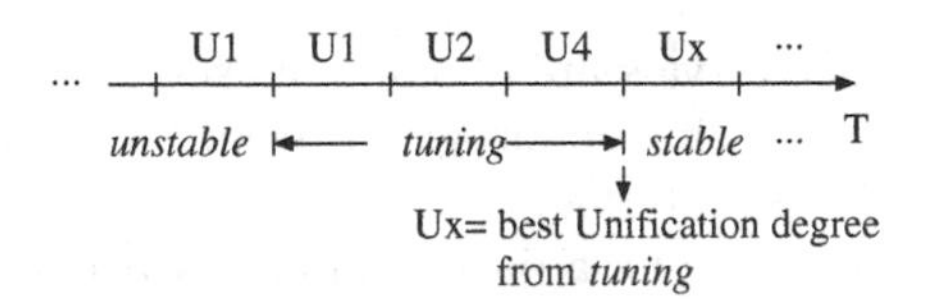

Fig. 3. Outline of execution under basic phase detection method

3.2 History Table Based Method

In order to use the feature that a phase will recur during the program execution, we designed the history table based method to keep the phase information in a history table. If the program comes into a phase that has appeared in the past, we can choose a suitable unification degree from the cached history information without starting a new tuning procedure.

Fig. 4 is the diagram of the hardware approach of this table method. Fig. 5 shows the detailed algorithm of history table based method.

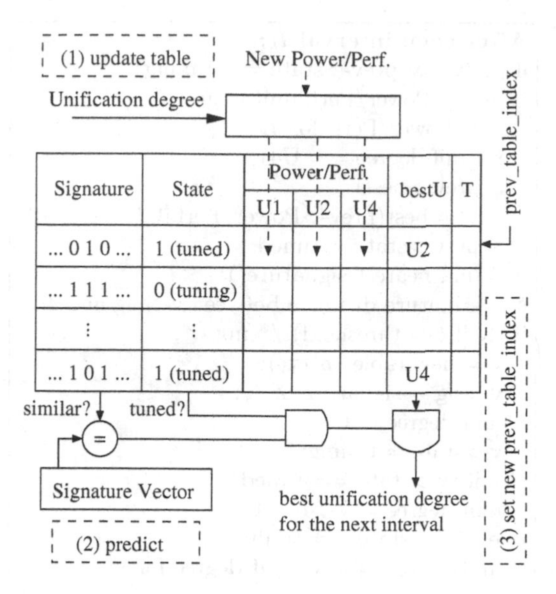

Fig. 4. Hardware approach for history table based method

The table employed in this algorithm is constructed as follows:

1. The signature field: Each different signature occupies one table entry so that we can use this field to index the table items. This field has a same storage size as the signature.
2. The state field: It denotes the state of the table entry. In this method, we define two states: *tuned* and *tuning*. A state of *tuning* means that this entry has just been added into the table and the best unification degree has not been tuned out. After all three unification degrees have been tried, we select a best unification degree from the tuned results (another field in the table) and set the state as *tuned*. One bit is used for this field.
3. The Power/Perf. field occupies three fixed point storage units for each entry. It holds the power/performance information for the interval represented by this signature. We keep the tuning information of different unification degrees in the three fields denoted as U1, U2 and U4, respectively. They are updated when the entry is under the *tuning* state.
4. The bestU field: It holds the best unification degree for this signature. This field is set after the tuning is complete. If this phase is observed again, we can predict the suitable unification degree from this field. Two bits are used for this field.
5. The T field: It records the time that the entry is touched. This field is referred to when replacing old entries. Several bits are used according to the table size.

At the point that we are about to predict a suitable unification degree for the coming interval I_{k+1}, the signature S_{k+1} that is needed as the index to look up

```
After each interval I_k:
if (prev && prev->state == tuning)
  prev->Power/Perf.[unif_degree]
     = Power/Perf. for I_k;
  if (unif_degree == U4)
     prev->bestU
        = best(prev->Power/Perf.[U1, U2, U4]);
     prev->state = tuned;
v = find_nearest_signature();
δ = signature distance between v->sig and I_k;
if (!v || δ > threshold) /* miss */
  v = new_table_entry();
  v->sig = signature of I_k;
  unif_degree = U1;
  v->state = tuning;
else if (v->state == tuned)
  unif_degree = v->bestU;
else /* v->state == tuning */
  unif_degree = next unif_degree for v;
prev = v;
```

Fig. 5. Algorithm of history table based method

the bestU field in the history table, has not been determined. To solve this problem, we engage a specific register named *prev_table_index* to store the table index of the previous interval. After each interval, we calculate the power/performance of current interval and store it in the entry which *prev_table_index* refers to ((1) in Fig. 4). Therefore the Power/Perf. field and bestU field of each current entry hold information for the next interval. After current interval is complete, we can look for the current signature in the history table. If there is a hit, the corresponding entry will probably carry the best unification degree for the next interval. And we can predict the best unification degree based on this entry ((2) in Fig. 4). The register *prev_table_index* is then updated to the current table index before starting the next interval ((3) in Fig. 4).

In Fig. 5, *prev* denotes *prev_table_index* and v denotes a temporary table index. Also the syntax like *prev->state* denotes the *state* field of the entry pointed by *prev*. *unif_degree* is the current unification degree.

There are two main actions which will be performed on the table:

1. Find the nearest signature. We simply look up the table, comparing the new signature with all cached signatures to find a smallest distance. If this smallest distance is larger than the threshold, we return a table miss and insert the new signature for the late tuning. Otherwise we report a table hit;
2. Replace the Least Recently Used (LRU) table entry when there is no sufficient room for the new signature, while we call new_table_entry() (in Fig. 5).

The performance of these two actions will greatly depend on the size of the table. As indicated in paper [7], a program will not show many different signatures during execution if the interval is set to be 100k instructions. We can set the table size at a small level, for example, 16 entries. Hence the overhead introduced by the process of looking up and replacing can be negligible. We will discuss this in more detail in Section 5.4.

In this method we have an assumption that if interval I_{k+1} happens once after interval I_k and I_k occurs again, the next interval will probably be I_{k+1}. It is similar to a simple history branch predictor. We can efficiently predict the best unification level for I_{k+1} if the next interval for I_k is always I_{k+1}, while we must endure some misprediction penalty if the next interval for I_k is variable. We will show the efficiency of this method in Section 5.

4 Simulation Methodology

We use a detailed cycle-accurate out-of-order execution simulator, SimpleScalar Tool Set [10], to measure energy and performance of different unification degrees. Table 1 lists the processor configuration. We assume a deep pipeline similar to the current processors. Table 2 summarizes the latencies and penalties in pipeline configuration of U1, U2 and U4, respectively.

We choose 8 integer benchmarks (gzip2, gcc, gzip, mcf, parser, perlbmk, vortex and vpr) from SPECint2000, with train inputs. 1.5 billion Instructions are simulated after skipping the first billion instructions.

To evaluate the energy and performance together in the tuning procedure, we can use PDP, EDP and EDDP as the metric, which can be calculated as $\frac{W}{MIPS}$, $\frac{W}{(MIPS)^2}$ and $\frac{W}{(MIPS)^3}$, respectively [11]. Each equation puts different emphasis on energy and performance and will show different efficiency according to the evaluated platforms. Basically, PDP is suitable for portable systems and EDP is used for high end systems such as workstations and laptop computers, while EDDP is good for server families. For simplicity, we apply one single metric during each program execution. The experiments and analyses in Section 5 are based on EDP because our PSU is targeted on high-performance mobile com-

Table 1. Processor configuration

Processor	8-way out-of-order issue, 128-entry RUU, 64-entry LSQ, 8 int ALU, 4 int mult/div, 8 fp ALU, 4 fp mult/div, 8 memory ports
Branch Prediction	32K-entry gshare, 13-bit history, 4K-entry BTB, 32-entry RAS
L1 Icache	64KB/32B line/2way
L1 Dcache	64KB/32B line/2way
L2 unified cache	2MB/64B line/4-way
Memory	64 cycles first hit, 2 cycles burst interval

Table 2. Assumptions of latencies and penalty

unification degree	U1	U2	U4
clock frequency rate	100%	50%	25%
branch mispred. resolution latency	20	10	5
L1 Icache hit latency	4	2	1
L1 Dcache hit latency	4	2	1
L2 cache hit latency	16	8	4
int Mult latency	3	2	1
fp ALU latency	2	1	1
fp Mult latency	4	2	1

puter. Our mechanisms can easily change to the metric of PDP or EDDP to fit for different platforms.

In this paper, we are considering the energy saving in the processor. Energy saving in U2 and U4 contains two parts (1) Energy saved by stopping clock drivers of some pipeline registers in order to inactivate and bypass them; (2) Pipeline stalls decreased by better IPC due to small latencies and penalties. We get Equation 2 from paper [1,11,12] to calculate the energy saving under different unification degrees.

$$\frac{E_{U_x}}{E_{normal}} = \frac{IPC_{normal}}{IPC_{U_x}} \times (1 - \beta) \tag{2}$$

Where E_{U_x} is energy in unification degree U_x and E_{normal} is energy in normal execution mode; IPC_{normal} is IPC in normal execution while IPC_{U_x} is IPC in U_x; β is the power saving part gained by inactivating pipeline registers. According to paper [1], we assume β to be 15% under U2 and 22.5% under U4.

5 Results and Analyses

5.1 Two Non-phase Based Methods for Comparison

Before we apply our algorithms on the PSU controller, we run the benchmarks under single unification degree method and optimal method. These two methods are used to measure the efficiency of the phase detection based algorithms.

(1) Single unification degree method

Use a fixed unification degree U1, U2 or U4 in the whole program execution and collect EDP data of each interval.

(2) Optimal method

Based on the data collected from single unification degree method, a best unification degree for each interval can be found. By using such profiling data the unification degree can be set to the best one at beginning of each instruction interval. This method is a theoretical optimal one and can not be achieved in real execution because it is based on the post-simulated trace analysis. It will have a smallest EDP result among all the mechanisms we have mentioned. Therefore if the EDP result of a mechanism is close to this optimal case, we can say the mechanism is efficient.

5.2 General Analysis Via Comparing Average EDP

In the following experiments, the signature size was chosen to be 1024 bits and the threshold to be 0.5. Each interval has 100k instructions. A simple hash function based on masking and shifting is used to lower the signature collection cost. Fig. 6 shows the EDP results for all 8 benchmarks. In Fig. 6, the horizontal axis denotes benchmarks and the average value, and the vertical axis denotes EDP value normalized by EDP of the optimal method for each benchmark. The columns in one benchmark represent the normalized EDPs of U1, U2, U4, basic phase detection and history table based method, from left to right. The average results of all benchmarks are also listed. The method of smaller EDP result is more efficient.

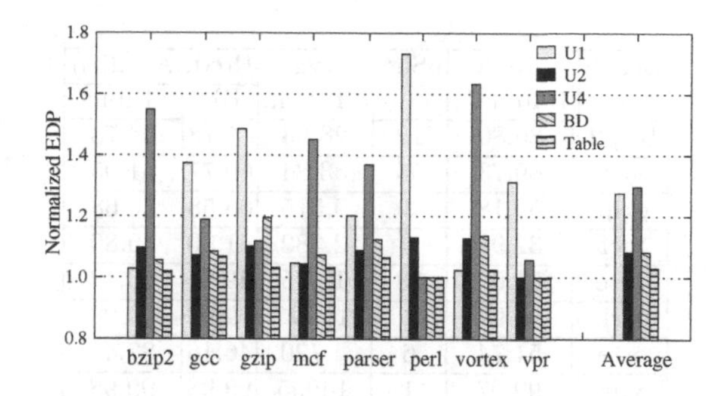

Fig. 6. Normalized EDP for SPECint2000 benchmarks

As shown in Fig. 6, we can see that not all of the benchmarks will show the smallest EDP results under a single unification degree. For benchmarks like bzip2, mcf and vortex, U1 is the most energy efficient unification degree. For benchmarks like perlbmk, degree U4 has the smallest EDP result. For other benchmarks, including gcc, gzip, parser, and vpr, U2 is better than U1 or U4. These results confirm our assumption that there is no fixed pipeline configurations which can always have best energy performance efficiency for all the programs and reconfigurations during the execution are necessary.

For the efficiency of our mechanisms, Fig. 6 shows that the basic phase detection method can achieve an average EDP of 108%, compared with the optimal method. This method has an EDP decrease of 15.1%, 1.41% and 16.4% when compared with single U1, U2 and U4, respectively.

The history table based method shows better average results, as compared with the basic phase detection method. It can achieve an average EDP of 103% of the optimal method. Compared with single U1, U2 and U4, it can gain a total EDP decrease of 19.2%, 4.82% and 20.5%.

From these results, it can be seen that both basic detection method and history table based method can have the ability to reduce the processor energy consumption by prediction the next suitable pipeline unification degree. As expected, the history table based method is more effective since it caches more history information which can reduce the tuning cost.

5.3 Prediction Accuracy

Since we are designing the dynamic mechanisms to predict a suitable unification degree for the next interval, the prediction accuracy is very important to the final energy saving result. To study the efficiency of the design methods in more detail, we list the prediction accuracy of the unification degrees in Table 3, together with some benchmark characteristics.

Table 3. Prediction accuracy of each benchmark, together with benchmark characteristics

Bench mark	Stable Rate (%)	nSigs	Avg. ST_Len.	Pred. Acc.(%)	
				BD	Table
bzip2	86.80	12	28.93	85.78	94.73
gcc	89.73	55	53.84	60.77	51.95
gzip	59.18	3	9.575	66.58	84.68
mcf	32.98	6	2.382	41.30	49.80
parser	67.63	33	11.75	49.24	60.12
perl.	99.97	1	14995	99.98	99.98
vortex	51.74	6	4.720	46.16	87.41
vpr	99.97	1	14995	99.98	99.98

In Table 3, the column of "stable rate" stands for the percentage of the total intervals that are in stable time. The "nSigs" column denotes the number of different signatures when the programs are under the stable time. We obtain this value by comparing the signatures of two stable phases. If the distance is larger than the predefined threshold, we increase this count by 1. It can be roughly used to represent the complexity of the benchmark. A higher value shows that the programs can be classified into more different stable phase groups and may require more tunings. It may potentially increase the complexity for dynamical prediction. The column of "Avg. ST_len" represents the average interval length of the stable phase for each benchmark. These three columns are the statistical results we got from the basic detection method. The accuracy of using working set signature to identify the program phase is important for further reconfiguration on processor and has been evaluated by Dhodapkar [7].

Another column named "Pred. Acc." in Table 3 is the ratio of the precise prediction of the unification degree for basic detection method and the history table based method, respectively. We obtained these two columns by calculating the similarity of predicted unification degrees with those theoretical precise unification degrees from the optimal method. In order to show the efficiency of history table based method optimally, we simply choose an infinite table size in Table 3. Fixed table size will be discussed in Section 5.4.

From Table 3, it is apparent that the prediction accuracy of the history table based method is better than the basic detection based method. This is similar to the conclusion obtained in Section 5.2.

It can also be seen from Table 3 that the prediction accuracy changes due to the program characteristics. For simple benchmarks like perlbmk and vpr, most intervals are of the same stable phase. For these two benchmarks, the prediction accuracy of both dynamical methods can reach nearly 100%. The prediction accuracy of the basic detection based method falls when the program becomes less stable. This may related with the simple design of the basic detection method. We compare only the signatures of consecutive intervals and start a tuning at each point the program goes toward stable. If the stability of program is low, the basic detection method will perform poorly because of the low energy saving rate in the unstable and tuning periods.

Unlike the basic detection method, the history table base method is less sensitive to the program stability as illustrated in the benchmarks gzip and vortex. Although the stability ratios for these two benchmarks are less than 60%, prediction accuracies of 84.68% and 87.41% are respectively attained. The increased prediction accuracies can be attributed to the historical tuned information recorded in the extra structures described by the history table based method. If the jump direction from one signature to another signature is stable, the prediction will be accurate. On the other hand, this method is sensitive to the number of phase groups. For example, gcc is quite stable but the number of different signatures during stable phase is large which lead to the uncertainty in the jump directions. More detailed results of the history table based method for gcc will be listed in Section 5.4.

5.4 Lower the Cost of History Table Based Method

The size of the history table is an important parameter for the table based method. We can see from Table 3 that the numbers of different signatures in stable phases are relatively small for most of the benchmarks and the majority of the benchmarks are quite stable. Therefore it is possible to set a small fixed table size, in order to lower the additional hardware cost introduced by the table structure, without degrading the ratio of prediction accuracy.

In this serial of experiments, we set the table size as a fixed number, from 16 entries to 2 entries. We use the LRU mechanism to replace the table entry when there is insufficient room for new signatures. The results are shown in Fig. 7. A signature size of 1024 bit and a threshold of 0.5 are used for the configuration.

In Fig. 7, the columns for each benchmark represent the results of infinite table size, 16-entry, 8-entry, 4-entry and 2-entry, respectively. We can see from the results that there is almost no degradation between the infinite-entry, 16-entry and 8-entry for all benchmarks. A sharp decrease of prediction accuracy occurs on 4-entry table size for gzip and vortex. Other benchmarks like bzip2,

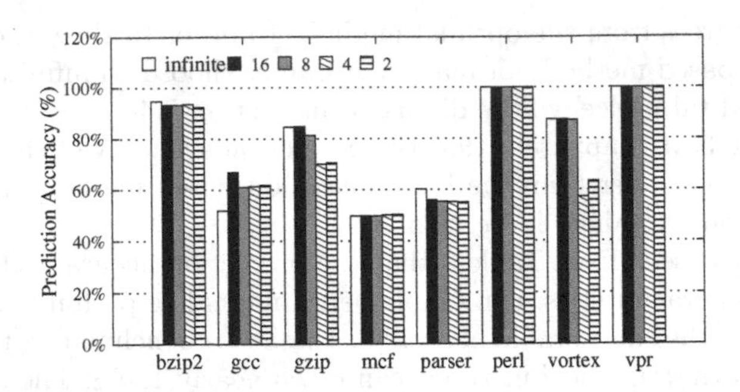

Fig. 7. Prediction accuracy of different table sizes for table based method

perlbmk and vpr show no loss of accuracy even when the size shrinks to 2-entry. For benchmark gcc, the accuracy actually increases after the table size is reducted from infinite to 16 entries. It appears the old historical information for gcc may sometimes cause misprediction of the best unification degree.

From these results, we can assume that an 8-entry table size will be sufficient for SPECint2000 benchmarks. With a small table size, the table can be accessed more quickly which introduces less overhead into the PSU control system.

6 Conclusions and Future Work

In this paper, we have designed two dynamic control mechanisms for PSU enabled processors in order to achieve good EDP. These two mechanisms are based on phase detection via working set signature. By using these two methods, we dynamically reconfigure the unification degree during the program execution according to the program behavior change. Our simulations show that the two methods can achieve an average EDP decreasing of 15.1% and 19.2%, compared to the original system without PSU enabling. Such results are about 8.34% and 3.02% larger than the optimal mode. Both methods can reduce energy consumption in the processor via dynamically predicting a unification degree for the coming interval. The two dynamical methods have different advantages. The basic detection method is simple and introduces less hardware complexity while the history table based method demonstrates greater efficiency at prediction.

Currently the energy consumption model in this paper is still very rough. We are planning to study the hardware approach so as to build a more accurate model, including the detailed overhead introduced by the dynamical prediction mechanisms. In addition, alternate program phase detection methods other than the working set signature will be tried on the PSU system.

Acknowledgments. This research is partially supported by Grant-in-Aid for Fundamental Scientific Research (S) #16100001 from Ministry of Education, Culture, Sports, Science and Technology Japan.

References

1. Shimada, H., Ando, H., Shimada, T.: Pipeline stage unification: a low-energy consumption technique for future mobile processors. In: Proceedings of the 2003 international symposium on Low power electronics and design, pp. 326–329. ACM Press, New York (2003)
2. Shimada, H., Ando, H., Shimada, T.: Pipelinig with variable depth for low power consumption (in Japanese). In: IPSJ Technical Report, 2001-ARC-145, Information Processing Society of Japan, pp.57–62 (2001)
3. Koppanalil, J., Ramrakhyani, P., Desai, S., Vaidyanathan, A., Rotenberg, E.: A case for dynamic pipeline scaling. In: Proceedings of the 2002 international conference on Compilers, architecture, and synthesis for embedded systems, pp. 1–8. ACM Press, New York (2002)
4. Hrishikesh, M.S., Burger, D., Jouppi, N.P., Keckler, S.W., Farkas, K.I., Shivakumar, P.: The optimal logic depth per pipeline stage is 6 to 8 FO4 inverter delays. In: Proceedings of the 29th annual international symposium on Computer architecture, Washington, DC, USA, pp. 14–24. IEEE Computer Society Press, Los Alamitos (2002)
5. Srinivasan, V., Brooks, D., Gschwind, M., Bose, P., Zyuban, V., Strenski, P.N., Emma, P.G.: Optimizing pipelines for power and performance. In: Proceedings of the 35th annual ACM/IEEE international symposium on Microarchitecture, pp. 333–344. IEEE Computer Society Press, Los Alamitos (2002)
6. Shimada, H., Ando, H., Shimada, T.: Power consumption reduction through combining pipeline stage unification and DVS (in Japanese). IPSJ Transactions on Advanced Computing Systems 48(3), 75–87 (2007)
7. Dhodapkar, A.S., Smith, J.E.: Managing multi-configuration hardware via dynamic working set analysis. In: Proceedings of the 29th annual international symposium on Computer architecture, Washington, DC, USA, pp. 233–244. IEEE Computer Society Press, Los Alamitos (2002)
8. Sherwood, T., Perelman, E., Hamerly, G., Sair, S., Calder, B.: Discovering and exploiting program phases. IEEE Micro 23(6), 84–93 (2003)
9. Balasubramonian, R., Albonesi, D., Buyuktosunoglu, A., Dwarkadas, S.: Memory hierarchy reconfiguration for energy and performance in general-purpose processor architectures. In: Proceedings of the 33rd annual ACM/IEEE international symposium on Microarchitecture, pp. 245–257. ACM Press, New York (2000)
10. Burger, D., Austin, T.M.: The simplescalar tool set, version 2.0. SIGARCH Computer Architecture News 25(3), 13–25 (1997)
11. Gonzalez, R., Horowitz, M.: Energy dissipation in general purpose microprocessors. IEEE Journal of Solid-State Circuits 31(9), 1277–1284 (1996)
12. Gowan, M.K., Biro, L.L., Jackson, D.B.: Power considerations in the design of the alpha 21264 microprocessor. In: Proceedings of the 35th annual conference on Design automation, pp. 726–731. ACM Press, New York (1998)

Reducing Energy in Instruction Caches by Using Multiple Line Buffers with Prediction

Kashif Ali[1], Mokhtar Aboelaze[2], and Suprakash Datta[2]

[1] School of Computing, Queens University, Kingston on Canada
[2] Department of Computer Science and Engineering, York University, Toronto on Canada

Abstract. Energy consumption plays a crucial role in the design of embedded processors especially for portable devices. Since memory access consumes a significant portion of the energy of a processor, the design of fast low-energy caches has become a very important aspect of modern processor design. In this paper, we present a novel cache architecture for reduced energy instruction caches. Our proposed cache architecture consists of the L1 cache, multiple line buffers, and a prediction mechanism to predict which line buffer, or L1 cache to access next. We used simulation to evaluate our proposed architecture and compare it with the HotSpot cache, Filter cache, Predictive line buffer cache and Way-Halting cache. Simulation results show that our approach can reduce instruction cache energy consumption, on average, by 75% without sacrificing performance

1 Introduction

On-chip caches can have a huge impact on the processor performance. Caches are faster than the main memory, and consume less power per access than the main memory. A well-designed cache results in a fast and energy efficient processor.

In modern processors, the cache takes a considerable portion of the chip area, for the DEC 21164 processor, 43% of the total energy consumed in the chip is consumed by the cache [3]. Therefore, reducing energy consumption in caches is a priority for computer architects. In [7] the authors showed how to use a unified cache to reduce the total area of the cache by 20-30% and maintain the same hit rate as a split cache. Albonesi in [1] proposed the selective way cache. In the selective way cache, preferred ways (a subset of all the ways) are accessed first; in case of a miss, the rest of the ways are accessed. The savings in energy (by not accessing all the ways) is accomplished at the expense of increasing the access time. Zhang et al [12] proposed a cache where by setting a configuration register they can reconfigure the cache size, the cache associativity, and the cache line size. By fine-tuning the cache parameters to the application, they achieved a power saving of up to 40%.

Way prediction was used in [13] to reduce cache energy. In order not to sacrifice the cache speed, they used a 2-level prediction scheme. First, they decide if they use way prediction or not; if not then all the ways in a set associative cache are accessed. However, if the decision is to use way prediction, the predicted way is

J. Labarta, K. Joe, and T. Sato (Eds.): ISHPC 2005 and ALPS 2006, LNCS 4759, pp. 508–521, 2008.

accessed first, in case of a miss, the rest of the ways are accessed. A non-uniform cache was introduced in [6]. In this design, the cache has different values for associativity. The optimal value for the number of ways is determined for each application. They also proposed some techniques in order to minimize the access to redundant cache way and cache tags to minimize energy consumption.

HotSpot cache was introduced in [10] where a small filter cache was used to store loops that are executed more than a specific threshold. The loops are detected by using the Branch Target Buffer (or BTB) and is promoted to the HotSpot cache when they reach their threshold values. Their design resulted in reducing the energy consumption of the cache. Jouppi in [5] showed how to use a small fully associative cache and prefetching to improve the performance of a direct-mapped cache without paying the price of a fully associative cache. Zhang et al introduced the way-halting cache in [11] where they used some bits from the tag in order to choose which way to access in a multi-way (set associative) cache.

In this paper, we introduce a new cache architecture that has a slightly better average cache access time than many existing architectures and consumes much less energy compared to the existing architectures. We use MediaBench and Mibench benchmark to compare our results with the standard cache without any line buffers, the Filter cache, the HotSpot cache, the Way-halting cache and the single predictive line buffer.

The organization of this paper is as follows. In Section 2, we discuss the motivation behind our architecture. In Section 3 we propose and explain our architecture. Section 4 gives details of our prediction and line placement algorithm. Section 5 presents the simulation setup and compares our architectures with the HotSpot cache, Filter cache and single line buffer cache. Section 6 concludes the paper.

2 Motivation

The overall occurrences of branches are application dependent e.g. multimedia application tends to have more structured conditional branches compared with others applications (SPEC2000 applications). But no matter what type of workload, conditional branches represents a significant fraction of total instructions executed by a program.

A loop block, sequence of instructions whose iterations is controlled by conditional instruction, may contain minimal 2 instructions and can be up to 32 instructions or more. Over 75% of such control instructions jumps a maximum of 16 instructions. For multimedia application (MediaBench/MiBench) on the average, almost 50% of the control instructions jumps no more then 6 instructions. For such applications, a loop block can be captured within 2-3 line buffer. By careful analysis of the programs in the MediaBench and SPEC2000 we found the following

- More than 75% of the loops in the MediaBench suite include 16 or less instructions.

- Almost 95% of the loops in the MediaBench suite contain 30 or less instructions
- Almost 90% of the loops in SPEC2000 contains 30 or less instructions.

In [2] we showed how to use a single line buffer with prediction in order to reduce energy consumption in a direct-mapped cache. While 16-instruction loops cannot be be captured using a single line buffer, they could be captured if 4-8 line buffers are used with a good cache organization to guarantee that the instructions in the loops are mapped to the entire set of line buffers instead of replacing each other in a small number of line buffers. Increasing the line size is not the solution since it affects the temporal locality and may reduce the hit ratio.

3 Proposed Architecture

Our single predictive line buffer (proposed in [2]) does not have the ability to capture most of the loop blocks i.e. cannot take advantage of temporal locality, hence require accessing lower level (level-1) cache more often. To fully utilize the temporal locality in a program, we now extend our single predictive line buffer scheme by adding multiple line buffers between the CPU and L1 cache. We are also proposing a new prediction scheme to control access to the buffers. During the fetch cycle only one of the line buffers is accessed. In case of a miss in the line buffer, the instruction will be fetched from L1 and the line containing the fetched instruction is placed in one of the line buffers. Figure 1 shows a schematic of the proposed architecture. with 4 line buffers, labeled as bb1, bb2, bb3 and bb4, and the L1 cache. The optimal number of line buffers depends on the application. In our simulations, we found that having anywhere between 4 to 8 line buffers achieves the best results for most of the applications in MediaBench and Mibench benchmark. We present detail analysis relating to number of line buffers in Section 5.2.

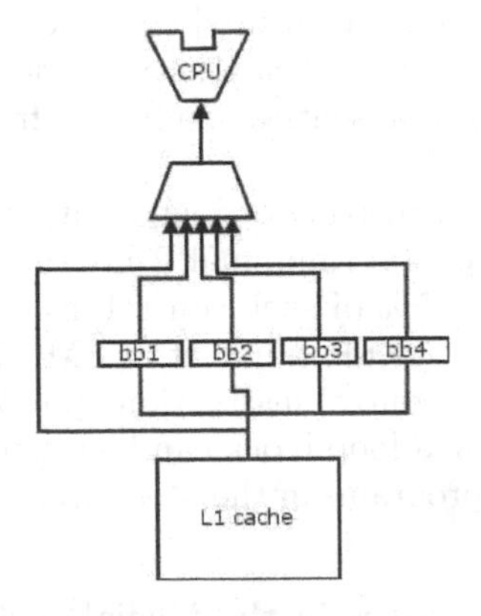

Fig. 1. Cache with Multiple Line Buffers

Our scheme dynamically selects either one of the line buffers or the L1 cache for fetching instructions. We assume the existence of BTB (branch target buffer), which is common in many modern processors. The added hardware for the selection mechanism is minimum – it involves some extra registers (called *tag-bit registers*) for storing a part of the tag for the cache lines in the line buffers.

3.1 Tag-Bit Registers

Our goal is to spread the loop(s) between the different line buffers and to keep the loop(s) that are being currently executed in the line buffers. If we are successful in doing that, the tags of the data in the line buffers are sequential, and they differ only in the low-order bits. This observation can be effectively used to predict the line buffer containing the instruction to be fetched. The main idea of our proposed architecture is to cache a few of the low-order bits of tags in special registers called tag-bit registers. The i low order bits of the tags in each line in line buffer are kept in tag-bit registers.

Our algorithm compares the contents of the tag-bit register with the corresponding bits of the instruction address. If one matches, this is the predicted line buffer and we access it to get the instruction. This requires much less energy than accessing the L1 cache. If our prediction is incorrect, then we have to go to the L1 cache to access the required instruction. Figure 2 shows the organization of the tag-bit register.

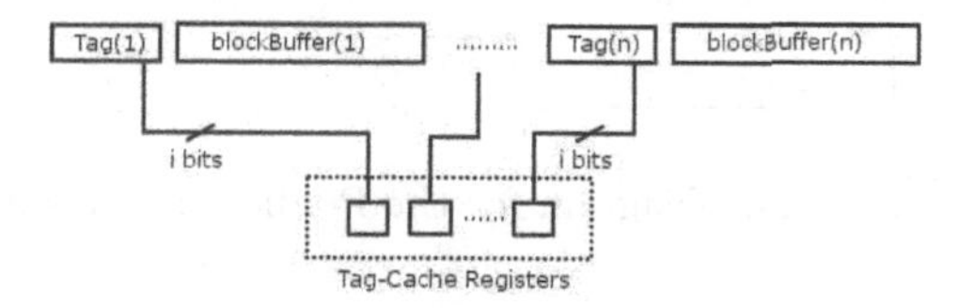

Fig. 2. Tag-Bit Registers

In the next section, we present the prediction mechanism and the placement mechanism used in order to increase the probability of placing the long loops successfully in the line buffers.

4 Prediction Scheme

In order to predict effectively between line buffers and the L1 cache we need to keep some state information. These state information are *fetch_mode* which could be either L1 cache, or line buffer. If *fetch_mode* points to the line buffer, then *curr_bb* is a pointer to the predicted line buffer that holds the instruction to be fetched. Finally, *TAG_bits* is an array which holds the low order i bits of every tag in the line buffers. This may be a physical register, or could be just the i bits in the TAG stored in every line buffer. Figure 3 shows the flow diagram for the prediction algorithm.

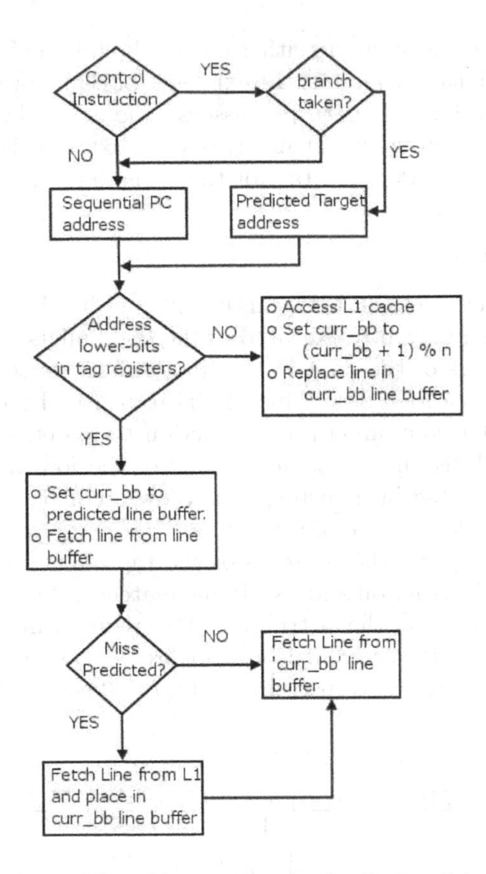

Fig. 3. Flow Diagram for the Prediction Scheme

The main idea of the prediction algorithm is as follows. Once the instruction is fetched, the program counter (PC) is checked against the BTB to see if that instruction is a branch and predicted taken or not. If the instruction is a branch and is predicted taken, the target address is loaded from the BTB into the PC, otherwise the next sequential address is loaded in the PC (PC is incremented according to the instruction length). The low order bits of the TAG part of the address is checked against the *TAG_bits* array. If there is a match, then the *fetch_mode* is set to that particular line buffer. Otherwise it is set to L1 cache, the instruction will be then fetched from the L1 cache and will be stored in line buffer *curr_bb*+1.

If there is a match, then the instruction is fetched from the predicted line buffer. The fetched instruction may or may not be the required instruction (we checked only the low order i bits). If our prediction is correct, then the instruction is sent to the CPU. If we miss-predicted, then the instruction is accessed from the cache, the line contains that word is sent to the *curr_bb*.

5 Experimental Results

In this section, we compare our proposed scheme, multiple predictive line with various other scheme for both performance and energy. We first show the effect of the number of bits in the tag-bit registers on the performance. We then evaluate the effectiveness of multiple predictive line buffer scheme by comparing it to Filter cache, HotSpot cache, single predictive line buffer and Way-halting cache. And finally we'll compare scheme's optimization goal of reducing energy without affecting performance.

5.1 Experimental Setup

We use SimpleScalar toolset [8] and CACTI 3.2 [9] to conduct our experiments. We modified SimpleScalar to simulate Filter Caches, HotSpot caches, Predictive Line buffer and Way-Halting cache. Our baseline architecture uses a 16KB direct-mapped cache or 16KB 4-way set-associative cache. Our Line buffer is 32 bytes. We have used a 512 bytes, direct-mapped L0 cache for Filter cache and HotSpot cache. The BTB is 4-way set-associative with 512 sets. We also used a 2-level branch predictor in our simulation. We evaluated energy consumption using $0.35\mu m$ process technology. For HotSpot cache, we used a value of 16 as candidate threshold as was suggested in [10]. As our proposed scheme is targeted toward embedded microprocessors, we have used multimedia benchmarks, MediaBench and Mibench but our scheme yields similar results for other types of workload. Each applications was limited to 500 million instructions using the data set included with the benchmark suites. We choose sets of encoder/decoder from different media types such as data, voice, speech, image, video and communication. (See Table-1). Energy per cache access, as is obtained from CACTI 3.2 and is shown in Table-2

Table 1. Benchmark Applications Summary

Application	Type	Benchmark
crc32/fft	Communication	Mibench
epic	Data	MediaBench
adpcm/g721/gsm	Voice/Speech	MediaBench
jpeg	Image	MediaBench
lame	Mp3	Mibench
mpeg2	Video	MediaBench

5.2 Optimal Number of Line Buffers

The ideal number of Line buffer depends on each application. Our experiments shows that 4-8 lines buffers are optimal for most applications.

Table 2. Energy per Access for Various Cache Configurations

Cache	Energy
512 L0 cache	0.69nJ
line buffer	0.12nJ
16KB direct-map	1.63nJ
16KB 4-way set-assoc	2.49nJ

By analyzing average normalized energy for various applications of benchmarks, we observed that number of line buffer to use depends upon the conditional branches relatives target address. Knowing, on the average, how far these instructions jumps to (i.e. size of loop block), we can relate them to how many line buffers is required for that particular application. For instance, for communication application crc32, using 6 line buffer is optimal and adding 7 or 8 line buffer does not improve the energy consumption. This is because for such application, almost none of loop block are greater then 14 instruction and can be easily captured using maximum of 6 line buffers. Similar observation can be made for other applications in MediaBench and MiBench applications.

5.3 Tag-Bit Registers Misprediction

The key component for effectiveness of multiple predictive line buffer is tag-bit registers. We experimented with numerous width of low-order tag bits. A near-ideal tag bit size is the smallest tag-bit that can find the right line buffer, hence less miss-prediction. By storing more bit in the register, that results in slightly more accurate prediction but will also increase overhead as the register size increases.

Table 3 and 4 show the result of using between 2 and 6 bits with 4 line buffer and 4-8 bits with 8 line buffer for direct map cache. From the average of these two tables we can see when using 2 tag bit with 4 line buffer, the miss prediction can be as high as 22.43%. As we stated in Section 2, that almost 90% of loops contains 30 or less instructions, using 2 bits we can only accurately distinguish between maximum of 4 instructions in any loop block. Therefore using more bits will help us predicts more efficiently. From the tables we can see that using 4 or 5 bits can gives satisfactory results for most of applications. In our experiment, we use 5 bits for the tag-bit register.

5.4 Energy

In this section we compare energy savings of multiple predictive line buffer with conventional line buffer, HotSpot Cache and single Predictive line buffer. Fig. 4 and Fig. 5 show normalized energy reduction using conventional direct-map and 4-way set-associative as base cache respectively. We can see that using multiple line buffers significantly reduces energy consumption. For some applications

Table 3. Miss Prediction Ratio (4 Line Buffer, 2-6 bits for tag-bit Register) using Direct-Map L1 cache

BenchMark	2-tag	3-tag	4-tag	5-tag	6-tag
apdcm-decode	26.59	0.18	0.14	0.00	0.00
apdcm-encode	25.88	3.86	0.06	0.05	0.01
crc32	20.02	0.02	0.01	0.00	0.00
epic	18.24	1.05	0.01	0.00	0.00
fft	27.70	8.63	3.37	1.47	1.20
fft-inv	27.70	8.63	3.37	1.47	1.20
g721-decode	19.18	3.97	1.88	0.90	0.47
g721-encode	18.47	2.66	1.15	0.40	0.17
gsm-decode	30.88	3.27	1.46	0.04	0.02
gsm-encode	26.56	0.57	0.17	0.10	0.00
jpeg2-decode	26.11	1.91	0.54	0.04	0.01
jpeg2-encode	19.64	3.90	0.45	0.08	0.03
lame	19.22	3.07	1.20	0.15	0.06
mpeg2-decode	15.25	4.32	2.34	2.24	2.20
mpeg2-encode	22.05	4.23	1.32	0.56	0.01
unepic	15.72	2.53	0.38	0.02	0.01
Average	22.45	3.30	1.12	0.47	0.34

Table 4. Miss Prediction Ratio (8 Line Buffer, 4-8 bits for tag-bit Register), using direct-Map L1 cache

BenchMark	4-tag	5-tag	6-tag	7-tag	8-tag
apdcm-decode	0.16	0.01	0.00	0.00	0.00
apdcm-encode	0.07	0.06	0.00	0.00	0.00
crc32	0.01	0.00	0.00	0.00	0.00
epic	0.02	0.01	0.00	0.00	0.00
fft	7.20	3.85	2.70	1.31	0.71
fft-inv	7.20	3.85	2.70	1.31	0.71
g721-decode	4.19	1.35	0.76	0.25	0.24
g721-encode	2.71	1.05	0.55	0.08	0.07
gsm-decode	2.38	0.61	0.58	0.55	0.00
gsm-encode	1.72	1.15	0.05	0.04	0.00
jpeg2-decode	3.00	0.17	0.08	0.02	0.00
jpeg2-encode	2.28	1.40	0.34	0.28	0.28
lame	2.49	0.67	0.31	0.17	0.02
mpeg2-decode	2.84	2.57	2.36	0.04	0.03
mpeg2-encode	3.76	2.03	0.04	0.00	0.00
unepic	1.23	0.08	0.05	0.01	0.00
Average	2.58	1.18	0.66	0.25	0.13

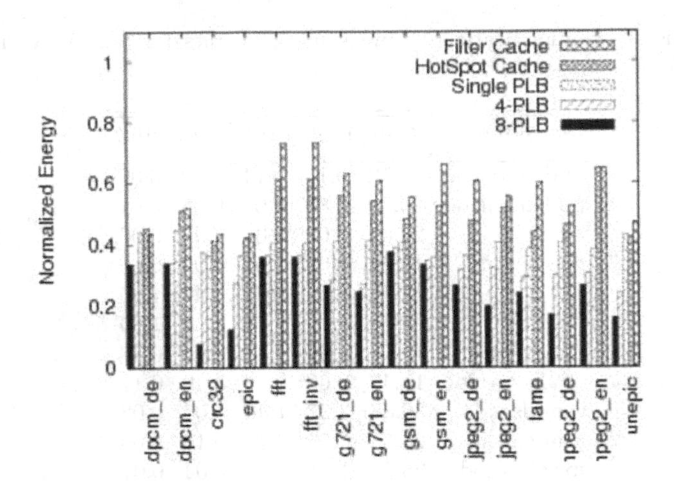

Fig. 4. Normalized energy reduction using direct-map L1 cache

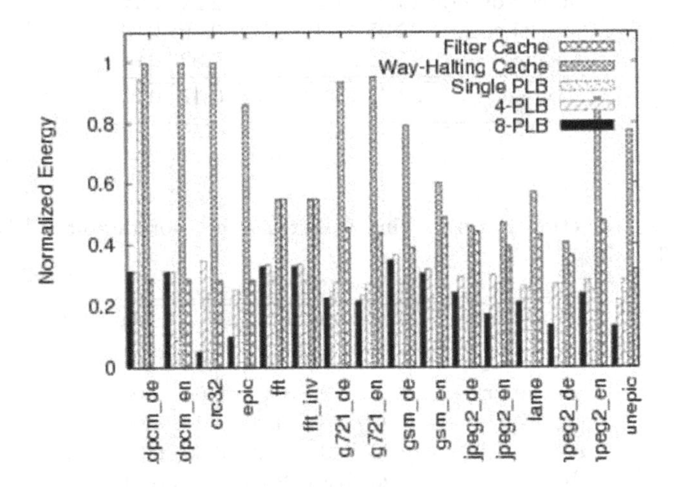

Fig. 5. Normalized energy using 4-way set-associative L1 cache

such as crc32, epic, mpeg2-decode and jpeg-encodes significantly reduces energy consumption, when 8 line buffer are used, over 4 line buffer.

For these application, significant portion of control instruction jump can't be contained using 4 line buffer. Therefore using 8 line buffer significantly reduces energy consumption for these application compared to others in benchmark. Using 8-multiple line buffer with line buffer reduces normalized energy consumption by up to 74%, comparing with HotSpot Cache 47%.

Table 5 shows the average normalized energy consumption of various schemes. From table, we can observe that using 8 multiple line buffer, on average, lower energy consumption significantly compared with others.

Table 5. Various Schemes Average Normalized Energy Using Direct-map And 4-way Set-associative Cache

Scheme	Direct Mapped	Set-associative
4 PLB	0.32	0.29
8 PLB	0.26	0.22
Filter Cache	0.57	0.46
HotSpot Cache	0.53	0.41
Single PLB	0.40	0.36
Way-Halting Cache	N/A	0.61

5.5 Delay

In this section we show that multiple predictive line buffer scheme does not sacrifice performance in order to reduce energy consumption. We compare the normalized delay for multiple predictive line buffer (using 4 and 8 line buffer), with HotSpot Cache and single predictive line buffer, with both direct-map L1 cache.

Normalized delay using conventional direct-map L1 cache as the base architecture is shown in Fig. 6, while Fig. 7 shows the same results for a 4-way set associative baseline cache. Using 4 predictive line buffer performs as good as single predictive line buffer. Using 8 multiple line buffer have higher delay compared to single predictive line buffer but still is significantly better then HotSpot Cache. As we mentioned in Section 5.3, if using 6 tag width, the delay for 8 line buffer can be improved, but improvement is still very minimal . Table 6 shows the average normalized delay for HotSpot Cache, single predictive line buffer, 4 and

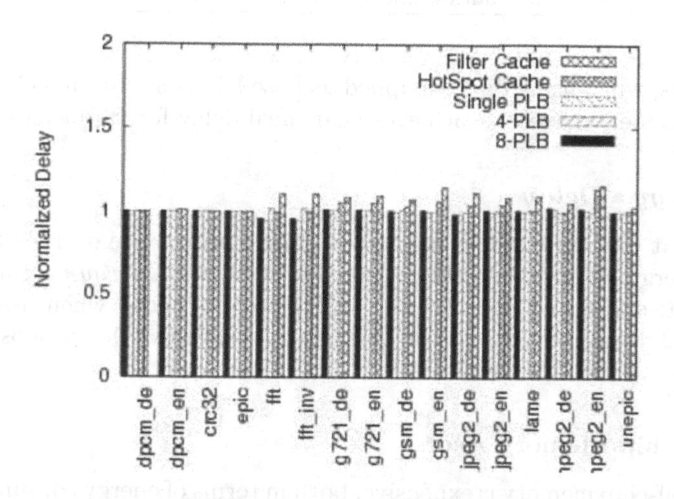

Fig. 6. Normalized Delay using direct-map L1 cache

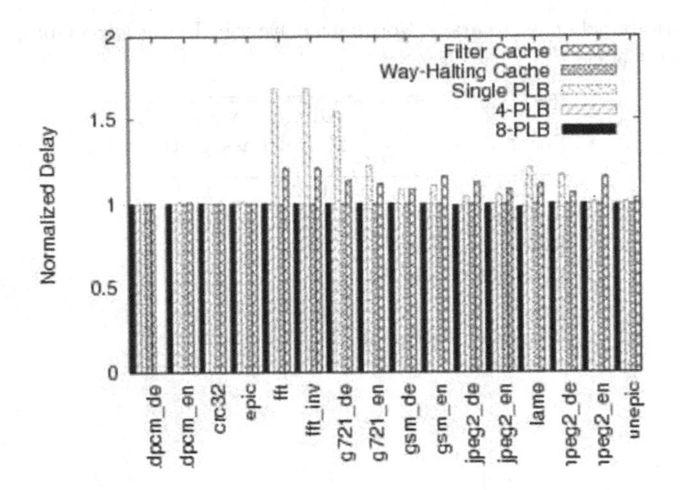

Fig. 7. Normalized Delay using 4-way set-associative cache

Table 6. Various Schemes Average Normalized Delay Using Direct-map And 4-way Set-associative Cache

Scheme	Direct Mapped	Set-associative
4 PLB	0.993	0.997
8 PLB	0.999	0.997
Filter Cache	1.070	1.091
HotSpot Cache	1.027	N/A
Single PLB	1.004	1.007
Way-Halting Cache	N/A	1.000

8 line buffers, when using direct-mapped as base L1 cache. Results clearly shows that our scheme, on average, achieves near-ideal delay for various applications.

5.6 *Energy ∗ Delay*

To show that our scheme does not impose any performance overhead while decreasing energy consumption, Fig. 8 and Fig. 9 shows *Energy ∗ Delay* product for HotSpot, single and multiple line buffer cache scheme, when using Direct-mapped and 4-way Set-associative L1 cache respectively. Our proposed scheme outperforms all the other schemes.

5.7 Off-Chip Memory Access

Accessing off-chip memory is expensive, both in terms of energy consumption and delay. Accessing 512KB 4-way set-associative off-chip cache is almost 6 times as

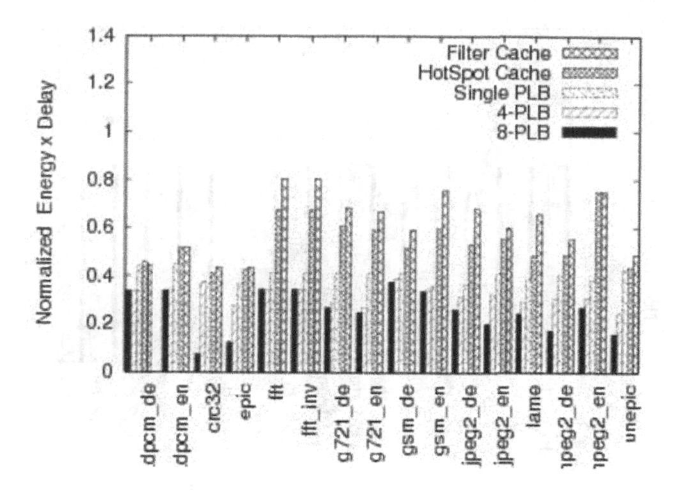

Fig. 8. Normalized *Energy* ∗ *Delay* using direct-map L1 cache

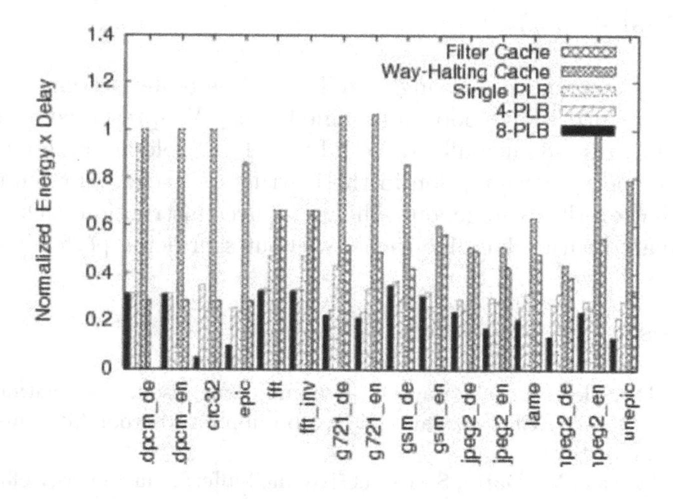

Fig. 9. Normalized *Energy* ∗ *Delay* using 4-way set-associative L1 cache

expensive as accessing L1 cache (512KB 4-way set-associative caches consumes $14.21nJ$ per access, see Table 2). Although direct-map although relatively has fast access, but can suffers from thrashing problem. Thrashing occurs when two memory lines maps to same line in the cache. Thrashing can cause performance issue as most of time is spend in moving data between memory and cache. Thrashing can be avoided if the loop block can be captured in the upper level cache hence avoiding conflicts. Our proposed scheme did not increase the off-chip access. For certain application, such as jpeg-decode, we can even lower memory access up to 50% comparing with HotSpot cache. On average, our scheme lowered off-chip memory access compared to HotSpot and Filter Cache.

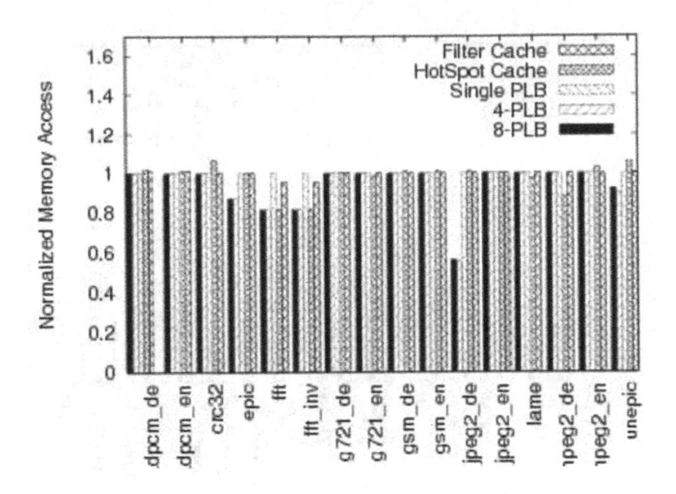

Fig. 10. Normalized Off-Chip Memory Access

6 Conclusion

In this paper, we extended our single predictive-line buffer scheme (proposed in [2]) in order to capture long loops in the line buffers. We presented a cache architecture that utilizes 4-8 line buffers, the BTB and a simple prediction mechanism to reduce the energy consumption in the instruction cache. Our Simulation results show that on the average, our scheme reduces instruction cache energy up to 75% compared with a baseline cache, without sacrificing performance.

References

1. Albonesi, D.: Selective cache ways: on-demand cache resource allocation. In: Proc. of the 32nd ACM/IEEE International Symposium on Microarchitecture, pp. 248–259 (November 1999)
2. Ali, K., Aboelaze, M., Datta, S.: Predictive line buffer: A fast energy efficient cache architecture (submitted) IEEE SoutheastCon 2006
3. Edmondson, J.F.: Internal organization of the Alpha 21164, a 300-MHz 64 bit quad-issue CMOS RISC microprocessor. Digital Technology J. 7(1), 119–135 (1995)
4. Hasegawa, A., Kawasaki, I., Yoshioka, S., Kawasaki, S., Biswas, P.: SH3: High code density, low power. IEEE Micro 15(6), 11–19 (1995)
5. Jouppi, N.P.: Improving direct-mapped cache performance by the addition of a small fully associative cache and prefetch buffers. In: The 17th Annual International Symposium on Computer Architecture ISCA, pp. 364–373 (May 1990)
6. Ishihara, T., Fallah, F.: A non-uniform cache architecture for low power system design. In: ISLPED 2005. Proceedings of the 2005 International Symposium on Low Power Electronics and Design (August 2005)
7. Mizuno, H., Ishibashi, K.: A separated bit-line unified cache: Conciliating small on-chip cache die-area and low miss ratio. IEEE Transactions on Very Large Scale Integration (VLSI) Systems 7(1), 139–144 (1999)

8. The Simplescalar simulator (May 2006), www.simplescalar.com
9. Shivakumar, P., Jouppi, N.: CACTI 3.0: An integrated cache timing, power, and area model. Technical Report 2001.2 Compaq Research Lab (2001)
10. Yang, C.-L., Lee, C.-H.: HotSpot cache: joint temporal and spatial locality exploitation for I-cache energy reduction. In: ISPLD 2004. Proc. of the 2004 International Symposium on Low Power Electronics and Design, pp. 114–119 (August 2004)
11. Zhang, C., Vahid, F., Yang, J., Najjar, W.: A way-halting cache for low-power high-performance systems. In: ISPLD 2004. Proc. of the 2004 International Symposium on Low Power Electronics and Design (August 2004)
12. Zhang, C., Vahid, F., Najjar, W.: A Highly Configurable Cache for Low Energy Embedded Systems. ACM Transactions on Embedded Computing Systems (TECS) 4(2), 363–387 (2005)
13. Zhu, Z., Zhang, X.: Access mode prediction for low-power cache design. IEEE Micro 22(2), 58–71 (2002)

Author Index

Lecture Notes in Computer Science

Sublibrary 1: Theoretical Computer Science and General Issues

For information about Vols. 1–4576
please contact your bookseller or Springer

Vol. 4917: P. Stenström, M. Dubois, M. Katevenis, R. Gupta, T. Ungerer (Eds.), High Performance Embedded Architectures and Compilers. XIII, 400 pages. 2008.

Vol. 4910: V. Geffert, J. Karhumäki, A. Bertoni, B. Preneel, P. Návrat, M. Bieliková (Eds.), SOFSEM 2008: Theory and Practice of Computer Science. XV, 792 pages. 2008.

Vol. 4904: S. Rao, M. Chatterjee, P. Jayanti, C.S.R. Murthy, S.K. Saha (Eds.), Distributed Computing and Networking. XVIII, 588 pages. 2007.

Vol. 4878: E. Tovar, P. Tsigas, H. Fouchal (Eds.), Principles of Distributed Systems. XIII, 457 pages. 2007.

Vol. 4875: S.-H. Hong, T. Nishizeki, W. Quan (Eds.), Graph Drawing. XIII, 402 pages. 2008.

Vol. 4873: S. Aluru, M. Parashar, R. Badrinath, V.K. Prasanna (Eds.), High Performance Computing – HiPC 2007. XXIV, 663 pages. 2007.

Vol. 4863: A. Bonato, F.R.K. Chung (Eds.), Algorithms and Models for the Web-Graph. X, 217 pages. 2007.

Vol. 4860: G. Eleftherakis, P. Kefalas, G. Păun, G. Rozenberg, A. Salomaa (Eds.), Membrane Computing. IX, 453 pages. 2007.

Vol. 4855: V. Arvind, S. Prasad (Eds.), FSTTCS 2007: Foundations of Software Technology and Theoretical Computer Science. XIV, 558 pages. 2007.

Vol. 4851: S. Boztaş, H.-F.(F.) Lu (Eds.), Applied Algebra, Algebraic Algorithms and Error-Correcting Codes. XII, 368 pages. 2007.

Vol. 4847: M. Xu, Y. Zhan, J. Cao, Y. Liu (Eds.), Advanced Parallel Processing Technologies. XIX, 767 pages. 2007.

Vol. 4846: I. Cervesato (Ed.), Advances in Computer Science – ASIAN 2007. XI, 313 pages. 2007.

Vol. 4838: T. Masuzawa, S. Tixeuil (Eds.), Stabilization, Safety, and Security of Distributed Systems. XIII, 409 pages. 2007.

Vol. 4835: T. Tokuyama (Ed.), Algorithms and Computation. XVII, 929 pages. 2007.

Vol. 4783: J. Holub, J. Žďárek (Eds.), Implementation and Application of Automata. XIII, 324 pages. 2007.

Vol. 4782: R. Perrott, B.M. Chapman, J. Subhlok, R.F. de Mello, L.T. Yang (Eds.), High Performance Computing and Communications. XIX, 823 pages. 2007.

Vol. 4771: T. Bartz-Beielstein, M.J. Blesa Aguilera, C. Blum, B. Naujoks, A. Roli, G. Rudolph, M. Sampels (Eds.), Hybrid Metaheuristics. X, 202 pages. 2007.

Vol. 4770: V.G. Ganzha, E.W. Mayr, E.V. Vorozhtsov (Eds.), Computer Algebra in Scientific Computing. XIII, 460 pages. 2007.

Vol. 4769: A. Brandstädt, D. Kratsch, H. Müller (Eds.), Graph-Theoretic Concepts in Computer Science. XIII, 341 pages. 2007.

Vol. 4763: J.-F. Raskin, P.S. Thiagarajan (Eds.), Formal Modeling and Analysis of Timed Systems. X, 369 pages. 2007.

Vol. 4759: J. Labarta, K. Joe, T. Sato (Eds.), High-Performance Computing. XV, 524 pages. 2008.

Vol. 4746: A. Bondavalli, F. Brasileiro, S. Rajsbaum (Eds.), Dependable Computing. XV, 239 pages. 2007.

Vol. 4743: P. Thulasiraman, X. He, T.L. Xu, M.K. Denko, R.K. Thulasiram, L.T. Yang (Eds.), Frontiers of High Performance Computing and Networking ISPA 2007 Workshops. XXIX, 536 pages. 2007.

Vol. 4742: I. Stojmenovic, R.K. Thulasiram, L.T. Yang, W. Jia, M. Guo, R.F. de Mello (Eds.), Parallel and Distributed Processing and Applications. XX, 995 pages. 2007.

Vol. 4739: R. Moreno Díaz, F. Pichler, A. Quesada Arencibia (Eds.), Computer Aided Systems Theory – EUROCAST 2007. XIX, 1233 pages. 2007.

Vol. 4736: S. Winter, M. Duckham, L. Kulik, B. Kuipers (Eds.), Spatial Information Theory. XV, 455 pages. 2007.

Vol. 4732: K. Schneider, J. Brandt (Eds.), Theorem Proving in Higher Order Logics. IX, 401 pages. 2007.

Vol. 4731: A. Pelc (Ed.), Distributed Computing. XVI, 510 pages. 2007.

Vol. 4728: S. Bozapalidis, G. Rahonis (Eds.), Algebraic Informatics. VIII, 291 pages. 2007.

Vol. 4726: N. Ziviani, R. Baeza-Yates (Eds.), String Processing and Information Retrieval. XII, 311 pages. 2007.

Vol. 4719: R. Backhouse, J. Gibbons, R. Hinze, J. Jeuring (Eds.), Datatype-Generic Programming. XI, 369 pages. 2007.

Vol. 4711: C.B. Jones, Z. Liu, J. Woodcock (Eds.), Theoretical Aspects of Computing – ICTAC 2007. XI, 483 pages. 2007.

Vol. 4710: C.W. George, Z. Liu, J. Woodcock (Eds.), Domain Modeling and the Duration Calculus. XI, 237 pages. 2007.

Vol. 4708: L. Kučera, A. Kučera (Eds.), Mathematical Foundations of Computer Science 2007. XVIII, 764 pages. 2007.

Vol. 4707: O. Gervasi, M.L. Gavrilova (Eds.), Computational Science and Its Applications – ICCSA 2007, Part III. XXIV, 1205 pages. 2007.

Vol. 4706: O. Gervasi, M.L. Gavrilova (Eds.), Computational Science and Its Applications – ICCSA 2007, Part II. XXIII, 1129 pages. 2007.

Vol. 4705: O. Gervasi, M.L. Gavrilova (Eds.), Computational Science and Its Applications – ICCSA 2007, Part I. XLIV, 1169 pages. 2007.

Vol. 4703: L. Caires, V.T. Vasconcelos (Eds.), CONCUR 2007 – Concurrency Theory. XIII, 507 pages. 2007.

Vol. 4700: C.B. Jones, Z. Liu, J. Woodcock (Eds.), Formal Methods and Hybrid Real-Time Systems. XVI, 539 pages. 2007.

Vol. 4699: B. Kågström, E. Elmroth, J. Dongarra, J. Waśniewski (Eds.), Applied Parallel Computing. XXIX, 1192 pages. 2007.

Vol. 4698: L. Arge, M. Hoffmann, E. Welzl (Eds.), Algorithms – ESA 2007. XV, 769 pages. 2007.

Vol. 4697: L. Choi, Y. Paek, S. Cho (Eds.), Advances in Computer Systems Architecture. XIII, 400 pages. 2007.

Vol. 4688: K. Li, M. Fei, G.W. Irwin, S. Ma (Eds.), Bio-Inspired Computational Intelligence and Applications. XIX, 805 pages. 2007.

Vol. 4684: L. Kang, Y. Liu, S. Zeng (Eds.), Evolvable Systems: From Biology to Hardware. XIV, 446 pages. 2007.

Vol. 4683: L. Kang, Y. Liu, S. Zeng (Eds.), Advances in Computation and Intelligence. XVII, 663 pages. 2007.

Vol. 4681: D.-S. Huang, L. Heutte, M. Loog (Eds.), Advanced Intelligent Computing Theories and Applications. XXVI, 1379 pages. 2007.

Vol. 4672: K. Li, C. Jesshope, H. Jin, J.-L. Gaudiot (Eds.), Network and Parallel Computing. XVIII, 558 pages. 2007.

Vol. 4671: V.E. Malyshkin (Ed.), Parallel Computing Technologies. XIV, 635 pages. 2007.

Vol. 4669: J.M. de Sá, L.A. Alexandre, W. Duch, D. Mandic (Eds.), Artificial Neural Networks – ICANN 2007, Part II. XXXI, 990 pages. 2007.

Vol. 4668: J.M. de Sá, L.A. Alexandre, W. Duch, D. Mandic (Eds.), Artificial Neural Networks – ICANN 2007, Part I. XXXI, 978 pages. 2007.

Vol. 4666: M.E. Davies, C.J. James, S.A. Abdallah, M.D. Plumbley (Eds.), Independent Component Analysis and Blind Signal Separation. XIX, 847 pages. 2007.

Vol. 4665: J. Hromkovič, R. Královič, M. Nunkesser, P. Widmayer (Eds.), Stochastic Algorithms: Foundations and Applications. X, 167 pages. 2007.

Vol. 4664: J. Durand-Lose, M. Margenstern (Eds.), Machines, Computations, and Universality. X, 325 pages. 2007.

Vol. 4661: U. Montanari, D. Sannella, R. Bruni (Eds.), Trustworthy Global Computing. X, 339 pages. 2007.

Vol. 4649: V. Diekert, M.V. Volkov, A. Voronkov (Eds.), Computer Science – Theory and Applications. XIII, 420 pages. 2007.

Vol. 4647: R. Martin, M.A. Sabin, J.R. Winkler (Eds.), Mathematics of Surfaces XII. IX, 509 pages. 2007.

Vol. 4646: J. Duparc, T.A. Henzinger (Eds.), Computer Science Logic. XIV, 600 pages. 2007.

Vol. 4644: N. Azémard, L. Svensson (Eds.), Integrated Circuit and System Design. XIV, 583 pages. 2007.

Vol. 4641: A.-M. Kermarrec, L. Bougé, T. Priol (Eds.), Euro-Par 2007 Parallel Processing. XXVII, 974 pages. 2007.

Vol. 4639: E. Csuhaj-Varjú, Z. Ésik (Eds.), Fundamentals of Computation Theory. XIV, 508 pages. 2007.

Vol. 4638: T. Stützle, M. Birattari, H. H. Hoos (Eds.), Engineering Stochastic Local Search Algorithms. X, 223 pages. 2007.

Vol. 4630: H.J. van den Herik, P. Ciancarini, H.H.L.M.(J.) Donkers (Eds.), Computers and Games. XII, 283 pages. 2007.

Vol. 4628: L.N. de Castro, F.J. Von Zuben, H. Knidel (Eds.), Artificial Immune Systems. XII, 438 pages. 2007.

Vol. 4627: M. Charikar, K. Jansen, O. Reingold, J.D.P. Rolim (Eds.), Approximation, Randomization, and Combinatorial Optimization. XII, 626 pages. 2007.

Vol. 4624: T. Mossakowski, U. Montanari, M. Haveraaen (Eds.), Algebra and Coalgebra in Computer Science. XI, 463 pages. 2007.

Vol. 4623: M. Collard (Ed.), Ontologies-Based Databases and Information Systems. X, 153 pages. 2007.

Vol. 4621: D. Wagner, R. Wattenhofer (Eds.), Algorithms for Sensor and Ad Hoc Networks. XIII, 415 pages. 2007.

Vol. 4619: F. Dehne, J.-R. Sack, N. Zeh (Eds.), Algorithms and Data Structures. XVI, 662 pages. 2007.

Vol. 4618: S.G. Akl, C.S. Calude, M.J. Dinneen, G. Rozenberg, H.T. Wareham (Eds.), Unconventional Computation. X, 243 pages. 2007.

Vol. 4616: A.W.M. Dress, Y. Xu, B. Zhu (Eds.), Combinatorial Optimization and Applications. XI, 390 pages. 2007.

Vol. 4614: B. Chen, M. Paterson, G. Zhang (Eds.), Combinatorics, Algorithms, Probabilistic and Experimental Methodologies. XII, 530 pages. 2007.

Vol. 4613: F.P. Preparata, Q. Fang (Eds.), Frontiers in Algorithmics. XI, 348 pages. 2007.

Vol. 4600: H. Comon-Lundh, C. Kirchner, H. Kirchner (Eds.), Rewriting, Computation and Proof. XVI, 273 pages. 2007.

Vol. 4599: S. Vassiliadis, M. Berekovič, T.D. Hämäläinen (Eds.), Embedded Computer Systems: Architectures, Modeling, and Simulation. XVIII, 466 pages. 2007.

Vol. 4598: G. Lin (Ed.), Computing and Combinatorics. XII, 570 pages. 2007.

Vol. 4596: L. Arge, C. Cachin, T. Jurdziński, A. Tarlecki (Eds.), Automata, Languages and Programming. XVII, 953 pages. 2007.

Vol. 4595: D. Bošnački, S. Edelkamp (Eds.), Model Checking Software. X, 285 pages. 2007.

Vol. 4590: W. Damm, H. Hermanns (Eds.), Computer Aided Verification. XV, 562 pages. 2007.

Vol. 4588: T. Harju, J. Karhumäki, A. Lepistö (Eds.), Developments in Language Theory. XI, 423 pages. 2007.

Vol. 4583: S.R. Della Rocca (Ed.), Typed Lambda Calculi and Applications. X, 397 pages. 2007.

Vol. 4580: B. Ma, K. Zhang (Eds.), Combinatorial Pattern Matching. XII, 366 pages. 2007.